Dendritic Cells in Cancer

Michael R. Shurin • Russell D. Salter
Editors

Dendritic Cells in Cancer

Springer

Editors
Michael R. Shurin
Departments of Pathology and Immunology
Clinical Immunopathology
University of Pittsburgh Medical Center and University of Pittsburgh Cancer Institute
5725 Children's Hospital of Pittsburgh Main Tower
200 Lothrop Street
Pittsburgh PA 15213
USA
shurinmr@upmc.edu

Russell D. Salter
Department of Immunology
Immunology Graduate Program
E1052 Biomedical Science Tower
200 Lothrop Street
University of Pittsburgh School of Medicine
Pittsburgh PA 15213
USA
rds@pitt.edu

ISBN 978-0-387-88610-7 e-ISBN 978-0-387-88611-4
DOI 10.1007/978-0-387-88611-4

Library of Congress Control Number: 2008940078

Printed on acid-free paper

springer.com

Preface

The role of dendritic cells in the immune system has been the subject of intense study for the past two decades, and it is now beyond question that these cells play a critical role in determining all aspects of the immune response, including its kinetics, magnitude, direction and character. During an acute challenge such as that presented by many infections, dendritic cells in peripheral sites must internalize antigens and migrate to secondary lymphoid tissues to initiate immune responses. With chronic exposure to antigens in states such as cancer, however, dendritic cells are bombarded with conflicting signals. A complex environment within the tumor or tumor-bearing host determines the magnitude and polarity of numerous changes in dendritic cells induced by surrounding cells and factors.

The goal of this book is to assemble and integrate, for the first time, our knowledge of how dendritic cells function in the setting of cancer, providing a comprehensive survey of the field in a single volume. To this end, chapters are organized within thematic sections, each addressing major areas of current research. The authors of each chapter were chosen for their expertise and standing in their respective fields and have provided up-to-date accounts of the latest research findings. Specific topics include analysis of dendritic cell behavior in the tumor microenvironment, including endogenous and exogenous dendritic cells, multiple dendritic cell populations, molecular pathways responsible for dendritic cell dysfunction, tumor-derived factors altering dendritic cell polarization and activation, mechanisms of dendritic cell alterations and the role of dendritic cells in tumor escape from immune recognition and elimination. Furthermore, additional chapters provide extensive analysis of the consequences of cancer therapy on dendritic cells and how aging impacts dendritic cell function in the tumor microenvironment. Finally, chapters are included examining strengths and pitfalls of current methodologies for generating dendritic cells from cancer patients for therapeutic purposes and on the role of tumor-mediated modulation of the dendritic cell system in cancer immunotherapy.

This book should prove to be an essential reference guide for researchers in the fields of tumor immunology, immunotherapy and vaccine development and will be highly useful for students and others entering the field and seeking an introduction to the exciting and dynamic topic of immunobiology of dendritic cells in cancer.

Contents

1 The Central Role of Dendritic Cells in Immunity 1
Jessica Chu and Russell D. Salter

2 Dendritic Cells in Cancer: Emergence of the Discipline 11
Michael R. Shurin and Michael T. Lotze

Part I Alterations of the Dendritic Cell System in Cancer

3 Tumor-Infiltrating Dendritic Cells: The Pathologist's Perspective 33
Dmitry W. Gutkin

4 Functional Defects of Dendritic Cells in Cancer 57
Mikhail M. Dikov and Irina S. Chekneva

5 General Properties of Dendritic Cell Populations in Cancer 77
Benjamin J. Daniel, Duane P. Jeansonne, Suzanne R. Thibodeaux, and Tyler J. Curiel

6 Elimination of Dendritic Cells in Cancer 89
Viktor Umansky and Dirk Schadendorf

Part II Mechanisms of Dendritic Cell Dysfunction in Cancer

7 Tumor-Derived Factors Responsible for Dendritic Cell Dysfunction 103
Alberto Pinzon-Charry and J. Alejandro López

8 Tumor-Derived Exosomes as Dendritic Cell Modulators 119
Roberta Valenti, Veronica Huber, Paola Filipazzi, Manuela Iero, Giorgio Parmiani, and Licia Rivoltini

9 **Signaling Pathways Mediating Dendritic Cell Dysfunction in Cancer** ... 129
Jing Yang and Qing Yi

10 **A Role for STAT3 in Dendritic Cell Regulation by Tumor-Derived Factors** ... 143
Dmitry I. Gabrilovich

11 **Tumor-Associated Inflammation and Impact on Dendritic Cell Function** ... 157
Zoltan Pos, Marianna Sabatino, Tara Spivey, Hui Liu, Andras Falus, and Francesco M. Marincola

12 **Changes in Dendritic Cells in Cancer and Aging** ... 171
Annabelle Grolleau-Julius and Raymond L. Yung

13 **Stress, Immunity and Dendritic Cells in Cancer** ... 181
Rachel Kohman and Alexander W. Kusnecov

14 **Cancer Therapy and Dendritic Cell Immunomodulation** ... 201
Galina V. Shurin, Neil Amina, and Michael R. Shurin

Part III Role of Dendritic Cells in Tumor Escape

15 **Role of IDO in Dendritic Cell Differentiation and Function in Cancer** ... 219
Alexey Popov and Joachim L. Schultze

16 **Dendritic Cells: From Inducers of Specific T-Cell Responses to Promoters of Angiogenesis** ... 231
George Coukos and Fabian Benencia

17 **Protumorigenic Function of Dendritic Cells** ... 243
Anjli Kukreja

18 **Dendritic Cell Maturation Versus Polarization in Tumor Escape** ... 257
Michael W. Lipscomb, Walter J. Storkus, and Amy K. Wesa

Part IV Dendritic Cells in Cancer: Clinical Aspects

19 **Trafficking of Dendritic Cells in the Tumor Environment** ... 271
Zoya R. Yurkovetsky and Ghena N. Yurkovetsky

20 **Dendritic Cells in Tumor-Draining Lymph Nodes** 291
Alistair J. Cochran, Rong-Rong Huang, Eijun Itakura, Jonathan H. Lee, and Barbara G. Molenkamp

21 **Dendritic Cell Vaccines in Cancer: Obstacles to Overcome** 309
Ezra D. Mirvish, Rebecca G. Pomerantz, Louis D. Falo, and Larisa J. Geskin

22 **How Best to Generate Dendritic Cells from Patients with Cancer and How Best to Use them for Immunotherapeutic Purposes** 331
Mark M. Aloysius, Richard A. Robins, and Oleg Eremin

23 **Genetically Modified Dendritic Cells in Cancer Immunotherapy** 347
Brian J. Morrison, Jason C. Steel, Melissa Gregory, John C. Morris, and Anatoli M. Malyguine

24 **Dendritic Cell Tumors** 365
Lourdes R. Ylagan

25 **Future Directions in Dendritic Cell Research in Cancer** 375
Madhav V. Dhodapkar

Index 385

Contributors

Mark M. Aloysius Section of Surgery, Division of Gastrointestinal Surgery, Nottingham University Hospitals, Queen's Medical Centre, Nottingham, NG7 2UH, UK

Neil Amina Departments of Pathology and Immunology, University of Pittsburgh Medical Center, S735 Scaife Hall, 3550 Terrace Street, Pittsburgh, PA 15261, USA

Fabian Benencia Department of Biomedical Sciences, College of Osteopathic Medicine and, BME, Russ College of Engineering and Technology, Ohio University, Edison Biotechnology Institute, Konneker Labs, The Ridges, Athens, OH 45701, USA

Irina S. Chekneva Division of Hematology and Oncology, Department of Medicine, Vanderbilt University, 2200 Pierce Ave., 648 PRB, Nashville, TN 37232-6838, USA

Jessica Chu Department of Immunology, Immunology Graduate Program, University of Pittsburgh School of Medicine, E1006 Biomedical Science Tower, 200 Lothrop Street, Pittsburgh, PA 15213, USA

Alistair J. Cochran Departments of Pathology and Laboratory Medicine, Department of Surgery, Jonsson Comprehensive Cancer Center, David Geffen School of Medicine at UCLA, John Wayne Cancer Institute, 10833 LeConte Avenue, Los Angeles, CA 90095, USA

George Coukos Center for Research on Ovarian Cancer Early Detection and Cure, Abramson Family Cancer Research Institute, and Division of Gynecologic Oncology, Department of Obstetrics and Gynecology, University of Pennsylvania, 1209 Biomedical Research Building II/III, 421 Curie Boulevard, Philadelphia, PA 19104-6142, USA

Tyler J. Curiel Cancer Therapy & Research Center at UTHSCSA, 7979 Wurzbach Road, suite U637, San Antonio, TX 78229, USA

Benjamin J. Daniel Cancer Therapy & Research Center at UTHSCSA, 7979 Wurzbach Road, suite U637, San Antonio, TX 78229, USA

Madhav V. Dhodapkar Section of Hematology;, Yale University, New Haven, 333 Cedar Street, Box 208021, New Haven, CT 06510, USA

Mikhail M. Dikov Division of Hematology and Oncology, Department of Medicine, Vanderbilt University, 2200 Pierce Ave., 648 PRB, Nashville, TN 37232-6838, USA

Oleg Eremin Director of R&D and Lead Clinician for Breast Services, Research & Development Department, United Lincolnshire Hospitals NHS Trust, Lincoln County Hospital, Greetwell Road, Lincoln LN2 5QY, UK; Department of Surgery, University Hospital Nottingham, Nottingham, UK

Louis D. Falo Department of Dermatology, University of Pittsburgh School of Medicine and University of Pittsburgh Cancer Institute, 145 Lothrop Hall, 190 Lothrop Street, Pittsburgh, PA 15213, USA

Andras Falus Department of Genetics, Cell and Immunobiology; Inflammation Biology and Immungenomics Research Group, Hungarian Academy of Sciences, Semmelweis University, Budapest, 1089, Hungary

Paola Filipazzi Unit of Immunotherapy of Human Tumors, Fondazione-IRCCS Istituto Nazionale per lo Studio e la Cura dei Tumori, Via Venezian 1, 20133 Milano, Italy

Dmitry I. Gabrilovich Division of Dendritic Cell Biology, H. Lee Moffitt Cancer Center, University of South Florida, MRC, Rm. 2067, 12902 Magnolia Dr., Tampa, FL 33612, USA

Larisa J. Geskin Department of Dermatology, Cutaneous Oncology and Photopheresis Center, University of Pittsburgh School of Medicine and University of Pittsburgh Cancer Institute, 190 Lothrop Street, Lothrop Hall, Suite 101, Pittsburgh, PA 15213, USA

Melissa Gregory Applied and Developmental Research Support Program, SAIC Frederick, Inc. NCI-Frederick, Laboratory of Cell-Mediated Immunity, 1050 Boyles St., Bldg 560, 12-77, Frederick, MD 21702, USA

Annabelle Grolleau-Julius Divisions of Geriatric Medicine and Rheumatology, Department of Internal Medicine, University of Michigan, Room 3023 BSRB, 109 Zina Pitcher Place, Ann Arbor, Michigan 48109-2200, USA

Dmitry W. Gutkin Department of Pathology and Lab Medicine, VAMC Pittsburgh, Department of Pathology, University of Pittsburgh School of Medicine, VA Pittsburgh Healthcare System, University Drive C, 132L-U, Pittsburgh, PA 15240, USA

Rong-Rong Huang Departments of Pathology and Laboratory Medicine, David Geffen School of Medicine at UCLA, 10833 LeConte Avenue, Los Angeles, CA 90095, USA

Veronica Huber Unit of Immunotherapy of Human Tumors, Fondazione-IRCCS Istituto Nazionale per lo Studio e la Cura dei Tumori, Via Venezian 1, 20133 Milano, Italy

Manuela Iero Unit of Immunotherapy of Human Tumors, Fondazione-IRCCS Istituto Nazionale per lo Studio e la Cura dei Tumori, Via Venezian 1, 20133 Milano, Italy

Eijun Itakura Departments of Pathology and Laboratory Medicine, David Geffen School of Medicine at UCLA, 10833 LeConte Avenue, Los Angeles, CA 90095, USA

Duane P. Jeansonne Cancer Therapy & Research Center at UTHSCSA, 7979 Wurzbach Road, suite U637, San Antonio, TX 78229, USA

Rachel Kohman School of Pharmacy, Joint Graduate Program in Toxicology, Behavioral Neuroscience Program, Rutgers University, 152 Frelinghuysen Road, Piscataway, NJ 08855, USA; University of Medicine and Dentistry of New Jersey, Piscataway, New Jersey, USA

Anjli Kukreja Section of Hematology, Yale University, 333 Cedar Street, Box 208021, New Haven, CT 06520-8021, USA

Alexander W. Kusnecov Behavioral Neuroscience Program, Department of Psychology, Rutgers University, 152 Frelinghuysen Road, Piscataway, NJ 08855, USA; University of Medicine and Dentistry of New Jersey, Piscataway, New Jersey, USA

Jonathan H. Lee Division of Surgical Oncology, UMDNJ/RWJ Medical School, The Cancer Institute of New Jersey, 195 Little Albany Street, New Brunswick, NJ 08903, USA

Michael W. Lipscomb Departments of Immunology and Dermatology, University of Pittsburgh Medical Center, W1045 Biomedical Sciences Tower, 200 Lothrop Street, Pittsburgh, PA 15213, USA

Hui Liu Immunogenetics Section, Department of Transfusion Medicine, Clinical Center, National Institutes of Health, Building 10, Room 1C-738, 10 Center Drive MSC 1502, Bethesda, Maryland 20892-1502, USA

José Alejandro López Dendritic Cell and Cancer Laboratory, Griffith University, School of Biomolecular and Physical Sciences, Queensland Institute of Medical Research, Royal Brisbane Hospital Post Office, Brisbane 4029, Australia

Michael T. Lotze Departments of Surgery and Bioengineering, UPSHS, University of Pittsburgh Medical Center and University of Pittsburgh Cancer Institute, Rm 405, Physician's Office Building 2, 5150 Centre Dr., Pittsburgh, PA 15232, USA

Anatoli Malyguine Laboratory of Cell Mediated Immunity, CSP, SAIC-Frederick, Inc., National Cancer Institute at Frederick, P.O. Box B, Frederick, MD 21702-1201, USA

Francesco M. Marincola Chief Infectious Disease and Immunogenetics Section, Department of Transfusion Medicine, Clinical Center, National Institutes of Health, Bldg 10 - Room 1N226, Bethesda - Maryland 20892, USA

Ezra D. Mirvish Department of Dermatology, University of Pittsburgh School of Medicine and University of Pittsburgh Cancer Institute, 190 Lothrop Street, Lothrop Hall, Suite 101, Pittsburgh, PA 15213, USA

Barbara G. Molenkamp Department of Surgical Oncology, VU University Medical Center, P.O. Box 7057, 1007 MB Amsterdam, The Netherlands

John C. Morris Metabolism Branch, National Cancer Institute, Mark O. Hatfield Clinical Research Center, Room 4-5330, 10 Center Drive, Bethesda, MD 20892-1457, USA

Brian J. Morrison Dendritic Cell and Cancer Laboratory, Queensland Institute of Medical Research, Griffith University, Royal Brisbane Hospital Post Office, Brisbane 4029, Australia; Metabolism Branch, Center for Cancer Research, National Cancer Institute, Bethesda, MD, USA

Giorgio Parmiani Unit of Immunotherapy of Human Tumors, Fondazione-Istituto Nazionale per lo Studio e la Cura dei Tumori, Via Venezian 1, 20133 Milan, Italy

Alberto Pinzon-Charry NHMRC Peter Doherty Fellow, Bancroft Centre, L floor, Queensland Institute of Medical Research, PO Royal Brisbane Hospital, Brisbane, Queensland, 4029, Australia

Rebecca G. Pomerantz Department of Dermatology, University of Pittsburgh School of Medicine and University of Pittsburgh Cancer Institute, 190 Lothrop Street, Lothrop Hall, Suite 101, Pittsburgh, PA 15213, USA

Alexey Popov Laboratory for Genomics and Immunoregulation, Program Unit Molecular Immune & Cell Biology, LIMES (Life and Medical Sciences), University of Bonn, Karlrobert-Kreiten Street 13, 53115 Bonn, Germany

Zoltan Pos Immunogenetics Section, Department of Transfusion Medicine, Clinical Center, National Institutes of Health, Building 10, Room 1C-738, 10 Center Drive MSC 1502, Bethesda, Maryland 20892-1502, USA

Licia Rivoltini Unit of Immunotherapy of Human Tumors, Fondazione-IRCCS Istituto Nazionale per lo Studio e la Cura dei Tumori, Via Venezian 1, 20133 Milano, Italy

Richard A. Robins Department of Immunology, University Hospital Nottingham, Nottingham, NG7 2UH, UK

Marrianna Sabatino Immunogenetics Section, Department of Transfusion Medicine, Clinical Center, National Institutes of Health, Building 10, Room 1C-738, 10 Center Drive MSC 1502, Bethesda, Maryland 20892-1502, USA

Russell D. Salter Department of Immunology, Immunology Graduate Program, E1052 Biomedical Science Tower, 200 Lothrop Street, University of Pittsburgh School of Medicine, Pittsburgh, PA 15213, USA

Dirk Schadendorf Clinical Cooperation Unit Dermato-Oncology (G300), German Cancer Research Center, Im Neuenheimer Feld 280, 69120 Heidelberg, Germany; Skin Cancer Unit, German Cancer Research Center, Heidelberg, Germany; University Hospital Mannheim, Germany

Joachim L. Schultze Molekulare Tumorimmunologie, Medizinische Klinik I, Cologne Center for Genomics, Kerpener Str. 62 / Hs 16 UG, 50924 Köln, Germany; Program Unit Molecular Immune & Cell Biology, Life and Medical Sciences, Laboratory for Genomics and Immunoregulation, University of Bonn, Bonn, Germany

Michael R. Shurin Departments of Pathology and Immunology, Clinical Immunopathology, University of Pittsburgh Medical Center and University of Pittsburgh Cancer Institute, 5725 Children's Hospital of Pittsburgh Main Tower, 200 Lothrop Street, Pittsburgh, PA 15213, USA

Galina V. Shurin Departments of Pathology and Immunology, University of Pittsburgh Medical Center, S730 Scaife Hall, 3550 Terrace Street, Pittsburgh, PA 15261, USA

Tara Spivey Immunogenetics Section, Department of Transfusion Medicine, Clinical Center, National Institutes of Health, Building 10, Room 1C-738, 10 Center Drive MSC 1502, Bethesda, Maryland 20892-1502, USA

Jason C. Steel Metabolism Branch, National Cancer Institute, Building 10, Room 4N115, 10 Center Drive, Bethesda, MD 20892-1374, USA

Walter J. Storkus Departments of Dermatology and Immunology, University of Pittsburgh School of Medicine, W1041.2 Biomedical Sciences Tower, 200 Lothrop Street, Pittsburgh, PA 15213, USA

Suzanne R. Thibodeaux Cancer Therapy & Research Center at UTHSCSA, 7979 Wurzbach Road, suite U637, San Antonio, TX 78229, USA

Viktor Umansky, PhD Clinical Cooperation Unit Dermato-Oncology (G300), German Cancer Research Center, Im Neuenheimer Feld 280, 69120 Heidelberg, Germany; Skin Cancer Unit, German Cancer Research Center, Heidelberg, Germany; University Hospital Mannheim, Germany

Roberta Valenti Unit of Immunotheapy of Human Tumors, Fondazione-IRCCS Istituto Nazionale per lo Studio e la Cura dei Tumori, via Venezian 1, 20133 Milano, Italy

Amy K. Wesa Department of Dermatology, University of Pittsburgh Medical Center, W1045 Biomedical Sciences Tower, 200 Lothrop Street, Pittsburgh, PA 15213, USA

Jing Yang Department of Lymphoma and Myeloma, and the Center for Cancer Immunology Research University of Texas M. D. Anderson Cancer Center, 1515 Holcombe Boulevard, Unit 0903, Houston, Texas 77030, USA

Qing Yi Department of Lymphoma and Myeloma, and the Center for Cancer Immunology Research University of Texas MD Anderson Cancer Center, 1515 Holcombe Boulevard, Unit 0903, Houston, Texas 77030, USA

Lourdes Ylagan Washington University Medical School, 660 South Euclid Ave Box 8118, St. Louis, Missouri 63110, USA

Raymond Yung Divisions of Geriatric Medicine and Rheumatology, Department of Internal Medicine, University of Michigan, Room 3023 BSRB, 109 Zina Pitcher Place, Ann Arbor, Michigan 48109-2200, USA; GRECC, Ann Arbor Veterans Affairs Health System, Ann Arbor, Michigan, USA

Ghena N. Yurkovetsky AoP Cancer Research Support Foundation, 1323 Babcock Blvd, Pittsburgh, PA 15209, USA

Zoya R. Yurkovetsky University of Pittsburgh Medical Center, University of Pittsburgh Cancer Institute, Hillman Cancer Center, Room 1.19d, 5117 Centre Ave, Pittsburgh, PA 15213, USA

Introduction

Every year, hundreds of papers bring new insights into the immunobiology of dendritic cells and their role in the regulation of immune homeostasis and immune responses in different diseases. However, the role of diverse dendritic cell subsets in controlling tumor development, growth, spreading and metastasis is still being debated in light of constantly emerging experimental data and novel concepts. In the first two chapters, the general role of dendritic cells in immunity is introduced; this is followed by a description of the dendritic cell system as it pertains to cancer, including a historical perspective on the dendritic cell's place in tumor immunology and the development of our understanding of how dendritic cells interact with tumor cells, resulting in either acceleration of tumor progression and spreading or induction of antitumor immune responses and tumor elimination.

Chapter 1
The Central Role of Dendritic Cells in Immunity

Jessica Chu and Russell D. Salter

Abstract Professional antigen-presenting cells play a major role in the initiation of immune responses against microbial and viral pathogens and are also critical for eliciting antitumor immunity. Dendritic cells are particularly potent antigen-presenting cells due to their ability to take up, process, and present antigen to both CD4+ and CD8+ T cells. They can promote inflammation at local sites of infection and act as carriers of antigen to lymphoid organs for T-cell priming. In this way, they bridge innate and adaptive immunity. The following section will review the many diverse subsets of both human and mouse dendritic cells and the phenotypic and functional changes they undergo following antigen exposure. Additionally, dendritic cell-directed T-cell differentiation will be discussed with some emphasis on dendritic cell plasticity and the role of dendritic cells in disease.

1.1 Dendritic Cell Subsets

Dendritic cell subsets can be characterized by their anatomical localization as well as their phenotypic markers. In terms of anatomical location, dendritic cells (DC) are found in the circulation and in lymphoid and non-lymphoid tissues. DC can be found in afferent lymph where they are known as veiled cells and in blood, in which approximately 1% of peripheral blood mononuclear cells are comprised of DC and their precursors (Sato and Fujita 2007). Blood DC precursors give rise to non-lymphoid tissue-resident DC like Langerhans cells (LC, Fig. 1.1) and dermal DC (DDC) in the epidermis, interstitial DC in organs such as the liver, kidney, heart, and lung, and mucosal DC that line the mucosa of the oral cavity, intestinal tract, and respiratory tract (Banchereau et al. 2000; Steinman 1991). Lymphoid tissues such as tonsil, spleen, thymus, and lymph nodes house germinal center DC (GCDC), follicular DC (FDC),

---

R.D. Salter (✉)
Department of Immunology, University of Pittsburgh School of Medicine, E1006 Biomedical Science Tower, 200 Lothrop Street, Pittsburgh, PA 15213, USA
e-mail: rds@pitt.edu

M.R. Shurin, R.D. Salter (eds.), *Dendritic Cells in Cancer*,
DOI 10.1007/978-0-387-88611-4_1, © Springer Science+Business Media, LLC 2009

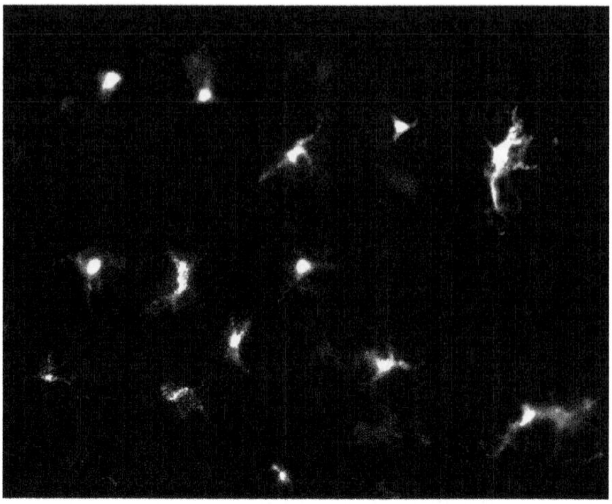

Fig. 1.1 Langerhans cells show a distinctive morphology with dendritic extensions. Mouse skin from ear was sectioned and stained with a fluoresceinated antibody against the class II MHC marker. A 400-fold magnification is shown (image provided by Dr. Adriana Larregina, UPMC)

and interdigitating DC (IDC) found in T-cell areas (Banchereau et al. 2000; Sato and Fujita 2007; Steinman 1991).

Phenotypically, mouse and human DC are defined by different sets of surface markers, with the added complexity of distinct subsets within each species. In mice, all DC express CD11c and MHC class II, but vary in their expression of CD4 and CD8α (Sato and Fujita 2007). Conventional DC consist of myeloid lineage CD4–CD8α^- and CD4+CD8α^- cells in the marginal zone of the spleen and lymph nodes and CD4–CD8α^{low} cells in skin-draining lymph nodes as well as lymphoid lineage CD4–CD8α^{high} cells in T-cell areas of the spleen and lymph nodes. The other major murine DC subset is the CD8α+CD205+B220+Gr-1+ plasmacytoid DC (pDC) found in the spleen, bone marrow, thymus, and lymph nodes. In humans, DC are distinguished by their expression of CD1a and CD11c (Sato and Fujita 2007). CD4+CD1a+CD11c^{high} and CD4+CD1a–CD11c^{low} comprise the conventional myeloid lineage human DC while the lymphoid pDC lineage consists of CD4+CD1a–CD11c– cells. Though these distinct subsets of DC have been clearly defined in mice and humans, it should be noted that DC exhibit great plasticity, and it is possible that these cells may exist as a continuum of cell types rather than as separate entities.

1.2 Dendritic Cells: Antigen Uptake and Processing

Initial recruitment of DC from the blood to peripheral sites occurs continuously in the steady state and is governed by chemokines, complement, cleavage products, defensins, and bacterial peptides (McWilliam et al. 1996; Sabatte et al. 2007; Sozzani 2005). Before activation, DC are considered immature with high endocytic capabilities and low expression of MHC and costimulatory molecules

(Tan and O'Neill 2005). In this state, they typically do not elicit T-cell responses in the absence of a pathogenic or stress-related threat. Immature DC in the body are found at sites of potential antigen exposure such as the skin, mucosal surfaces, and the spleen, where they carry out an important immunosurveillance function.

DC can interact with antigen through intracellular infection or can engulf exogenous antigen through three main processes: macropinocytosis, phagocytosis, and receptor-mediated endocytosis. Macropinocytosis involves receptor-independent sampling of extracellular fluid that is captured in 1–3 μm endocytic vesicles (Sabatte et al. 2007) while phagocytosis often employs Fc receptors, complement receptors, and other receptors (i.e., mannose receptor (Syme et al. 2002), DC-SIGN (Cambi and Figdor 2003; Tailleux et al. 2003)) to take up opsonized or non-opsonized microbes, dead or dying cells, and other antigenic particles (Banchereau et al. 2000; Banchereau and Steinman 1998; Sabatte et al. 2007). Receptor-mediated endocytosis of antigens is typically clathrin dependent and involves C-type lectins (i.e., mannose receptor, DEC-205 (Mahnke et al. 2000)) targeting carbohydrate structures on pathogens, scavenger receptors that bind polyanionic ligands, and FcγR types I and II that bind immune complexes (Banchereau et al. 2000; Sabatte et al. 2007). These endocytic receptors cluster in clathrin-coated pits to facilitate the endocytic process. Following antigen uptake, DC can process and present antigen using distinct classical and non-classical pathways described below.

Two classically defined pathways exist for the presentation of intracellular antigen to CD8+ T cells and exogenous antigen to CD4+ T cells. Intracellularly infected DC may contain cytosolic antigens that can be ubiquitinated and degraded to peptide fragments by the proteasome (Banchereau et al. 2000). These peptides can then be translocated into the endoplasmic reticulum (ER) via TAP1/2 transporters where they are loaded onto MHC class I molecules before transport to the cell surface, where they can stimulate CD8+ T-cell responses. A second quite distinct pathway involves the uptake of soluble and particulate extracellular antigens internalized into endosomes where they are degraded into peptides following vesicle acidification (Banchereau et al. 2000). These antigen-containing endosomes fuse with other vesicles rich in MHC class II to form a unique compartment known as the MIIC (MHC class II compartment). MIIC also contains HLA-DM, which is critical in the removal of the CLIP fragment of the class II-associated invariant chain that shields the peptide groove. Following removal of the CLIP, peptide-receptive MHC class II molecules can bind peptides with appropriate sequences, forming complexes which are then translocated to the cell surface. Once at the cell surface, MHC class II molecules present peptides to antigen-specific CD4+ T cells.

DC also have a unique capacity in vivo to process exogenous antigens for presentation to CD8+ T cells. This ability, termed cross-presentation or cross-priming, allows DC to prime CD8+ T-cell responses against pathogens which do not infect DC themselves. Multiple mechanisms have been demonstrated to allow for cross-presentation and are listed below (Groothuis and Neefjes 2005).

They include but are not limited to (1) gap junctions connecting infected and non-infected cells allowing transfer of antigenic peptides; (2) loading of antigens into MHC class I present in recycling pathways; (3) fusion of phagosomal membranes with endoplasmic reticulum, allowing for loading of exogenous antigens into nascent class I molecules; (4) export of antigens from endosomes into the cytosol; (5) transfer of antigens contained within microvesicles secreted by antigen-positive cells to bystander DC; and (6) transfer of membrane proteins between cells through nanotube structures (Onfelt et al. 2004; Rustom et al. 2004; Watkins and Salter 2005). Lastly, CD1 molecules are used in another non-classical antigen presentation method for the presentation of endogenous and exogenous microbial lipids and glycolipid-containing antigens (Banchereau et al. 2000).

1.3 Dendritic Cell Activation and Maturation

Simultaneously with antigen capture, processing, and presentation, DC also undergo an activation and maturation process that will ultimately enable them to migrate to lymphoid organs for the priming and activation of antigen-specific T cells. At the sites of local infection, DC encounter specific molecules that initiate their maturation. These molecules include pathogen-associated molecular patterns (PAMP) from infecting pathogens and damage-associated molecular patterns (DAMP) from damaged host tissues that bind to toll-like receptors (TLR) on the DC surface or within endosomes (Bianchi 2007; Sabatte et al. 2007; van Vliet et al. 2007). Ligation of these receptors results in the generation of pro-inflammatory cytokines such as IL-1, TNF-α, and IL-6 (Sabatte et al. 2007) that enable the recruitment of innate immune cells, like neutrophils, to the site of infection. This shift toward a pro-inflammatory cytokine environment further enhances DC activation. Additionally, the presence of T-cell-derived signals (CD40L (Caux et al. 1994; Schulz et al. 2000)), innate immune cells (NKT cells (Hermans et al. 2003), $\gamma\delta$ T cells (Conti et al. 2005), neutrophils (Ludwig et al. 2006)), and other stress signals (extracellular acidosis (Vermeulen et al. 2004), kinins (Aliberti et al. 2003), complement components (Soruri et al. 2003), oxidative stress (Rutault et al. 1999)) influence the activation and maturation of DC.

During maturation, DC lose their endocytic and phagocytic receptors; upregulate costimulatory molecules (CD40, CD80, CD86); change morphology by downregulating adhesion molecules, undergoing cytoskeletal reorganization, and obtaining high motility; and rearrange MIIC for higher peptide–MHC class II expression at the cell surface (Banchereau et al. 2000; Banchereau and Steinman 1998; Reis e Sousa et al. 1999; Sabatte et al. 2007; Tan and O'Neill 2005). These changes facilitate DC in their preparation to migrate to lymphoid tissues where they will present specific antigen and costimulate T cells for a robust immune response. Migration occurs when DC upregulate the chemokine receptor CCR7 (Sallusto et al. 1998), which allows them to chemotax to the chemokines CCL19 and CCL21 via afferent lymphatics to lymphoid tissues (Randolph et al. 2005). Once DC reach lymphoid tissues, they probe the circulating naïve T-cell pool for antigen-specific T cells. In order for DC to activate these antigen-specific T cells, three signals are required. Signal 1 consists of TCR cross-linking to peptide–MHC, signal 2 is the interaction of costimulatory molecules CD80 and CD86 on DC with CD28 on T cells, and signal 3 involves factors such as cytokines to promote the differentiation of T cells toward a distinct effector cell type (Diebold 2008; Sabatte et al. 2007).

1.4 Dendritic Cells in Directing T-Cell Responses

DC-directed T-cell fate can be influenced by many factors including the nature of signal 3, pathogen type, route of pathogen delivery, tissue-derived environmental factors, other innate immune cells, and feedback signals from activated T cells (Diebold 2008). Typically, in the presence of all three signals, myeloid DC that interact with intracellular microbes such as bacteria, viruses, and parasites produce IL-12 and type I interferons that program naïve CD4+ T cells to become IFN-γ-producing Th1 cells (Kadowaki 2007). DC licensed by T helper cells then activate macrophages and CD8+ cytotoxic T lymphocytes (CTL) for the killing of intracellular microbes infecting host cells. In addition to cellular immunity, neutralizing antibody activity is also generated in the context of intracellular infection (Diebold 2008). On the other hand, CTL responses are not useful during infection with extracellular bacteria, fungi, protozoa, and parasites. Instead, DC exposed to these extracellular microbes skew naïve CD4+ T cells to Th2-type cells, which secrete IL-4, IL-5, and IL-13 (Diebold 2008; Kadowaki 2007). This leads to antibody class switching in B cells for the production of IgE, which activates mast cells and eosinophils to eliminate the extracellular pathogens. The induction of a humoral response is important for opsonization, phagocyte activation, complement induction, and toxin neutralization in this type of infection. Th2 immune responses are also important for the maintenance of gut homeostasis. Mucosal DC continuously sample both pathogens and commensal bacteria from the gut lumen and must avoid detrimental responses to commensals while mounting immune responses to harmful bacteria. Thus, when mucosal DC encounter pathogenic microbes, they elicit a Th2 response that leads to the generation of IgA while keeping a non-inflammatory environment (Diebold 2008). Dysregulation of Th2 polarization in the gut can result in inflammatory diseases such as Crohn's disease (Rimoldi et al. 2005). Currently, it is unknown if the Th2 pathway is the natural default pathway in the absence of IL-12 or if specific instructive signals are also required.

Aside from infecting pathogen type and nature of signal 3, other factors also play a role in skewing Th responses. Most notably, prostaglandin E_2 (PgE_2) (Kalinski et al. 2001) and histamine (Idzko et al. 2002) elicit TNF-α-dependent DC maturation but inhibit IL-12p70 production such that naïve CD4+ T cells become Th2 polarized. IgE-activated mast cells have also been shown to lead to DC maturation in the absence of IL-12. Another major factor involved in DC-instructed Th2 responses is thymic stromal-derived lymphopoietin (TSLP) (Ito et al. 2005; Reche et al. 2001), which has also been observed at high levels in skin lesions of patients with atopic dermatitis (Soumelis et al. 2002) and allergic asthma (Al-Shami et al. 2005). Lastly, Notch ligands Delta and Jagged play a role in the generation of Th1 and Th2 responses, respectively (Amsen et al. 2004).

For a long time, the paradigm of Th1 versus Th2 skewing has been well accepted. However, in recent years it has become apparent that IL-17-producing Th17 cells also play a role in the control of infection by extracellular and intracellular bacteria, fungi, and parasites. The differentiation of this helper T-cell subtype relies on different instructive signals depending on the species. In mice, it has been determined that TGF-β and IL-6 are required during the initial differentiation stage toward the Th17 lineage (Matsushita and Higashi 2008). On the other hand, in humans, TGF-β has been shown to inhibit IL-17 production (Acosta-Rodriguez et al. 2007). Instead, human CD4+ naïve T cells require IL-1β in order to become Th17 cells and IL-6 and IL-23 enhance this process (Acosta-Rodriguez et al. 2007; Matsushita and Higashi 2008). It has been shown that human DC exposed to *Escherichia coli* and nucleotide adenosine triphosphate (ATP) produce IL-23, suggesting the involvement of DC in the instruction of Th17 cells (Schnurr et al. 2005). Moreover, van Beelen and colleagues have determined that bacterial NOD2-ligand MDP primes DC to produce IL-23 and IL-1 for the promotion of IL-17 expression in T cells (van Beelen et al. 2007). It should be noted that uncontrolled Th17 responses, resulting in excessive IL-17 production, have been implicated in the inflammatory autoimmune diseases such as rheumatoid arthritis and experimental autoimmune encephalomyelitis (EAE), a model for multiple sclerosis (Weaver et al. 2007).

1.5 Tolerogenic Dendritic Cells

It is evident that fully mature DC, requiring the presence of all three T-cell stimulation signals, prime naïve T cells to elicit strong immune responses against invading pathogens. However, in the absence of the third signal, namely cytokines and other factors to skew T-cell differentiation, naïve T cells undergo anergy, deletion, or conversion to a regulatory T-cell type (Diebold 2008; Tan and O'Neill 2005). This process is important both for the control of inflammation after an infection has resolved and for the maintenance of tolerance to self-antigens in the absence of an antigenic threat. During the steady state, tissue-resident immature DC induce T-cell anergy due to their low expression of costimulatory molecules, surface MHC, and pro-inflammatory cytokines (Lutz and Schuler 2002). Additionally, another type of steady-state DC has been described. They are known as semi-mature DC, which have high levels of costimulatory molecules and surface MHC class I and II loaded with self-peptides due to continuous sampling of apoptotic cells in the surrounding environment (Lutz and Schuler 2002; Tan and O'Neill 2005). By upregulating CCR7, they have the capacity to migrate to secondary lymphoid organs where they present self-antigen to circulating autoreactive T cells. However, because of their inability to produce IL-12 and other pro-inflammatory cytokines, DC either program T cells to become CD4+ IL-10-producing T regulatory (Treg) cells or induce the deletion of self-reactive T cells (Lutz and Schuler 2002). These tolerogenic DC are important for the maintenance of tolerance in the periphery and the elimination of autoreactive T cells for the prevention of autoimmune diseases. Factors such as TNF-α have been shown to produce tolerogenic DC in vitro and intravenous injection of these DC into mice has been shown to prevent EAE due to the generation of CD4+IL-10+ Treg cells (Menges et al. 2002). IL-10 (De Smedt et al. 1997) and PgE_2 (Diebold 2008; Sa-Nunes et al. 2007) may also be involved in alternative DC maturation.

The generation of tolerogenic DC and thus T regulatory cells can also occur in the presence of infection or during the resolution stage. It has been suggested that DC exhaustion may lead to tolerogenic DC as a way to resolve an immune response after an infection has been eliminated (Diebold 2008). In the presence of infection, particularly viral, pDC have been shown to induce IFN-γ, IL-10-producing cytotoxic T cells with an immunoregulatory role (Kawamura et al. 2006). Similarly, under conditions of exposure to IL-3 and CD40L, such as parasitic infection, pDC have been shown to differentiate CD8+ T cells into IL-10-producing suppressor cells (Gilliet and Liu 2002). CpG-ODN has also been implicated in programming pDC to differentiate naïve CD4+ T cells into CD4+CD25+Foxp3+ IL-10-producing suppressor cells (Moseman et al. 2004).

1.6 Conclusion

Dendritic cells have important functions at multiple stages during immune responses. After interacting with pathogens they can stimulate innate immune cells through secretion of cytokines and chemokines. Following capture of antigens and subsequent migration to secondary lymphoid tissues, they orchestrate the adaptive immune response by stimulating naïve T cells. In addition, they can acquire tolerogenic functions for the resolution of infection and in the absence of a pathogenic threat where maintaining peripheral tolerance to self-antigens is crucial. Dysregulation of tolerogenic DC can result in the formation of autoimmune pathologies and on the flip side, impairment of immunogenic DC function during pathogen invasion can lead to uncontrolled infection. Therefore, a fine balance must be achieved for DC to perform the appropriate functions for each given circumstance. Because of their plasticity, DC could be useful tools for the control of overactive immune responses, as in the cases of transplantation and autoimmune diseases, and for the elicitation of strong immunity against cancer, which will be discussed extensively in the remainder of this book.

References

Acosta-Rodriguez, E. V., Napolitani, G., Lanzavecchia, A. and Sallusto, F. 2007. Interleukins 1beta and 6 but not transforming growth factor-beta are essential for the differentiation of interleukin 17-producing human T helper cells. Nat Immunol 8:942–949.

Al-Shami, A., Spolski, R., Kelly, J., Keane-Myers, A. and Leonard, W. J. 2005. A role for TSLP in the development of inflammation in an asthma model. J Exp Med 202:829–839.

Aliberti, J., Viola, J. P., Vieira-de-Abreu, A., Bozza, P. T., Sher, A. and Scharfstein, J. 2003. Cutting edge: bradykinin induces IL-12 production by dendritic cells: a danger signal that drives Th1 polarization. J Immunol 170:5349–5353.

Amsen, D., Blander, J. M., Lee, G. R., Tanigaki, K., Honjo, T. and Flavell, R. A. 2004. Instruction of distinct CD4 T helper cell fates by different notch ligands on antigen-presenting cells. Cell 117:515–526.

Banchereau, J., Briere, F., Caux, C., Davoust, J., Lebecque, S., Liu, Y. J., Pulendran, B. and Palucka, K. 2000. Immunobiology of dendritic cells. Annu Rev Immunol 18:767–811.

Banchereau, J. and Steinman, R. M. 1998. Dendritic cells and the control of immunity. Nature 392:245–252.

Bianchi, M. E. 2007. DAMPs, PAMPs and alarmins: all we need to know about danger. J Leukoc Biol 81:1–5.

Cambi, A. and Figdor, C. G. 2003. Dual function of C-type lectin-like receptors in the immune system. Curr Opin Cell Biol 15:539–546.

Caux, C., Massacrier, C., Vanbervliet, B., Dubois, B., Van Kooten, C., Durand, I. and Banchereau, J. 1994. Activation of human dendritic cells through CD40 cross-linking. J Exp Med 180:1263–1272.

Conti, L., Casetti, R., Cardone, M., Varano, B., Martino, A., Belardelli, F., Poccia, F. and Gessani, S. 2005. Reciprocal activating interaction between dendritic cells and pamidronate-stimulated gammadelta T cells: role of CD86 and inflammatory cytokines. J Immunol 174:252–260.

De Smedt, T., Van Mechelen, M., De Becker, G., Urbain, J., Leo, O. and Moser, M. 1997. Effect of interleukin-10 on dendritic cell maturation and function. Eur J Immunol 27:1229 1235.

Diebold, S. S. 2008. Determination of T-cell fate by dendritic cells. Immunol Cell Biol 86:389–397.

Gilliet, M. and Liu, Y. J. 2002. Generation of human CD8 T regulatory cells by CD40 ligand-activated plasmacytoid dendritic cells. J Exp Med 195:695–704.

Groothuis, T. A. and Neefjes, J. 2005. The many roads to cross-presentation. J Exp Med 202:1313–1318.

Hermans, I. F., Silk, J. D., Gileadi, U., Salio, M., Mathew, B., Ritter, G., Schmidt, R., Harris, A. L., Old, L. and Cerundolo, V. 2003. NKT cells enhance CD4+ and CD8+ T cell responses to soluble antigen in vivo through direct interaction with dendritic cells. J Immunol 171:5140–5147.

Idzko, M., la Sala, A., Ferrari, D., Panther, E., Herouy, Y., Dichmann, S., Mockenhaupt, M., Di Virgilio, F., Girolomoni, G. and Norgauer, J. 2002. Expression and function of histamine receptors in human monocyte-derived dendritic cells. J Allergy Clin Immunol 109:839–846.

Ito, T., Wang, Y.-H., Duramad, O., Hori, T., Delespesse, G. J., Watanabe, N., Qin, F. X.-F., Yao, Z., Cao, W. and Liu, Y.-J. 2005. TSLP-activated dendritic cells induce an inflammatory T helper type 2 cell response through OX40 ligand. J. Exp. Med. 202:1213–1223.

Kadowaki, N. 2007. Dendritic cells: a conductor of T cell differentiation. Allergol Int 56:193–199.

Kalinski, P., Vieira, P. L., Schuitemaker, J. H., de Jong, E. C. and Kapsenberg, M. L. 2001. Prostaglandin E(2) is a selective inducer of interleukin-12 p40 (IL-12p40) production and an inhibitor of bioactive IL-12p70 heterodimer. Blood 97:3466–3469.

Kawamura, K., Kadowaki, N., Kitawaki, T. and Uchiyama, T. 2006. Virus-stimulated plasmacytoid dendritic cells induce CD4+ cytotoxic regulatory T cells. Blood 107:1031–1038.
Ludwig, I. S., Geijtenbeek, T. B. and van Kooyk, Y. 2006. Two way communication between neutrophils and dendritic cells. Curr Opin Pharmacol 6:408–413.
Lutz, M. B. and Schuler, G. 2002. Immature, semi-mature and fully mature dendritic cells: which signals induce tolerance or immunity? Trends Immunol 23:445–449.
Mahnke, K., Guo, M., Lee, S., Sepulveda, H., Swain, S. L., Nussenzweig, M. and Steinman, R. M. 2000. The dendritic cell receptor for endocytosis, DEC-205, can recycle and enhance antigen presentation via major histocompatibility complex class II-positive lysosomal compartments. J Cell Biol 151:673–684.
Matsushita, S. and Higashi, T. 2008. Human Th17 cell clones and natural immune responses. Allergol Int 57:135–140.
McWilliam, A. S., Napoli, S., Marsh, A. M., Pemper, F. L., Nelson, D. J., Pimm, C. L., Stumbles, P. A., Wells, T. N. C. and Holt, P. G. 1996. Dendritic cells are recruited into the airway epithelium during the inflammatory response to a broad spectrum of stimuli. J Exp Med 184:2429–2432.
Menges, M., Rossner, S., Voigtlander, C., Schindler, H., Kukutsch, N. A., Bogdan, C., Erb, K., Schuler, G. and Lutz, M. B. 2002. Repetitive injections of dendritic cells matured with tumor necrosis factor alpha induce antigen-specific protection of mice from autoimmunity. J Exp Med 195:15–21.
Moseman, E. A., Liang, X., Dawson, A. J., Panoskaltsis-Mortari, A., Krieg, A. M., Liu, Y. J., Blazar, B. R. and Chen, W. 2004. Human plasmacytoid dendritic cells activated by CpG oligodeoxynucleotides induce the generation of CD4+CD25+ regulatory T cells. J Immunol 173:4433–4442.
Onfelt, B., Nedvetzki, S., Yanagi, K. and Davis, D. M. 2004. Cutting edge: Membrane nanotubes connect immune cells. J Immunol 173:1511–1513.
Randolph, G. J., Angeli, V. and Swartz, M. A. 2005. Dendritic-cell trafficking to lymph nodes through lymphatic vessels. Nat Rev Immunol 5:617–628.
Reche, P. A., Soumelis, V., Gorman, D. M., Clifford, T., Liu, M., Travis, M., Zurawski, S. M., Johnston, J., Liu, Y. J., Spits, H., de Waal Malefyt, R., Kastelein, R. A. and Bazan, J. F. 2001. Human thymic stromal lymphopoietin preferentially stimulates myeloid cells. J Immunol 167:336–343.
Reis e Sousa, C., Sher, A. and Kaye, P. 1999. The role of dendritic cells in the induction and regulation of immunity to microbial infection. Curr Opin Immunol 11:392–399.
Rimoldi, M., Chieppa, M., Salucci, V., Avogadri, F., Sonzogni, A., Sampietro, G. M., Nespoli, A., Viale, G., Allavena, P. and Rescigno, M. 2005. Intestinal immune homeostasis is regulated by the crosstalk between epithelial cells and dendritic cells. Nat Immunol 6:507–514.
Rustom, A., Saffrich, R., Markovic, I., Walther, P. and Gerdes, H. H. 2004. Nanotubular highways for intercellular organelle transport. Science 303:1007–1010.
Rutault, K., Alderman, C., Chain, B. M. and Katz, D. R. 1999. Reactive oxygen species activate human peripheral blood dendritic cells. Free Radic Biol Med 26:232–238.
Sabatte, J., Maggini, J., Nahmod, K., Amaral, M. M., Martinez, D., Salamone, G., Ceballos, A., Giordano, M., Vermeulen, M. and Geffner, J. 2007. Interplay of pathogens, cytokines and other stress signals in the regulation of dendritic cell function. Cytokine Growth Factor Rev 18:5–17.
Sallusto, F., Schaerli, P., Loetscher, P., Schaniel, C., Lenig, D., Mackay, C. R., Qin, S. and Lanzavecchia, A. 1998. Rapid and coordinated switch in chemokine receptor expression during dendritic cell maturation. Eur J Immunol 28:2760–2769.
Sato, K. and Fujita, S. 2007. Dendritic cells: nature and classification. Allergol Int 56:183–191.

Schnurr, M., Toy, T., Shin, A., Wagner, M., Cebon, J. and Maraskovsky, E. 2005. Extracellular nucleotide signaling by P2 receptors inhibits IL-12 and enhances IL-23 expression in human dendritic cells: a novel role for the cAMP pathway. Blood 105:1582–1589.

Schulz, O., Edwards, A. D., Schito, M., Aliberti, J., Manickasingham, S., Sher, A. and Reis e Sousa, C. 2000. CD40 triggering of heterodimeric IL-12 p70 production by dendritic cells in vivo requires a microbial priming signal. Immunity 13:453–462.

Soruri, A., Riggert, J., Schlott, T., Kiafard, Z., Dettmer, C. and Zwirner, J. 2003. Anaphylatoxin C5a induces monocyte recruitment and differentiation into dendritic cells by TNF-alpha and prostaglandin E2-dependent mechanisms. J Immunol 171:2631–2636.

Soumelis, V., Reche, P. A., Kanzler, H., Yuan, W., Edward, G., Homey, B., Gilliet, M., Ho, S., Antonenko, S., Lauerma, A., Smith, K., Gorman, D., Zurawski, S., Abrams, J., Menon, S., McClanahan, T., de Waal-Malefyt Rd, R., Bazan, F., Kastelein, R. A. and Liu, Y. J. 2002. Human epithelial cells trigger dendritic cell mediated allergic inflammation by producing TSLP. Nat Immunol 3:673–680.

Sozzani, S. 2005. Dendritic cell trafficking: More than just chemokines. Cytokine & Growth Factor Reviews 16:581–592.

Steinman, R. M. 1991. The dendritic cell system and its role in immunogenicity. Annu Rev Immunol 9:271–296.

Syme, R. M., Spurrell, J. C., Amankwah, E. K., Green, F. H. and Mody, C. H. 2002. Primary dendritic cells phagocytose Cryptococcus neoformans via mannose receptors and Fcgamma receptor II for presentation to T lymphocytes. Infect Immun 70:5972–5981.

Tailleux, L., Schwartz, O., Herrmann, J. L., Pivert, E., Jackson, M., Amara, A., Legres, L., Dreher, D., Nicod, L. P., Gluckman, J. C., Lagrange, P. H., Gicquel, B. and Neyrolles, O. 2003. DC-SIGN is the major Mycobacterium tuberculosis receptor on human dendritic cells. J Exp Med 197:121–127.

Tan, J. K. and O'Neill, H. C. 2005. Maturation requirements for dendritic cells in T cell stimulation leading to tolerance versus immunity. J Leukoc Biol 78:319–324.

van Beelen, A. J., Zelinkova, Z., Taanman-Kueter, E. W., Muller, F. J., Hommes, D. W., Zaat, S. A., Kapsenberg, M. L. and de Jong, E. C. 2007. Stimulation of the intracellular bacterial sensor NOD2 programs dendritic cells to promote interleukin-17 production in human memory T cells. Immunity 27:660–669.

van Vliet, S. J., den Dunnen, J., Gringhuis, S. I., Geijtenbeek, T. B. and van Kooyk, Y. 2007. Innate signaling and regulation of Dendritic cell immunity. Curr Opin Immunol 19:435–440.

Vermeulen, M. E., Gamberale, R., Trevani, A. S., Martinez, D., Ceballos, A., Sabatte, J., Giordano, M. and Geffner, J. R. 2004. The impact of extracellular acidosis on dendritic cell function. Crit Rev Immunol 24:363–384.

Watkins, S. C. and Salter, R. D. 2005. Functional connectivity between immune cells mediated by tunneling nanotubules. Immunity 23:309–318.

Weaver, C. T., Hatton, R. D., Mangan, P. R. and Harrington, L. E. 2007. IL-17 family cytokines and the expanding diversity of effector T cell lineages. Annu Rev Immunol 25:821–852.

Chapter 2
Dendritic Cells in Cancer: Emergence of the Discipline

Michael R. Shurin and Michael T. Lotze

Abstract Immunologic research, following the discovery of dendritic cells radically changed our understanding of the induction, maintenance, and emergence of immune-mediated inflammatory disorders, including malignant diseases. The mechanisms central to the etiology and pathogenesis of many of these chronic inflammatory conditions involve dendritic cells. Identification of dendritic cells in tumors as well as clinical evaluation of dendritic cell vaccines led to the realization that very complex interactions between dendritic cells and other cellular and extracellular components of the tumor microenvironment dictated clinical outcome. Dendritic cells interestingly either induce antitumor immune response or promote a wound repair phenotype including reparative epithelial tumor proliferation, resumption of "barrier function", promotion of the premetastatic niche, and metastases. The limited success of dendritic cell-based therapies suggests the need for a deeper understanding of immunobiology of these key cells of the immune system as they develop within the complex tumor microenvironment. Reanalyzing and reexamining the accumulated data and concepts in the field, as done in this chapter and the book overall, serve this important goal.

2.1 Discovery of Dendritic Cells

Dendritic cells (DC) are a critical component of immunity, previously underappreciated in a discipline dictated by the specificity and charm of specific antibodies and T cells. Immunologic research following the discovery of DC significantly changed our understanding of the induction and emergence of mechanisms central to the etiology and pathogenesis of many inflammatory disorders, including malignant diseases. DC were first described in the skin by

M.R. Shurin (✉)
Clinical Immunopathology, University of Pittsburgh Medical Center and University of Pittsburgh Cancer Institute, 5725 CHP-MT, 200 Lothrop Street, Pittsburgh, Pennsylvania, USA
e-mail: shurinmr@upmc.edu

M.R. Shurin, R.D. Salter (eds.), *Dendritic Cells in Cancer*,
DOI 10.1007/978-0-387-88611-4_2, © Springer Science+Business Media, LLC 2009

German pathologist Paul Langerhans (1847–1888) and provided the eponymic, Langerhans cells (LC). In 1868, still as an undergraduate, Paul Langerhans stained a sample of human skin with gold chloride and analyzed epidermal skin cells as part of an open competition organized by Berlin University. He identified the cells, which from their appearance, Langerhans believed were nerve cells. The "branched skin cells resembling neuron", described in his 1868 paper entitled "On the nerves of the human skin" (Langerhans 1868), remained an enigma for over a century before their immunological function was recognized. Even after this, the biological and clinical significance of LC has been a matter of conjecture for many years.

In the nineteenth century, in discussing the nature of the cells recently described by Langerhans, Ranvier in 1875 expressed the opinion that they were neither nervous elements nor branched pigmentary cells, but rather lymphatic cells that had migrated into the epidermis (Ranvier 1875). As quoted by Ranvier, Eberth was the first to contest the neural origin of the cell described 2 years earlier by Paul Langerhans. He thought in 1870 that these cells were either branched pigmentary cells or migratory cells (Eberth 1870). It is, indeed, surprising that that opinion has remained ignored for many years.

A hundred years after the discovery of LC in the skin, these cells were identified in the lymph nodes and thymus (Olah et al. 1968; Jimbow et al. 1969; Kondo 1969; Van Haelst 1969). In the 1960s, some investigators believed that LC were skin macrophages (Tarnowski and Hashimoto 1967; Hashimoto and Tarnowski 1968; Prunieras 1969; Hashimoto 1971) and their capability to uptake foreign particles was proven (Wolff and Schreiner 1970; Sagebiel 1972). Interestingly, LC were also suggested in the 1960s – early 1970s to be lymphoid cells which are capable of producing antibody (Billingham and Silvers 1965; Kuwahara 1971). Based on these and other results, it has been proposed that LC being the skin "dendritic cells" (as in (Prunieras 1969)) may capture antigenic material and thus play a role in the primary immune response (Prunieras 1969; Hashimoto 1971; Vernon et al. 1973). In the early 1970s, Silberberg et al. based on the electron microscopy analyses postulated that if "LC are shown to possess antigen on or near their surface and in the course of the contact allergic reaction reach lymph nodes, they may very well participate in immune responses (i.e., immunoproliferative processes) in the lymph nodes as well as locally in the skin" and concluded that "Langerhans cells are a previously unrecognized cell population of immunologic importance" (Silberberg 1971, 1972, 1973; Silberberg et al. 1974, 1976). It is interesting to note that it was also suggested by the same group that LC might interact with sensitized lymphocytes and "release substances which lead to further inflammatory changes in the skin".

LC thus were known to be DC, but the term was only based on "the dendritic nature": for example, according to Michael Birbeck in 1961, "the perikarya of these cells appear characteristically dendritic" (Birbeck et al. 1961). There are also several interesting terminological issues. Although "dendritic cells" were described in the skin in the 1960s (Zelickson 1965; Zelickson and Mottaz 1968;

Prunieras 1969; Mishima and Kawasaki 1970) (e.g., Michael Birbeck's et al. paper from 1961 (Birbeck et al. 1961)), they included different cell populations. In addition to using the term "dendritic cells" for LC, it was used for melanocytes and an unusual cell population described as non-keratinocytic cells (α-cells) that have neither premelanosomes nor Langerhans (Birbeck) granules (Mishima and Kawasaki 1970; Kidd et al. 1971). These cells were also known as α-DC (Mishima et al. 1972).

It was not until 1973, however, that the term "dendritic cells" was used by Ralph M. Steinman and Zanvil A. Cohn for a new class of white blood cells with a number of distinctive features and functions (Steinman and Cohn 1973). Their publication is commonly accepted as the beginning of the modern era of dendritic cell science. Today, the term *dendritic cell* defines a diverse and multi-functional group of cells that serve as sentinels, adjuvants, and conductors of many immune functions, including the host defense against pathogens, both infectious and neoplastic.

2.2 Dendritic Cells in Cancer: Recognition of Their Functional Significance

Understanding of the functional significance of DC in initiating and maintenance of antitumor immunity and in cancer immunosurveillance began with the identification of LC at the tumor site, first in the skin tumors and later in other solid malignancies. From the early 1970s, an increased number of LC has been reported in benign epidermal tumors, whereas contradictory findings of the presence and the number of LC have been reported in squamous cell carcinomas both of cutaneous and mucosal origin (Lisi 1973; Wilborn et al. 1978; Loning et al. 1982; Fernandez-Bussy et al. 1983; Gatter et al. 1984). Basset et al. (1974) reported "Large numbers of dendritic cells similar in structure to Langerhans cells of normal epidermis and other epithelia were observed within nodules of a tumour identified as a bronchiolar-alveolar tumour of primary alveolar origin". Interestingly, although the overall function of LC was not yet proven at that time, the authors speculated that their presence at the tumor site "might be regarded as part of a defense process, since their occurrence in lymph nodes, thymus, and spleen could suggest they may have some function in this connection".

Functional studies on LC have always been hampered by the fact that these cells represent only a very small subpopulation of epidermal cells and it was therefore not possible to separate them from other epidermal cells. Only in the late 1970s, the functional role for LC was reported in direct in vitro studies and the ability of LC to present antigen was directly proven. Based on the contemporary identification of Fc receptors on LC in 1977 (Stingl et al. 1977), LC were isolated by rosetting and gradient centrifugation. Stingl et al. asked whether LC could act as antigen-presenting cells for either soluble protein antigens (PPD, OVA) or for simple chemical haptens (TNP). The results clearly demonstrated that LC-induced antigen-specific T-cell proliferative responses comparable in magnitude to that induced by macrophages (Stingl et al. 1978a). Following the identification of Ia molecules on LC (Klareskog et al. 1977; Rowden et al. 1977), their ability to stimulate allogeneic T-cell proliferation in MLR assay was also tested and compared to that of macrophages (Stingl et al. 1978a). The authors reported that Ia-dependent ability of LC to stimulate allogeneic T cells was similar to that of macrophages in the same assay. Although this study has dealt exclusively with epidermal LC, the authors concluded "It is conceivable that all nonlymphoid organs contain a small percentage of cells with similar immunologic properties" (Stingl et al. 1978a).

Based on morphological, ultrastructural, immunohistochemical and biochemical analyses, "a relationship between Langerhans cells and the monocyte–macrophage system" was repeatedly suggested (Stingl et al. 1978b; Sterry and Steigleder 1979). Thus, by the late 1970s, it was accepted that "Langerhans cells represent specific granule-containing dendritic cells . . . , occur in the squamous epithelium, and also in the corium, lymph node and thymus. . ., are able to phagocytize antigens. . . as well as to migrate through the lymph vessels into the regional lymph nodes. . ., [and display] the antigen-presenting and lymphocyte stimulating functions. . ." (Haustein 1979). At the same time, in ongoing studies utilizing the in vitro MLR assay as a model of T-cell recognition and response to cell surface alloantigens, it was shown that splenic adherent cells, called SAC or A-cells, and "lymphoid dendritic cells", represent the same cell population (Steinman and Witmer 1978; Ahmann et al. 1979, 1981). Furthermore, the ability of LC to induce CTL responses similarly to SAC has been also reported (Pehamberger et al. 1983). Based on various similarities, LC of the skin and "indeterminate dendritic cells" as well as interdigitating cells (IDC), first described by Veldman (Veldman et al. 1978), of the normal tonsils, spleen, and thymic medulla were suggested to be related (Heusermann et al. 1974; Kelly et al. 1978; Hoefsmit et al. 1979). The latter cell subset circulates and was referred to as "veiled cells" in the lymph and "macrophages with ruffled membrane" in the peritoneal exudates (Kelly et al. 1978; Spry et al. 1980). Thus, several types of irregularly shaped *dendritic cells* have been identified, including dendritic cells and interdigitating cells in lymphoid tissues, epidermal Langerhans cells, follicular or germinal center dendritic cells, and veiled cells in lymph (Van Voorhis et al. 1983b). Interestingly, in 1982, Stella C. Knight suggested that veiled cells resemble DC described by Steinman or may be the precursors of DC (Balfour et al. 1982; Knight et al. 1982). At that time, the "DC of Steinman" were considered to be in vitro equivalents of LC and IDC (Hoefsmit 1982). While LC, IDC, and veiled cells were believed to belong to a subpopulation of the macrophages and could be developed from monocytes, DC described by Steinman and Cohn were considered to be bone marrow derived as a separate cell subset (Steinman and Nussenzweig 1980). DC that were originally identified in rodents were soon seen in and isolated from human tissues and their role "as inducer cells in the immune response" was also confirmed (Van Voorhis et al. 1983a; Richtsmeier et al. 1984).

In relation to the tumor, clusters of Langerhans and lymphoid cells in the inflammatory peritumoral infiltrate in basal cell carcinoma, for instance, were described in the 1970s and based on the similarity between "Langerhans' and interdigitating reticulum cells" it was thought that this specific microenvironment might be "favourable to certain immunological activities of T-lymphocyte populations" (Macadam 1978; Porfiri et al. 1979). "Dendritic cells similar to Langerhans' cells of normal epidermis" were recognized at the same time in specimens from cutaneous T-cell lymphoma (CTCL), oral squamous cell carcinoma (SCC), salivary gland adenoma, malignant melanoma, and other tumor types (Rowden et al. 1979; David and Buchner 1980; Schenk 1980; Szekeres and Daroczy 1981; Thomas et al. 1984). For instance, HLA-DR positive DC were shown in the infiltrate and between the melanoma cells (Poppema et al. 1983). Analyzing individual mononuclear cell populations at the tumor site, including antigen-presenting DC, CD4+ T helpers, and CD8+ cytotoxic T lymphocytes, the authors concluded that they might play a role in immune defenses against malignant cells.

In 1981, LC were shown to express S-100 antigen (Cocchia et al. 1981) and soon S-100-expressing DC were identified and characterized in various neoplastic tissues (Nakajima et al. 1982). For instance, S-100 antigen-containing cells with dendritic features, recognizable by morphological and immunohistochemical criteria as belonging to the Langerhans' cell type, have been found in undifferentiated nasopharyngeal carcinoma and in lymph node metastases. S-100+ DC appeared to be few or absent in most SCC of both mucosal and epidermal origin. The presence of these cells, which have a special function of antigen presentation in immune responses, was speculated to be involved in host–tumor interactions (Lauriola et al. 1984). Similar results were reported for S-100+ DC in many different tumor types, including lung, oral, and cervical cancer as well as others (Kurihara and Hashimoto 1985; Nakajima et al. 1985; Tay et al. 1987).

Thus, soon after identification of DC at the tumor site, it was accepted that they may play an important role in tumor–host interactions and may be involved in initiation of antitumor immune response (Fig. 2.1A). This concept was primarily supported by functional studies modulating DC activity at the tumor site and by direct experiments demonstrating antitumor potential of DC in pre-clinical models. For example, in the middle 1980s it was shown that DC can pick up tumor antigens and modulate antitumor immunity in vivo (Knight et al. 1985): DC isolated from the spleen and pulsed with tumor antigens caused inhibition of tumor growth when injected into tumor-bearing mice, thereby serving as antitumor vaccines. At the same time involvement of LC in antitumor immunity was also suggested: Muller at al. (1985) showed that the chemical carcinogen DMBA induced skin tumors and depleted LC from treated skin, which however repopulated the skin upon cessation of the DMBA treatment (Muller et al. 1985). During this repopulation of the skin by LC, the tumors decreased in size. The authors concluded that since LC function as local cutaneous APC, their depletion during tumor induction may allow DMBA-transformed cells to circumvent the immune system and form tumors. Their reappearance associated with tumor regression suggested that the LC are

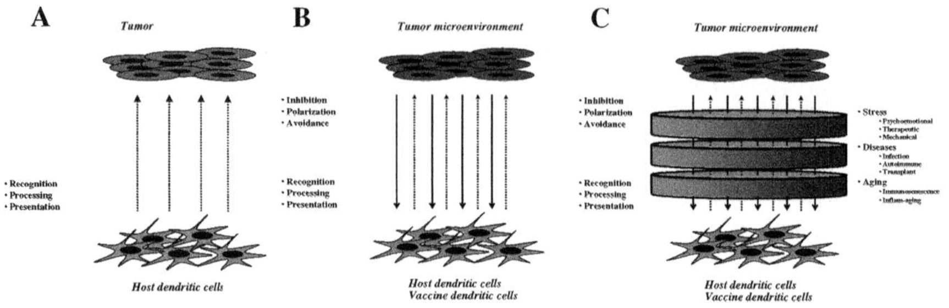

Fig. 2.1 Emergence of understanding of dendritic cell–tumor cell interaction. Dendritic cells (DC) were recognized as important mediators, inducing specific immune responses. Identification of DC within the tumor mass supported the notion that they played a consequential involvement in initiating antitumor immunity (**A**). This view has received additional support in numerous reports demonstrating a direct correlation between the number of tumor-infiltrating DC and prognosis. However, from the very early reports it became clear that the morphological appearance of DC in the tumor environment, as well as phenotypic maturation and function of tumor-associated DC or DC isolated from patients with cancer could be compromised or deficient. Tumor-derived factors suppress differentiation, activation, and longevity of DC precursors and DC themselves in vitro and in vivo provided sustaining evidence of tumor-mediated repression of DC. These findings suggested that tumor-induced inhibition of DC differentiation (dendropoiesis), functioning, and survival represents pathophysiological mechanisms by which tumors survive immunological recognition and elimination (**B**). Nevertheless, novel experimental and clinical data reveal additional layers of complexity of existing concept of DC functioning in the setting of cancer (**C**). First of all, common psychologic stressors in patients diagnosed with malignant diseases, as well as those associated with therapy (e.g., chemotherapy and radiation) or mechanical disruption (e.g., surgery and biopsy), strongly affect dendropoiesis and DC activation via release of glucocorticoids, catecholamines, neuropeptides, and inflammatory cytokines and chemokines. Secondly, infectious, allergic, and autoimmune diseases, especially in the pediatric and elderly populations of cancer patients drastically modulate the state of the DC and its ability to respond to tumor-derived or therapy-associated stimuli. Cell and organ transplantation and related immunosuppressive treatments may both account for unusual behavior of DC within the tumor milieu. Thirdly, as has been recently proven, immunosenescence of the elderly arises in part due to the altered differentiation and function of individual subpopulations of DC and, thus, might also compromise development and sustaining of DC-mediated antitumor immunity in cancer patients. Most of these factors that modify the DC in patients with malignancies should be applied not only to endogenous host DC, but also to exogenously administered DC vaccines. The interaction between injected DC and resident DC may compromise the induction of effective antitumor immune responses. Understanding of DC behavior in the tumor environment in a particular patient is crucial for designing and applying efficient therapy for cancer treatment

involved in an immune response against the tumors (Muller et al. 1985). Moreover, DC receiving TA were also shown to induce protective immunity against a subsequent tumor challenge (Gyure et al. 1987).

Furthermore, the role of DC in the development of antitumor immunity (Fig. 2.1A) was supported by a growing number of clinical and pathological data analyzing a correlation between the levels of tumor-infiltrating DC and the patients' outcomes. For instance, Igisu et al. (1983) found in CTCL, the fewer the number of intratumoral LC, the poorer the prognosis. Survival time in patients with the advanced stage of gastric carcinoma correlated well with the density of LC. In patients with a marked infiltration of LC, survival time was longer than in cases of only a slight infiltration (Tsujitani et al. 1987). Patients with many S-100+ DC in colorectal adenocarcinomas survived longer than did those with few, most often in those patients with no metastases (Ambe et al. 1989).

In addition, Nomori et al. (1986) found that the degree of density of T-zone histiocytes (i.e., DC) was significantly related to prognosis in primary site of biopsy specimens of nasopharyngeal carcinoma. Furukawa et al. found similar results in stage 1a adenocarcinoma of the lung (Furukawa et al. 1985) and reported that LC were present more frequently in moderately to well-differentiated adenocarcinoma than in poorly differentiated adenocarcinoma of the lung. Cochran et al. (1987) also showed that DC were infrequent in lymph nodes infiltrated by melanoma or located near melanoma but were numerous in nodes located farther from tumor. Interestingly the appearance of LC seemed to be related to the presence of chronic inflammation: Very few LC were observed in carcinomas in which inflammatory cells were rare (Nakajima et al. 1985). Many subsequent reports confirmed the conclusion that survival of patients whose tumors contain high density of DC is more favorable when compared to those whose tumors have a low density of these cells in different types of cancer. Altogether these and similar results allowed wide acceptance of the prevalent concept that LC/DC in immunological defense mechanisms of the host against the tumor may be clinically effective during the early phases of tumor development.

2.3 Dendritic Cells in Cancer: Recognition of Their Role in Tumor Escape

By the mid-end 1980s, with only a few exceptions, analyzing S-100+ and HLA-DR+ DC and LC in human tumors, most investigators commonly concluded that there was a significant decrease of these cells in tumors when compared with non-malignant or surrounding tissues. Also, the distribution of DC in the tumor tissues and regional lymph nodes could be considered as a reference indicator of tumor histologic grade and clinical prognosis of patients with different tumor types, and thus could reflect the degree of tumor immunity induced in the tumor-bearing host. However, pathogenetic and immunologic mechanisms regulating DC distribution and homing at the tumor site were still unknown. Soon a clue appeared with the accumulation of data suggesting first morphologic then functional abnormalities of LC/DC in the tumor mass. In characterizing distribution of LC in human skin tumors, "A striking finding to emerge was that in benign skin lesions Langerhans' cells were increased, whereas in malignant tumours they were not only markedly depleted or absent but also grossly stunted and deformed in outline" (Gatter et al. 1984). This is probably the first report describing tumor-associated abnormalities in DC. Similar alterations were seen in LC in the skin of patients with basal cell carcinoma (BCC): "Perturbations of ATPase-positive, dendritic LC were evident in all specimens. These perturbations included various degrees of disruption of the usually uniform LC network and alterations in the morphology of LC. Many LC had rounded, deformed cell bodies, dendrites that were shortened or completely absent" (Azizi et al. 1987). The authors speculated that the perturbations could reflect an effect of the tumor cells on LC morphology. It took, however, almost a decade before direct experimental evidence of tumor-mediated inhibition of DC functioning was reported.

Early attempts to identify potential alterations of function of DC isolated from patients with cancer or from growing tumors were inconclusive (Stene et al. 1992; Chaux et al. 1993; Tas et al. 1993), in spite of well thought-out predictions of tumor cell capability of abrogating the anticancer activity of the DC (Becker 1992) through, for instance, tumor-derived IL-10 (Becker 1993b). In fact IL-10 production by tumor cells has been soon shown to represent one of the mechanisms involved in the modulation of the antigen-presenting cell function of tumor-associated DC and in tumor immunological escape (Chaux 1995). In the early 1990s, a growing number of studies reported deficiency of DC function in the tumor microenvironment, whereas only a few reports failed to detect DC abnormalities in cancer. For instance, the capability of DC prepared from the blood to form cellular clusters with allogeneic cells was shown to be impaired in 26/44 patients with head and neck squamous cell carcinoma (HNSCC) (Stoger et al. 1993), which could be explained by defects in expression of various molecules on tumor-associated DC (Chaux et al. 1995) or small-Rho GTPase-mediated inhibition of actin polymerization and associated cell functions including motility and adhesiveness (Shurin et al. 2005b; Tourkova et al. 2007). In 1996, Gabrilovich et al. in a series of in vivo and in vitro experiments demonstrated that DC from tumor-bearing mice had a reduced ability to present antigens and stimulate T-cell proliferation (Gabrilovich et al. 1996a, b). Decreased expression of CD80 and CD86 on tumor-infiltrating DC was also reported (Chaux et al. 1996). Enk et al. (1997) revealed that melanoma-derived factors convert DC-antigen-presenting cell function to tolerance induction against tumor tissue, changing antitumor DC to "silencers" of antitumor immune responses.

These and other early reports supported the notion that inadequate presentation of tumor antigens by host DC is one potential mechanism of tumor-mediated inhibition of DC function as antigen-presenting cells (Fig. 2.1B). At the same time, it was shown that tumor-derived factors might suppress differentiation and functional maturation of DC (Gabrilovich et al. 1996a) resulting in a decreased pool of active DC capable of inducing antitumor immune responses in the tumor environment. Progressing skin tumors produce factors that inhibit LC migration from the epidermis to lymph nodes, which may enable tumors to evade the activation of protective immunity (Lucas and

Halliday 1999). Furthermore, in 1999, we reported that not only may tumors suppress DC differentiation and function, but they could also actively induce apoptotic death of DC and DC precursors both in vitro and in vivo (Esche et al. 1999; Shurin et al. 1999; Katsenelson et al. 2001). All of these early findings were confirmed later in several reports (Kiertscher et al. 2000; Pirtskhalaishvili et al. 2000, 2001; Shurin et al. 2001; Peguet-Navarro et al. 2003; Pinzon-Charry et al. 2006). Because conventional DC are the primary cell population responsible for initiating antitumor immunity, it is not surprising that tumors develop mechanisms for inhibiting DC function and longevity. The important interplay between DC and tumor cells in regulating tumor progression and antitumor immunity was first predicted (Becker 1993a) and then proven in numerous pre-clinical and clinical studies (Fig. 2.1B). Thus tumor-mediated interference with DC generation, function, and survival represents one pathway by which tumors escape host immune mechanisms via inhibiting the activity of DC (Fig. 2.1B).

Other than inhibiting DC, tumor cells are able to redirect or re-polarize differentiation of hematopoietic precursors from the conventional DC lineage to regulatory or tolerogenic DC subpopulations, protumorigenic endothelial-like DC, or to myeloid-derived suppressor cells (MDSC) and monocyte/macrophages. For instance, several reports demonstrated accumulation of tumor-supporting plasmacytoid DC (pDC or DC2) in tumor ascites in patients with ovarian cancer (Zou et al. 2001; Curiel et al. 2004; Wertel et al. 2006). Melanoma cells also induced immunosuppressive DC (McCarter et al. 2007). Interestingly, a low circulating pDC count in cancer patients, as well as low number of tumor-infiltrating pDC, was reported as a good prognostic sign (Vakkila et al. 2004; O'Donnell et al. 2007). There is evidence that tumor cells may also skew monocyte differentiation from DC to macrophage-like cells: DC generated in the presence of soluble factors produced by lung SCC and adenocarcinoma were phenotypically and functionally more similar to macrophages than to untreated control DC (Avila-Moreno et al. 2006). The accumulation of MDSC, which could precede their development into immature DC (Rossner et al. 2005) or be a separate differentiation pathway (Auffray et al. 2008; Geissmann et al. 2008), is also associated with immune suppression in tumor-bearing mice and in cancer patients. These cells exhibit a protumorigenic role due to their pleiotropic activities that include, in addition to well-defined immunosuppression and tolerance, support of mutagenesis in the tumor microenvironment, promotion of angiogenesis and metastasis, as well as directly sustaining both neoplastic growth and inflammatory reaction (Marigo et al. 2008). Thus, re-polarization of myeloid progenitor cells or macrophage DC progenitors (MDP) (Fogg et al. 2006) induced by the presence of tumor-derived factors is a second mechanism of tumor escape from immune recognition limiting DC (Fig. 2.1B).

A recently described third mechanism of tumor evasion, so-called avoidance, is the loss of expression of DC chemokines in tumor cells. DC chemoattractants are produced by all normal tissues, but their expression may be down-regulated upon transition to malignancy. Absence of the chemokine CXCL14 is the best known factor enabling the avoidance pathway. Constitutive CXCL14 (BRAK) expression was described in all tested non-malignant tissues and cell lines, but not in tumor cell lines or tumor tissues (Hromas et al. 1999; Frederick et al. 2000). Thus CXCL14 is a strong chemoattractant for immature DC and an activator of DC maturation. Expression of CXCL14 correlates with DC attraction and homing, and its absence could make tumor cells "invisible" for recruitment of DC and should be considered as an additional mechanism of tumor escape from the immunological recognition (Shurin et al. 2005a) (Fig. 2.1B).

Conventional or myeloid DC (DC1) are a fundamental part of the immune defense mechanism, which can promote specific immunity by inducing T-cell activation, expansion, and ultimately recruitment to the tumor site. Despite this, their presence within the tumor microenvironment has been associated with enhanced antitumor immunity, while tolerogenic regulatory DC populations have been shown to promote tumor cell growth and spreading, angiogenesis, and immunosuppression. This paradoxical role of DC in cancer can be explained by their functional plasticity. It may result in the polarized expression of either pro- or anti-tumorigenic functions. Key players in the setting of DC phenotype are the tumor microenvironmental signals to which DC are exposed and which selectively tune their functions within a functional spectrum encompassing the DC1 and DC2 extremes (Shurin et al. 2006).

2.4 Dendritic Cells in Cancer: Recognition of the Diversity of Regulatory Roles

That an impaired or re-directed functionality of DC in the tumor environment plays a crucial role in tumor proliferation, spreading, and metastasis, i.e., in tumor escape mechanisms (Shurin and Gabrilovich 2001; Shurin et al. 2003; Shurin and Chatta 2008) is no longer in doubt. Specifically, abnormalities of DC that allow tumor progression include defective MHC class I (Tourkova et al. 2005) and class II (Gerner et al. 2008) presentation of tumor antigens; immaturity of DC and low level of expression of co-stimulatory molecules (Chaux et al. 1996; Shurin et al. 2001); induction of regulatory DC producing IL-10 and/or TGF-β that promote regulatory T cells (Ghiringhelli et al. 2005; Lan et al. 2006); low levels of IL-12, IL-15, and IL-18 production by DC (Shurin et al. 2002; Bellone et al. 2006; Capobianco et al. 2006); expression of immunosuppressive molecules HLA-G and B7-H1 (Curiel et al. 2003; Lemaoult et al. 2007) and IDO (Hou et al. 2007); production of TNF-α and IL-8 and induction of neovascularization (Conejo-Garcia et al. 2004; Curiel et al. 2004); impaired migration (Shurin et al. 2005b) and endocytic activity (Tourkova et al. 2007); and, probably, inhibited killing activity of DC (Taieb et al. 2006) in the tumor environment. Multiple tumor-derived and stroma-derived factors are responsible for altering the DC found at the tumor site and systemically. The list of these molecules includes but is not limited to VEGF, M-CSF, IL-6, IL-10, IL-8, TGF-β, CCL2, CCL20, SDF-1, prostaglandins, gangliosides, neuropeptides, tumor antigens (e.g., PSA and MUC1), lactic acid, NO, spermine, hyaluronan, reactive oxygen species, and other largely unknown factors (Shurin and Chatta 2008). These factors could be produced by various cells in the tumor microenvironment, including not only tumor cells themselves but also fibroblasts, endothelial cells, macrophages, lymphocytes, neutrophils, and other tissue-specific cells depending on the tumor location. This suggests that cellular interactions within the tissue at the tumor site are primarily responsible for the state of DC maturation, activation, functioning, and survival and, thus, for supporting tumor progression or antitumor immunity. Furthermore, other signals generated during the course of tumor process also alter the activation of DC and include extracellular acidosis, oxidative stress, and fever-like temperature. Understanding of how DC are regulated in the tumor microenvironment and how this impacts the efficacy of DC vaccines and other immunotherapeutic means is far from complete. Only a few clinical trials focusing on protection of DC from the harmful effects of the tumor environment have been tested.

In addition to a direct regulation of DC functionality by local factors in the tumor milieu, important factors and associated conditions may influence the functional state of DC. The activity of administered DC vaccines in patients with cancer may depend on these factors. Psychological stress associated with the diagnosis of cancer, selection of the treatment options, and the prognosis of individual diseases (Fig. 2.1C). Stress-associated hormones, neuropeptides and neuromediators, such as glucocorticoids, ACTH, substance P, and catecholamines, are well documented to alter differentiation, maturation, and activity of DC. In addition to the psychoemotional stress, patients with cancer commonly experience a variety of physical/mechanical stresses linked to the treatment procedures, such as surgery, radiation, and chemotherapy. These stressors also cause an influx of common stress hormones, as well as mediators related to pain, trauma, tissue destruction, and specific effects of radiation and chemotherapy, which all strongly affect the functional state of DC (Saint-Mezard et al. 2003; Elftman et al. 2007; Goyarts et al. 2008; Kawasaki et al. 2008; Kleyn et al. 2008). Thus analyzing the behavior of DC in the tumor milieu is contingent on recognition of the acute and chronic stress conditions, which might substantially limit or exacerbate DC response to tumor/stroma-mediated signaling.

Another level of complexity is due to the pathogenic modifications of DC function in infectious, autoimmune, allergic, and other immune-mediated diseases sometimes associated with tumor development or tumor therapy (Fig. 2.1C). Alterations in DC homeostasis have been implicated in various human inflammatory, autoimmune, and allergic diseases (Adler and Steinbrink 2007; Blanco et al. 2008). Both in vitro and in vivo studies have shown that tolerogenic DC contribute to prevention of autoimmunity and allergic reactions. These cells might thus worsen the course of cancer and certain infectious diseases. Furthermore, immunosuppressive drugs used the following organ transplantation impair DC function, resulting in an increased incidence of viral associated malignancies (Sebelin et al. 2006). Mixed DC populations with a diverse spectrum of activities might be a common feature in cancer patients with mixed diseases. On the other hand, prevalence of "pro-dendritic" or "anti-dendritic" stimuli, i.e., immunogenic vs tolerogenic pathways, may determine the course of tumor progression and its response to therapy.

More than half of cancers arise in individuals older than 65 years of age. Age-related alterations of DC should be considered as a third layer of complexity in understanding the DC–tumor cell interplay in patients with cancer. It seems reasonable to suggest the existence of both pathways: (i) DC in the elderly are involved in the increased incidence of cancer and (ii) tumor-induced modulation of DC might direct appearance of various immune-mediated diseases in the elderly (Shurin et al. 2007; Agrawal et al. 2008). For instance, increased levels of IL-6 and IL-10 reported in older individuals (Caruso et al. 2004; Maggio et al. 2006) may directly affect dendropoiesis and function of DC and thus their potential to induce and maintain antitumor immunity. Thus, design of DC vaccines for elderly patients with cancer might differ from DC vaccines prepared for the younger patient population and should take into account specific behavior and immunological state of DC in aged individuals, especially if they suffer from other diseases. This simple conclusion sounds logical and reasonable, but unfortunately it is hard to transfer into modern clinical trials. Hopefully, new experimental and pre-clinical studies will demonstrate an undeniable demand for designing combinatorial DC vaccine-based therapy that accounts for specific medical conditions of cancer patients, including their age, psychoemotional status, and the presence of other immune-mediated diseases.

Cancer, arising in adults in the setting of chronic inflammation, thus can be envisioned as a network of interacting cells, emerging as a property of tissues with continuous autocrine growth as well as perpetual stress signaling. DC, as integrants of such signals arising from epithelia, stroma, endothelia, and recruited inflammatory cells, can thus play varying roles during the evolution of a tumor, in some instances impeding their development and in others actively promoting immune tolerance and impotence in the tumor microenvironment, responding to ambient microenvironmental nutrients, redox status, and oxygenation. Together, these observations suggest the existence of multiple means by which the activation, suppression, or polarization of DC can be induced in patients with cancer, supporting the view that in spite of a multiplicity of roles, the coherent manipulation of endogenous DC and rational preparation of DC vaccines could elicit robust antitumor immune responses leading to clinically significant outcomes. Full explication and development of DC-based strategies will require the emergence of a more mature perception of the tumor microenvironment in which they, and interacting inflammatory/immune cells, play their role.

References

Adler, H. S. and Steinbrink, K. 2007. Tolerogenic dendritic cells in health and disease: friend and foe! Eur J Dermatol 17:476–491

Agrawal, A., Agrawal, S., Tay, J. and Gupta, S. 2008. Biology of dendritic cells in aging. J Clin Immunol 28:14–20

Ahmann, G. B., Nadler, P. I., Birnkrant, A. and Hodes, R. J. 1979. T cell recognition in the mixed lymphocyte response. I. Non-T, radiation-resistant splenic adherent cells are the predominant stimulators in the murine mixed lymphocyte reaction. J Immunol 123:903–909

Ahmann, G. B., Nadler, P. I., Birnkrant, A. and Hodes, R. J. 1981. T cell recognition in the mixed lymphocyte response. II. Ia-positive splenic adherent cells are required for non-I region-induced stimulation. J Immunol 127:2308–2313

Ambe, K., Mori, M. and Enjoji, M. 1989. S-100 protein-positive dendritic cells in colorectal adenocarcinomas. Distribution and relation to the clinical prognosis. Cancer 63:496–503

Auffray, C., Emre, Y. and Geissmann, F. 2008. Homeostasis of dendritic cell pool in lymphoid organs. Nat Immunol 9:584–586

Avila-Moreno, F., Lopez-Gonzalez, J. S., Galindo-Rodriguez, G., Prado-Garcia, H., Bajana, S. and Sanchez-Torres, C. 2006. Lung squamous cell carcinoma and adenocarcinoma cell lines use different mediators to induce comparable phenotypic and functional changes in human monocyte-derived dendritic cells. Cancer Immunol Immunother 55:598–611

Azizi, E., Bucana, C., Goldberg, L. and Kripke, M. L. 1987. Perturbation of epidermal Langerhans cells in basal cell carcinomas. Am J Dermatopathol 9:465–473

Balfour, B., O'Brien, J. A., Perera, M., Clarke, J., Sumerska, T. and Knight, S. C. 1982. The effect of veiled cells on lymphocyte function. Adv Exp Med Biol 149:447–454

Basset, F., Soler, P., Wyllie, L., Abelanet, R., Le Charpentier, M., Kreis, B. and Breathnach, A. S. 1974. Langerhans cells in a bronchiolar-alveolar tumour of lung. Virchows Arch A Pathol Anat Histol 362:315–330

Becker, Y. 1992. Anticancer role of dendritic cells (DC) in human and experimental cancers – a review. Anticancer Res 12:511–520

Becker, Y. 1993a. Dendritic cell activity against primary tumors: an overview. In Vivo 7:187–191

Becker, Y. 1993b. Success and failure of dendritic cell (DC) anticancer activity may be modulated by nitric oxide synthetase (NOS) gene expression: a hypothesis. In Vivo 7:285–288

Bellone, G., Carbone, A., Smirne, C., Scirelli, T., Buffolino, A., Novarino, A., Stacchini, A., Bertetto, O., Palestro, G., Sorio, C., Scarpa, A., Emanuelli, G. and Rodeck, U. 2006. Cooperative induction of a tolerogenic dendritic cell phenotype by cytokines secreted by pancreatic carcinoma cells. J Immunol 177:3448–3460

Billingham, R. E. and Silvers, W. K. 1965. Some unsolved problems in the biology of the skin. In: Biology of the Skin and Hair Growth. Lyne and Short (Eds.), Angus and Robertson Publ., Sydney, pp. 1–24.

Birbeck, M. S., Breathnach, A. S. and Everall, J. D. 1961. An electron microscopy study of basal melanocytes and high-level clear cells (Langerhans cells) in vitiligo. J Invest Dermatol 37:51–64

Blanco, P., Palucka, A. K., Pascual, V. and Banchereau, J. 2008. Dendritic cells and cytokines in human inflammatory and autoimmune diseases. Cytokine Growth Factor Rev 19:41–52

Capobianco, A., Rovere-Querini, P., Rugarli, C. and Manfredi, A. A. 2006. Melanoma cells interfere with the interaction of dendritic cells with NK/LAK cells. Int J Cancer 119:2861–2869

Caruso, C., Lio, D., Cavallone, L. and Franceschi, C. 2004. Aging, longevity, inflammation, and cancer. Ann N Y Acad Sci 1028:1–13

Chaux, P. 1995. [Dendritic cells and immune function in cancer]. Pathol Biol (Paris) 43:897–903

Chaux, P., Hammann, A., Martin, F. and Martin, M. 1993. Surface phenotype and functions of tumor-infiltrating dendritic cells: CD8 expression by a cell subpopulation. Eur J Immunol 23:2517–2525

Chaux, P., Martin, M. S. and Martin, F. 1995. Defect of the CTLA4-Ig ligands on tumor-infiltrating dendritic cells. Adv Exp Med Biol 378:389–392

Chaux, P., Moutet, M., Faivre, J., Martin, F. and Martin, M. 1996. Inflammatory cells infiltrating human colorectal carcinomas express HLA class II but not B7-1 and B7-2 costimulatory molecules of the T-cell activation. Lab Invest 74:975–983
Cocchia, D., Michetti, F. and Donato, R. 1981. Immunochemical and immuno-cytochemical localization of S-100 antigen in normal human skin. Nature 294:85–87
Cochran, A. J., Pihl, E., Wen, D. R., Hoon, D. S. and Korn, E. L. 1987. Zoned immune suppression of lymph nodes draining malignant melanoma: histologic and immunohistologic studies. J Natl Cancer Inst 78:399–405
Conejo-Garcia, J. R., Benencia, F., Courreges, M. C., Kang, E., Mohamed-Hadley, A., Buckanovich, R. J., Holtz, D. O., Jenkins, A., Na, H., Zhang, L., Wagner, D. S., Katsaros, D., Caroll, R. and Coukos, G. 2004. Tumor-infiltrating dendritic cell precursors recruited by a beta-defensin contribute to vasculogenesis under the influence of Vegf-A. Nat Med 10:950–958
Curiel, T. J., Cheng, P., Mottram, P., Alvarez, X., Moons, L., Evdemon-Hogan, M., Wei, S., Zou, L., Kryczek, I., Hoyle, G., Lackner, A., Carmeliet, P. and Zou, W. 2004. Dendritic cell subsets differentially regulate angiogenesis in human ovarian cancer. Cancer Res 64:5535–5538
Curiel, T. J., Wei, S., Dong, H., Alvarez, X., Cheng, P., Mottram, P., Krzysiek, R., Knutson, K. L., Daniel, B., Zimmermann, M. C., David, O., Burow, M., Gordon, A., Dhurandhar, N., Myers, L., Berggren, R., Hemminki, A., Alvarez, R. D., Emilie, D., Curiel, D. T., Chen, L. and Zou, W. 2003. Blockade of B7-H1 improves myeloid dendritic cell-mediated antitumor immunity. Nat Med 9:562–567
David, R. and Buchner, A. 1980. Langerhans' cells in a pleomorphic adenoma of submandibular salivary gland. J Pathol 131:127–135
Eberth, C. I. 1870. Die endigung der Hautnerven. Arch. f. Mikr. Anat. 6:225–228
Elftman, M. D., Norbury, C. C., Bonneau, R. H. and Truckenmiller, M. E. 2007. Corticosterone impairs dendritic cell maturation and function. Immunology 122:279–290
Enk, A. H., Jonuleit, H., Saloga, J. and Knop, J. 1997. Dendritic cells as mediators of tumor-induced tolerance in metastatic melanoma. Int J Cancer 73:309–316
Esche, C., Lokshin, A., Shurin, G. V., Gastman, B. R., Rabinowich, H., Watkins, S. C., Lotze, M. T. and Shurin, M. R. 1999. Tumor's other immune targets: dendritic cells. J Leukoc Biol 66:336–344
Fernandez-Bussy, R., Cambazard, F., Mauduit, G., Schmitt, D. and Thivolet, J. 1983. T cell subsets and Langerhans cells in skin tumours. Eur J Cancer Clin Oncol 19:907–913
Fogg, D. K., Sibon, C., Miled, C., Jung, S., Aucouturier, P., Littman, D. R., Cumano, A. and Geissmann, F. 2006. A clonogenic bone marrow progenitor specific for macrophages and dendritic cells. Science 311:83–87
Frederick, M. J., Henderson, Y., Xu, X., Deavers, M. T., Sahin, A. A., Wu, H., Lewis, D. E., El-Naggar, A. K. and Clayman, G. L. 2000. In vivo expression of the novel CXC chemokine BRAK in normal and cancerous human tissue. Am J Pathol 156:1937–1950
Furukawa, T., Watanabe, S., Kodama, T., Sato, Y., Shimosato, Y. and Suemasu, K. 1985. T-zone histiocytes in adenocarcinoma of the lung in relation to postoperative prognosis. Cancer 56:2651–2656
Gabrilovich, D. I., Chen, H. L., Girgis, K. R., Cunningham, H. T., Meny, G. M., Nadaf, S., Kavanaugh, D. and Carbone, D. P. 1996a. Production of vascular endothelial growth factor by human tumors inhibits the functional maturation of dendritic cells. Nat Med 2:1096–1103
Gabrilovich, D. I., Ciernik, I. F. and Carbone, D. P. 1996b. Dendritic cells in antitumor immune responses. I. Defective antigen presentation in tumor-bearing hosts. Cell Immunol 170:101–110
Gatter, K. C., Morris, H. B., Roach, B., Mortimer, P., Fleming, K. A. and Mason, D. Y. 1984. Langerhans' cells and T cells in human skin tumours: an immunohistological study. Histopathology 8:229–244

Geissmann, F., Auffray, C., Palframan, R., Wirrig, C., Ciocca, A., Campisi, L., Narni-Mancinelli, E. and Lauvau, G. 2008. Blood monocytes: distinct subsets, how they relate to dendritic cells, and their possible roles in the regulation of T-cell responses. Immunol Cell Biol 86:398–408

Gerner, M. Y., Casey, K. A. and Mescher, M. F. 2008. Defective MHC class II presentation by dendritic cells limits CD4 T cell help for antitumor CD8 T cell responses. J Immunol 181:155–164

Ghiringhelli, F., Puig, P. E., Roux, S., Parcellier, A., Schmitt, E., Solary, E., Kroemer, G., Martin, F., Chauffert, B. and Zitvogel, L. 2005. Tumor cells convert immature myeloid dendritic cells into TGF-beta-secreting cells inducing CD4+CD25+ regulatory T cell proliferation. J Exp Med 202:919–929

Goyarts, E., Matsui, M., Mammone, T., Bender, A. M., Wagner, J. A., Maes, D. and Granstein, R. D. 2008. Norepinephrine modulates human dendritic cell activation by altering cytokine release. Exp Dermatol 17:188–196

Gyure, L. A., Barfoot, R., Denham, S. and Hall, J. G. 1987. Immunity to a syngeneic sarcoma induced in rats by dendritic lymph cells exposed to the tumour either in vivo or in vitro. Br J Cancer 55:17–20

Hashimoto, K. 1971. Langerhans' cell granule. An endocytotic organelle. Arch Dermatol 104:148–160

Hashimoto, K. and Tarnowski, W. M. 1968. Some new aspects of the Langerhans cell. Arch Dermatol 97:450–464

Haustein, U. F. 1979. [The Langerhans cell–its macrophages-analogous function in the triggering of the allergic contact eczema (author's transl)]. Allerg Immunol (Leipz) 25:116–131

Heusermann, U., Stutte, H. J. and Muller-Hermelink, H. K. 1974. Interdigitating cells in the white pulp of the human spleen. Cell Tissue Res 153:415–417

Hoefsmit, E. C. 1982. Macrophages, Langerhans cells, interdigitating and dendritic accessory cells: a summary. Adv Exp Med Biol 149:463–468

Hoefsmit, E. C., Balfour, B. M., Kamperdijk, E. W. and Cvetanov, J. 1979. Cells containing Birbeck granules in the lymph and the lymph node. Adv Exp Med Biol 114:389–394

Hou, D. Y., Muller, A. J., Sharma, M. D., DuHadaway, J., Banerjee, T., Johnson, M., Mellor, A. L., Prendergast, G. C. and Munn, D. H. 2007. Inhibition of indoleamine 2,3-dioxygenase in dendritic cells by stereoisomers of 1-methyl-tryptophan correlates with antitumor responses. Cancer Res 67:792–801

Hromas, R., Broxmeyer, H. E., Kim, C., Nakshatri, H., Christopherson, K., 2nd, Azam, M. and Hou, Y. H. 1999. Cloning of BRAK, a novel divergent CXC chemokine preferentially expressed in normal versus malignant cells. Biochem Biophys Res Commun 255:703–706

Igisu, K., Watanabe, S., Shimosato, Y. and Kukita, A. 1983. Langerhans cells and their precursors with S100 protein in mycosis fungoides. Jpn J Clin Oncol 13:693–702

Jimbow, K., Sato, S. and Kukita, A. 1969. Cells containing Langerhans granules in human lymph nodes of dermatopathic lymphadenopathy. J Invest Dermatol 53:295–299

Katsenelson, N. S., Shurin, G. V., Bykovskaia, S. N., Shogan, J. and Shurin, M. R. 2001. Human small cell lung carcinoma and carcinoid tumor regulate dendritic cell maturation and function. Mod Pathol 14:40–45

Kawasaki, T., Choudhry, M. A., Schwacha, M. G., Fujimi, S., Lederer, J. A., Bland, K. I. and Chaudry, I. H. 2008. Trauma-hemorrhage inhibits splenic dendritic cell proinflammatory cytokine production via a mitogen-activated protein kinase process. Am J Physiol Cell Physiol 294:C754–764

Kelly, R. H., Balfour, B. M., Armstrong, J. A. and Griffiths, S. 1978. Functional anatomy of lymph nodes. II. Peripheral lymph-borne mononuclear cells. Anat Rec 190:5–21

Kidd, R. L., Krawczyk, W. S. and Wilgram, G. F. 1971. The Merkel cell in human epidermis: its differentiation from other dendritic cells. Arch Dermatol Forsch 241:374–384

Kiertscher, S. M., Luo, J., Dubinett, S. M. and Roth, M. D. 2000. Tumors promote altered maturation and early apoptosis of monocyte-derived dendritic cells. J Immunol 164:1269–1276
Klareskog, L., Tjernlund, U., Forsum, U. and Peterson, P. A. 1977. Epidermal Langerhans cells express Ia antigens. Nature 268:248–250
Kleyn, C. E., Schneider, L., Saraceno, R., Mantovani, C., Richards, H. L., Fortune, D. G., Cumberbatch, M., Dearman, R. J., Terenghi, G., Kimber, I. and Griffiths, C. E. 2008. The effects of acute social stress on epidermal Langerhans' cell frequency and expression of cutaneous neuropeptides. J Invest Dermatol 128:1273–1279
Knight, S. C., Balfour, B. M., O'Brien, J., Buttifant, L., Sumerska, T. and Clarke, J. 1982. Role of veiled cells in lymphocyte activation. Eur J Immunol 12:1057–1060
Knight, S. C., Hunt, R., Dore, C. and Medawar, P. B. 1985. Influence of dendritic cells on tumor growth. Proc Natl Acad Sci USA 82:4495–4497
Kondo, Y. 1969. Macrophages containing Langerhans cell granules in normal lymph nodes of the rabbit. Z Zellforsch Mikrosk Anat 98:506–511
Kurihara, K. and Hashimoto, N. 1985. The pathological significance of Langerhans cells in oral cancer. J Oral Pathol 14:289–298
Kuwahara, H. 1971. Fine structures of Langerhans cell with special references to their function. Jap J Dermatol Series B 80:195–210
Lan, Y. Y., Wang, Z., Raimondi, G., Wu, W., Colvin, B. L., de Creus, A. and Thomson, A. W. 2006. "Alternatively activated" dendritic cells preferentially secrete IL-10, expand Foxp3 + CD4 + T cells, and induce long-term organ allograft survival in combination with CTLA4-Ig. J Immunol 177:5868–5877
Langerhans, P. 1868. Uber die Nerven der menschlichen Haut. . Virhouv Arch. [Pathol of Anat] 44:325–337
Lauriola, L., Michetti, F., Sentinelli, S. and Cocchia, D. 1984. Detection of S-100 labelled cells in nasopharyngeal carcinoma. J Clin Pathol 37:1235–1238
Lemaoult, J., Caumartin, J., Daouya, M., Favier, B., Rond, S. L., Gonzalez, A. and Carosella, E. D. 2007. Immune regulation by pretenders: cell-to-cell transfers of HLA-G make effector T cells act as regulatory cells. Blood 109:2040–2048
Lisi, P. 1973. Investigation on Langerhans cells in pathological human epidermis. Acta Derm Venereol 53:425–428
Loning, T., Caselitz, J., Seifert, G., Weber, K. and Osborn, M. 1982. Identification of Langerhans cells: simultaneous use of sera to intermediate filaments, T6 and HLA-DR antigens on oral mucosa, human epidermis and their tumours. Virchows Arch A Pathol Anat Histopathol 398:119–128
Lucas, A. D. and Halliday, G. M. 1999. Progressor but not regressor skin tumours inhibit Langerhans' cell migration from epidermis to local lymph nodes. Immunology 97:130–137
Macadam, R. F. 1978. An electron-microscopic study of basal cell carcinoma. J Pathol 126:149–156
Maggio, M., Guralnik, J. M., Longo, D. L. and Ferrucci, L. 2006. Interleukin-6 in aging and chronic disease: a magnificent pathway. J Gerontol A Biol Sci Med Sci 61:575–584
Marigo, I., Dolcetti, L., Serafini, P., Zanovello, P. and Bronte, V. 2008. Tumor-induced tolerance and immune suppression by myeloid derived suppressor cells. Immunol Rev 222:162–179
McCarter, M. D., Baumgartner, J., Escobar, G. A., Richter, D., Lewis, K., Robinson, W., Wilson, C., Palmer, B. E. and Gonzalez, R. 2007. Immunosuppressive dendritic and regulatory T cells are upregulated in melanoma patients. Ann Surg Oncol 14:2854–2860
Mishima, Y. and Kawasaki, H. 1970. Dendritic cell dynamics in progressive depigmentation. Read before the VII. International Pigment Cell Conference, Seattle, September 2–6, 1969. J Invest Derm 54:93
Mishima, Y., Kawasaki, H. and Pinkus, H. 1972. Dendritic cell dynamics in progressive depigmentations. Distinctive cytokinetics of -dendritic cells revealed by electron microscopy. Arch Dermatol Forsch 243:67–87

Muller, H. K., Halliday, G. M. and Knight, B. A. 1985. Carcinogen-induced depletion of cutaneous Langerhans cells. Br J Cancer 52:81–85
Nakajima, T., Kodama, T., Tsumuraya, M., Shimosato, Y. and Kameya, T. 1985. S-100 protein-positive Langerhans cells in various human lung cancers, especially in peripheral adenocarcinomas. Virchows Arch A Pathol Anat Histopathol 407:177–189
Nakajima, T., Watanabe, S., Sato, Y., Kameya, T., Hirota, T. and Shimosato, Y. 1982. An immunoperoxidase study of S-100 protein distribution in normal and neoplastic tissues. Am J Surg Pathol 6:715–727
Nomori, H., Watanabe, S., Nakajima, T., Shimosato, Y. and Kameya, T. 1986. Histiocytes in nasopharyngeal carcinoma in relation to prognosis. Cancer 57:100–105
O'Donnell, R. K., Mick, R., Feldman, M., Hino, S., Wang, Y., Brose, M. S. and Muschel, R. J. 2007. Distribution of dendritic cell subtypes in primary oral squamous cell carcinoma is inconsistent with a functional response. Cancer Lett 255:145–152
Olah, I., Dunay, C., Rohlich, P. and Toro, I. 1968. A special type of cells in the medulla of the rat thymus. Acta Biol Acad Sci Hung 19:97–113
Peguet-Navarro, J., Sportouch, M., Popa, I., Berthier, O., Schmitt, D. and Portoukalian, J. 2003. Gangliosides from human melanoma tumors impair dendritic cell differentiation from monocytes and induce their apoptosis. J Immunol 170:3488–3494
Pehamberger, H., Stingl, L. A., Pogantsch, S., Steiner, G., Wolff, K. and Stingl, G. 1983. Epidermal cell-induced generation of cytotoxic T-lymphocyte responses against alloantigens or TNP-modified syngeneic cells: requirement for Ia-positive Langerhans cells. J Invest Dermatol 81:208–211
Pinzon-Charry, A., Maxwell, T., McGuckin, M. A., Schmidt, C., Furnival, C. and Lopez, J. A. 2006. Spontaneous apoptosis of blood dendritic cells in patients with breast cancer. Breast Cancer Res 8:R5
Pirtskhalaishvili, G., Shurin, G. V., Esche, C., Trump, D. L. and Shurin, M. R. 2001. TNF-alpha protects dendritic cells from prostate cancer-induced apoptosis. Prostate Cancer Prostatic Dis 4:221–227
Pirtskhalaishvili, G., Shurin, G. V., Gambotto, A., Esche, C., Wahl, M., Yurkovetsky, Z. R., Robbins, P. D. and Shurin, M. R. 2000. Transduction of dendritic cells with Bcl-xL increases their resistance to prostate cancer-induced apoptosis and antitumor effect in mice. J Immunol 165:1956–1964
Poppema, S., Brocker, E. B., de Leij, L., Terbrack, D., Visscher, T., Ter Haar, A., Macher, E., The, T. H. and Sorg, C. 1983. In situ analysis of the mononuclear cell infiltrate in primary malignant melanoma of the skin. Clin Exp Immunol 51:77–82
Porfiri, B., Nigro, M. and Garcovich, A. 1979. [Presence of clusters of Langerhans cells and lymphoid cells in the peritumoral infiltrate of a case of basal cell naevus syndrome. Ultrastructural study (author's transl)]. Ann Dermatol Venereol 106:237–240
Prunieras, M. 1969. Interactions between keratinocytes and dendritic cells. J Invest Dermatol 52:1–17
Ranvier, L. 1875. Traite technique d'Histologie. Savy Edit. Paris.
Richtsmeier, W. J., Bowers, W. E., Ellsworth, C. A., Sorge, K. and Berkowitz, M. 1984. Dendritic cell identification in head and neck lymphoid tissue. Newly recognized cells control T-lymphocyte functions. Arch Otolaryngol 110:701–706
Rossner, S., Voigtlander, C., Wiethe, C., Hanig, J., Seifarth, C. and Lutz, M. B. 2005. Myeloid dendritic cell precursors generated from bone marrow suppress T cell responses via cell contact and nitric oxide production in vitro. Eur J Immunol 35:3533–3544
Rowden, G., Lewis, M. G. and Sullivan, A. K. 1977. Ia antigen expression on human epidermal Langerhans cells. Nature 268:247–248
Rowden, G., Phillips, T. M., Lewis, M. G. and Wilkinson, R. D. 1979. Target role of Langerhans cells in mycosis fungoides: transmission and immuno-electron microscopic studies. J Cutan Pathol 6:364–382

Sagebiel, R. W. 1972. In vivo and in vitro uptake of ferritin by Langerhans cells of the epidermis. J Invest Dermatol 58:47–54
Saint-Mezard, P., Chavagnac, C., Bosset, S., Ionescu, M., Peyron, E., Kaiserlian, D., Nicolas, J. F. and Berard, F. 2003. Psychological stress exerts an adjuvant effect on skin dendritic cell functions in vivo. J Immunol 171:4073–4080
Schenk, P. 1980. [Langerhans cells in invasive laryngeal carcinoma (author's transl)]. Laryngol Rhinol Otol (Stuttg) 59:232–237
Sebelin, K., Schulzki, A., Kloetzel, P. M., Dorken, B., Pezzutto, A. and Subklewe, M. 2006. Impairment of circulating myeloid dendritic cells in immunosuppressed renal/pancreas transplant recipients. Transplantation 82:779–787
Shurin, M. R. and Chatta, G. 2008. Immunobiology of dendritic cells in cancer. In: Mechanisms of Therapeutic Reversal of Immune Suppression in Cancer. D. I. Gabrilovich and A. Hurtwiz (Eds.), pp. 101–130.
Shurin, M. R., Esche, C., Lokshin, A. and Lotze, M. T. 1999. Apoptosis of dendritic cells. In: Dendritic Cells. M .T. Lotze and A. W. Thomson (Eds.), Academic Press, San Diego, CA, pp. 673–692.
Shurin, G. V., Ferris, R. L., Tourkova, I. L., Perez, L., Lokshin, A., Balkir, L., Collins, B., Chatta, G. S. and Shurin, M. R. 2005a. Loss of new chemokine CXCL14 in tumor tissue is associated with low infiltration by dendritic cells (DC), while restoration of human CXCL14 expression in tumor cells causes attraction of DC both in vitro and in vivo. J Immunol 174:5490–5498
Shurin, G. V., Shurin, M. R., Bykovskaia, S., Shogan, J., Lotze, M. T. and Barksdale, E. M., Jr. 2001. Neuroblastoma-derived gangliosides inhibit dendritic cell generation and function. Cancer Res 61:363–369
Shurin, G. V., Tourkova, I. L., Chatta, G. S., Schmidt, G., Wei, S., Djeu, J. Y. and Shurin, M. R. 2005b. Small rho GTPases regulate antigen presentation in dendritic cells. J Immunol 174:3394–3400
Shurin, M. R. and Gabrilovich, D. I. 2001. Regulation of dendritic cell system by tumor. Cancer Res, Ther Control 11:65–78
Shurin, M. R., Shurin, G. V. and Chatta, G. S. 2007. Aging and the dendritic cell system: implications for cancer. Crit Rev Oncol Hematol 64:90–105
Shurin, M. R., Shurin, G. V., Lokshin, A., Yurkovetsky, Z. R., Gutkin, D. W., Chatta, G., Zhong, H., Han, B. and Ferris, R. L. 2006. Intratumoral cytokines/chemokines/growth factors and tumor infiltrating dendritic cells: friends or enemies? Cancer Metastasis Rev 25:333–356
Shurin, G. V., Yurkovetsky, Z. R. and Shurin, M. R. 2003. Tumor-induced dendritic cell dysfunction. In: Mechanisms of Tumor Escape from the Immune Response. A. C. Ochoa (Ed.), Taylor & Francis Publ., New York, pp. 112–138.
Shurin, M. R., Yurkovetsky, Z. R., Tourkova, I. L., Balkir, L. and Shurin, G. V. 2002. Inhibition of CD40 expression and CD40-mediated dendritic cell function by tumor-derived IL-10. Int J Cancer 101:61–68
Silberberg, I. 1971. Ultrastructural studies of Langerhans cells in contact sensitive and irritant reactions to mercuric chloride (abstr.). Clin Res 19:715
Silberberg, I. 1972. Ultrastructural features of Langerhans cells at sites of negative and positive contact allergic patch test reactions (abstr.). Clin Res 20:419.
Silberberg, I. 1973. Apposition of mononuclear cells to langerhans cells in contact allergic reactions. An ultrastructural study. Acta Derm Venereol 53:1–12
Silberberg, I., Baer, R. L. and Rosenthal, S. A. 1974. Circulating Langerhans cells in a dermal vessel. Acta Derm Venereol 54:81–85
Silberberg, I., Baer, R. L. and Rosenthal, S. A. 1976. The role of Langerhans cells in allergic contact hypersensitivity. A review of findings in man and guinea pigs. J Invest Dermatol 66:210–217

Spry, C. J., Pflug, A. J., Janossy, G. and Humphrey, J. H. 1980. Large mononuclear (veiled) cells like 'Ia-like' membrane antigens in human afferent lympn. Clin Exp Immunol 39:750–755
Steinman, R. M. and Cohn, Z. A. 1973. Identification of a novel cell type in peripheral lymphoid organs of mice. I. Morphology, quantitation, tissue distribution. J Exp Med 137:1142–1162
Steinman, R. M. and Nussenzweig, M. C. 1980. Dendritic cells: features and functions. Immunol Rev 53:127–147
Steinman, R. M. and Witmer, M. D. 1978. Lymphoid dendritic cells are potent stimulators of the primary mixed leukocyte reaction in mice. Proc Natl Acad Sci USA 75:5132–5136
Stene, M. A., Holthoj-Asnong, C. and Cochran, A. J. 1992. Effect of epidermal Langerhans cells from melanoma patients on lymphoproliferative responses. Melanoma Res 2:57–62
Sterry, W. and Steigleder, G. K. 1979. Acid nonspecific esterase in human Langerhans cells. J Cutan Pathol 6:476–478
Stingl, G., Katz, S. I., Clement, L., Green, I. and Shevach, E. M. 1978a. Immunologic functions of Ia-bearing epidermal Langerhans cells. J Immunol 121:2005–2013
Stingl, G., Katz, S. I., Shevach, E. M., Rosenthal, A. S. and Green, I. 1978b. Analogous functions of macrophages and Langerhans cells in the initiation in the immune response. J Invest Dermatol 71:59–64
Stingl, G., Wolff-Schreiner, E. C., Pichler, W. J., Gschnait, F., Knapp, W. and Wolff, K. 1977. Epidermal Langerhans cells bear Fc and C3 receptors. Nature 268:245–246
Stoger, H., Wilders-Truschnig, M., Samonigg, H., Schmid, M., Bauernhofer, T., Tiran, A., Tas, M. and Drexhage, H. A. 1993. The presence of immunosuppressive 'p15E-like' factors in the serum and urine of patients suffering from malign and benign breast tumours. Clin Exp Immunol 93:437–441
Szekeres, L. and Daroczy, J. 1981. Electron microscopic investigation on the local cellular reaction to primary malignant melanoma. Dermatologica 163:137–144
Taieb, J., Chaput, N., Menard, C., Apetoh, L., Ullrich, E., Bonmort, M., Pequignot, M., Casares, N., Terme, M., Flament, C., Opolon, P., Lecluse, Y., Metivier, D., Tomasello, E., Vivier, E., Ghiringhelli, F., Martin, F., Klatzmann, D., Poynard, T., Tursz, T., Raposo, G., Yagita, H., Ryffel, B., Kroemer, G. and Zitvogel, L. 2006. A novel dendritic cell subset involved in tumor immunosurveillance. Nat Med 12:214–219
Tarnowski, W. M. and Hashimoto, K. 1967. Langerhans' cell granules in histiocytosis X. The epidermal Langerhans' cell as a macrophage. Arch Dermatol 96:298–304
Tas, M. P., Simons, P. J., Balm, F. J. and Drexhage, H. A. 1993. Depressed monocyte polarization and clustering of dendritic cells in patients with head and neck cancer: in vitro restoration of this immunosuppression by thymic hormones. Cancer Immunol Immunother 36:108–114
Tay, S. K., Jenkins, D., Maddox, P., Campion, M. and Singer, A. 1987. Subpopulations of Langerhans' cells in cervical neoplasia. Br J Obstet Gynaecol 94:10–15
Thomas, J. A., Iliescu, V., Crawford, D. H., Ellouz, R., Cammoun, M. and de-The, G. 1984. Expression of HLA-DR antigens in nasopharyngeal carcinoma: an immunohistological analysis of the tumour cells and infiltrating lymphocytes. Int J Cancer 33:813–819
Tourkova, I. L., Shurin, G. V., Chatta, G. S., Perez, L., Finke, J., Whiteside, T. L., Ferrone, S. and Shurin, M. R. 2005. Restoration by IL-15 of MHC class I antigen-processing machinery in human dendritic cells inhibited by tumor-derived gangliosides. J Immunol 175:3045–3052
Tourkova, I. L., Shurin, G. V., Wei, S. and Shurin, M. R. 2007. Small rho GTPases mediate tumor-induced inhibition of endocytic activity of dendritic cells. J Immunol 178:7787–7793
Tsujitani, S., Furukawa, T., Tamada, R., Okamura, T., Yasumoto, K. and Sugimachi, K. 1987. Langerhans cells and prognosis in patients with gastric carcinoma. Cancer 59:501–505

Vakkila, J., Thomson, A. W., Vettenranta, K., Sariola, H. and Saarinen-Pihkala, U. M. 2004. Dendritic cell subsets in childhood and in children with cancer: relation to age and disease prognosis. Clin Exp Immunol 135:455–461
Van Haelst, U. J. 1969. Light and electron microscopic study of the normal and pathological thymus of the rat. 3. A mesenchymal histiocytic type of cell. Z Zellforsch Mikrosk Anat 99:198–209
Van Voorhis, W. C., Valinsky, J., Hoffman, E., Luban, J., Hair, L. S. and Steinman, R. M. 1983a. Relative efficacy of human monocytes and dendritic cells as accessory cells for T cell replication. J Exp Med 158:174–191
Van Voorhis, W. C., Witmer, M. D. and Steinman, R. M. 1983b. The phenotype of dendritic cells and macrophages. Fed Proc 42:3114–3118
Veldman, J. E., Molenaar, I. and Keuning, F. J. 1978. Electron microscopy of cellular immunity reactions in B-cell deprived rabbits. Thymus derived antigen reactive cells, their micro-environment and progeny in the lymph node. Virchows Arch B Cell Pathol 28:217–228
Vernon, M. L., Fountain, L., Krebs, H. M., Horta-Barbosa, L., Fuccillo, D. A. and Sever, J. L. 1973. Birbeck granules (Langerhans' cell granules) in human lymph nodes. Am J Clin Pathol 60:771–779
Wertel, F., Polak, G., Rolinski, J., Barczynski, B. and Kotarski, J. 2006. Myeloid and lymphoid dendritic cells in the peritoneal fluid of women with ovarian cancer. Adv Med Sci 51:174–177
Wilborn, W. H., Dismukes, D. E. and Montes, L. F. 1978. Ultrastructural identification of Langerhans cells in seborrheic keratoses. J Cutan Pathol 5:368–372
Wolff, K. and Schreiner, E. 1970. Uptake, intracellular transport and degradation of exogenous protein by Langerhans cells. An electron microscopic-cytochemical study using peroxidase as tracer substance. J Invest Dermatol 54:37–47
Zelickson, A. S. 1965. The Langerhans Cell. J Invest Dermatol 44:201–212
Zelickson, A. S. and Mottaz, J. H. 1968. Epidermal dendritic cells. A quantitative study. Arch Dermatol 98:652–659
Zou, W., Machelon, V., Coulomb-L'Hermin, A., Borvak, J., Nome, F., Isaeva, T., Wei, S., Krzysiek, R., Durand-Gasselin, I., Gordon, A., Pustilnik, T., Curiel, D. T., Galanaud, P., Capron, F., Emilie, D. and Curiel, T. J. 2001. Stromal-derived factor-1 in human tumors recruits and alters the function of plasmacytoid precursor dendritic cells. Nat Med 7:1339–1346

Part I
Alterations of the Dendritic Cell System in Cancer

From the very first publications examining dendritic cells in solid tumors it has been evident that intratumoral dendritic cells are morphologically and functionally altered. These observations have been confirmed by functional analyses of dendritic cells isolated or generated from tumor-bearing animals or patients with cancer. The following chapters in this section describe, respectively, where and how dendritic cells localize within different tumor types and which immunohistochemical markers are useful for characterization of tumor-associated dendritic cells; abnormalities in key functions of dendritic cells and dendritic cell subsets that develop as tumors progress both in experimental animals and patients with cancer; and pathways involved in dendritic cell elimination in the tumor microenvironment, including inhibition of dendropoiesis, i.e., dendritic cell differentiation and maturation, polarization of dendritic cell development and stimulation of apoptotic death of dendritic cells. These processes are described without emphasis on the molecular mechanisms or signaling pathways involved, as these topics will be discussed in the next section of the book.

Chapter 3
Tumor-Infiltrating Dendritic Cells: The Pathologist's Perspective

Dmitry W. Gutkin

Abstract In order to perform immunosurveillance and initiate antitumor immune response, dendritic cells must be physically present in malignant tissues. This chapter summarizes data on the density and functional status of tumor-infiltrating dendritic cells in different types of human malignant neoplasms, generated by immunohistochemistry methods. The results show that density and maturation of tumor-associated dendritic cells vary in specific types of tumors and correlate with histologic grade and clinical stage of the disease. In majority of cases there is positive correlation between intratumoral dendritic cell density and favorable disease prognosis; however, certain exceptions are present. Possible mechanisms of tumor-related alterations of dendritic cell homing, survival, and maturation at the tumor site are discussed.

3.1 Dendritic Cells at the Tumor Site: Introduction

According to the cancer immune surveillance paradigm, the immune system is able to detect and eliminate emerging malignant cells to prevent their uncontrolled proliferation. The key element of this surveillance system is a dendritic cell (DC) that is capable of recognizing a malignant cell, uptake cancer-related antigens, process them, and present to T lymphocytes. In order to complete this task, DC must be able to enter the tumor, survive in the tumor microenvironment, follow maturation steps, and emigrate from the tumor site (Fig. 3.1).

Conventional DC emerge in the tissue from hematopoietic blood-born precursors. In their immature state they express specific proteins that allow them to uptake and process antigens. Immature DC (iDC) are characterized by high endocytic activity and low T-cell activation potential. They express chemokine receptors (i.e., CCR6) that allow their migration into tissue. After uptaking antigenic materials, DC undergo maturation steps, up-regulate the

D.W. Gutkin (✉)
Program Leader, Pathology and Lab Medicine, VAMC Pittsburgh, University Drive C, 132L-U, Pittsburgh, PA 15240, USA
e-mail: dwgutkin@hotmail.com

M.R. Shurin, R.D. Salter (eds.), *Dendritic Cells in Cancer*,
DOI 10.1007/978-0-387-88611-4_3, © Springer Science+Business Media, LLC 2009

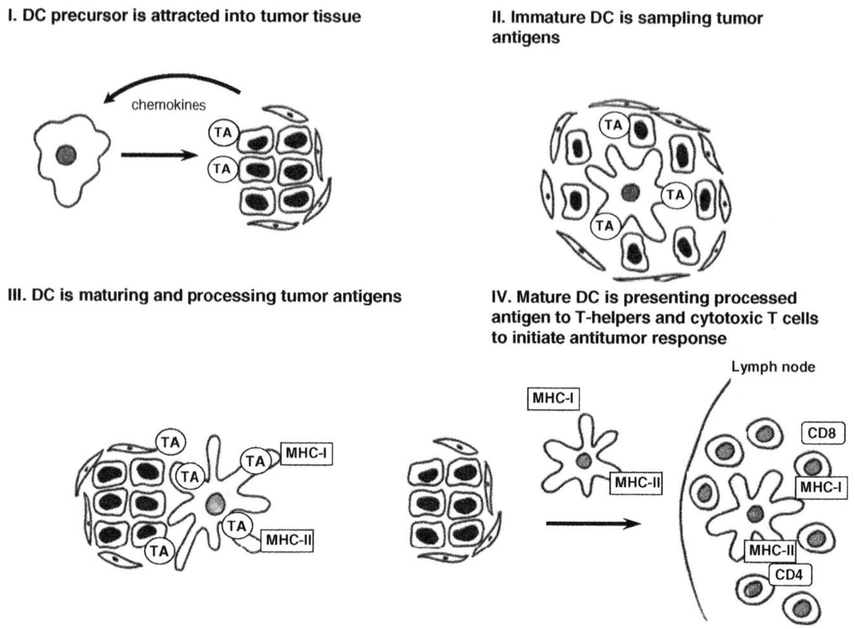

Fig. 3.1 Dendritic cell role in initiating antitumor immune response. (**A**) Tumor cells attract DC precursors by releasing chemokines. (**B**) Immature DC sample neoplastic cells, engulf, and process tumor antigens. (**C**) While maturing, TIDC up-regulate expression of MHC class I and class II molecules for antigen presentation. (**D**) DC emigrate from the tumor tissue to the regional lymph nodes to present processed tumor antigens in the context of MHC class I and class II complexes to antigen-specific CD8 and CD4 T cells. DC, dendritic cells; TA, tumor antigen(s)

proteins of antigen-processing machinery (APM), and travel to the regional lymph nodes. Simultaneously, they up-regulate cell surface receptors, such as CD80, CD86, and CD40, that act as co-receptors in T-cell activation (Hashimoto et al. 2000). These processes are regulated by cytokines, some of which, like IL-4 and GM-CSF, direct differentiation of DC, while others, like TNF-α, induce their maturation (Caux et al. 2000; Shurin et al. 2006). DC also up-regulate expression of CCR7, a chemotactic receptor that directs their trafficking through the lymphatic system to a lymph node, in response to chemokines CCL19 and CCL21. Finally, mature DC (mDC) present processed antigens to CD8 and CD4 T cells in the context of MHC I and MHC II complexes, respectively (Fig. 3.1). In addition, DC can perform an effector cell functions, triggering tumor cell death by various mechanisms (Manna and Mohanakumar 2002; Huang et al. 2005; Nicolas et al. 2007). A different type of DC – plasmacytoid DC, is predominantly found in peripheral lymph nodes (Cox et al. 2005). These cells can be detected in the tumor tissue; however, their role remains largely unknown.

In the last 25 years, numerous studies were performed to clarify the role of DC in neoplastic process. While describing tumor-infiltrating DC (TIDC), researchers usually answered the following questions:

1. Are these cells present in the tumor tissue and in what numbers?
2. In what stage of maturation are they present (what is the relative number of iDC and mDC in the tumor tissue)?
3. What is the spatial distribution of TIDC (are they located in tumor epithelium, advancing tumor margin, or peritumoral stroma)?
4. Is there a correlation of number, state of maturation, or spatial distribution of TIDC with tumor grade, stage, and prognosis?

As it will be shown below, today there are still no solid answers to these questions, and in many instances the results of various studies are controversial. Additionally, there are no uniformly acceptable methods of studying TIDC, therefore careful attention to techniques is warranted to compare results of different studies.

3.2 TIDC: Technical Aspects

Evaluation of DC in tumors is commonly performed by microscopic examination of frozen or fixed tumor tissue. However, routine histochemical stains (i.e., H&E) do not allow recognizing TIDC with certainty. In addition, common histology cannot discriminate between different subpopulations of DC or determine their state of maturation. Thus, evaluation of TIDC requires utilization of methods that can determine not only morphology of the cells but also their molecular phenotype. These methods include immunohistochemistry that detects the specific protein expression in the cells of interest and in situ hybridization that detects the expression of specific mRNAs. Specific proteins that are tested most often are listed in Table 3.1. For evaluation of total number of TIDC the most often used marker is S-100 protein. Specific proteins usually utilized to identify iDC include CD1a, CD209/DC-SIGN, and CD207/Langerin. For mDC this list consists of CD83, CD86, CD208/LAMP, and HLA–DR. The specific marker of plasmacytoid DC is CD123. Novel immunostains are being constantly introduced (Table 3.1).

Immunohistochemistry and in situ hybridization can be performed on fresh and fixed tissues. Formalin-fixed paraffin-embedded tissue is used most often, since it is the main type of tissue available for the pathology research. Tumor tissue is cut into slices 5–10 μm thick and specially treated to make cellular proteins more accessible and detectable (antigen retrieval) (Fig. 3.2). Specific, usually monoclonal, antibodies against proteins of interest are applied to the tissue and the resulting antigen–antibody complexes are visualized with fluorescent or color stains. In situ hybridization, instead of looking for specific proteins, is detecting specific mRNAs by using complementary

Table 3.1 Commonly used immunohistochemical markers for TIDC analysis

Antibody	Properties	Target
S-100 protein	Low-molecular weight protein characterized by two calcium binding sites of the helix-loop-helix conformation	All types of DC (mature and immature DC). Cells derived from neural crest (Schwann cells, melanocytes, glial cells), chondrocytes, adipocytes, and myoepithelial cells
CD1a	49 kDa cell surface glycoprotein expressed in association with β2-microglobulin. Expressed predominantly in early steps of DC maturation	Myeloid immature DC
CD209/DC-SIGN (DC-specific ICAM-3-grabbing non-integrin)	DC-specific adhesion receptor that mediates DC binding to ICAM-3. Presumably mediates the recognition of non-self and the presentation of foreign antigens. Can regulate important adhesion processes	Myeloid immature DC
CD207/Langerin	C-type lectin responsible for the formation of Birbeck granules, a typical hallmark for Langerhans cells	Myeloid immature DC
CD83	40–45 kDa glycoprotein expressed predominantly in the late steps of DC maturation. CD83+ DC co-express the highest levels of HLA II molecules	Myeloid mature DC
HLA–DR	Major histocompatibility complex class II molecules, cell surface receptor	Mature DC, B lymphocytes, monocytes/macrophages
CD86	Membrane protein of the immunoglobulin superfamily, which provides a co-stimulatory signal for T-cell activation and survival	Myeloid mature DC
CD208/DC–LAMP (Dendritic cell–lysosomal-associated membrane protein),	Member of the lysosomal-associated membrane protein (LAMP) family. Plays an important role in antigen processing and MHC class II-restricted antigen presentation	Myeloid mature DC
CD123	IL-3 receptor α-chain involved in cell growth and differentiation	Plasmacytoid DC
CD11c	Part of an adhesion molecule of integrin type (integrin gp150/95)	Different types of DC and some other leukocytes

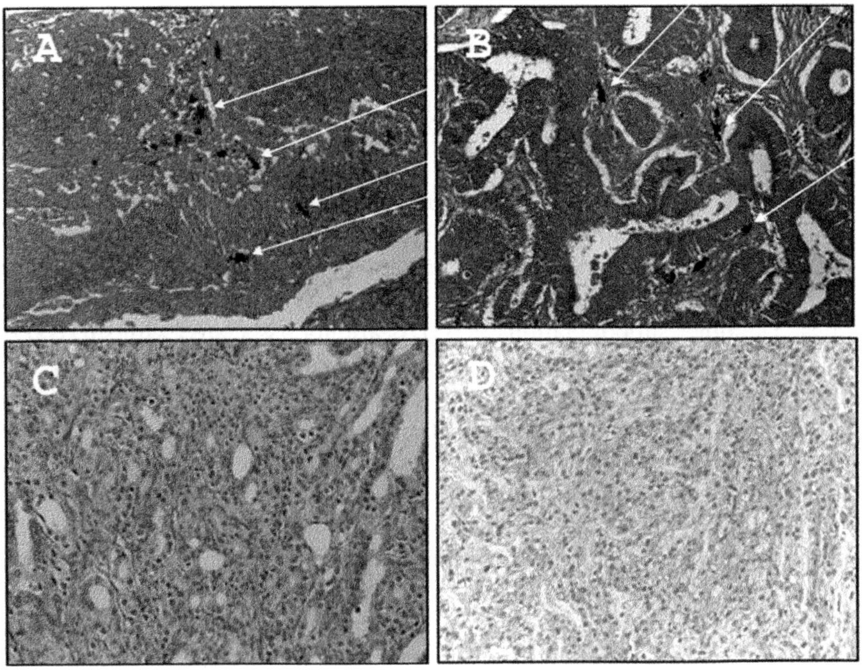

Fig. 3.2 Identification of tumor-infiltrating dendritic cells in different types of cancer by using S-100 protein immunostain. Formalin-fixed paraffin-embedded tissue samples were cut at 4 μm thick. Monoclonal antibody against S-100 protein was applied, followed by secondary biotinylated antibody, avidin–biotin complex, and diaminobenzidine as a chromogen. The slides were counterstained with hematoxilin. **(A)** Pulmonary squamous cell carcinoma. **(B)** Colonic adenocarcinoma. **(C)** Prostatic adenocarcinoma. **(D)** Renal cell carcinoma. *Arrows* indicate TIDC. Notice the absence of TIDC in prostatic and renal cell carcinoma

oligonucleotides. Cells, expressing proteins or mRNAs of interest, are defined as positive and can be counted under the light or fluorescent microscope.

Several DC characteristics seem to be influenced by the nature (cell origin) of tumors and thus immunopathological results generated in tumors of one type (i.e., colonic adenocarcinoma) cannot be directly applied to other types of tumor. As of today only limited number of malignancies has been evaluated for the presence of TIDC. Studies performed on the most common types of tumors are summarized below.

3.3 TIDC in Head and Neck Malignancy

High frequency of head and neck cancers and availability of tumor tissue for research made these tumors a favorite target for investigative pathologists. One of the earliest studies of TIDC was performed in 1985 by Kurihara and

Hasimoto, who found an increased number of S-100+ DC in squamous cell carcinoma (SCC) of the tongue (Kurihara and Hashimoto 1985). However, when a large series of oral SCC was evaluated it was found that S-100+DC infiltrates were low in 20% of specimens, intermediate in 42% of specimens, and high in 37% of specimens. TIDC density strongly correlated with a prognosis: a low number of infiltrating S-100+DC were more predictive of poor survival than lymph node involvement or tumor stage (Reichert et al. 2001). As expected, different subpopulations of TIDC have dissimilar tissue distribution and prognostic significance. In SCC of the tongue, patients with greater numbers of iDC adjacent to tumor have improved survival and decreased recurrence rates. The other subpopulations of TIDC are not associated with survival or recurrence. In addition, the number of iDC in peritumoral epithelium decreases as the tumor stage rises and nodal metastases develop (Goldman et al. 1998). When comparing density of CD1a+ iDC and CD83+ mDC, Kikuchi and co-workers found that the number of iDC in tissue adjacent to the primary tumor was greater in patients with lower stage of the disease (without metastasis to the regional lymph nodes), compared with those who had lymph node metastases. Interestingly, mDC density showed the reverse correlation: CD83+ DC in the primary tumors were found in patients with lymph node metastases (Kikuchi et al. 2002). With regard to tumor grade, the number of iDC is significantly higher in well-differentiated tumors than in moderately and poorly differentiated tumors (Chen et al. 2005). A recent study utilizing novel DC markers (CD207/Langerin and CD209/DC-SIGN for immature myeloid DC, CD208/DC–LAMP for mature myeloid DC, and CD123 for plasmacytoid DC) revealed that CD207/Langerin+ iDC were present predominantly inside the tumor nests, whereas CD209/DC-SIGN+ iDC were located in bands of tissue surrounding nests of malignant cells. Mature DC–LAMP+ DC were rare and predominantly extranestal. Plasmacytoid DC were also identified and their presence was associated with poor prognosis, i.e., decreased patients' survival (O'Donnell et al. 2007). Thus, in regard to *oral SSC* all of the studies led to a similar conclusion: iDC are located in the tumor nests, mDC are present around the tumor nests, and higher density of TIDC (specifically iDC) indicates better prognosis.

Much less is known about TIDC in *laryngeal SCC*. In a study of Gallo and co-workers, S-100+ DC were identified in malignant tissue, and although no correlation with histologic grade or stage was found, high TIDC density was associated with longer survival (Gallo et al. 1991). Karakok and co-workers also identified S-100+ TIDC in tumoral and peritumoral tissue of laryngeal SCC, but did not find any correlation of TIDC density with tumor grade, stage, or patients' survival (Karakok et al. 2003). In another type of head and neck cancer, *nasopharyngeal carcinoma*, S-100+ TIDC are identified in 50% of cases, typically within the cancer nests (Lauriola et al. 1984; Giannini et al. 1991). There is a strong positive correlation of TIDC density with the patients' survival (Giannini et al. 1991; Ma et al. 1995)

In *thyroid carcinoma*, TIDC infiltration strongly depends on the type of tumor. Several groups reported that papillary carcinoma contains significantly higher numbers of TIDC than follicular carcinoma, poorly differentiated, and undifferentiated carcinoma (Yamakawa et al. 1995; Tsuge et al. 2005; Ugolini et al. 2007). Majority of TIDC are immature, expressing CD1a or DC-SIGN. They are primarily located in cancer nodules and in peritumoral tissue. Correlation of TIDC density with clinical outcome is uncertain, although numbers of TIDC are higher in low-grade tumors.

3.4 TIDC in Digestive System Malignancy

Majority of studies on TIDC in *esophageal SCC* revealed a positive association with tumor prognosis. Dense TIDC infiltrates were found in 20% of tumors (Furihata et al. 1992). TIDC density correlates well with pathologic stage: tumors are more superficial and the disease is at an earlier clinical stage in patients with high DC density (Matsuda et al. 1990; Furihata et al. 1992). In terms of patients survival the findings are less clear: some groups reported longer survival time in patients with high TIDC numbers (Matsuda et al. 1990; Furihata et al. 1992; Ishigami et al. 2003), while others did not find any correlation between TIDC density and prognosis (Cao et al. 2005).

In *gastric adenocarcinoma*, S-100+ TIDC are mainly interspersed among the tumor cells (Tsujitani et al. 1987). The density of CD 83+ mDC density is significantly lower in gastric cancer tissue than in normal gastric tissue and also significantly lower in advanced stage gastric cancer than in its early stage (Tsukayama et al. 2005). The prognosis for patients with high density of TIDC was significantly better than that for patients with low TIDC density (Ishigami et al. 2000; Takahashi et al. 2002; Tsukayama et al. 2005). Specifically, patients with high TIDC density have less lymph nodal involvement (Tsujitani et al. 1987; Tsujitani et al. 1993) and lower peritoneal recurrence rate (Tsujitani et al. 1992). Not only the total number of TIDC but also the number of CD83+ mDC positively correlates with the survival (Tsukayama et al. 2005).

In *colonic adenocarcinoma*, iDC are located predominantly in the tumor epithelium, whereas mDC are observed predominantly in the cancer invasive margin and cancer stroma (Suzuki et al. 2002; Dadabayev et al. 2004; Miyagawa et al. 2004). Total TIDC density correlates well with favorable prognosis: patients with high density of TIDC survive longer, often with no metastases (Ambe et al. 1989; Nakayama et al. 2003; Nagorsen et al. 2007). Several studies addressed correlation between maturation status of TIDC and clinical outcomes, but the results are controversial. Schwaab and co-workers found that density of CD 83+ mDC in colorectal cancer was three times lower than that seen in normal colonic mucosa and that mDC were rarely observed in metastatic tumors (mDC density in metastases was sixfold lower than in primary tumors) (Schwaab et al. 2001). It was also reported that high amounts of HLA II+ mDC in the tumor stroma as well as high amounts of LAMP+ mDC in tumor epithelium correlated with an adverse survival outcome (Dadabayev et al. 2004; Sandel et al. 2005). Conversely, Inoue and co-workers determined that patients with advanced stage tumors and lymph node metastases have significantly less CD83+ mDC (Inoue et al. 2005). Interestingly, DC density is also significantly reduced in non-neoplastic mucosa adjacent to colorectal adenocarcinoma, compared to normal colonic mucosa (Cui et al. 2007).

In *hepatocellular carcinoma*, high TIDC density is closely related to improved clinical outcome and represents an independent prognostic factor (Yin et al. 2003; Cai et al. 2006). In contrast, the number of DC in pericancerous tissues does not correlate with patient's prognosis (Cai et al. 2006).

In *biliary carcinoma*, CD1a+ iDC are found in cancer epithelium and peritumoral areas, while CD83+ mDC are located predominantly at the invasive margin of carcinoma (Takagi et al. 2004; Furihata et al. 2005). High numbers of TIDC correlate with low frequency of lymph node metastases and with prolonged patients survival (Nakakubo et al. 2003; Furihata et al. 2005).

Pancreatic adenocarcinoma is characterized by a paucity of TIDC: significant numbers of S100+ TIDC and CD1a+ iDC were found in only 4% of tumors. When present, DC were located mainly in the peritumoral area. Other immune cells, including T cells, B cells, and macrophages, were rare or absent in the majority of the specimens (Dallal et al. 2002).

3.5 TIDC in Pulmonary Malignancy

The data on TIDC density and distribution in pulmonary carcinoma are complex due to the heterogeneity of lung cancers. In two main types of pulmonary non-small cell carcinoma, adenocarcinoma, and SCC, TIDC are found in 60–80% of tumors (Nakajima et al. 1985; Colasante et al. 1993, 1995; Inoshima et al. 2002). TIDC are present in low-grade tumors (bronchioloalveolar adenocarcinoma, well-moderately differentiated SCC), but are absent in high-grade tumors (poorly differentiated SCC) (Zeid and Muller 1993). Interestingly, TIDC density is significantly higher in female patients than in males, and in non-smokers than in smokers (Inoshima et al. 2002). The majority of TIDC are CD1a+ iDC; only a few CD83+ mDC are observed in peritumoral stroma, thus indicating a negative impact of tumor on DC maturation (Bergeron et al. 2006; Baleeiro et al. 2008). Recently, Perrot and co-workers showed that TIDC display "semi-mature" phenotype in non-small cell carcinoma, which can be associated with a compromised tumor-specific immune response (Perrot et al. 2007). TIDC are located predominantly in cancer nests and their number correlates with the extent of cancer cell apoptosis: in areas of scattered DC distribution, only a few apoptotic tumor cells can be detected, while in the areas of DC aggregations, apoptotic tumor cells are significantly more abundant (Kurabayashi et al. 2004). High numbers of TIDC in non-small cell carcinoma is a favorable prognostic factor and is associated with lower disease stage and longer survival (Zeid and Muller 1993; Inoshima et al. 2002).

TIDC density in two types of pulmonary neuroendocrine tumors (small cell carcinoma and carcinoid tumor) is usually very low (Nakajima et al. 1985; Coli et al. 1990; Zeid and Muller 1993; Katsenelson et al. 2001). Katsenelson and co-workers found different populations of TIDC, including CD1a+ iDC and CD83+ mDC, in small cell carcinoma, but samples of carcinoid tumor were devoid of DC. Generation and differentiation of DC were completely abrogated by carcinoid-conditioned medium and significantly reduced by small cell carcinoma-conditioned medium. The authors suggested that pulmonary neuroendocrine carcinoma produces factors that inhibit DC generation, maturation, or induce DC apoptosis (Katsenelson et al. 2001).

In summary, TIDC density is higher in well differentiated than in poorly differentiated pulmonary carcinoma. Immature DC are mainly located in tumor nests, especially in areas of tumor cell apoptosis. Mature DC are sparse and are functionally impaired. TIDC are rarely observed in small cell neuroendocrine carcinoma and are absent in carcinoid tumor. High TIDC density correlates with a favorable prognosis.

3.6 TIDC in Genitourinary Malignancy

In *prostate carcinoma*, S-100+ TIDC are found mainly in low-grade tumors (grade 1–2). TIDC were inconspicuous in grade 4 and virtually absent in grade 5 cancers (Bigotti et al. 1991). The authors concluded that the presence of DC in carcinoma of the prostate represents a good prognostic indicator, being constantly associated with the clinically less aggressive low-grade tumors. Troy et al. (1998a) compared the number of TIDC in prostate carcinoma and adjacent normal prostatic tissue and found that DC represent a small subset of leucocytes in both benign and malignant prostatic tissues. There were significantly less CD1a+ iDC in prostate cancer compared with normal prostatic tissue and only a small subset of DC expressed markers of activation, such as CD83, CD86, or CMRF44. The authors concluded that there is no active recruitment of DC into prostate cancer and those DC present are only minimally activated.

In *transitional cell carcinoma of the urinary bladder*, a dense infiltrate of S-100+ TIDC is detected in 50% of cases (Inoue et al. 1993). Immature CD1a+ TIDC are found mainly interspersed among the epithelial tumor cells (Ioachim-Velogianni et al. 1995; Troy et al. 1999), while mDC are present in the lamina propria adjacent to the tumor (Troy et al. 1999). The data on correlation of TIDC density with tumor grade, stage, and prognosis are controversial: some authors reported a positive correlation (Inoue et al. 1993), whereas others did not find any correlation (Ioachim-Velogianni et al. 1995), or even a negative correlation (Troy et al. 1999).

In *renal cell carcinoma*, a population of TIDC is sparce and is mainly comprised of CD1a + iDC (Troy et al. 1998b), while CD83 + mDC are more evident in peritumoral tissue and are incompletely activated (Troy et al. 1998b; Schwaab et al. 1999; Aso et al. 2004). It appears that renal cell carcinoma recruits few DC into the tumor substance and that the tumor environment fails to initiate their activation. In patients with renal cell carcinoma who received cytokine treatment for metastases, higher numbers of S100 + TIDC and CD83 + mDC were associated with favorable treatment response (Kobayashi et al. 2007).

3.7 TIDC in Breast Carcinoma

In breast carcinoma, TIDC are detected in 30–50% of tumors (Bell et al. 1999). Immature CD1a + DC, mostly of Langerhans type (Langerin +), are retained predominantly within the tumor epithelium, whereas CD83 + and LAMP + mDC are confined to peritumoral areas (Hillenbrand et al. 1999; Lespagnard et al. 1999; Coventry et al. 2002). The number of S100 + TIDC is higher in high-grade tumors (Lespagnard et al. 1999). In regards to prognosis, no correlation was found between total TIDC and iDC density and clinical outcome (Iwamoto et al. 2003; Treilleux et al. 2004). At the same time, high mDC density is a favorable prognostic marker: there is a significant association of increased number of mDC with longer relapse-free and overall survival (Iwamoto et al. 2003; Treilleux et al. 2004). CD123 + plasmacytoid DC infiltration was found in 13% of the breast carcinoma. Their presence was strongly associated with shorter overall survival and relapse-free survival and was found to be an independent adverse prognostic factor (Treilleux et al. 2004).

3.8 TIDC in Gynecologic Malignancy

SCC of the uterine cervix is a type of malignancy, which has a strong proven connection with infectious etiology (Human papilloma virus, HPV). Thus, it is on a crossroad of antiviral and antitumor immunity and presents a special interest for immunopathologists. It is generally accepted that cervical SCC progresses through stages of intraepithelial dysplasia (low-grade dysplasia, high-grade dysplasia, carcinoma in situ). Thus it is feasible to monitor DC involvement in different steps of neoplastic process (the task that is difficult to achieve in other types of tumor). Majority of studies showed that HPV infection is associated with decreased density of DC in cervical epithelium (McArdle and Muller 1986; Hawthorn et al. 1988; Jimenez-Flores et al. 2006). When dysplasia is followed by carcinoma, the total number of DC appears to increase (McArdle and Muller 1986; Hawthorn et al. 1988; Levi et al. 2005). However, there is a significant decline of CD1a+ iDC (Hubert et al. 2005; Hayati and Zulkarnaen 2007) and a simultaneous increase of CD83+ mDC (Hayati and Zulkarnaen 2007). Recently, plasmacytoid DC were detected in the majority of the cervical carcinoma cases, primarily in the stroma (Bontkes et al. 2005). Taken together these data point toward impaired immune response in the presence of HPV infection. An increase of mDC in invasive carcinoma may be related to associated inflammation, as Zijlmans et al. (2007) showed that mDC density in cervical carcinoma correlated positively with the expression levels of proinflammatory cytokines (GM-CSF and TNF-α) in the tumor microenvironment, whereas iDC density did not.

A peculiar correlation of TIDC density with the degree of squamous cell dysplasia in vulvar lesions was reported in a study of Brustmann (2006). Density of CD1a+ iDC was determined in normal squamous epithelium, vulvar condylomas, high-grade vulvar intraepithelial neoplasias (VIN), and invasive keratinizing SCC. The frequency of iDC increased from normal and condyloma to VIN, but was lowest in SCC. Interestingly, it did not correlate with stage and grade of cancer, but did correlate with recurrence of the disease.

In *endometrial carcinoma*, high S100+ TIDC density seems to be associated with good prognosis. In a study of Coppola et al. (1998), most low-grade endometrial carcinomas were infiltrated by numerous S-100+ DC, while most high-grade carcinomas were DC depleted. Honig et al. (2005) reported that compared to tumors with low TIDC infiltration, samples with high numbers of TIDC showed a higher degree of cancer differentiation, lower overexpression of p53, and lower proliferation index.

3.9 TIDC in Malignant Melanoma

Malignant melanoma (MM) is a popular object of immunopathologic research for a number of reasons. First, the disease has a significant morbidity and mortality, and current treatments (especially in aggressive cases) remain suboptimal. Second, pathologic tissue is easily available for examination and special testing. Third, there is a convincing evidence of spontaneous regression in some cases of MM, which may be related to successful antitumor immune attack. However, evaluation of TIDC in MM is complicated by the fact that distribution of DC in skin is quite complex. Immature DC are mainly located in epidermis and belong to two different subtypes: Langerhans type and non-Langerhans type. While both subtypes express CD1a, only Langerhans cell expresses Langerin (protein essential for the development of Birbeck's granules). Dermal DC comprise a mixture of iDC and mDC. Due to complexity of the DC system in the skin and the fact that every research group utilized different markers of DC identification, results are rather controversial.

Stene et al. (1988) found that compared with histologically normal skin, epidermal DC of Langerhans type were depleted above "deeply invasive" melanomas but were relatively unchanged above "early invasive" melanomas. Dermal DC were significantly increased around in situ and "early invasive" melanomas but not around "deeply invasive" melanomas. Other groups found a substantial reduction in total DC and a lesser decline in CD1 + iDC in the epidermis overlying melanoma with a simultaneous increase in the frequency of cells expressing these phenotypes in the dermis deep to tumor (Toriyama et al. 1993; Garcia-Plata et al. 1995). In contrast, Vermi and co-workers reported an increase of TIDC in the epidermis and the peritumoural area. Phenotypically intraepidermal DC were mostly CD1a+/Langerin+ iDC, whereas peritumoural DC included a large population of DC-SIGN+/CD1a- mDC, a small subset of CD1a+ iDCs, and CD123+ plasmacytoid DC. All DC subsets were predominantly immature. The authors concluded that the paucity of intratumoral DC and the predominantly immature phenotype of peritumoral dermal DC indicate defective maturation of TIDC resulting in lack of T-cell priming (Vermi et al. 2003).

The density of iDC and mDC shows strong inverse correlation with the stage (thickness) of melanomas (Toriyama et al. 1993; Ladanyi et al. 2007). In terms of prognosis, high peritumoral density of mDC correlates with significantly longer survival (Ladanyi et al. 2007; Simonetti et al. 2007).

3.10 TIDC in Pediatric Malignancy

Very interesting study was performed by Vakkila and co-workers to compare TIDC density in pediatric and adult tumors (Vakkila et al. 2006). Unlike in adult tumors (colon carcinoma, breast carcinoma, esophageal carcinoma), TIDC were virtually absent in pediatric malignancies (Ewing's sarcoma, rhabdomyosarcoma, hepatoblastoma, neuroblastoma, Wilms' tumor). Inflammatory infiltrate in pediatric tumors was composed mainly of macrophages, whereas in adult tumors, TIDC formed 37% of leukocytes within the tumor islands and 25% around the tumors. The reason for this striking difference merits further investigation.

3.11 Conclusions

Although some of the data collected to date are controversial, certain conclusions can be drawn. In the majority of solid tumors, TIDC density inversely correlates with tumor pathologic grade and stage (more TIDC are present in well-differentiated and less-invasive tumors) and positively correlates with favorable prognostic features, such as the absence of lymph node metastases, distant metastases, and overall survival. This conclusion is well aligned with the concept of the important role of DC in antitumor immune surveillance. On the basis of this concept an established malignant tumor can be viewed as a failure of the immune system to launch a successful antitumor attack, and at the same time, as a success of an emerging tumor to escape such an attack. Therefore, the absolute numbers of TIDC and their functional and maturation characteristics should be interpreted with caution and with the clear understanding that cases of the immune system success in controlling tumor growth rarely reach a pathologist's desk.

It becomes evident that different tumors have varying ways of interaction with TIDC. In certain cancer types, i.e., pancreatic carcinoma and urothelial carcinoma, there are very few TIDC, while others have a substantial number of TIDC, which appear to be functionally impaired. There are several mechanisms explaining a paucity of TIDC and their dysfunctional status. These mechanisms will be reviewed in detail in the following chapters; however, it is worth outlining them here (Fig. 3.3).

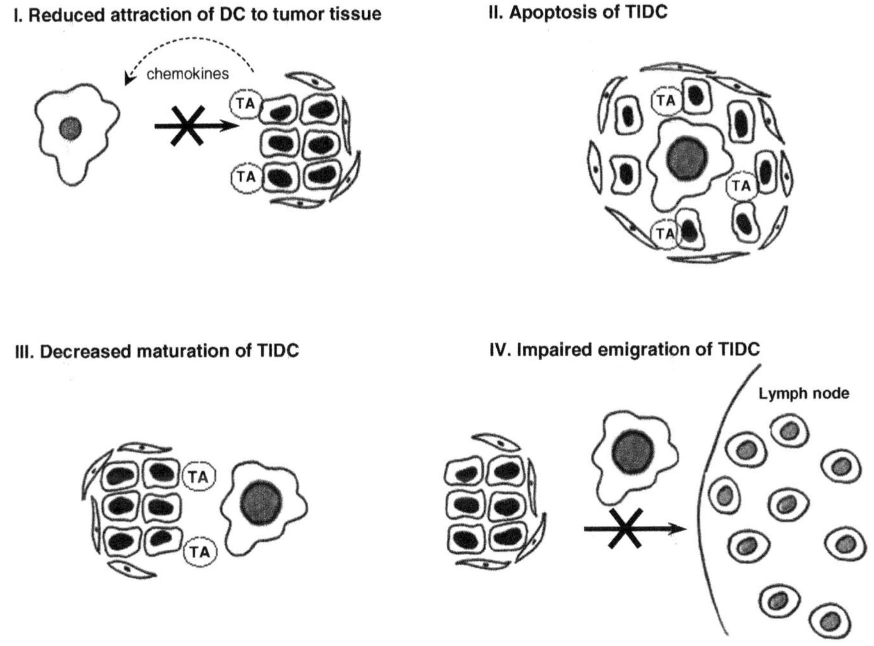

Fig. 3.3 Putative mechanisms of dendritic cell alteration in the local tumor microenvironment. **(A)** Tumor cells down-regulate expression of DC attractive chemokines, thus diminishing DC infiltration and homing at the tumor site. **(B)** Tumor cells induce apoptosis of TIDC decreasing their numbers at the tumor site. **(C)** Tumor cells interfere with TIDC maturation and function, thus preventing tumor antigen uptake, processing, and presentation. **(D)** Incompletely mature TIDC fail to emigrate and present processed tumor antigens to T lymphocytes. DC, dendritic cells; TA, tumor antigen(s); TIDC, tumor-infiltrating dendritic cells

1. *Insufficient chemoattraction of DC to the tumor tissue.* Homing of DC depends on the action of chemokines, "inviting" DC to survey the tissue microenvironment (Vicari et al. 2004). Although numerous chemokines may be involved in DC attraction (i.e., CCL20 acting on CCR6), one of them, called BRAK (CXCL14), has recently become a focus of a particular interest. CXCL14 is steadily expressed in normal tissue of different types (Hromas et al. 1999; Frederick et al. 2000) and has been shown to be a potent iDC chemoattractant and activator (Shellenberger et al. 2004; Schaerli et al. 2005; Shurin et al. 2005). Notably, its expression is markedly reduced in several malignant neoplasms (Frederick et al. 2000; Shurin et al. 2005) suggesting a deficient attraction and homing of DC. This phenomenon presents a simple and effective mechanism of tumor escape and explains a paucity of TIDC in advanced tumors.
2. *Tumor-induced apoptosis of DC.* It has been extensively demonstrated that DC undergo apoptosis in vitro and in vivo after contact with tumors (Esche et al. 1999; Pirtskhalaishvili et al. 2001; Balkir et al. 2004). This phenomenon can result in depletion of TIDC, especially in more aggressive high-grade tumors (see Shurin et al. (2006) for review and Chapter 6).
3. *Blockade of DC maturation.* In the absence of appropriate maturation and activation cascades, high density of TIDC is insufficient to induce an effective immune response. In fact, immature or semi-mature TIDC can cause immunologic tolerance (Adema et al. 2005; Ghiringhelli et al. 2007). Tumors can prevent DC differentiation and maturation (dendropoiesis) by releasing such cytokines as IL-6, IL-10, TGF-β, M-CSF, VEGF, and others (see Fricke and Gabrilovich (2006) and Shurin et al. (2006) for review). Specifically, VEGF has been shown to interfere with TIDC maturation and function in lung cancer (Inoshima et al. 2002), gastric carcinoma (Saito et al. 1998; Takahashi et al. 2002), oral SCC (Kikuchi et al. 2006), breast carcinoma (Iwamoto et al. 2003), and head and neck SCC (Strauss et al. 2005). Other factors, specific to certain tumors can also exert such an effect. For example, prostate cancer microenvironment inhibits DC maturation by direct action of prostate-specific antigen (Aalamian et al. 2003). Lung cancer inhibits DC maturation and function by releasing bombesin-like peptides (Makarenkova et al. 2003). Neuroblastoma inhibits DC generation and function by producing gangliosides (Shurin et al. 2001). Undoubtedly, other tumor-specific factors will soon be discovered. This phenomenon can explain disproportional increase of immature or partially mature TIDC seen in some tumors.
4. *"Capturing" DC in the tumor tissue.* Factors of the tumor microenvironment can prevent emigration of TIDC from the tumor site to the draining lymph nodes [see Adema et al. (2005) for review]. DC captured in tumor are inefficient in initiating T-cell responses, and this phenomenon can explain an apparent discordance between dense TIDC infiltration and a lack of efficient antitumor response in some tumors.
5. *"Reassignment" of DC differentiation.* Tumor-released substances can also cause a dramatic change in DC function by altering local DC subtypes. For

instance, it was recently shown that pro-angiogenic factors, like VEGF and oncostatin M, can lead to trans-differentiation of TIDC into endothelial-like cells, thus promoting tumor vascularization (Coukos et al. 2005; Gottfried et al. 2007). This phenomenon additionally explains a discrepancy between TIDC number and intensity of immune response.

3.12 Future Directions

As it is presented in this chapter, the initial phase of TIDC study has generated vast data on the number of TIDC in different types of cancer and its relation to clinical prognosis. The focus of research is shifting now toward identification of mechanisms related to TIDC abnormality and finding the ways of therapeutic interference. Following questions seem to be the most important for future studies:

1. What is a functional status of TIDC at different stages of maturation? It has become evident that TIDC (both immature and mature) are not fully functional; however, which molecular pathways are altered remains largely unknown. The main reasons for this are technical limitations of traditional methods, specifically, inability to characterize the expression of multiple proteins in the cell simultaneously. New morphometric and molecular techniques can provide a significant breakthrough in this field. For example, QuantumDot Immunochemistry can analyze numerous cellular constituents in a single TIDC and thus more precisely characterize its functional and maturation status. Individual TIDC can be isolated from fresh or fixed tumor tissue by Laser Capture Microscopy (LCM) and analyzed by DNA and RNA microarrays (Fend et al. 2000; Dong et al. 2004; Buckanovich et al. 2006; Espina et al. 2006).
2. Which factors of the tumor microenvironment are responsible for TIDC alterations? Several comprehensive reviews addressed this issue recently (Gabrilovich et al. 1997; Shurin et al. 2006; Whiteside 2006). However, details of tumor cell–DC interactions need to be significantly better clarified. Analysis of intratumoral soluble factors will identify those that are able to influence DC homing, maturation, longevity, and emigration. A promising approach is a microsampling of the tumor environment in animal models followed by RNA microarrays and protein expression profiling (Dabrosin 2005) or intratumoral microdialysis in live freely moving animals (Zhong et al. 2007). Direct evaluation of in vivo DC trafficking can be performed by using magnetic resonance tracking (de Vries et al. 2005).
3. What are the ways to prevent/reverse TIDC disfunctionality? Disclosing of molecular details of TIDC alteration will ultimately lead to new methods of therapeutic intervention. A good example of such an approach is a study of Shurin et al. (2005) showing that restoration of BRAK expression in head

and neck SCC leads to DC recruitment into the tumor and induction of antitumor immunity. In a study of Furumoto and co-workers, transduction of tumor cells (melanoma and colonic adenocarcinoma cell lines) with iDC-attracting chemokine CCL20 and injection of DC-activating CG-rich motifs led to increase of TIDC density and restoration of their activity (Furumoto et al. 2004).

In conclusion, TIDC appear to play a major role in inducing antitumor immunity while their dysfunction may represent an important mechanism of tumor escape from immune surveillance. Analysis of intratumoral DC and molecular pathways responsible for their dysfunction is important for understanding immunobiology of the DC system and its role in controlling tumor progression and metastasis. New methods to protect/restore DC function will have to be found to improve the efficacy of DC vaccines and other immunotherapeutic approaches for cancer treatment.

References

Aalamian, M., Tourkova, I. L., Chatta, G. S., Lilja, H., Huland, E., Huland, H., Shurin, G. V. and Shurin, M. R. 2003. Inhibition of dendropoiesis by tumor derived and purified prostate specific antigen. J Urol 170:2026–2030.

Adema, G. J., de Vries, I. J., Punt, C. J. and Figdor, C. G. 2005. Migration of dendritic cell based cancer vaccines: in vivo veritas? Curr Opin Immunol 17:170–174.

Ambe, K., Mori, M. and Enjoji, M. 1989. S-100 protein-positive dendritic cells in colorectal adenocarcinomas. Distribution and relation to the clinical prognosis. Cancer 63:496–503.

Aso, T., Ogawa, Y., Naoe, M., Fukagai, T., Yoshida, H. and Kushima, M. 2004. [Immunohistochemical analysis of CD83, CD8 and CD4 positive cells in renal cell carcinoma]. Nippon Hinyokika Gakkai Zasshi 95:645–650.

Baleeiro, R. B., Anselmo, L. B., Soares, F. A., Pinto, C. A., Ramos, O., Gross, J. L., Haddad, F., Younes, R. N., Tomiyoshi, M. Y., Bergami-Santos, P. C. and Barbuto, J. A. 2008. High frequency of immature dendritic cells and altered in situ production of interleukin-4 and tumor necrosis factor-alpha in lung cancer. Cancer Immunol Immunother 57(9):1335–1345.

Balkir, L., Tourkova, I. L., Makarenkova, V. P., Shurin, G. V., Robbins, P. D., Yin, X. M., Chatta, G. and Shurin, M. R. 2004. Comparative analysis of dendritic cells transduced with different anti-apoptotic molecules: sensitivity to tumor-induced apoptosis. J Gene Med 6:537–544.

Bell, D., Chomarat, P., Broyles, D., Netto, G., Harb, G. M., Lebecque, S., Valladeau, J., Davoust, J., Palucka, K. A. and Banchereau, J. 1999. In breast carcinoma tissue, immature dendritic cells reside within the tumor, whereas mature dendritic cells are located in peritumoral areas. J Exp Med 190:1417–1426.

Bergeron, A., El-Hage, F., Kambouchner, M., Lecossier, D. and Tazi, A. 2006. Characterisation of dendritic cell subsets in lung cancer micro-environments. Eur Respir J 28:1170–1177.

Bigotti, G., Coli, A. and Castagnola, D. 1991. Distribution of Langerhans cells and HLA class II molecules in prostatic carcinomas of different histopathological grade. Prostate 19:73–87.

Bontkes, H. J., Ruizendaal, J. J., Kramer, D., Meijer, C. J. and Hooijberg, E. 2005. Plasmacytoid dendritic cells are present in cervical carcinoma and become activated by human papillomavirus type 16 virus-like particles. Gynecol Oncol 96:897–901.

Brustmann, H. 2006. Galectin-3 and CD1a-positive dendritic cells are involved in the development of an invasive phenotype in vulvar squamous lesions. Int J Gynecol Pathol 25:30–37.

Buckanovich, R. J., Sasaroli, D., O'Brien-Jenkins, A., Botbyl, J., Conejo-Garcia, J. R., Benencia, F., Liotta, L. A., Gimotty, P. A. and Coukos, G. 2006. Use of immuno-LCM to identify the in situ expression profile of cellular constituents of the tumor microenvironment. Cancer Biol Ther 5:635–642.

Cai, X. Y., Gao, Q., Qiu, S. J., Ye, S. L., Wu, Z. Q., Fan, J. and Tang, Z. Y. 2006. Dendritic cell infiltration and prognosis of human hepatocellular carcinoma. J Cancer Res Clin Oncol 132:293–301.

Cao, F. M., Zhang, X. H., Yan, X., Wang, J. L., Wang, X. L., Zhang, Z. G., Shen, H. T. and Wang, F. R. 2005. [Prognostic significances of natural killer cells and dendritic cells infiltrations in esophageal squamous cell carcinoma]. Ai Zheng 24:232–236.

Caux, C., Ait-Yahia, S., Chemin, K., de Bouteiller, O., Dieu-Nosjean, M. C., Homey, B., Massacrier, C., Vanbervliet, B., Zlotnik, A. and Vicari, A. 2000. Dendritic cell biology and regulation of dendritic cell trafficking by chemokines. Springer Semin Immunopathol 22:345–369.

Chen, W. K., Chen, F. J., Zeng, Z. Y., Wu, G. H., Guo, Z. M., Wei, M. W., Yang, A. K., Zhang, Q., He, J. H. and Hou, J. H. 2005. [Expression and clinical significance of CD1a + dendritic cells in glottic squamous cell carcinoma]. Ai Zheng 24:1124–1126.

Colasante, A., Castrilli, G., Aiello, F. B., Brunetti, M. and Musiani, P. 1995. Role of cytokines in distribution and differentiation of dendritic cell/Langerhans' cell lineage in human primary carcinomas of the lung. Hum Pathol 26:866–872.

Colasante, A., Poletti, V., Rosini, S., Ferracini, R. and Musiani, P. 1993. Langerhans cells in Langerhans cell histiocytosis and peripheral adenocarcinomas of the lung. Am Rev Respir Dis 148:752–759.

Coli, A., Bigotti, G. and Cione, T. 1990. Langerhans cell and HLA class II molecules distribution in lung carcinoma. J Nucl Med Allied Sci 34:185–189.

Coppola, D., Fu, L., Nicosia, S. V., Kounelis, S. and Jones, M. 1998. Prognostic significance of p53, bcl-2, vimentin, and S100 protein-positive Langerhans cells in endometrial carcinoma. Hum Pathol 29:455–462.

Coukos, G., Benencia, F., Buckanovich, R. J. and Conejo-Garcia, J. R. 2005. The role of dendritic cell precursors in tumour vasculogenesis. Br J Cancer 92:1182–1187.

Coventry, B. J., Lee, P. L., Gibbs, D. and Hart, D. N. 2002. Dendritic cell density and activation status in human breast cancer – CD1a, CMRF-44, CMRF-56 and CD-83 expression. Br J Cancer 86:546–551.

Cox, K., North, M., Burke, M., Singhal, H., Renton, S., Aqel, N., Islam, S. and Knight, S. C. 2005. Plasmacytoid dendritic cells (PDC) are the major DC subset innately producing cytokines in human lymph nodes. J Leukoc Biol 78:1142–1152.

Cui, G., Yuan, A., Goll, R., Olsen, T., Husebekk, A., Vonen, B. and Florholmen, J. 2007. Distinct changes of dendritic cell number and IL-12 mRNA level in adjacent mucosa throughout the colorectal adenoma-carcinoma sequence. Cancer Immunol Immunother 56:1993–2001.

Dabrosin, C. 2005. Microdialysis - an in vivo technique for studies of growth factors in breast cancer. Front Biosci 10:1329–1335.

Dadabayev, A. R., Sandel, M. H., Menon, A. G., Morreau, H., Melief, C. J., Offringa, R., van der Burg, S. H., Janssen-van Rhijn, C., Ensink, N. G., Tollenaar, R. A., van de Velde, C. J. and Kuppen, P. J. 2004. Dendritic cells in colorectal cancer correlate with other tumor-infiltrating immune cells. Cancer Immunol Immunother 53:978–986.

Dallal, R. M., Christakos, P., Lee, K., Egawa, S., Son, Y. I. and Lotze, M. T. 2002. Paucity of dendritic cells in pancreatic cancer. Surgery 131:135–138.

de Vries, I. J., Lesterhuis, W. J., Barentsz, J. O., Verdijk, P., van Krieken, J. H., Boerman, O. C., Oyen, W. J., Bonenkamp, J. J., Boezeman, J. B., Adema, G. J., Bulte, J. W., Scheenen, T. W.,

Punt, C. J., Heerschap, A. and Figdor, C. G. 2005. Magnetic resonance tracking of dendritic cells in melanoma patients for monitoring of cellular therapy. Nat Biotechnol 23:1407–1413.

Dong, P., Li, X., Zhu, Z., Yu, Z., Lu, G., Sun, Z. and Wang, S. 2004. Application of tissue microarray: evaluation of the expression of S-100-positive dendritic cells, tumor suppressor gene p63 and tissue inhibitor of metalloproteinase-1 in laryngeal carcinoma. Acta Otolaryngol 124:1204–1207.

Esche, C., Lokshin, A., Shurin, G. V., Gastman, B. R., Rabinowich, H., Watkins, S. C., Lotze, M. T. and Shurin, M. R. 1999. Tumor's other immune targets: dendritic cells. J Leukoc Biol 66:336–344.

Espina, V., Wulfkuhle, J. D., Calvert, V. S., VanMeter, A., Zhou, W., Coukos, G., Geho, D. H., Petricoin, E. F., 3rd and Liotta, L. A. 2006. Laser-capture microdissection. Nat Protoc 1:586–603.

Fend, F., Kremer, M. and Quintanilla-Martinez, L. 2000. Laser capture microdissection: methodical aspects and applications with emphasis on immuno-laser capture microdissection. Pathobiology 68:209–214.

Frederick, M. J., Henderson, Y., Xu, X., Deavers, M. T., Sahin, A. A., Wu, H., Lewis, D. E., El-Naggar, A. K. and Clayman, G. L. 2000. In vivo expression of the novel CXC chemokine BRAK in normal and cancerous human tissue. Am J Pathol 156:1937–1950.

Fricke, I. and Gabrilovich, D. I. 2006. Dendritic cells and tumor microenvironment: a dangerous liaison. Immunol Invest 35:459–483.

Furihata, M., Ohtsuki, Y., Ido, E., Iwata, J., Sonobe, H., Araki, K., Ogoshi, S. and Ohmori, K. 1992. HLA-DR antigen- and S-100 protein-positive dendritic cells in esophageal squamous cell carcinoma–their distribution in relation to prognosis. Virchows Arch B Cell Pathol Incl Mol Pathol 61:409–414.

Furihata, M., Ono, Y., Ichikawa, K., Tomita, S., Fujimori, T. and Kubota, K. 2005. Prognostic significance of CD83 positive, mature dendritic cells in the gallbladder carcinoma. Oncol Rep 14:353–356.

Furumoto, K., Soares, L., Engleman, E. G. and Merad, M. 2004. Induction of potent antitumor immunity by in situ targeting of intratumoral DCs. J Clin Invest 113:774–783.

Gabrilovich, D. I., Corak, J., Ciernik, I. F., Kavanaugh, D. and Carbone, D. P. 1997. Decreased antigen presentation by dendritic cells in patients with breast cancer. Clin Cancer Res 3:483–490.

Gallo, O., Libonati, G. A., Gallina, E., Fini-Storchi, O., Giannini, A., Urso, C. and Bondi, R. 1991. Langerhans cells related to prognosis in patients with laryngeal carcinoma. Arch Otolaryngol Head Neck Surg 117:1007–1010.

Garcia-Plata, D., Lessana-Leibowitch, M., Palangie, A., Gulliemette, J., Sedel, D., Mendez, L. and Mozos, E. 1995. Immunophenotype analysis of dendritic cells and lymphocytes associated with cutaneous malignant melanomas. Invasion Metastasis 15:125–134.

Ghiringhelli, F., Apetoh, L., Housseau, F., Kroemer, G. and Zitvogel, L. 2007. Links between innate and cognate tumor immunity. Curr Opin Immunol 19:224–231.

Giannini, A., Bianchi, S., Messerini, L., Gallo, O., Gallina, E., Asprella Libonati, G., Olmi, P. and Zampi, G. 1991. Prognostic significance of accessory cells and lymphocytes in nasopharyngeal carcinoma. Pathol Res Pract 187:496–502.

Goldman, S. A., Baker, E., Weyant, R. J., Clarke, M. R., Myers, J. N. and Lotze, M. T. 1998. Peritumoral CD1a-positive dendritic cells are associated with improved survival in patients with tongue carcinoma. Arch Otolaryngol Head Neck Surg 124:641–646.

Gottfried, E., Kreutz, M., Haffner, S., Holler, E., Iacobelli, M., Andreesen, R. and Eissner, G. 2007. Differentiation of human tumour-associated dendritic cells into endothelial-like cells: an alternative pathway of tumour angiogenesis. Scand J Immunol 65:329–335.

Hashimoto, S. I., Suzuki, T., Nagai, S., Yamashita, T., Toyoda, N. and Matsushima, K. 2000. Identification of genes specifically expressed in human activated and mature dendritic cells through serial analysis of gene expression. Blood 96:2206–2214.

Hawthorn, R. J., Murdoch, J. B., MacLean, A. B. and MacKie, R. M. 1988. Langerhans' cells and subtypes of human papillomavirus in cervical intraepithelial neoplasia. BMJ 297:643–646.
Hayati, A. R. and Zulkarnaen, M. 2007. An immunohistochemical study of CD1a and CD83-positive infiltrating dendritic cell density in cervical neoplasia. Int J Gynecol Pathol 26:83–88.
Hillenbrand, E. E., Neville, A. M. and Coventry, B. J. 1999. Immunohistochemical localization of CD1a-positive putative dendritic cells in human breast tumours. Br J Cancer 79:940–944.
Honig, A., Schaller, N., Dietl, J., Backe, J. and Kammerer, U. 2005. S100 as an immunohistochemically-detected marker with prognostic significance in endometrial carcinoma. Anticancer Res 25:1747–1753.
Hromas, R., Broxmeyer, H. E., Kim, C., Nakshatri, H., Christopherson, K., 2nd, Azam, M. and Hou, Y. H. 1999. Cloning of BRAK, a novel divergent CXC chemokine preferentially expressed in normal versus malignant cells. Biochem Biophys Res Commun 255:703–706.
Huang, J., Tatsumi, T., Pizzoferrato, E., Vujanovic, N. and Storkus, W. J. 2005. Nitric oxide sensitizes tumor cells to dendritic cell-mediated apoptosis, uptake, and cross-presentation. Cancer Res 65:8461–8470.
Hubert, P., Caberg, J. H., Gilles, C., Bousarghin, L., Franzen-Detrooz, E., Boniver, J. and Delvenne, P. 2005. E-cadherin-dependent adhesion of dendritic and Langerhans cells to keratinocytes is defective in cervical human papillomavirus-associated (pre)neoplastic lesions. J Pathol 206:346–355.
Inoshima, N., Nakanishi, Y., Minami, T., Izumi, M., Takayama, K., Yoshino, I. and Hara, N. 2002. The influence of dendritic cell infiltration and vascular endothelial growth factor expression on the prognosis of non-small cell lung cancer. Clin Cancer Res 8:3480–3486.
Inoue, K., Furihata, M., Ohtsuki, Y. and Fujita, Y. 1993. Distribution of S-100 protein-positive dendritic cells and expression of HLA-DR antigen in transitional cell carcinoma of the urinary bladder in relation to tumour progression and prognosis. Virchows Arch A Pathol Anat Histopathol 422:351–355.
Inoue, Y., Nakayama, Y., Minagawa, N., Katsuki, T., Nagashima, N., Matsumoto, K., Shibao, K., Tsurudome, Y., Hirata, K., Nagata, N. and Itoh, H. 2005. Relationship between interleukin-12-expressing cells and antigen-presenting cells in patients with colorectal cancer. Anticancer Res 25:3541–3546.
Ioachim-Velogianni, E., Stavropoulos, N. E., Kitsiou, E., Stefanaki, S. and Agnantis, N. J. 1995. Distribution of CD1a-positive Langerhans cells and lymphocyte subsets in transitional cell carcinoma of the urinary bladder. An immunohistological study on frozen sections. J Pathol 177:401–406.
Ishigami, S., Natsugoe, S., Matsumoto, M., Okumura, H., Sakita, H., Nakashima, S., Takao, S. and Aikou, T. 2003. Clinical implications of intratumoral dendritic cell infiltration in esophageal squamous cell carcinoma. Oncol Rep 10:1237–1240.
Ishigami, S., Natsugoe, S., Tokuda, K., Nakajo, A., Xiangming, C., Iwashige, H., Aridome, K., Hokita, S. and Aikou, T. 2000. Clinical impact of intratumoral natural killer cell and dendritic cell infiltration in gastric cancer. Cancer Lett 159:103–108.
Iwamoto, M., Shinohara, H., Miyamoto, A., Okuzawa, M., Mabuchi, H., Nohara, T., Gon, G., Toyoda, M. and Tanigawa, N. 2003. Prognostic value of tumor-infiltrating dendritic cells expressing CD83 in human breast carcinomas. Int J Cancer 104:92–97.
Jimenez-Flores, R., Mendez-Cruz, R., Ojeda-Ortiz, J., Munoz-Molina, R., Balderas-Carrillo, O., de la Luz Diaz-Soberanes, M., Lebecque, S., Saeland, S., Daneri-Navarro, A., Garcia-Carranca, A., Ullrich, S. E. and Flores-Romo, L. 2006. High-risk human papilloma virus infection decreases the frequency of dendritic Langerhans' cells in the human female genital tract. Immunology 117:220–228.
Karakok, M., Bayazit, Y. A., Ucak, R., Ozer, E., Kanlikama, M., Mumbuc, S. and Sari, I. 2003. Langerhans cell related inflammatory reaction in laryngeal squamous cell carcinoma. Auris Nasus Larynx 30:81–84.

Katsenelson, N. S., Shurin, G. V., Bykovskaia, S. N., Shogan, J. and Shurin, M. R. 2001. Human small cell lung carcinoma and carcinoid tumor regulate dendritic cell maturation and function. Mod Pathol 14:40–45.

Kikuchi, K., Kusama, K., Sano, M., Nakanishi, Y., Ishige, T., Ohni, S., Oinuma, T. and Nemoto, N. 2006. Vascular endothelial growth factor and dendritic cells in human squamous cell carcinoma of the oral cavity. Anticancer Res 26:1833–1848.

Kikuchi, K., Kusama, K., Taguchi, K., Ishikawa, F., Okamoto, M., Shimada, J., Sakashita, H. and Yamamo, Y. 2002. Dendritic cells in human squamous cell carcinoma of the oral cavity. Anticancer Res 22:545–557.

Kobayashi, M., Suzuki, K., Yashi, M., Yuzawa, M., Takayashiki, N. and Morita, T. 2007. Tumor infiltrating dendritic cells predict treatment response to immunotherapy in patients with metastatic renal cell carcinoma. Anticancer Res 27:1137–1141.

Kurabayashi, A., Furihata, M., Matsumoto, M., Hayashi, H. and Ohtsuki, Y. 2004. Distribution of tumor-infiltrating dendritic cells in human non-small cell lung carcinoma in relation to apoptosis. Pathol Int 54:302–310.

Kurihara, K. and Hashimoto, N. 1985. The pathological significance of Langerhans cells in oral cancer. J Oral Pathol 14:289–298.

Ladanyi, A., Kiss, J., Somlai, B., Gilde, K., Fejos, Z., Mohos, A., Gaudi, I. and Timar, J. 2007. Density of DC–LAMP(+) mature dendritic cells in combination with activated T lymphocytes infiltrating primary cutaneous melanoma is a strong independent prognostic factor. Cancer Immunol Immunother 56:1459–1469.

Lauriola, L., Michetti, F., Sentinelli, S. and Cocchia, D. 1984. Detection of S-100 labelled cells in nasopharyngeal carcinoma. J Clin Pathol 37:1235–1238.

Lespagnard, L., Gancberg, D., Rouas, G., Leclercq, G., de Saint-Aubain Somerhausen, N., Di Leo, A., Piccart, M., Verhest, A. and Larsimont, D. 1999. Tumor-infiltrating dendritic cells in adenocarcinomas of the breast: a study of 143 neoplasms with a correlation to usual prognostic factors and to clinical outcome. Int J Cancer 84:309–314.

Levi, G., Feldman, J., Holman, S., Salarieh, A., Strickler, H. D., Alter, S. and Minkoff, H. 2005. Relationship between HIV viral load and Langerhans cells of the cervical epithelium. J Obstet Gynaecol Res 31:178–184.

Ma, C. X., Jia, T. C., Li, X. R., Zhand, Z. F. and Yiao, C. B. 1995. Langerhans cells in nasopharyngeal carcinoma in relation to prognosis. In Vivo 9:225–229.

Makarenkova, V. P., Shurin, G. V., Tourkova, I. L., Balkir, L., Pirtskhalaishvili, G., Perez, L., Gerein, V., Siegfried, J. M. and Shurin, M. R. 2003. Lung cancer-derived bombesin-like peptides down-regulate the generation and function of human dendritic cells. J Neuroimmunol 145:55–67.

Manna, P. P. and Mohanakumar, T. 2002. Human dendritic cell mediated cytotoxicity against breast carcinoma cells in vitro. J Leukoc Biol 72:312–320.

Matsuda, H., Mori, M., Tsujitani, S., Ohno, S., Kuwano, H. and Sugimachi, K. 1990. Immunohistochemical evaluation of squamous cell carcinoma antigen and S-100 protein-positive cells in human malignant esophageal tissues. Cancer 65:2261–2265.

McArdle, J. P. and Muller, H. K. 1986. Quantitative assessment of Langerhans' cells in human cervical intraepithelial neoplasia and wart virus infection. Am J Obstet Gynecol 154:509–515.

Miyagawa, S., Soeda, J., Takagi, S., Miwa, S., Ichikawa, E. and Noike, T. 2004. Prognostic significance of mature dendritic cells and factors associated with their accumulation in metastatic liver tumors from colorectal cancer. Hum Pathol 35:1392–1396.

Nagorsen, D., Voigt, S., Berg, E., Stein, H., Thiel, E. and Loddenkemper, C. 2007. Tumor-infiltrating macrophages and dendritic cells in human colorectal cancer: relation to local regulatory T cells, systemic T-cell response against tumor-associated antigens and survival. J Transl Med 5:62.

Nakajima, T., Kodama, T., Tsumuraya, M., Shimosato, Y. and Kameya, T. 1985. S-100 protein-positive Langerhans cells in various human lung cancers, especially in peripheral adenocarcinomas. Virchows Arch A Pathol Anat Histopathol 407:177–189.

Nakakubo, Y., Miyamoto, M., Cho, Y., Hida, Y., Oshikiri, T., Suzuoki, M., Hiraoka, K., Itoh, T., Kondo, S. and Katoh, H. 2003. Clinical significance of immune cell infiltration within gallbladder cancer. Br J Cancer 89:1736–1742.

Nakayama, Y., Inoue, Y., Minagawa, N., Katsuki, T., Nagashima, N., Onitsuka, K., Tsurudome, Y., Sako, T., Hirata, K., Nagata, N. and Itoh, H. 2003. Relationships between S-100 protein-positive cells and clinicopathological factors in patients with colorectal cancer. Anticancer Res 23:4423–4426.

Nicolas, A., Cathelin, D., Larmonier, N., Fraszczak, J., Puig, P. E., Bouchot, A., Bateman, A., Solary, E. and Bonnotte, B. 2007. Dendritic cells trigger tumor cell death by a nitric oxide-dependent mechanism. J Immunol 179:812–818.

O'Donnell, R. K., Mick, R., Feldman, M., Hino, S., Wang, Y., Brose, M. S. and Muschel, R. J. 2007. Distribution of dendritic cell subtypes in primary oral squamous cell carcinoma is inconsistent with a functional response. Cancer Lett 255:145–152.

Perrot, I., Blanchard, D., Freymond, N., Isaac, S., Guibert, B., Pacheco, Y. and Lebecque, S. 2007. Dendritic cells infiltrating human non-small cell lung cancer are blocked at immature stage. J Immunol 178:2763–2769.

Pirtskhalaishvili, G., Shurin, G. V., Esche, C., Trump, D. L. and Shurin, M. R. 2001. TNF-alpha protects dendritic cells from prostate cancer-induced apoptosis. Prostate Cancer Prostatic Dis 4:221–227.

Reichert, T. E., Scheuer, C., Day, R., Wagner, W. and Whiteside, T. L. 2001. The number of intratumoral dendritic cells and zeta-chain expression in T cells as prognostic and survival biomarkers in patients with oral carcinoma. Cancer 91:2136–2147.

Saito, H., Tsujitani, S., Ikeguchi, M., Maeta, M. and Kaibara, N. 1998. Relationship between the expression of vascular endothelial growth factor and the density of dendritic cells in gastric adenocarcinoma tissue. Br J Cancer 78:1573–1577.

Sandel, M. H., Dadabayev, A. R., Menon, A. G., Morreau, H., Melief, C. J., Offringa, R., van der Burg, S. H., Janssen-van Rhijn, C. M., Ensink, N. G., Tollenaar, R. A., van de Velde, C. J. and Kuppen, P. J. 2005. Prognostic value of tumor-infiltrating dendritic cells in colorectal cancer: role of maturation status and intratumoral localization. Clin Cancer Res 11:2576–2582.

Schaerli, P., Willimann, K., Ebert, L. M., Walz, A. and Moser, B. 2005. Cutaneous CXCL14 targets blood precursors to epidermal niches for Langerhans cell differentiation. Immunity 23:331–342.

Schwaab, T., Schned, A. R., Heaney, J. A., Cole, B. F., Atzpodien, J., Wittke, F. and Ernstoff, M. S. 1999. In vivo description of dendritic cells in human renal cell carcinoma. J Urol 162:567–573.

Schwaab, T., Weiss, J. E., Schned, A. R. and Barth Jr, R. J. 2001. Dendritic cell infiltration in colon cancer. J Immunother 24:130–137.

Shellenberger, T. D., Wang, M., Gujrati, M., Jayakumar, A., Strieter, R. M., Burdick, M. D., Ioannides, C. G., Efferson, C. L., El-Naggar, A. K., Roberts, D., Clayman, G. L. and Frederick, M. J. 2004. BRAK/CXCL14 is a potent inhibitor of angiogenesis and a chemotactic factor for immature dendritic cells. Cancer Res 64:8262–8270.

Shurin, G. V., Ferris, R. L., Tourkova, I. L., Perez, L., Lokshin, A., Balkir, L., Collins, B., Chatta, G. S. and Shurin, M. R. 2005. Loss of new chemokine CXCL14 in tumor tissue is associated with low infiltration by dendritic cells (DC), while restoration of human CXCL14 expression in tumor cells causes attraction of DC both in vitro and in vivo. J Immunol 174:5490–5498.

Shurin, G. V., Shurin, M. R., Bykovskaia, S., Shogan, J., Lotze, M. T. and Barksdale, E. M., Jr. 2001. Neuroblastoma-derived gangliosides inhibit dendritic cell generation and function. Cancer Res 61:363–369.

Shurin, M. R., Shurin, G. V., Lokshin, A., Yurkovetsky, Z. R., Gutkin, D. W., Chatta, G., Zhong, H., Han, B. and Ferris, R. L. 2006. Intratumoral cytokines/chemokines/growth factors and tumor infiltrating dendritic cells: friends or enemies? Cancer Metastasis Rev 25:333–356.

Simonetti, O., Goteri, G., Lucarini, G., Rubini, C., Stramazzotti, D., Lo Muzio, L., Biagini, G. and Offidani, A. 2007. In melanoma changes of immature and mature dendritic cell expression correlate with tumor thickness: an immunohistochemical study. Int J Immunopathol Pharmacol 20:325–333.

Stene, M. A., Babajanians, M., Bhuta, S. and Cochran, A. J. 1988. Quantitative alterations in cutaneous Langerhans cells during the evolution of malignant melanoma of the skin. J Invest Dermatol 91:125–128.

Strauss, L., Volland, D., Kunkel, M. and Reichert, T. E. 2005. Dual role of VEGF family members in the pathogenesis of head and neck cancer (HNSCC): possible link between angiogenesis and immune tolerance. Med Sci Monit 11:BR280–292.

Suzuki, A., Masuda, A., Nagata, H., Kameoka, S., Kikawada, Y., Yamakawa, M. and Kasajima, T. 2002. Mature dendritic cells make clusters with T cells in the invasive margin of colorectal carcinoma. J Pathol 196:37–43.

Takagi, S., Miyagawa, S., Ichikawa, E., Soeda, J., Miwa, S., Miyagawa, Y., Iijima, S., Noike, T., Kobayashi, A. and Kawasaki, S. 2004. Dendritic cells, T-cell infiltration, and Grp94 expression in cholangiocellular carcinoma. Hum Pathol 35:881–886.

Takahashi, A., Kono, K., Itakura, J., Amemiya, H., Feng Tang, R., Iizuka, H., Fujii, H. and Matsumoto, Y. 2002. Correlation of vascular endothelial growth factor-C expression with tumor-infiltrating dendritic cells in gastric cancer. Oncology 62:121–127.

Toriyama, K., Wen, D. R., Paul, E. and Cochran, A. J. 1993. Variations in the distribution, frequency, and phenotype of Langerhans cells during the evolution of malignant melanoma of the skin. J Invest Dermatol 100:269S–273S.

Treilleux, I., Blay, J. Y., Bendriss-Vermare, N., Ray-Coquard, I., Bachelot, T., Guastalla, J. P., Bremond, A., Goddard, S., Pin, J. J., Barthelemy-Dubois, C. and Lebecque, S. 2004. Dendritic cell infiltration and prognosis of early stage breast cancer. Clin Cancer Res 10:7466–7474.

Troy, A., Davidson, P., Atkinson, C. and Hart, D. 1998a. Phenotypic characterisation of the dendritic cell infiltrate in prostate cancer. J Urol 160:214–219.

Troy, A. J., Davidson, P. J., Atkinson, C. H. and Hart, D. N. 1999. CD1a dendritic cells predominate in transitional cell carcinoma of bladder and kidney but are minimally activated. J Urol 161:1962–1967.

Troy, A. J., Summers, K. L., Davidson, P. J., Atkinson, C. H. and Hart, D. N. 1998b. Minimal recruitment and activation of dendritic cells within renal cell carcinoma. Clin Cancer Res 4:585–593.

Tsuge, K., Takeda, H., Kawada, S., Maeda, K. and Yamakawa, M. 2005. Characterization of dendritic cells in differentiated thyroid cancer. J Pathol 205:565–576.

Tsujitani, S., Furukawa, T., Tamada, R., Okamura, T., Yasumoto, K. and Sugimachi, K. 1987. Langerhans cells and prognosis in patients with gastric carcinoma. Cancer 59:501–505.

Tsujitani, S., Kakeji, Y., Maehara, Y., Sugimachi, K. and Kaibara, N. 1993. Dendritic cells prevent lymph node metastasis in patients with gastric cancer. In Vivo 7:233–237.

Tsujitani, S., Kakeji, Y., Watanabe, A., Kohnoe, S., Maehara, Y. and Sugimaghi, K. 1992. Infiltration of S-100 protein positive dendritic cells and peritoneal recurrence in advanced gastric cancer. Int Surg 77:238–241.

Tsukayama, S., Omura, K., Yoshida, K., Tanaka, Y. and Watanabe, G. 2005. Prognostic value of CD83-positive mature dendritic cells and their relation to vascular endothelial growth factor in advanced human gastric cancer. Oncol Rep 14:369–375.

Ugolini, C., Basolo, F., Proietti, A., Vitti, P., Elisei, R., Miccoli, P. and Toniolo, A. 2007. Lymphocyte and immature dendritic cell infiltrates in differentiated, poorly differentiated, and undifferentiated thyroid carcinoma. Thyroid 17:389–393.

Vakkila, J., Jaffe, R., Michelow, M. and Lotze, M. T. 2006. Pediatric cancers are infiltrated predominantly by macrophages and contain a paucity of dendritic cells: a major nosologic difference with adult tumors. Clin Cancer Res 12:2049–2054.

Vermi, W., Bonecchi, R., Facchetti, F., Bianchi, D., Sozzani, S., Festa, S., Berenzi, A., Cella, M. and Colonna, M. 2003. Recruitment of immature plasmacytoid dendritic cells (plasmacytoid monocytes) and myeloid dendritic cells in primary cutaneous melanomas. J Pathol 200:255–268.

Vicari, A. P., Treilleux, I. and Lebecque, S. 2004. Regulation of the trafficking of tumour-infiltrating dendritic cells by chemokines. Semin Cancer Biol 14:161–169.

Whiteside, T. L. 2006. The role of immune cells in the tumor microenvironment. Cancer Treat Res 130:103–124.

Yamakawa, M., Yamada, K., Orui, H., Tsuge, T., Ogata, T., Dobashi, M. and Imai, Y. 1995. Immunohistochemical analysis of dendritic/Langerhans cells in thyroid carcinomas. Anal Cell Pathol 8:331–343.

Yin, X. Y., Lu, M. D., Lai, Y. R., Liang, L. J. and Huang, J. F. 2003. Prognostic significances of tumor-infiltrating S-100 positive dendritic cells and lymphocytes in patients with hepatocellular carcinoma. Hepatogastroenterology 50:1281–1284.

Zeid, N. A. and Muller, H. K. 1993. S100 positive dendritic cells in human lung tumors associated with cell differentiation and enhanced survival. Pathology 25:338–343.

Zhong, H., Han, B., Tourkova, I. L., Lokshin, A., Rosenbloom, A., Shurin, M. R. and Shurin, G. V. 2007. Low-dose Paclitaxel prior to intratumoral dendritic cell vaccine modulates intratumoral cytokine network and lung cancer growth. Clin Cancer Res 13:5455–5462.

Zijlmans, H. J., Fleuren, G. J., Baelde, H. J., Eilers, P. H., Kenter, G. G. and Gorter, A. 2007. Role of tumor-derived proinflammatory cytokines GM-CSF, TNF-alpha, and IL-12 in the migration and differentiation of antigen-presenting cells in cervical carcinoma. Cancer 109:556–565.

Chapter 4
Functional Defects of Dendritic Cells in Cancer

Mikhail M. Dikov and Irina S. Chekneva

Abstract Altered dendritic cell function is one of the most fundamental mechanisms of tumor escape from immune surveillance. Functional characteristics of dendritic cells evolve with their differentiation and are tightly controlled by the cytokine network. Tumors actively interfere with dendritic cell differentiation and their acquisition of functional properties and render them defective. This chapter reviews the data characterizing the effect of tumor on the key functions of dendritic cells, including antigen uptake, antigen presentation, expression of cell surface molecules, motility and cytokine production.

4.1 Introduction

Dendritic cells (DC) are strategically positioned as a key element of both innate and adaptive immunity. They constitute a complex network of antigen-presenting cells that have an essential role in the modulation of primary immune responses. DC are crucial for the induction and maintenance of antitumor immunity, as they can uniquely initiate T-cell responses against tumors by stimulating naïve T cells and are also the most effective cells for the activation of secondary T-cell responses. It is now evident that DC in patients with malignancy or tumor-bearing animals demonstrate functional deficiencies that inhibit their capacity to mount an effective antitumor response.

Induction of antitumor immune response is a multistep process in which DC have important functions. Functional characteristics of DC evolve with their stage of maturation, which is finely controlled by the cytokine network and the microenvironmental features. DC originate from hematopoietic stem cells within the bone marrow and their development occurs in distinct stages. DC precursors differentiate into immature DC that circulate in the blood and home to a large variety of tissues where they act as sentinel for immune monitoring.

M.M. Dikov (✉)
Division of Hematology and Oncology, Department of Medicine, Vanderbilt University, 2200 Pierce Ave., 648 PRB, Nashville, TN 37232-6838, USA
e-mail: Mikhail.Dikov@Vanderbilt.edu

M.R. Shurin, R.D. Salter (eds.), *Dendritic Cells in Cancer*,
DOI 10.1007/978-0-387-88611-4_4, © Springer Science+Business Media, LLC 2009

The efficient immune responses occur only if antigens reach secondary lymphoid organs, and are expressed there for a definite period and at a certain level (Zinkernagel 2000). The physiological function of immature DC is to capture antigens. Following their encounter with antigens and pro-inflammatory signals, immature DC initiate their maturation process and migrate to the lymphoid organs. Interaction of CD40 with its ligand expressed on T helper cells, inflammatory cytokines, bacterial components or immune complexes induces maturation of DC. Maturing DC lose endocytic activity, increase surface expression and stability of MHC–peptide complexes, up-regulate the surface expression of adhesion and co-stimulatory molecules, while secreting inflammatory cytokines that are essential for T-cell activation. Coinciding with the phenotypic maturation, DC increase their migratory capacity. Mature DC migrate to secondary lymphoid organs where they prime T cells by interacting and presenting antigens in the context of cell surface MHC class I and II molecules. DC produce a variety of pro-inflammatory cytokines and chemokines. Interaction with $CD4^+$ T cells induces DC to produce IL-12, a critical Th1-polarizing cytokine also supporting cytotoxic T lymphocytes. DC produce pro-inflammatory cytokines, chemokines and IL-12 only for brief periods of time, and at defined stages of maturation; this is the important feature that might influence the outcome of the immune response (Langenkamp et al. 2000; Schuurhuis et al. 2006).

Myeloid and plasmocytoid DC are the two subsets of DC identified in the human blood based on the expression of the β_2 integrin, CD11c (O'Doherty et al. 1994; Ueno et al. 2007). However, the combination of the abilities to pick up antigens, migrate from peripheral tissues to the draining lymph nodes, and efficiently stimulate naïve T cells is a unique property of myeloid DC (Kalinski et al. 1999; Steinman and Inaba 1999), which places them at the pivotal role in the initiation and regulation of antitumor immune responses. Plasmocytoid DC produce large amounts of IFN-α but are inefficient in antigen capture at all maturation stages, although they have a proven role in regulation of T-cell responses (Grouard et al. 1997; Schuurhuis et al. 2006).

Clinical and experimental data clearly demonstrate that growing tumor not only decreases the number of DC but also affects all their basic functions (Shurin and Gabrilovich 2001). Tumors suppress DC antigen uptake, expression of cell surface molecules, decrease migration to draining lymph nodes and alter the profile of secreted cytokines (reviewed in Gabrilovich 2004; Pinzon-Charry et al. 2005b; Bennaceur et al. 2008). Both circulating and tumor-infiltrating DC from cancer patients appear to be phenotypically and functionally defective (Steinman and Inaba 1999; Zhang et al. 2003; Dave et al. 2004; Galon et al. 2006; Hiraoka et al. 2006; Gottfried et al. 2008). The available data link the defective DC functions with their aberrant differentiation and altered phenotype. A few studies in which analysis of different DC characteristics was performed revealed that in cancer all DC functions are affected at the same time (Dong et al. 2003; Pockaj et al. 2004). This seems natural, as specific DC characteristics and the efficacy of DC in T-cell activation are not attributed to a single specific molecule, but are the result of quantitative changes associated with their differentiation and maturation. However, the effects of tumor-derived factors on specific functions of mature DC were also reported (Beckebaum et al. 2004b).

4.2 Uptake and Processing of Antigens

Immature DC are specialized in capturing and processing antigen to form MHC–peptide complexes. Macropinocytosis has emerged as a key mechanism of antigen capture by DC, allowing substantial volumes of the extracellular milieu to be engulfed and processed (Sallusto et al. 1995; West et al. 1999). In peripheral tissues, DC constitutively macropinocytose extracellular fluid and take up antigen via phagocytosis and receptor-mediated endocytosis. DC express several receptors that facilitate the internalization and presentation of antigens, including C-type lectin receptors such as the mannose receptor (Sallusto et al. 1995) and DEC205 (Swiggard et al. 1995), as well as receptors for the Fc domain of immunoglobulins, FcγR and FcεR. Specific receptors for heat shock proteins (HSP) mediate the internalization of HSP–peptide complexes (Singh-Jasuja et al. 2000; Kuppner et al. 2001).

Reduced endocytic capacity of circulating DC in patients with malignancy was observed. For instance, DC isolated from patients with renal cell carcinoma showed a reduced capacity to take up antigen (Thurnher et al. 1996). The studies with DC generated from monocytes of cancer patients revealed defective antigen uptake and processing. A detailed investigation by Dong et al. (2003) described differences in the organization of the actin cytoskeleton in BCR-ABL oncogen-expressing DC generated from chronic myeloid leukemia (CML) patients compared with normal DC. Cells had a decreased ability to spread and polarize and had fewer podosomes compared with normal DC. They were also defective in processing and presentation of exogenous antigens such as tetanus toxoid. The antigen-processing defect was attributed to the reduced capacity of DC from CML patients to capture antigen via macropinocytosis or via mannose receptors. Monocyte-derived DC from breast cancer patients showed significantly reduced expression of co-stimulatory molecules (CD80 and CD40) and demonstrated reduced phagocytic ability, reduced antigen presentation to T cells and reduced capacity to mature in response to lipopolysaccharide (Pockaj et al. 2004). Aberrant antigen processing by DC was associated with the increased COX-2 expression and PgE_2 levels within the tumor milieu and in the circulation. Sharma et al. (2003) assessed the capability of DC generated from mouse bone marrow progenitors in the presence of tumor supernatant to process and present the ovalbumin antigens. Consistent with the reduction in the number of tumor supernatant-treated DC expressing transporter associated with antigen processing (TAP), these DC had a reduced capacity to process and present the ovalbumin antigens. Tumor COX-2 inhibition resulted in maintenance of TAP and antigen-presenting properties of DC.

Another report revealed an association of inhibited phagocytosis with the secretion of vascular endothelial growth factor (VEGF), an important immunosuppressive cytokine produced by tumor cells (Ishida et al. 1998). In this mouse model, blockade of VEGF led to an increase in antigen uptake and migration of tumor-associated DC. In another murine model, Tourkova et al. revealed a decreased uptake of dextran 40 and polystyrene beads by DC generated in the presence of different tumor cell lines, including prostate, colon, lung and oral squamous cell carcinomas in vitro and by DC prepared from tumor-bearing mice ex vivo. Impaired endocytic activity of DC co-cultured with tumor cells was associated with decreased levels of active small Rho GTPases Cdc42 and Rac1 (Tourkova et al. 2007).

Tumor-derived molecules can directly interfere with the DC antigen-capture and antigen-presenting pathways. For example, mucin-1 (MUC-1), secreted by breast cancer cells, is endocytosed by DC but mostly retained in early endosomes, leading to its inefficient processing and presentation to T cells and lower frequency of MUC-1-specific effector cells (Hiltbold et al. 2000; Vlad et al. 2004).

In some cases, DC purified from the peripheral blood of patients with cancer seem to retain normal endocytic ability but have low T-cell stimulatory capacity (Hasebe et al. 2000). Of interest is that co-culture of immature or mature monocyte-derived DC with glioma cells did not affect incorporation of dextran by DC but affected their surface phenotype and IL-12 production, suggesting that certain DC functions may be regulated independently (Kikuchi et al. 2002).

A few studies that specifically addressed the function of DC antigen-processing machinery (AMP) in cancer demonstrated impairment of AMP component expression induced by tumor-derived factors (Whiteside et al. 2004; Tourkova et al. 2005). Human DC generated in the presence of oral squamous cell carcinoma cell lines exhibited a profound down-regulation of several APM components, including LMP2, LMP10, MB1, TAP1/2 and other antigen-processing pathway proteins. The attenuated expression of these proteins was associated with a decreased presentation of an antigen to antigen-specific autologous $CD8^+$ T cells and was induced, at least in part, by tumor-derived gangliosides (Tourkova et al. 2005). Similar results were obtained using mouse DC generated in the presence of culture supernatants from COX-2-expressing tumors (Sharma et al. 2003). Furthermore, expression of several MHC class I APM components, including delta, MB-1, LMP-10, ERp57 and tapasin, was significantly decreased in murine DC generated in the presence of prostate cancer cells. APM component down-regulation was associated with decreased ability of DC to present model antigen and was mediated by IRF-8, a member of the interferon regulatory factor family (Tourkova et al. 2008).

4.3 Migration

Upon activation, DC migrate to the lymphoid organs such as spleen and lymph nodes. There, DC attract T and B cells and maintain the viability of re-circulating T lymphocytes by releasing chemokines and cytokines (Banchereau et al.

2000; Clark et al. 2000; Caux et al. 2002). Migration of DC is tightly regulated as a function of maturation (Parlato et al. 2001; De Vries et al. 2003). A variety of cytokines and chemokines (GM-CSF, TNF-α, IL-1, CCL21, CCL20 and CCL19), but also non-chemokine chemotactic agonists, lipid mediators and membrane proteins modulate DC movement and maturation (Sozzani 2005). Maturing DC down-regulate expression of CC chemokine receptors (CCR) 1 and 2 and up-regulate CCR7 that directs DC to the lymphoid vessels and T-cell areas of secondary lymphoid organs (Yanagihara et al. 1998; Chen et al. 2001; Jarrossay et al. 2001; Caux et al. 2002). CCR7 ligands CCL21 and CCL19 potently induce migration of mature DC (Kellermann et al. 1999). Switch in the expression of chemokine receptors is accompanied by up-regulation of MCH class II and co-stimulatory molecules (HLA-DR, CD80, CD83 and CD86). Experiments with intradermal injection of DC showed that mature DC have, on average, a six- to eightfold higher migratory capacity than immature DC (Ridolfi et al. 2004).

Within the tumor microenvironment several factors contribute to the sequestration of DC within tumor tissues and the subsequent inhibition of their migration. The observation that in vitro-generated DC injected in human tumors were not able to migrate to draining lymph nodes has suggested that this crucial step may be impaired in cancer (Triozzi et al. 2000). It has been reported that melanoma cell lines can effectively chemoattract DC, modulate their phenotype and eventually damage DC motility. Melanoma-conditioned DC exhibited an increased adhesion capacity to melanoma cells in vitro and did not migrate in response to lymphoid chemokines (Remmel et al. 2001). In vitro-generated DC failed to home into draining lymph nodes upon intratumoral injection in patients with metastatic melanomas (Triozzi et al. 2000). Neuroblastoma has been shown to hinder chemokine-mediated DC migration by interfering with CCR7-CCL19 intracellular signal transduction pathways (Walker et al. 2006). IL-8, which is produced by different types of solid tumors, including hepatocellular carcinoma, colorectal and pancreatic cancers was implicated in the retention of DC inside malignant lesions and the impairment of DC migration toward CCR7 ligands (Feijoo et al. 2005). Melanoma-derived gangliosides can also impair DC migratory function through the down-regulation of CCR7 expression (Bennaceur et al. 2006). Hypoxia, which is a characteristic of solid tumors, inhibits monocyte-derived DC migration through the suppression of matrix metalloproteinase (MMP) production (Zhao et al. 2005; Darmanin et al. 2007). Altered actin organization in DC derived from CML patients results in that the capacity of these DC to migrate is impaired (Dong et al. 2003).

In an interesting study, Yang et al. (2003) reported that PgE_2 signaling via EP2 receptor negatively regulates DC migration in the tumor-bearing animals and contributes to tumor-induced immune suppression. DC from $EP2^{-/-}$ knockout tumor-bearing mice showed a significantly increased ability to home to axillary and inguinal lymph nodes, compared with that of tumor-bearing wild-type animals. This was associated with significantly higher expression of CCR7 and improved chemotaxis in response to CCL19 and CCL21 by DC derived from tumor-bearing knockout mice.

Of many tumor-derived factors affecting DC function TGF-β and IL-10 are also implied in regulation of DC motility. TGF-β increases expression of chemokine receptor such as CCR1, CCR2, CCR3 and CCR6 in immature DC but down-regulates CCR7 expression induced by TNF-α and prevents DC migration toward lymph nodes maintaining them immature (Sato et al. 2000). IL-10 has also been shown to modulate expression of chemokines and chemokine receptors by DC resulting in down-regulation of CCR7 and up-regulation of CCR5 (Takayama et al. 2001).

4.4 Maturation, Antigen Presentation and Co-Stimulation

Maturation of DC is a key event in the induction of the immune response; it is required for eliciting primary T-cell responses (Banchereau and Steinman 1998). DC maturation is a series of several coordinated events such as loss of endocytic/phagocytic receptors, up-regulation of MHC molecules, alteration in expression of adhesion molecules, cytokine receptors and cytokine production, changes in morphology, lysosomal and MHC class II-enriched compartments (Banchereau et al. 2000; Turley et al. 2000; Kleijmeer et al. 2001). DC maturation is associated with the expression of many accessory molecules (CD40, CD54, CD58, CD80, CD83, CD86, CD70) that interact with receptors on T cells to enhance adhesion and signaling (co-stimulation). Depending on the conditions, DC can stimulate the outgrowth and activation of a variety of T cells. The balance between pro-inflammatory and anti-inflammatory signals in the local microenvironment, including TNF-α, IL-1, IL-6, IL-10, TGF-β and prostanoids, plays a role in the maturation process (Kalinski et al. 1998; Banchereau et al. 2000). These signals, as well as many tumor-derived factors, can regulate transition of immature DC to mature and presentation of captured antigen to lymphocytes.

Systemic tumor-induced defects of DC differentiation and maturation result in the decreased production of mature functionally competent DC and the accumulation of immature DC that have characteristics of lineage-committed DC but cannot up-regulate MHC class I and class II and co-stimulatory molecules or produce appropriate cytokines (Fig. 4.1). A number of tumor types have been shown to imbalance DC maturation and function, including lung cancer, breast cancer, melanoma, prostate cancer, hepatoma, renal cell carcinoma, gastric cancer and neuroblastoma via production of a variety of systemic and local factors and cytokines (reviewed in Shurin and Gabrilovich 2001; Gabrilovich 2004; Yang and Carbone 2004; Pinzon-Charry et al. 2005b; Shurin et al. 2006; Bennaceur et al. 2008; Gottfried et al. 2008).

DC isolated from cancer patients exhibit quantitative and functional deficiencies (Gabrilovich et al. 1996b; Troy et al. 1998b; Brown et al. 2001;

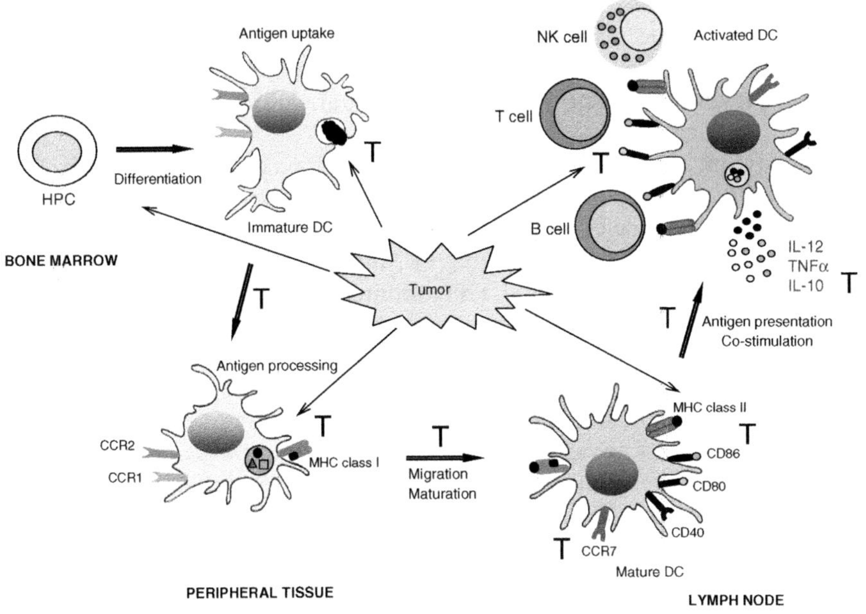

Fig. 4.1 Tumor affects both dendritic cell differentiation and function. Functional characteristics of DC evolve with their differentiation. DC originate from hematopoietic stem cells within the bone marrow. Their precursors differentiate into immature DC that circulate in the blood and home to a variety of tissues. In the peripheral tissue immature DC take up and process tumor antigens. Following their encounter with the antigens and pro-inflammatory signals, immature DC initiate their maturation process, increase their migratory capacity and migrate to the lymphoid organs. There, they stimulate effector cells by interacting and presenting antigens in the context of cell surface MHC class I and II molecules. DC also produce a variety of pro-inflammatory cytokines and chemokines. Interaction with $CD4^+$ T cells induces DC activation and up-regulates their IL-12 production. Tumors actively interfere with DC differentiation and their acquisition of functional properties. Tumors suppress DC antigen uptake and processing, expression of co-stimulatory and adhesion cell surface molecules, decrease migration to draining lymph nodes, and alter the profile of secreted cytokines and chemokines. *Black arrows* and letter “T” point to the major functions of DC affected by tumor

Hoffmann et al. 2002). Tumors inhibit the host immune system at least in part by interfering with the effectiveness of antigen presentation (Gabrilovich et al. 1996b; Kudela et al. 2001). Gabrilovich et al. (1997) evaluated T-cell stimulatory activity in patients with breast cancer and described defective responses to allogeneic and defined antigens in patients with advanced disease. Defective antigen presentation by DC also appears to be a major determinant of cytotoxic T lymphocyte non-responsiveness to peptide antigens in tumor-bearing mice (Gabrilovich et al. 1996b). DC from patients with hepatocellular carcinoma had significantly lower capacity to stimulate allogeneic T-cell proliferation, compared with DC isolated from patients with liver cirrhosis and normal controls

(Ninomiya et al. 1999). In these patients, DC expressed significantly lower levels of HLA-DR and decreased induction of IL-12. Overall, the DC compartment has been shown to exhibit increased number of immature cells with impaired antigen-presenting capacity (Pinzon-Charry et al. 2005a; Pinzon-Charry et al. 2005b).

Studies that assessed the effects of tumors on blood or tumor-infiltrating DC revealed that progression of cancer negatively correlates with the expression of co-stimulatory molecules, in particular CD80 and CD86, on both myeloid DC and plasmocytoid DC. All circulating DC subsets exhibited low MHC class II expression. Thus, Bergeron et al. (2006) found that various types of DC infiltrate lung carcinomas and display immature phenotype and low IL-12 expression. Circulating DC in hepatocellular carcinoma patients consisted mainly of immature, non-activated cell subsets (Troy et al. 1998b; Beckebaum et al. 2004a,b). Minimal number of DC was recruited to the tumor bed in patients with renal and prostate cancer. Those present had low levels of co-stimulatory molecules and a decreased capacity to stimulate allogeneic T-cell proliferation (Troy et al. 1998a,b). These DC express no or low levels of the co-stimulatory molecules CD80 and CD86. Furthermore, DC derived from progressing melanoma metastases do not express CD86 (Enk et al. 1997). Similar data were obtained for DC isolated from basal-cell carcinomas (Nestle et al. 1997); these DC also had reduced antigen-presenting function. In breast carcinoma, intratumoral DC appeared immature, whereas peritumoral DC showed a mature phenotype (Wilson et al. 2003). The presence of high numbers of infiltrating $CD1a^+$ cells seems to be associated with an improved prognosis in breast carcinoma (del Hoyo et al. 2002; Mende et al. 2006). Blood DC of multiple myeloma patients had lower expression of HLA-DR, CD40 and CD80 molecules, as compared with that of healthy controls (Ratta et al. 2002).

A study in patients with breast cancer has shown that freshly isolated DC although exhibited a more mature phenotype were markedly reduced in number and ability to stimulate antigen-specific T-cell responses (Della Bella et al. 2003). These findings possibly reflect the distinct mechanisms by which the tumors interfere with the immune system of the host. Chaux et al. (1997) reported that only a small proportion of DC isolated from colon carcinomas express CD80 or CD86. Application of maturation stimulus (CD40L or TNF-α) in vitro did not induce CD80 expression by tumor-infiltrating DC, which indicates that lack of co-stimulatory molecules expression is a result of defective cell differentiation (Chaux et al. 1997). Consistent with these observations, an increased proportion of immature DC with reduced expression of co-stimulatory molecules was found in the peripheral blood of patients with breast, head and neck, lung and esophageal cancers (Gabrilovich et al. 1997; Ishida et al. 1998), and similar data have been obtained using several mouse tumor models (Gabrilovich et al. 1996b; Ishida et al. 1998; Cheng et al. 2003).

In a number of studies, DC derived from monocyte of cancer patients showed lower expression of HLA-DR, CD80 and CD86 and accordingly impaired potency to stimulate T cells (Hasebe et al. 2000; Onishi et al. 2002;

Bellone et al. 2006). However, other investigations with monocyte-derived patients' DC and experiments with animal DC generated from hematopoietic progenitor cells demonstrated normal allostimulatory capacity of DC (Gabrilovich et al. 1996a; Gabrilovich et al. 1997; Ratta et al. 2002). Some reports indicated that tumor antigens are also responsible for impaired DC maturation and function. For instance, purified, tumor-derived and circulating prostate-specific antigen noticeably inhibited differentiation and maturation of DC in vitro (Aalamian et al. 2003; Aalamian-Matheis et al. 2007). Polyamines putrescine, spermidine and spermine, nutrients constantly produced in the tumor microenvironment, have also been implicated in impaired DC maturation (Della Bella et al. 2003), as well as tumor-derived gangliosides (Shurin et al. 2001) and bombesin-like peptides (Makarenkova et al. 2003).

Most recent investigations demonstrated that DC may cover dual functions in the tumor microenvironment. Immature or partially differentiated myeloid DC cannot induce antitumor immune responses but can function as regulatory DC and are an important component of the immunosuppressive networks in the tumor microenvironment (Zou et al. 2001; Curiel et al. 2004b; Mellor and Munn 2004; Fricke and Gabrilovich 2006; Dhodapkar et al. 2008; Muthuswamy et al. 2008). This regulatory capacity of inappropriately matured DC is a result of the influence of the tumor microenvironment and tumor-derived factor and represents a mean of how tumors escape immune recognition by interfering with the process of DC maturation/activation.

4.5 Cytokine Production

Generation of an appropriate cytokine milieu is a critical feature of DC and is essential for efficient antigen presentation and potent activation of T-cell-mediated immune responses. Depending on DC subtype and maturation/activation status, DC cytokine repertoire comprises a variety of immunomodulating molecules, including pro-inflammatory cytokines and chemokines such as TNF-α and IL-8, the major Th1-promoting factor IL-12, T-cell inhibitory cytokine IL-10 and tolerogenic TGF-β (Inoue et al. 1993; Troy et al. 1998b; Schwaab et al. 1999; Sandel et al. 2005).

Cytokine production by DC is tightly regulated by the factors in the microenvironment, which is particularly relevant in the case of IL-12 expression. IL-12 signal is a key regulator of the induction of adaptive Th1-type immunity (Del Vecchio et al. 2007) and supports clonal outgrowth of antigen-specific $CD8^+$ cytotoxic T lymphocytes (Kalinski et al. 1999; Curtsinger et al. 2003). Recently, it was also found that it promotes the reactivation and survival of memory $CD4^+$ T cells (Yoo et al. 2002) and can contribute to the re-polarization of dysfunctional antitumor Th2 $CD4^+$ T cells into Th1 cells (Wesa et al. 2007). IL-12 production by DC is restricted to a narrow temporal window after maturation (Langenkamp et al. 2000), whereas the capacity to induce Th2 responses is a property of DC that do not produce Th1-polarizing cytokines.

Tumors dramatically affect IL-12 production by DC. Thus, DC in patients with hepatocellular carcinoma exhibited immature phenotype and expressed low levels of IL-12 but had increased production of IL-10 and TNF-α (Kakumu et al. 2000; Beckebaum et al. 2004b). A significant negative correlation has been found between plasma levels of spermine and DC production of IL-12 in patients with breast cancer (Della Bella et al. 2003). DC from chronic lymphocytic leukemia patients were severely defective and had a reduced ability to release IL-12 to drive a type 1 T-cell response (Orsini et al. 2003).

In vitro experiments also confirm that tumor-derived factor can interfere with IL-12 secretion by DC. MUC-1 produced by breast cancer cells was shown to inhibit the capacity of DC to secrete IL-12, thereby skewing the development of T-cell responses toward type 2 (Carlos et al. 2005). Using in vitro co-culture model of tumor spheroid, Gottfried et al. (2006) found that lactic acid produced by melanoma or prostate carcinoma cells significantly reduced IL-12 secretion by DC.

Secretion of several other functionally important DC cytokines and chemokines is also affected by tumor. In colorectal carcinoma, DC density has shown a direct correlation with levels of TNF-α expressed by DC (Schwaab et al. 2001). DC at the tumor sites skew $CD4^+$ T-cell differentiation toward T cells secreting high levels of Th2 (IL-4 and IL-13) cytokines, which promote early tumor progression. IL-13 secreted from such T cells appears to be responsible for the tumor growth, as blocking of IL-13 partially inhibits the tumor growth in a humanized mouse model of breast cancer (Aspord et al. 2007). Productions of tumoricidal IFN-α by plasmocytoid DC were inhibited in patients with head and neck squamous cell carcinoma (Hartmann et al. 2003). Leukemic plasmocytoid DC, but not leukemic myeloid DC, had impaired capacity for maturation, decreased allostimulatory activity, and altered ability to secrete IFN-α (Mohty et al. 2001). It was reported that PgE_2, a factor overproduced in chronic inflammation and cancer, induces stable regulatory T-cell-attracting properties in maturing DC mediated by expression of CCL22 (Muthuswamy et al. 2008). At the same time, PgE_2-exposed DC secreted less Th1-attracting chemokines CXCL9, CXCL10, CXCL11 and CCL5. Interestingly, IFN-α was able to reverse the effect of PgE_2 (Muthuswamy et al. 2008).

Results of the recent studies shed light on yet poorly understood molecular mechanisms of tumor-induced down-regulation of IL-12 expression by DC. IL-10 produced by colon adenocarcinoma was shown to inhibit CD40-dependent IL-12 production by DC in addition to the fact that CD40 ligation on DC obtained from tumor bearers did not result in inducible expression of IL-12 protein or IL-12 p40 mRNA (Shurin et al. 2002). Melanoma lysates suppressed IL-12p70 secretion in TLR4-stimulated DC, reducing the generation of Th1 responses through activation of p44/42 MAP kinase. Blockade of melanoma-dependent p44/42 activation with a MEK1/2 inhibitor restored IL-12p70 production and the generation of Th1 cells from naïve $CD4^+$ cells (Jackson et al. 2008). Constitutive Stat3 activity in tumors inhibits the production of pro-inflammatory cytokines, while promoting the release of soluble factors that suppress

DC functions (Wang et al. 2004). Furthermore, these factors up-regulate Stat3 expression in DC, resulting in the induction of antitumor tolerance rather than immunity (Kortylewski et al. 2005).

It has now become apparent that in addition to aberrant immune responses tumor can induce DC to promote tumor angiogenesis by secretion of pro-angiogenic factors (Curiel et al. 2004a; Murdoch et al. 2008; Ryzhov et al. 2008). For example, high numbers of plasmacytoid DC in ovarian cancer attracted by CXCL-12 induced angiogenesis in vivo through the production of TNF-α and IL-8. By contrast, myeloid DC, which could suppress angiogenesis in vivo through the production of IL-12, were absent from malignant ascites (Curiel et al. 2004a). Moreover, human immature DC also release osteopontin, which triggers the release of IL-1β by monocytes, a potent angiogenic cytokine involved in neovascularization (Naldini et al. 2006). Several tumor-derived factors including HGF (Okunishi et al. 2005), TGF-α (Alard et al. 2004), PgE_2 (Pockaj et al. 2004), lactate (Puig-Kroger et al. 2003), adenosine (Novitskiy et al. 2008; Ryzhov et al. 2008) and osteopontin (Konno et al. 2006) have also been shown to suppress DC maturation resulting in cells with a pro-angiogenic phenotype.

4.6 Concluding Remarks

Recent studies have demonstrated a remarkable functional plasticity of DC, which is regulated by the cytokine network and local microenvironment. This unique property enables them to respond to a variety of pathogenic signals and to induce different types of immune responses. The same feature makes them vulnerable to the regulation by multiple tumor-derived factors, which potently affect DC differentiation and functionality. The reviewed data suggest that tumors can affect every major DC function from recruitment to the tumor site and antigen uptake to co-stimulation of effector cells and cytokine secretion. Most often, the impairment of DC functions correlates with their aberrant differentiation and maturation.

During immunological reactions, pro-inflammatory and anti-inflammatory mechanisms and polarization of T-cell responses are regulated in a finely tuned manner. Pro-inflammatory and Th1-promoting signals are followed by the signals inducing type 2-polarized response, immune tolerance and angiogenesis to provide for wound healing after inflammatory damage. DC have an important role in orchestrating these events by their ability to activate both mechanisms. It seems that tumors force DC to skip the first part of the immune response but instead assume their immune inhibitory and tissue protective role. Understanding the molecular mechanisms controlling the switch in DC functions would be essential for their successful therapeutic applications in cancer.

References

Aalamian-Matheis, M., Chatta, G. S., Shurin, M. R., Huland, E., Huland, H. and Shurin, G. V. 2007. Inhibition of dendritic cell generation and function by serum from prostate cancer patients: correlation with serum-free PSA. Adv Exp Med Biol 601:173–182.

Aalamian, M., Tourkova, I. L., Chatta, G. S., Lilja, H., Huland, E., Huland, H., Shurin, G. V. and Shurin, M. R. 2003. Inhibition of dendropoiesis by tumor derived and purified prostate specific antigen. J Urol 170:2026–2030.

Alard, P., Clark, S. L. and Kosiewicz, M. M. 2004. Mechanisms of tolerance induced by TGF beta-treated APC: CD4 regulatory T cells prevent the induction of the immune response possibly through a mechanism involving TGF beta. Eur J Immunol 34:1021–1030.

Aspord, C., Pedroza-Gonzalez, A., Gallegos, M., Tindle, S., Burton, E. C., Su, D., Marches, F., Banchereau, J. and Palucka, A. K. 2007. Breast cancer instructs dendritic cells to prime interleukin 13-secreting CD4+ T cells that facilitate tumor development. J Exp Med 204:1037–1047.

Banchereau, J., Briere, F., Caux, C., Davoust, J., Lebecque, S., Liu, Y. J., Pulendran, B. and Palucka, K. 2000. Immunobiology of dendritic cells. Annu Rev Immunol 18:767–811.

Banchereau, J. and Steinman, R. M. 1998. Dendritic cells and the control of immunity. Nature 392:245–252.

Beckebaum, S., Cicinnati, V. R. and Broelsch, C. E. 2004a. Future directions in immunosuppression. Transplant Proc 36:574S–576S.

Beckebaum, S., Zhang, X., Chen, X., Yu, Z., Frilling, A., Dworacki, G., Grosse-Wilde, H., Broelsch, C. E., Gerken, G. and Cicinnati, V. R. 2004b. Increased levels of interleukin-10 in serum from patients with hepatocellular carcinoma correlate with profound numerical deficiencies and immature phenotype of circulating dendritic cell subsets. Clin Cancer Res 10:7260–7269.

Bellone, G., Carbone, A., Smirne, C., Scirelli, T., Buffolino, A., Novarino, A., Stacchini, A., Bertetto, O., Palestro, G., Sorio, C., Scarpa, A., Emanuelli, G. and Rodeck, U. 2006. Cooperative induction of a tolerogenic dendritic cell phenotype by cytokines secreted by pancreatic carcinoma cells. J Immunol 177:3448–3460.

Bennaceur, K., Chapman, J., Brikci-Nigassa, L., Sanhadji, K., Touraine, J. L. and Portoukalian, J. 2008. Dendritic cells dysfunction in tumour environment. Cancer Lett 272(2):186–196.

Bennaceur, K., Popa, I., Portoukalian, J., Berthier-Vergnes, O. and Peguet-Navarro, J. 2006. Melanoma-derived gangliosides impair migratory and antigen-presenting function of human epidermal Langerhans cells and induce their apoptosis. Int Immunol 18:879–886.

Bergeron, A., El-Hage, F., Kambouchner, M., Lecossier, D. and Tazi, A. 2006. Characterisation of dendritic cell subsets in lung cancer micro-environments. Eur Respir J 28:1170–1177.

Brown, R. D., Pope, B., Murray, A., Esdale, W., Sze, D. M., Gibson, J., Ho, P. J., Hart, D. and Joshua, D. 2001. Dendritic cells from patients with myeloma are numerically normal but functionally defective as they fail to up-regulate CD80 (B7-1) expression after huCD40LT stimulation because of inhibition by transforming growth factor-beta1 and interleukin-10. Blood 98:2992–2998.

Carlos, C. A., Dong, H. F., Howard, O. M., Oppenheim, J. J., Hanisch, F. G. and Finn, O. J. 2005. Human tumor antigen MUC1 is chemotactic for immature dendritic cells and elicits maturation but does not promote Th1 type immunity. J Immunol 175:1628–1635.

Caux, C., Vanbervliet, B., Massacrier, C., Ait-Yahia, S., Vaure, C., Chemin, K., Dieu-Nosjean, M-C. and Vicari, A. 2002. Regulation of dendritic cell recruitment by chemokines. Transplantation 73:S7–S11.

Chaux, P., Favre, N., Martin, M. and Martin, F. 1997. Tumor-infiltrating dendritic cells are defective in their antigen-presenting function and inducible B7 expression in rats. Int J Cancer 72:619–624.

Chen, Z., Dehm, S., Bonham, K., Kamencic, H., Juurlink, B., Zhang, X., Gordon, J. R. and Xiang, J. 2001. DNA array and biological characterization of the impact of the maturation status of mouse dendritic cells on their phenotype and antitumor vaccination efficacy. Cell Immunol 214:60–71.

Cheng, P., Nefedova, Y., Miele, L., Osborne, B. A. and Gabrilovich, D. 2003. Notch signaling is necessary but not sufficient for differentiation of dendritic cells. Blood 102:3980–3988.

Clark, G. J., Angel, N., Kato, M., Lopez, J. A., MacDonald, K., Vuckovic, S. and Hart, D. N. 2000. The role of dendritic cells in the innate immune system. Microbes Infect 2:257–272.

Curiel, T. J., Cheng, P., Mottram, P., Alvarez, X., Moons, L., Evdemon-Hogan, M., Wei, S., Zou, L., Kryczek, I., Hoyle, G., Lackner, A., Carmeliet, P. and Zou, W. 2004a. Dendritic cell subsets differentially regulate angiogenesis in human ovarian cancer. Cancer Res 64:5535–5538.

Curiel, T. J., Coukos, G., Zou, L., Alvarez, X., Cheng, P., Mottram, P., Evdemon-Hogan, M., Conejo-Garcia, J. R., Zhang, L., Burow, M., Zhu, Y., Wei, S., Kryczek, I., Daniel, B., Gordon, A., Myers, L., Lackner, A., Disis, M. L., Knutson, K. L., Chen, L. and Zou, W. 2004b. Specific recruitment of regulatory T cells in ovarian carcinoma fosters immune privilege and predicts reduced survival. Nat Med 10:942–949.

Curtsinger, J. M., Lins, D. C. and Mescher, M. F. 2003. Signal 3 determines tolerance versus full activation of naive CD8 T cells: dissociating proliferation and development of effector function. J Exp Med 197:1141–1151.

Darmanin, S., Chen, J., Zhao, S., Cui, H., Shirkoohi, R., Kubo, N., Kuge, Y., Tamaki, N., Nakagawa, K., Hamada, J., Moriuchi, T. and Kobayashi, M. 2007. All-trans retinoic acid enhances murine dendritic cell migration to draining lymph nodes via the balance of matrix metalloproteinases and their inhibitors. J Immunol 179:4616–4625.

Dave, S. S., Wright, G., Tan, B., Rosenwald, A., Gascoyne, R. D., Chan, W. C., Fisher, R. I., Braziel, R. M., Rimsza, L. M., Grogan, T. M., Miller, T. P., LeBlanc, M., Greiner, T. C., Weisenburger, D. D., Lynch, J. C., Vose, J., Armitage, J. O., Smeland, E. B., Kvaloy, S., Holte, H., Delabie, J., Connors, J. M., Lansdorp, P. M., Ouyang, Q., Lister, T. A., Davies, A. J., Norton, A. J., Muller-Hermelink, H. K., Ott, G., Campo, E., Montserrat, E., Wilson, W. H., Jaffe, E. S., Simon, R., Yang, L., Powell, J., Zhao, H., Goldschmidt, N., Chiorazzi, M. and Staudt, L. M. 2004. Prediction of survival in follicular lymphoma based on molecular features of tumor-infiltrating immune cells. N Engl J Med 351:2159–2169.

del Hoyo, G. M., Martin, P., Vargas, H. H., Ruiz, S., Arias, C. F. and Ardavin, C. 2002. Characterization of a common precursor population for dendritic cells. Nature 415:1043–1047.

Del Vecchio, M., Bajetta, E., Canova, S., Lotze, M. T., Wesa, A., Parmiani, G. and Anichini, A. 2007. Interleukin-12: biological properties and clinical application. Clin Cancer Res 13:4677–4685.

Della Bella, S., Gennaro, M., Vaccari, M., Ferraris, C., Nicola, S., Riva, A., Clerici, M., Greco, M. and Villa, M. L. 2003. Altered maturation of peripheral blood dendritic cells in patients with breast cancer. Br J Cancer 89:1463–1472.

De Vries, I. J., Krooshoop, D. J., Scharenborg, N. M., Lesterhuis, W. J., Diepstra, J. H., Van Muijen, G. N., Strijk, S. P., Ruers, T. J., Boerman, O. C., Oyen, W. J., Adema, G. J., Punt, C. J. and Figdor, C. G. 2003. Effective migration of antigen-pulsed dendritic cells to lymph nodes in melanoma patients is determined by their maturation state. Cancer Res 63:12–17.

Dhodapkar, M. V., Dhodapkar, K. M. and Palucka, A. K. 2008. Interactions of tumor cells with dendritic cells: balancing immunity and tolerance. Cell Death Differ 15:39–50.

Dong, R., Cwynarski, K., Entwistle, A., Marelli-Berg, F., Dazzi, F., Simpson, E., Goldman, J. M., Melo, J. V., Lechler, R. I., Bellantuono, I., Ridley, A. and Lombardi, G. 2003. Dendritic cells from CML patients have altered actin organization, reduced antigen processing, and impaired migration. Blood 101:3560–3567.

Enk, A. H., Jonuleit, H., Saloga, J. and Knop, J. 1997. Dendritic cells as mediators of tumor-induced tolerance in metastatic melanoma. Int J Cancer 73:309–316.

Feijoo, E., Alfaro, C., Mazzolini, G., Serra, P., Penuelas, I., Arina, A., Huarte, E., Tirapu, I., Palencia, B., Murillo, O., Ruiz, J., Sangro, B., Richter, J. A., Prieto, J. and Melero, I. 2005. Dendritic cells delivered inside human carcinomas are sequestered by interleukin-8. Int J Cancer 116:275–281.

Fricke, I. and Gabrilovich, D. I. 2006. Dendritic cells and tumor microenvironment: a dangerous liaison. Immunol Invest 35:459–483.

Gabrilovich, D. 2004. Mechanisms and functional significance of tumour-induced dendritic-cell defects. Nat Rev Immunol 4:941–952.

Gabrilovich, D. I., Chen, H. L., Girgis, K. R., Cunningham, H. T., Meny, G. M., Nadaf, S., Kavanaugh, D. and Carbone, D. P. 1996a. Production of vascular endothelial growth factor by human tumors inhibits the functional maturation of dendritic cells. Nat Med 2:1096–1103.

Gabrilovich, D. I., Ciernik, I. F. and Carbone, D. P. 1996b. Dendritic cells in antitumor immune responses. I. Defective antigen presentation in tumor-bearing hosts. Cell Immunol 170:101–110.

Gabrilovich, D. I., Corak, J., Ciernik, I. F., Kavanaugh, D. and Carbone, D. P. 1997. Decreased antigen presentation by dendritic cells in patients with breast cancer. Clin Cancer Res 3:483–490.

Galon, J., Costes, A., Sanchez-Cabo, F., Kirilovsky, A., Mlecnik, B., Lagorce-Pages, C., Tosolini, M., Camus, M., Berger, A., Wind, P., Zinzindohoue, F., Bruneval, P., Cugnenc, P. H., Trajanoski, Z., Fridman, W. H. and Pages, F. 2006. Type, density, and location of immune cells within human colorectal tumors predict clinical outcome. Science 313:1960–1964.

Gottfried, E., Kreutz, M. and Mackensen, A. 2008. Tumor-induced modulation of dendritic cell function. Cytokine Growth Factor Rev 19:65–77.

Gottfried, E., Kunz-Schughart, L. A., Ebner, S., Mueller-Klieser, W., Hoves, S., Andreesen, R., Mackensen, A. and Kreutz, M. 2006. Tumor-derived lactic acid modulates dendritic cell activation and antigen expression. Blood 107:2013–2021.

Grouard, G., Rissoan, M. C., Filgueira, L., Durand, I., Banchereau, J. and Liu, Y. J. 1997. The enigmatic plasmacytoid T cells develop into dendritic cells with interleukin (IL)-3 and CD40-ligand. J Exp Med 185:1101–1111.

Hartmann, E., Wollenberg, B., Rothenfusser, S., Wagner, M., Wellisch, D., Mack, B., Giese, T., Gires, O., Endres, S. and Hartmann, G. 2003. Identification and functional analysis of tumor-infiltrating plasmacytoid dendritic cells in head and neck cancer. Cancer Res 63:6478–6487.

Hasebe, H., Nagayama, H., Sato, K., Enomoto, M., Takeda, Y., Takahashi, T. A., Hasumi, K. and Eriguchi, M. 2000. Dysfunctional regulation of the development of monocyte-derived dendritic cells in cancer patients. Biomed Pharmacother 54:291–298.

Hiltbold, E. M., Vlad, A. M., Ciborowski, P., Watkins, S. C. and Finn, O. J. 2000. The mechanism of unresponsiveness to circulating tumor antigen MUC1 is a block in intracellular sorting and processing by dendritic cells. J Immunol 165:3730–3741.

Hiraoka, K., Miyamoto, M., Cho, Y., Suzuoki, M., Oshikiri, T., Nakakubo, Y., Itoh, T., Ohbuchi, T., Kondo, S. and Katoh, H. 2006. Concurrent infiltration by CD8 + T cells and CD4 + T cells is a favourable prognostic factor in non-small-cell lung carcinoma. Br J Cancer 94:275–280.

Hoffmann, T. K., Muller-Berghaus, J., Ferris, R. L., Johnson, J. T., Storkus, W. J. and Whiteside, T. L. 2002. Alterations in the frequency of dendritic cell subsets in the peripheral circulation of patients with squamous cell carcinomas of the head and neck. Clin Cancer Res 8:1787–1793.

Inoue, K., Furihata, M., Ohtsuki, Y. and Fujita, Y. 1993. Distribution of S-100 protein-positive dendritic cells and expression of HLA-DR antigen in transitional cell carcinoma of the urinary bladder in relation to tumour progression and prognosis. Virchows Arch A Pathol Anat Histopathol 422:351–355.

Ishida, T., Oyama, T., Carbone, D. P. and Gabrilovich, D. I. 1998. Defective function of Langerhans cells in tumor-bearing animals is the result of defective maturation from hemopoietic progenitors. J Immunol 161:4842–4851.
Jackson, A. M., Mulcahy, L. A., Zhu, X. W., O'Donnell, D. and Patel, P. M. 2008. Tumour-mediated disruption of dendritic cell function: inhibiting the MEK1/2-p44/42 axis restores IL-12 production and Th1-generation. Int J Cancer 123:623–632.
Jarrossay, D., Napolitani, G., Colonna, M., Sallusto, F. and Lanzavecchia, A. 2001. Specialization and complementarity in microbial molecule recognition by human myeloid and plasmacytoid dendritic cells. Eur J Immunol 31:3388–3393.
Kakumu, S., Ito, S., Ishikawa, T., Mita, Y., Tagaya, T., Fukuzawa, Y. and Yoshioka, K. 2000. Decreased function of peripheral blood dendritic cells in patients with hepatocellular carcinoma with hepatitis B and C virus infection. J Gastroenterol Hepatol 15:431–436.
Kalinski, P., Hilkens, C. M., Wierenga, E. A. and Kapsenberg, M. L. 1999. T-cell priming by type-1 and type-2 polarized dendritic cells: the concept of a third signal. Immunol Today 20:561–567.
Kalinski, P., Schuitemaker, J. H., Hilkens, C. M. and Kapsenberg, M. L. 1998. Prostaglandin E2 induces the final maturation of IL-12-deficient CD1a + CD83 + dendritic cells: the levels of IL-12 are determined during the final dendritic cell maturation and are resistant to further modulation. J Immunol 161:2804–2809.
Kellermann, S. A., Hudak, S., Oldham, E. R., Liu, Y. J. and McEvoy, L. M. 1999. The CC chemokine receptor-7 ligands 6Ckine and macrophage inflammatory protein-3 beta are potent chemoattractants for in vitro- and in vivo-derived dendritic cells. J Immunol 162:3859–3864.
Kikuchi, T., Abe, T. and Ohno, T. 2002. Effects of glioma cells on maturation of dendritic cells. J Neurooncol 58:125–130.
Kleijmeer, M., Ramm, G., Schuurhuis, D., Griffith, J., Rescigno, M., Ricciardi-Castagnoli, P., Rudensky, A. Y., Ossendorp, F., Melief, C. J., Stoorvogel, W. and Geuze, H. J. 2001. Reorganization of multivesicular bodies regulates MHC class II antigen presentation by dendritic cells. J Cell Biol 155:53–63.
Konno, S., Eckman, J. A., Plunkett, B., Li, X., Berman, J. S., Schroeder, J. and Huang, S. K. 2006. Interleukin-10 and Th2 cytokines differentially regulate osteopontin expression in human monocytes and dendritic cells. J Interferon Cytokine Res 26:562–567.
Kortylewski, M., Kujawski, M., Wang, T., Wei, S., Zhang, S., Pilon-Thomas, S., Niu, G., Kay, H., Mule, J., Kerr, W. G., Jove, R., Pardoll, D. and Yu, H. 2005. Inhibiting Stat3 signaling in the hematopoietic system elicits multicomponent antitumor immunity. Nat Med 11:1314–1321.
Kudela, P., Schwarczova, Z., Sedlak, J. and Bizik, J. 2001. Conditioned medium from HeLa cells enhances motility of human monocyte-derived dendritic cells but abrogates their maturation and endocytic activity. Neoplasma 48:382–388.
Kuppner, M. C., Gastpar, R., Gelwer, S., Nossner, E., Ochmann, O., Scharner, A. and Issels, R. D. 2001. The role of heat shock protein (hsp70) in dendritic cell maturation: hsp70 induces the maturation of immature dendritic cells but reduces DC differentiation from monocyte precursors. Eur J Immunol 31:1602–1609.
Langenkamp, A., Messi, M., Lanzavecchia, A. and Sallusto, F. 2000. Kinetics of dendritic cell activation: impact on priming of TH1, TH2 and nonpolarized T cells. Nat Immunol 1:311–316.
Makarenkova, V. P., Shurin, G. V., Tourkova, I. L., Balkir, L., Pirtskhalaishvili, G., Perez, L., Gerein, V., Siegfried, J. M. and Shurin, M. R. 2003. Lung cancer-derived bombesin-like peptides down-regulate the generation and function of human dendritic cells. J Neuroimmunol 145:55–67.
Mellor, A. L. and Munn, D. H. 2004. IDO expression by dendritic cells: tolerance and tryptophan catabolism. Nat Rev Immunol 4:762–774.
Mende, I., Karsunky, H., Weissman, I. L., Engleman, E. G. and Merad, M. 2006. Flk2+ myeloid progenitors are the main source of Langerhans cells. Blood 107:1383–1390.

Mohty, M., Jarrossay, D., Lafage-Pochitaloff, M., Zandotti, C., Briere, F., de Lamballeri, X. N., Isnardon, D., Sainty, D., Olive, D. and Gaugler, B. 2001. Circulating blood dendritic cells from myeloid leukemia patients display quantitative and cytogenetic abnormalities as well as functional impairment. Blood 98:3750–3756.

Murdoch, C., Muthana, M., Coffelt, S. B. and Lewis, C. E. 2008. The role of myeloid cells in the promotion of tumour angiogenesis. Nat Rev Cancer 8:618–631.

Muthuswamy, R., Urban, J., Lee, J. J., Reinhart, T. A., Bartlett, D. and Kalinski, P. 2008. Ability of mature dendritic cells to interact with regulatory T cells is imprinted during maturation. Cancer Res 68:5972–5978.

Naldini, A., Leali, D., Pucci, A., Morena, E., Carraro, F., Nico, B., Ribatti, D. and Presta, M. 2006. Cutting edge: IL-1beta mediates the proangiogenic activity of osteopontin-activated human monocytes. J Immunol 177:4267–4270.

Nestle, F. O., Burg, G., Fah, J., Wrone-Smith, T. and Nickoloff, B. J. 1997. Human sunlight-induced basal-cell-carcinoma-associated dendritic cells are deficient in T cell co-stimulatory molecules and are impaired as antigen-presenting cells. Am J Pathol 150:641–651.

Ninomiya, T., Akbar, S. M., Masumoto, T., Horiike, N. and Onji, M. 1999. Dendritic cells with immature phenotype and defective function in the peripheral blood from patients with hepatocellular carcinoma. J Hepatol 31:323–331.

Novitskiy, S. V., Ryzhov, S., Zaynagetdinov, R., Goldstein, A. E., Huang, Y., Tikhomirov, O. Y., Blackburn, M. R., Biaggioni, I., Carbone, D. P., Feoktistov, I. and Dikov, M. M. 2008. Adenosine receptors in regulation of dendritic cell differentiation and function. Blood 112(5):1822–1831.

O'Doherty, U., Peng, M., Gezelter, S., Swiggard, W. J., Betjes, M., Bhardwaj, N. and Steinman, R. M. 1994. Human blood contains two subsets of dendritic cells, one immunologically mature and the other immature. Immunology 82:487–493.

Okunishi, K., Dohi, M., Nakagome, K., Tanaka, R., Mizuno, S., Matsumoto, K., Miyazaki, J., Nakamura, T. and Yamamoto, K. 2005. A novel role of hepatocyte growth factor as an immune regulator through suppressing dendritic cell function. J Immunol 175:4745–4753.

Onishi, H., Morisaki, T., Baba, E., Kuga, H., Kuroki, H., Matsumoto, K., Tanaka, M. and Katano, M. 2002. Dysfunctional and short-lived subsets in monocyte-derived dendritic cells from patients with advanced cancer. Clin Immunol 105:286–295.

Orsini, E., Guarini, A., Chiaretti, S., Mauro, F. R. and Foa, R. 2003. The circulating dendritic cell compartment in patients with chronic lymphocytic leukemia is severely defective and unable to stimulate an effective T-cell response. Cancer Res 63:4497–4506.

Parlato, S., Santini, S. M., Lapenta, C., Di Pucchio, T., Logozzi, M., Spada, M., Giammarioli, A. M., Malorni, W., Fais, S. and Belardelli, F. 2001. Expression of CCR-7, MIP-3beta, and Th-1 chemokines in type I IFN-induced monocyte-derived dendritic cells: importance for the rapid acquisition of potent migratory and functional activities. Blood 98:3022–3029.

Pinzon-Charry, A., Ho, C. S., Laherty, R., Maxwell, T., Walker, D., Gardiner, R. A., O'Connor, L., Pyke, C., Schmidt, C., Furnival, C. and Lopez, J. A. 2005a. A population of HLA-DR+ immature cells accumulates in the blood dendritic cell compartment of patients with different types of cancer. Neoplasia 7:1112–1122.

Pinzon-Charry, A., Maxwell, T. and Lopez, J. A. 2005b. Dendritic cell dysfunction in cancer: a mechanism for immunosuppression. Immunol Cell Biol 83:451–461.

Pockaj, B. A., Basu, G. D., Pathangey, L. B., Gray, R. J., Hernandez, J. L., Gendler, S. J. and Mukherjee, P. 2004. Reduced T-cell and dendritic cell function is related to cyclooxygenase-2 overexpression and prostaglandin E2 secretion in patients with breast cancer. Ann Surg Oncol 11:328–339.

Puig-Kroger, A., Pello, O. M., Selgas, R., Criado, G., Bajo, M. A., Sanchez-Tomero, J. A., Alvarez, V., del Peso, G., Sanchez-Mateos, P., Holmes, C., Faict, D., Lopez-Cabrera, M., Madrenas, J. and Corbi, A. L. 2003. Peritoneal dialysis solutions inhibit the differentiation and maturation of human monocyte-derived dendritic cells: effect of lactate and glucose-degradation products. J Leukoc Biol 73:482–492.

Ratta, M., Fagnoni, F., Curti, A., Vescovini, R., Sansoni, P., Oliviero, B., Fogli, M., Ferri, E., Della Cuna, G. R., Tura, S., Baccarani, M. and Lemoli, R. M. 2002. Dendritic cells are functionally defective in multiple myeloma: the role of interleukin-6. Blood 100:230–237.

Remmel, E., Terracciano, L., Noppen, C., Zajac, P., Heberer, M., Spagnoli, G. C. and Padovan, E. 2001. Modulation of dendritic cell phenotype and mobility by tumor cells in vitro. Hum Immunol 62:39–49.

Ridolfi, R., Riccobon, A., Galassi, R., Giorgetti, G., Petrini, M., Fiammenghi, L., Stefanelli, M., Ridolfi, L., Moretti, A., Migliori, G. and Fiorentini, G. 2004. Evaluation of in vivo labelled dendritic cell migration in cancer patients. J Transl Med 2:27.

Ryzhov, S., Novitskiy, S. V., Zaynagetdinov, R., Goldstein, A. E., Carbone, D. P., Biaggioni, I., Dikov, M. M. and Feoktistov, I. 2008. Host A2B Adenosine Receptors Promote Carcinoma Growth. Neoplasia 10.

Sallusto, F., Cella, M., Danieli, C. and Lanzavecchia, A. 1995. Dendritic cells use macropinocytosis and the mannose receptor to concentrate macromolecules in the major histocompatibility complex class II compartment: downregulation by cytokines and bacterial products. J Exp Med 182:389–400.

Sandel, M. H., Dadabayev, A. R., Menon, A. G., Morreau, H., Melief, C. J., Offringa, R., van der Burg, S. H., Janssen-van Rhijn, C. M., Ensink, N. G., Tollenaar, R. A., van de Velde, C. J. and Kuppen, P. J. 2005. Prognostic value of tumor-infiltrating dendritic cells in colorectal cancer: role of maturation status and intratumoral localization. Clin Cancer Res 11:2576–2582.

Sato, K., Kawasaki, H., Nagayama, H., Enomoto, M., Morimoto, C., Tadokoro, K., Juji, T. and Takahashi, T. A. 2000. TGF-beta 1 reciprocally controls chemotaxis of human peripheral blood monocyte-derived dendritic cells via chemokine receptors. J Immunol 164:2285–2295.

Schuurhuis, D. H., Fu, N., Ossendorp, F. and Melief, C. J. 2006. Ins and outs of dendritic cells. Int Arch Allergy Immunol 140:53–72.

Schwaab, T., Schned, A. R., Heaney, J. A., Cole, B. F., Atzpodien, J., Wittke, F. and Ernstoff, M. S. 1999. In vivo description of dendritic cells in human renal cell carcinoma. J Urol 162:567–573.

Schwaab, T., Weiss, J. E., Schned, A. R. and Barth Jr, R. J. 2001. Dendritic Cell Infiltration in Colon Cancer. J Immunother 24:130–137.

Sharma, S., Stolina, M., Yang, S. C., Baratelli, F., Lin, J. F., Atianzar, K., Luo, J., Zhu, L., Lin, Y., Huang, M., Dohadwala, M., Batra, R. K. and Dubinett, S. M. 2003. Tumor cyclooxygenase 2-dependent suppression of dendritic cell function. Clin Cancer Res 9:961–968.

Shurin, G. V., Shurin, M. R., Bykovskaia, S., Shogan, J., Lotze, M. T. and Barksdale, E. M., Jr. 2001. Neuroblastoma-derived gangliosides inhibit dendritic cell generation and function. Cancer Res 61:363–369.

Shurin, M. R. and Gabrilovich, D. I. 2001. Regulation of dendritic cell system by tumor. Cancer Res Ther Control 11:65–78.

Shurin, M. R., Shurin, G. V., Lokshin, A., Yurkovetsky, Z. R., Gutkin, D. W., Chatta, G., Zhong, H., Han, B. and Ferris, R. L. 2006. Intratumoral cytokines/chemokines/growth factors and tumor infiltrating dendritic cells: friends or enemies? Cancer Metastasis Rev 25:333–356.

Shurin, M. R., Yurkovetsky, Z. R., Tourkova, I. L., Balkir, L. and Shurin, G. V. 2002. Inhibition of CD40 expression and CD40-mediated dendritic cell function by tumor-derived IL-10. Int J Cancer 101:61–68.

Singh-Jasuja, H., Toes, R. E., Spee, P., Munz, C., Hilf, N., Schoenberger, S. P., Ricciardi-Castagnoli, P., Neefjes, J., Rammensee, H. G., Arnold-Schild, D. and Schild, H. 2000. Cross-presentation of glycoprotein 96-associated antigens on major histocompatibility complex class I molecules requires receptor-mediated endocytosis. J Exp Med 191:1965–1974.

Sozzani, S. 2005. Dendritic cell trafficking: more than just chemokines. Cytokine Growth Factor Rev 16:581–592.

Steinman, R. M. and Inaba, K. 1999. Myeloid dendritic cells. J Leukoc Biol 66:205–208.

Swiggard, W. J., Mirza, A., Nussenzweig, M. C. and Steinman, R. M. 1995. DEC-205, a 205-kDa protein abundant on mouse dendritic cells and thymic epithelium that is detected by the monoclonal antibody NLDC-145: purification, characterization, and N-terminal amino acid sequence. Cell Immunol 165:302–311.

Takayama, T., Morelli, A. E., Onai, N., Hirao, M., Matsushima, K., Tahara, H. and Thomson, A. W. 2001. Mammalian and viral IL-10 enhance C-C chemokine receptor 5 but down-regulate C-C chemokine receptor 7 expression by myeloid dendritic cells: impact on chemotactic responses and in vivo homing ability. J Immunol 166:7136–7143.

Thurnher, M., Radmayr, C., Ramoner, R., Ebner, S., Bock, G., Klocker, H., Romani, N. and Bartsch, G. 1996. Human renal-cell carcinoma tissue contains dendritic cells. Int J Cancer 68:1–7.

Tourkova, I. L., Shurin, G. V., Chatta, G. S., Perez, L., Finke, J., Whiteside, T. L., Ferrone, S. and Shurin, M. R. 2005. Restoration by IL-15 of MHC class I antigen-processing machinery in human dendritic cells inhibited by tumor-derived gangliosides. J Immunol 175:3045–3052.

Tourkova, I. L., Shurin, G. V., Ferrone, S. and Shurin, M. R. 2009. Interferon regulatory factor 8 mediates tumor-induced inhibition of antigen processing and presentation by dendritic cells. Cancer Immunol Immunother 58:121–133.

Tourkova, I. L., Shurin, G. V., Wei, S. and Shurin, M. R. 2007. Small rho GTPases mediate tumor-induced inhibition of endocytic activity of dendritic cells. J Immunol 178:7787–7793.

Triozzi, P. L., Khurram, R., Aldrich, W. A., Walker, M. J., Kim, J. A. and Jaynes, S. 2000. Intratumoral injection of dendritic cells derived in vitro in patients with metastatic cancer. Cancer 89:2646–2654.

Troy, A., Davidson, P., Atkinson, C. and Hart, D. 1998a. Phenotypic characterisation of the dendritic cell infiltrate in prostate cancer. J Urol 160:214–219.

Troy, A. J., Summers, K. L., Davidson, P. J., Atkinson, C. H. and Hart, D. N. 1998b. Minimal recruitment and activation of dendritic cells within renal cell carcinoma. Clin Cancer Res 4:585–593.

Turley, S. J., Inaba, K., Garrett, W. S., Ebersold, M., Unternaehrer, J., Steinman, R. M. and Mellman, I. 2000. Transport of peptide–MHC class II complexes in developing dendritic cells. Science 288:522–527.

Ueno, H., Klechevsky, E., Morita, R., Aspord, C., Cao, T., Matsui, T., Di Pucchio, T., Connolly, J., Fay, J. W., Pascual, V., Palucka, A. K. and Banchereau, J. 2007. Dendritic cell subsets in health and disease. Immunol Rev 219:118–142.

Vlad, A. M., Kettel, J. C., Alajez, N. M., Carlos, C. A. and Finn, O. J. 2004. MUC1 immunobiology: from discovery to clinical applications. Adv Immunol 82:249–293.

Walker, S. R., Ogagan, P. D., DeAlmeida, D., Aboka, A. M. and Barksdale, E. M., Jr. 2006. Neuroblastoma impairs chemokine-mediated dendritic cell migration in vitro. J Pediatr Surg 41:260–265.

Wang, T., Niu, G., Kortylewski, M., Burdelya, L., Shain, K., Zhang, S., Bhattacharya, R., Gabrilovich, D., Heller, R., Coppola, D., Dalton, W., Jove, R., Pardoll, D. and Yu, H. 2004. Regulation of the innate and adaptive immune responses by Stat-3 signaling in tumor cells. Nat Med 10:48–54.

Wesa, A., Kalinski, P., Kirkwood, J. M., Tatsumi, T. and Storkus, W. J. 2007. Polarized type-1 dendritic cells (DC1) producing high levels of IL-12 family members rescue patient TH1-type antimelanoma CD4+ T cell responses in vitro. J Immunother 30:75–82.

West, M. A., Antoniou, A. N., Prescott, A. R., Azuma, T., Kwiatkowski, D. J. and Watts, C. 1999. Membrane ruffling, macropinocytosis and antigen presentation in the absence of gelsolin in murine dendritic cells. Eur J Immunol 29:3450–3455.

Whiteside, T. L., Stanson, J., Shurin, M. R. and Ferrone, S. 2004. Antigen-processing machinery in human dendritic cells: up-regulation by maturation and down-regulation by tumor cells. J Immunol 173:1526–1534.

Wilson, N. S., El-Sukkari, D., Belz, G. T., Smith, C. M., Steptoe, R. J., Heath, W. R., Shortman, K. and Villadangos, J. A. 2003. Most lymphoid organ dendritic cell types are phenotypically and functionally immature. Blood 102:2187–2194.

Yanagihara, S., Komura, E., Nagafune, J., Watarai, H. and Yamaguchi, Y. 1998. EBI1/CCR7 is a new member of dendritic cell chemokine receptor that is up-regulated upon maturation. J Immunol 161:3096–3102.

Yang, L. and Carbone, D. P. 2004. Tumor-host immune interactions and dendritic cell dysfunction. Adv Cancer Res 92:13–27.

Yang, L., Yamagata, N., Yadav, R., Brandon, S., Courtney, R. L., Morrow, J. D., Shyr, Y., Boothby, M., Joyce, S., Carbone, D. P. and Breyer, R. M. 2003. Cancer-associated immunodeficiency and dendritic cell abnormalities mediated by the prostaglandin EP2 receptor. J Clin Invest 111:727–735.

Yoo, J. K., Cho, J. H., Lee, S. W. and Sung, Y. C. 2002. IL-12 provides proliferation and survival signals to murine CD4+ T cells through phosphatidylinositol 3-kinase/Akt signaling pathway. J Immunol 169:3637–3643.

Zhang, L., Conejo-Garcia, J. R., Katsaros, D., Gimotty, P. A., Massobrio, M., Regnani, G., Makrigiannakis, A., Gray, H., Schlienger, K., Liebman, M. N., Rubin, S. C. and Coukos, G. 2003. Intratumoral T cells, recurrence, and survival in epithelial ovarian cancer. N Engl J Med 348:203–213.

Zhao, W., Darmanin, S., Fu, Q., Chen, J., Cui, H., Wang, J., Okada, F., Hamada, J., Hattori, Y., Kondo, T., Hamuro, J., Asaka, M. and Kobayashi, M. 2005. Hypoxia suppresses the production of matrix metalloproteinases and the migration of human monocyte-derived dendritic cells. Eur J Immunol 35:3468–3477.

Zinkernagel, R. M. 2000. Localization dose and time of antigens determine immune reactivity. Semin Immunol 12:163–171; discussion 257–344.

Zou, W., Machelon, V., Coulomb-L'Hermin, A., Borvak, J., Nome, F., Isaeva, T., Wei, S., Krzysiek, R., Durand-Gasselin, I., Gordon, A., Pustilnik, T., Curiel, D. T., Galanaud, P., Capron, F., Emilie, D. and Curiel, T. J. 2001. Stromal-derived factor-1 in human tumors recruits and alters the function of plasmacytoid precursor dendritic cells. Nat Med 7:1339–1346.

Chapter 5
General Properties of Dendritic Cell Populations in Cancer

Benjamin J. Daniel, Duane P. Jeansonne, Suzanne R. Thibodeaux, and Tyler J. Curiel

Abstract Various, distinct populations of dendritic cells (DC) populate different normal, quiescent tissues in differing amounts, where they subserve a variety of functions generally for the benefit of the host. In cancers, the types, relative proportions, and functions of DC are altered, often to the detriment of the host. This chapter reviews the general issues related to imbalanced DC populations and their causes and dysfunctional consequences in the tumor microenvironment. Potential applications of this knowledge to novel anti-cancer immunotherapy strategies are outlined. Human and mouse disease models suitable for the study of tumor microenvironmental DC are discussed.

5.1 Introduction

Normal tissues have specific architectural features that include defined types and amounts of dendritic cells (DC) in specific ratios. By contrast, far less is known about the types, amounts, distribution patterns, and functionality of the various DC subsets residing in the tumor microenvironment and in the circulation of cancer patients. Reasons for this deficit in knowledge include the rarity of DC generally, and specifically in certain cancers; technical, logistical, and (in human studies) legal issues related to obtaining sufficient numbers of intact or viable cells for studies; definitions of what constitutes the tumor microenvironment; and the only recent ability to identify specific and distinct DC subsets, among other factors.

This chapter gives a high-level overview of the major issues that relate to the altered numbers, types, proportions, and functions of DC in tumor-bearing hosts. Examples of animal models and human diseases useful to study various aspects of tumor-associated DC are discussed. Details of the DC and of the factors that alter their normal characteristics in tumor-bearing hosts are found elsewhere in this book.

T.J. Curiel (✉)
Cancer Therapy & Research Center at UTHSCSA, 7979 Wurzbach Road, suite U637, San Antonio, TX 78229, USA
e-mail: curielt@uthscsa.edu

M.R. Shurin, R.D. Salter (eds.), *Dendritic Cells in Cancer*,
DOI 10.1007/978-0-387-88611-4_5, © Springer Science+Business Media, LLC 2009

5.2 Factors Affecting Dendritic Cell Distribution in the Tumor Microenvironment

Accumulation of DC in the tumor microenvironment owes to a multiplicity of factors including their recruitment from circulation in blood, differentiation in the solid tumor mass, or tumor stroma, or from trafficking from draining lymph nodes or other adjacent tissues or body fluids such as malignant effusions. Each tissue or anatomic compartment could contain differing numbers and types of DC with differing functionalities and phenotypic properties, leading to heterogeneous DC populations in the tumor microenvironment.

Tumors secrete chemotactic factors, and these can favor attraction of one DC type over another, such as selective recruitment of plasmacytoid DC (PDC) into ovarian cancer through a CXCR4/CXCL12 interaction (Zou et al. 2001). Tumor MUC1 can preferentially recruit immature DC into tumors (Carlos et al. 2005). Additional factors that tumors secrete to attract DC include β-defensins (Conejo-Garcia et al. 2004), CCL20 (Thomachot et al. 2004) and other chemokines, pro-inflammatory molecules, and immune modulators, the details of which are presented elsewhere. Breast cancers can preferentially exclude mature DC from the solid tumor mass, which is infiltrated primarily with immature DC (Bell et al. 1999). Further, breast cancers can preferentially induce local differentiation of immature DC, which can be functionally defective (Thomachot et al. 2004). Tumors can produce factors that prevent DC ingress, such as C-reactive protein in squamous cell carcinoma of the head and neck (Frenzel et al. 2007). Some tumors produce IL-8 that can trap DC and potentially prevent egress to draining lymph nodes (Feijoo et al. 2005). CXCL12 actively repels T cells (Poznansky et al. 2000), the reverse of chemoattraction. Whether such a mechanism rids a specific DC subset from the tumor microenvironment remains to be shown. Local factors can boost survival of one DC type over another (Zou et al. 2001) or could encourage the differentiation of a specific DC subset. Tumor microvesicles can impede myeloid DC (MDC) differentiation (Valenti et al. 2006). Much of the work on DC in the tumor microenvironment is on MDC. Factors specifically affecting PDC function or differentiation in cancer are less studied, and poorly understood.

All major types of DC have been described in the tumor microenvironment, including MDC, PDC, and Langerhans DC. Novel DC subsets have also been identified in the tumor microenvironment. For example, vascular leukocytes were originally described in tumors (Conejo-Garcia et al. 2004). Interferon-producing killer DC have been identified in the tumor microenvironment (Chan et al. 2006; Taieb et al. 2006). Langerhans DC are not generally found in tumors aside from those of cutaneous origin and are thus not addressed further here. The DC content and distribution in tumors can convey prognostic information (Byrne and Halliday, 2002; Iwamoto et al. 2003; O'Donnell et al. 2007), underscoring their clinical relevance.

5.3 Factors Affecting Dendritic Cell Function in the Tumor Microenvironment

DC functionality or phenotype can be altered upon exposure to factors specific to, or concentrated in the tumor microenvironment. Thus, DC in the tumor microenvironment likely are heterogeneous populations reflecting their varying sources of origin, and the differing factors that have affected them.

Tumors can secrete factors or signal through surface molecules to alter DC differentiation, activation, maturation or function (Bell et al. 1999; Curiel et al. 2003; Gabrilovich et al. 1996; Kusmartsev and Gabrilovich, 2006). Distinct tumors produce different factors with varying effects on DC. For example, vascular endothelial growth factor is considered a major factor inhibiting DC maturation in many cancers (Gabrilovich et al. 1996). Nonetheless, other factors, such as macrophage colony-stimulating factor, IL-6, IL-10, hepatocyte growth factor (Okunishi et al. 2005), transforming growth factor-β (Alard et al. 2004), prostaglandin E2 (Pockaj et al. 2004), lactate (Gottfried et al. 2006), or gangliosides (Wolfl et al. 2002) can also inhibit DC differentiation, activation or maturation in tumors. In pancreatic cancer, vascular endothelial growth factor may not be as important as IL-10, transforming growth factor-β or IL-6 in this regard (Bellone et al. 2006). Tumors can also induce DC to secrete transforming growth factor-β that fosters regulatory T-cell development (Ghiringhelli et al. 2005).

Aside from direct effects of the tumor, other tumor microenvironmental cells can also degrade the function of local DC. For example, regulatory T cells (Zou 2006), $\gamma\delta$ T cells (Peng et al. 2007), or natural killer cells (Ebata et al. 2006) can inhibit DC function in tumors.

Effects on tumor microenvironmental DC may not necessarily be irreversible. For example, MDC taken directly from murine melanoma and matured ex vivo can act as fully competent antigen-presenting cells (Preynat-Seauve et al. 2006). Potentially reversible functional defects in tumor microenvironmental DC suggest novel therapeutic strategies, some of which are discussed below.

5.4 Characteristics of Tumor Microenvironmental Dendritic Cells

Recent studies are now beginning to define the specific DC subsets and the functional attributes in various cancers (Fiore et al. 2006; Gerlini et al. 2007; Kovarova et al. 2007; Lee et al. 2006; Nagorsen et al. 2007; O'Donnell et al. 2007; Perrot et al. 2007; Sakakura et al. 2006). While it is unwise to speak in sweeping generalities, on the whole, tumor microenvironmental DC differ from other tumor-associated DC (such as those circulating in blood), and these both generally differ from DC obtained from normal, quiescent tissues or blood under homeostatic conditions. The most commonly observed difference is over-representation of immature DC in the tumor microenvironment (Fricke and Gabrilovich 2006; Troy et al. 1998). These tumor microenvironmental DC defects can be apparent even in the very earliest stages of malignancy (Lee et al. 2006).

DC were initially described for their capacity to prime helper function from naïve $CD4^+$ T cells (Banchereau and Steinman 1998), typically leading to protective immunity. By contrast, a hallmark of tumor microenvironmental DC is their ability to generate dysfunctional, instead of protective T-cell-mediated immunity. Microenvironmental DC induce T-cell IL-10 (Zou et al. 2001) and the generation of regulatory T cells, particularly $CD4^+CD25^{hi}$ $FOXP3^+$ regulatory T cells (Zou 2005). Mechanisms include DC B7-H1 signals (Curiel et al. 2003) and DC-transforming growth factor-β (Ghiringhelli et al. 2005), among other factors (Zou 2006). DC-mediated regulatory T-cell generation is also a function of the relative immaturity of tumor microenvironmental DC. Tumor DC can exert deleterious effects on non-T lymphocyte populations, such as the recent report that MDC in blood of cancer patients can alter the function of natural killer T cells (van der Vliet et al. 2008). Tumor microenvironmental DC produce factors such as CCL22 that could attract regulatory T cells, although that specific mechanism has not yet been demonstrated in tumor.

DC instruct naive $CD4^+$ T cells in T-cell helper (Th) polarization. Thus, much can be understood of T-cell polarization in tumors by studying the tumor microenvironmental DC. The tumor microenvironment does not generally favor Th1-polarized immunity, and local DC tend not to induce the Th1-polarized immunity that is considered beneficial to anti-tumor immunity. The role of dysfunctional tumor microenvironmental MDC has been extensively described (Curiel et al. 2003; Zou 2005). For example, tumor MDC can induce T-cell IL-10, which suppresses anti-tumor immunity, instead of inducing protective Th1-polarized immunity characterized by secretion of T-cell interferon-γ (Curiel et al. 2003). The dysfunction of tumor microenvironmental PDC has received less attention. PDC in human ovarian carcinoma generate $CD8^+$ regulatory T cells that secrete IL-10 and have suppressive functions (Wei et al. 2005). PDC from draining lymph nodes of lung cancer patients induce Tc1 responses (Faith et al. 2007). PDC appear to be over-represented in melanoma-positive sentinel lymph nodes, where they appear incompletely matured (Gerlini et al. 2007). Recently, a novel Th lineage, the Th17 T-cell population, was described (Steinman 2007). Th17 cells have been identified in cancer (Kryczek et al. 2007), although their immunopathologic significance is uncertain. Whereas a putative DC17, that is, a DC tending to induce Th17-polarized immune response, has been identified (Iwamoto et al. 2007), such a DC has not yet been reported in cancer. This knowledge deficit likely will be corrected soon.

Aside from effects on T-cell differentiation and function, DC can also affect tumor vascularity. For example, in ovarian cancer, PDC secrete proangiogenic factors such as tumor necrosis factor-α and IL-8, whereas MDC secrete anti-angiogenic factors such as IL-12 (Curiel et al. 2004). Also, in ovarian cancer, a specialized DC (the vascular leukocyte) serves as a vascular endothelial cell progenitor cell (Conejo-Garcia et al. 2004). Tumors can potentially convert microenvironmental DC into vascular endothelial cells (Gottfried et al. 2007). Immature DC, or DC exposed to hypoxia can have numerous potential other effects on angiogenesis (or T-cell function), but these mechanisms have not yet been shown actually to occur in the tumor microenvironment. The importance of DC contributions to controlling tumor neoangiogenesis compared to non-DC myeloid cells remains to be fully defined.

5.5 Strategies to Improve Tumor Microenvironmental Dendritic Cell Numbers or Function

Because DC dysfunction underlies much of the immune dysfunction in cancer, many strategies have attempted to correct these DC defects. One straightforward, although by no means easy, method to improve DC numbers or function in cancer is to infuse ex vivo-derived DC. However, adoptively transfused DC will be subjected to the same factors affecting resident DC upon entry into the tumor microenvironment and could therefore be rendered dysfunctional. In support, injection of MDC into cancer patients can induce FOXP3$^+$ regulatory T cells (Banerjee et al. 2006).

Additional strategies to improve DC numbers or function in the tumor microenvironment have been investigated, including administration of recombinant granulocyte–macrophage colony-stimulating factor (Waller 2007). All-trans retinoic acid reduced the numbers of circulating immature MDC in a pilot study of cancer patients (Mirza et al. 2006). In vitro work suggested that blocking vascular endothelial cell growth factor could improve functional DC differentiation in patients with acute myeloid leukemia (Kang et al. 2006). However, pharmacologic blockade of vascular endothelial cell growth factor signals did not improve DC maturation (of either MDC or PDC) in a small clinical study (van Cruijsen et al. 2007). On the other hand, the anti-vascular endothelial cell growth factor antibody bevacizumab boosted the function of circulating DC in cancer patients (Osada et al. 2008). VEGF-trap improved the phenotypic maturation of DC but did not lead to improved anti-tumor immunity in a small clinical study (Fricke et al. 2007). Other strategies to improve microenvironmental DC have also been attempted (Xu et al. 2007).

5.6 Murine Tumor Models to Study Plasmacytoid Dendritic Cells in Cancer

Because of the difficulties in obtaining sufficient numbers of viable DC from human tumors, mouse models are invaluable tools to study DC biology in tumors, despite some differences in mouse versus human DC, which are discussed in detail elsewhere in this book.

Because essentially any mouse tumor model is useful to study tumor microenvironmental MDC, little will be said on that topic here. Tumors that induce ascites are particularly useful in that the ascites can be sampled over time without killing the animal to obtain viable microenvironmental DC for studies. Models useful in this regard include intraperitoneal injection of ID8 epithelial ovarian carcinoma into syngeneic BL6 mice, or MMV breast carcinoma cells into syngeneic FVB mice.

The B16 melanoma model is commonly used to study tumor PDC, where they exhibit immunostimulatory or tolerogenic properties in distinct settings (Liu et al. 2008; Preynat-Seauve et al. 2006; Salio et al. 2004; Sharma et al. 2007). When the K17-35 melanoma model was compared side-by-side with B16F10 melanoma, both tumors elicited the same two predominant DC populations: MDC ($CD11c^{+}B220^{-}CD8\alpha^{-}$) and PDC ($CD11c^{+}B220^{+}CD8\alpha^{-}$), although B16F10 tumors contained fewer PDC than K17-35, and some PDC expressed CD8α (Preynat-Seauve et al. 2006).

Using the B16 sublines B16F10, B16-OVA, or B78H1-GM-CSF, PDC were found to activate regulatory T cells through the expression of indoleamine 2,3-dioxygenase (Munn et al. 2005, 2004; Sharma et al. 2007). In the EG.7 T-cell lymphoma tumor model, PDC augmented the ability of MDC to prime naive T cells in vivo (Lou et al. 2007) in a contact-dependent manner. In another study, the EL-4/HHD tumor was used to investigate the effect of immunization with a modified TERT tumor antigen-derived peptide in the presence of oligodeoxynucleotide-CpG on PDC and other DC subsets (Cornet et al. 2006). We have shown that ID8 epithelial ovarian carcinoma boosts PDC numbers in syngeneic BL6 mice (our unpublished data). As PDC likely play differing roles in different tumors, it will be worthwhile to examine additional tumor models for additional insights.

Murine PDC can be identified using combinations of cell surface markers or can be identified using PDC-specific antibodies. The 120G8 antibody recognizes an antigen expressed on PDC that is also up-regulated by type I interferon on B cells and DC (Asselin-Paturel et al. 2003). The PDC-specific antibodies 120G8 (Asselin-Paturel et al. 2003) or mPDCA-1 (Krug et al. 2004) will deplete PDC in vivo. The PDC-specific antibody 440c does not deplete PDC, but selectively reduces PDC type I interferon production in vitro and in vivo (Blasius et al. 2004).

5.7 Human Cancers Amenable to Studies of Microenvironmental Dendritic Cells

A significant obstacle to the study of human tumor microenvironmental DC remains the ability to obtain sufficient numbers of viable, functional cells for detailed studies. Blood remains the most studied tumor-associated tissue because of the ease of accession, but in the case of epithelial carcinomas and sarcomas, blood generally represents a tumor-associated rather than a proper tumor microenvironmental tissue. In the case of leukemias and the leukemic phase of lymphomas, blood can be considered a tumor microenvironmental tissue, although bone marrow or involved lymph nodes may be superior in this regard.

Malignant effusions provide an excellent opportunity to collect large numbers of tumor microenvironmental DC. Ascites in ovarian cancer has been exploited for studies of tumor microenvironmental DC, where we routinely and safely collect fluid for study on an up to weekly basis, obtaining up to 100 million viable, functional immune cells from a single collection. Malignant ascites in ovarian cancer could mirror the microenvironment of the solid tumor closely. Ascites from other cancers including that of colon or pancreas is also useful to study tumor microenvironmental DC, although in these instances the malignant ascites may not be as representative of events and cells in the solid tumor mass as in ovarian cancer, a speculation that remains to be confirmed. Bronchoalveolar lavage fluid is useful to study DC in lung cancer and may represent what occurs in the solid tumor mass. Draining lymph nodes represent another source of sufficient DC for detailed phenotypic and functional studies, where they have been most extensively studied in melanoma. We obtain up to 1 million functional and viable DC from a single lymph node in ovarian cancer. Skin biopsies are useful sources of tumor microenvironmental DC, although quantities are generally small. Of course, mechanical disruption of solid tumors can be used to study microenvironmental DC of any histology.

5.8 Conclusions

The altered proportions and functionality of tumor microenvironmental DC contribute to the immunopathogenesis of cancers. Mechanisms include lack of ability to prime a protective immune response, active induction of dysfunctional immune cells (such as regulatory T cells), potential to kill immune beneficial immune cells, and potential to augment tumor neoangiogenesis. Strategies to prevent DC dysfunction or correct imbalance proportions of functionalities are novel means to undertake tumor immunotherapy. Study of functional tumor microenvironmental DC remains a challenge owing to logistical issues related to obtaining sufficient numbers of cells for study. Nonetheless, ingenious investigators have found appropriate models and techniques to undertake important new studies.

Acknowledgments Supported by CA105207, CA100425, CA054174, the Rippel Foundation, the Voelcker Trust, and UTHSCSA endowments. We apologize that space constraints prevented the citation of many important works from our colleagues.

References

Alard, P., Clark, S. L., and Kosiewicz, M. M. 2004. Mechanisms of tolerance induced by TGF beta-treated APC: CD4 regulatory T cells prevent the induction of the immune response possibly through a mechanism involving TGF beta. Eur J Immunol 34, 1021–1030.

Asselin-Paturel, C., Brizard, G., Pin, J. J., Briere, F., and Trinchieri, G. 2003. Mouse strain differences in plasmacytoid dendritic cell frequency and function revealed by a novel monoclonal antibody. J Immunol 171, 6466–6477.

Banchereau, J., and Steinman, R. M. 1998. Dendritic cells and the control of immunity. Nature 392, 245–252.

Banerjee, D. K., Dhodapkar, M. V., Matayeva, E., Steinman, R. M., and Dhodapkar, K. M. 2006. Expansion of FOXP3 high regulatory T cells by human dendritic cells (DCs) in vitro and after injection of cytokine-matured DCs in myeloma patients. Blood 108, 2655–2661.
Bell, D., Chomarat, P., Broyles, D., Netto, G., Harb, G. M., Lebecque, S., Valladeau, J., Davoust, J., Palucka, K. A., and Banchereau, J. 1999. In breast carcinoma tissue, immature dendritic cells reside within the tumor, whereas mature dendritic cells are located in peritumoral areas. J Exp Med 190, 1417–1426.
Bellone, G., Carbone, A., Smirne, C., Scirelli, T., Buffolino, A., Novarino, A., Stacchini, A., Bertetto, O., Palestro, G., Sorio, C., et al. 2006. Cooperative induction of a tolerogenic dendritic cell phenotype by cytokines secreted by pancreatic carcinoma cells. J Immunol 177, 3448–3460.
Blasius, A., Vermi, W., Krug, A., Facchetti, F., Cella, M., and Colonna, M. 2004. A cell-surface molecule selectively expressed on murine natural interferon-producing cells that blocks secretion of interferon-alpha. Blood 103, 4201–4206.
Byrne, S. N., and Halliday, G. M. 2002. Dendritic cells: making progress with tumour regression? Immunol Cell Biol 80, 520–530.
Carlos, C. A., Dong, H. F., Howard, O. M., Oppenheim, J. J., Hanisch, F. G., and Finn, O. J. 2005. Human tumor antigen MUC1 is chemotactic for immature dendritic cells and elicits maturation but does not promote Th1 type immunity. J Immunol 175, 1628–1635.
Chan, C. W., Crafton, E., Fan, H. N., Flook, J., Yoshimura, K., Skarica, M., Brockstedt, D., Dubensky, T. W., Stins, M. F., Lanier, L. L., et al. 2006. Interferon-producing killer dendritic cells provide a link between innate and adaptive immunity. Nat Med 12, 207–213.
Conejo-Garcia, J. R., Benencia, F., Courreges, M. C., Kang, E., Mohamed-Hadley, A., Buckanovich, R. J., Holtz, D. O., Jenkins, A., Na, H., Zhang, L., et al. 2004. Tumor-infiltrating dendritic cell precursors recruited by a beta-defensin contribute to vasculogenesis under the influence of Vegf-A. Nat Med 10, 950–958.
Cornet, S., Menez-Jamet, J., Lemonnier, F., Kosmatopoulos, K., and Miconnet, I. 2006. CpG oligodeoxynucleotides activate dendritic cells in vivo and induce a functional and protective vaccine immunity against a TERT derived modified cryptic MHC class I-restricted epitope. Vaccine 24, 1880–1888.
Curiel, T. J., Cheng, P., Mottram, P., Alvarez, X., Moons, L., Evdemon-Hogan, M., Wei, S., Zou, L., Kryczek, I., Hoyle, G., et al. 2004. Dendritic cell subsets differentially regulate angiogenesis in human ovarian cancer. Cancer Res 64, 5535–5538.
Curiel, T. J., Wei, S., Dong, H., Alvarez, X., Cheng, P., Mottram, P., Krzysiek, R., Knutson, K. L., Daniel, B., Zimmermann, M. C., et al. 2003. Blockade of B7-H1 improves myeloid dendritic cell-mediated antitumor immunity. Nat Med 9, 562–567.
Ebata, K., Shimizu, Y., Nakayama, Y., Minemura, M., Murakami, J., Kato, T., Yasumura, S., Takahara, T., Sugiyama, T., and Saito, S. 2006. Immature NK cells suppress dendritic cell functions during the development of leukemia in a mouse model. J Immunol 176, 4113–4124.
Faith, A., Peek, E., McDonald, J., Urry, Z., Richards, D. F., Tan, C., Santis, G., and Hawrylowicz, C. 2007. Plasmacytoid dendritic cells from human lung cancer draining lymph nodes induce Tc1 responses. Am J Respir Cell Mol Biol 36, 360–367.
Feijoo, E., Alfaro, C., Mazzolini, G., Serra, P., Penuelas, I., Arina, A., Huarte, E., Tirapu, I., Palencia, B., Murillo, O., et al. 2005. Dendritic cells delivered inside human carcinomas are sequestered by interleukin-8. Int J Cancer 116, 275–281.
Fiore, F., Von Bergwelt-Baildon, M. S., Drebber, U., Beyer, M., Popov, A., Manzke, O., Wickenhauser, C., Baldus, S. E., and Schultze, J. L. 2006. Dendritic cells are significantly reduced in non-Hodgkin's lymphoma and express less CCR7 and CD62L. Leuk Lymphoma 47, 613–622.
Frenzel, H., Pries, R., Brocks, C. P., Jabs, W. J., Wittkopf, N., and Wollenberg, B. 2007. Decreased migration of myeloid dendritic cells through increased levels of C-reactive protein. Anticancer Res 27, 4111–4115.
Fricke, I., and Gabrilovich, D. I. 2006. Dendritic cells and tumor microenvironment: a dangerous liaison. Immunol Invest 35, 459–483.

Fricke, I., Mirza, N., Dupont, J., Lockhart, C., Jackson, A., Lee, J. H., Sosman, J. A., and Gabrilovich, D. I. 2007. Vascular endothelial growth factor-trap overcomes defects in dendritic cell differentiation but does not improve antigen-specific immune responses. Clin Cancer Res 13, 4840–4848.

Gabrilovich, D. I., Chen, H. L., Girgis, K. R., Cunningham, H. T., Meny, G. M., Nadaf, S., Kavanaugh, D., and Carbone, D. P. 1996. Production of vascular endothelial growth factor by human tumors inhibits the functional maturation of dendritic cells [published erratum appears in Nat Med 1996 Nov;2(11):1267]. Nat Med 2, 1096–1103.

Gerlini, G., Urso, C., Mariotti, G., Di Gennaro, P., Palli, D., Brandani, P., Salvadori, A., Pimpinelli, N., Reali, U. M., and Borgognoni, L. 2007. Plasmacytoid dendritic cells represent a major dendritic cell subset in sentinel lymph nodes of melanoma patients and accumulate in metastatic nodes. Clin Immunol 125, 184–193.

Ghiringhelli, F., Puig, P. E., Roux, S., Parcellier, A., Schmitt, E., Solary, E., Kroemer, G., Martin, F., Chauffert, B., and Zitvogel, L. 2005. Tumor cells convert immature myeloid dendritic cells into TGF-beta-secreting cells inducing CD4+CD25+ regulatory T cell proliferation. J Exp Med 202, 919–929.

Gottfried, E., Kreutz, M., Haffner, S., Holler, E., Iacobelli, M., Andreesen, R., and Eissner, G. 2007. Differentiation of human tumour-associated dendritic cells into endothelial-like cells: an alternative pathway of tumour angiogenesis. Scand J Immunol 65, 329–335.

Gottfried, E., Kunz-Schughart, L. A., Ebner, S., Mueller-Klieser, W., Hoves, S., Andreesen, R., Mackensen, A., and Kreutz, M. 2006. Tumor-derived lactic acid modulates dendritic cell activation and antigen expression. Blood 107, 2013–2021.

Iwamoto, M., Shinohara, H., Miyamoto, A., Okuzawa, M., Mabuchi, H., Nohara, T., Gon, G., Toyoda, M., and Tanigawa, N. 2003. Prognostic value of tumor-infiltrating dendritic cells expressing CD83 in human breast carcinomas. Int J Cancer 104, 92–97.

Iwamoto, S., Iwai, S., Tsujiyama, K., Kurahashi, C., Takeshita, K., Naoe, M., Masunaga, A., Ogawa, Y., Oguchi, K., and Miyazaki, A. 2007. TNF-alpha drives human CD14+ monocytes to differentiate into CD70+ dendritic cells evoking Th1 and Th17 responses. J Immunol 179, 1449–1457.

Kang, H. K., Park, J. S., Kim, S. K., Choi, B. H., Pham, T. N., Zhu, X. W., Cho, D., Nam, J. H., Kim, Y. J., Rhee, J. H., et al. 2006. Down-regulation of cellular vascular endothelial growth factor levels induces differentiation of leukemic cells to functional leukemic-dendritic cells in acute myeloid leukemia. Leuk Lymphoma 47, 2224–2233.

Kovarova, L., Buchler, T., Pour, L., Zahradova, L., Ocadlikova, D., Svobodnik, A., Penka, M., Vorlicek, J., and Hajek, R. 2007. Dendritic cell counts and their subsets during treatment of multiple myeloma. Neoplasma 54, 297–303.

Krug, A., French, A. R., Barchet, W., Fischer, J. A., Dzionek, A., Pingel, J. T., Orihuela, M. M., Akira, S., Yokoyama, W. M., and Colonna, M. 2004. TLR9-dependent recognition of MCMV by IPC and DC generates coordinated cytokine responses that activate antiviral NK cell function. Immunity 21, 107–119.

Kryczek, I., Wei, S., Zou, L., Altuwaijri, S., Szeliga, W., Kolls, J., Chang, A., and Zou, W. 2007. Cutting edge: Th17 and regulatory T cell dynamics and the regulation by IL-2 in the tumor microenvironment. J Immunol 178, 6730–6733.

Kusmartsev, S., and Gabrilovich, D. I. 2006. Effect of tumor-derived cytokines and growth factors on differentiation and immune suppressive features of myeloid cells in cancer. Cancer Metastasis Rev 25, 323–331.

Lee, B. N., Follen, M., Rodriquez, G., Shen, D. Y., Malpica, A., Shearer, W. T., and Reuben, J. M. 2006. Deficiencies in myeloid antigen-presenting cells in women with cervical squamous intraepithelial lesions. Cancer 107, 999–1007.

Liu, C., Lou, Y., Lizee, G., Qin, H., Liu, S., Rabinovich, B., Kim, G. J., Wang, Y. H., Ye, Y., Sikora, A. G., et al. 2008. Plasmacytoid dendritic cells induce NK cell-dependent, tumor antigen-specific T cell cross-priming and tumor regression in mice. J Clin Invest 118, 1165–1175.

Lou, Y., Liu, C., Kim, G. J., Liu, Y. J., Hwu, P., and Wang, G. 2007. Plasmacytoid dendritic cells synergize with myeloid dendritic cells in the induction of antigen-specific antitumor immune responses. J Immunol 178, 1534–1541.

Mirza, N., Fishman, M., Fricke, I., Dunn, M., Neuger, A. M., Frost, T. J., Lush, R. M., Antonia, S., and Gabrilovich, D. I. 2006. All-trans-retinoic acid improves differentiation of myeloid cells and immune response in cancer patients. Cancer Res 66, 9299–9307.

Munn, D. H., Sharma, M. D., Baban, B., Harding, H. P., Zhang, Y., Ron, D., and Mellor, A. L. 2005. GCN2 kinase in T cells mediates proliferative arrest and anergy induction in response to indoleamine 2,3-dioxygenase. Immunity 22, 633–642.

Munn, D. H., Sharma, M. D., Hou, D., Baban, B., Lee, J. R., Antonia, S. J., Messina, J. L., Chandler, P., Koni, P. A., and Mellor, A. L. 2004. Expression of indoleamine 2,3-dioxygenase by plasmacytoid dendritic cells in tumor-draining lymph nodes. J Clin Invest 114, 280–290.

Nagorsen, D., Voigt, S., Berg, E., Stein, H., Thiel, E., and Loddenkemper, C. 2007. Tumor-infiltrating macrophages and dendritic cells in human colorectal cancer: relation to local regulatory T cells, systemic T-cell response against tumor-associated antigens and survival. J Transl Med 5, 62.

O'Donnell, R. K., Mick, R., Feldman, M., Hino, S., Wang, Y., Brose, M. S., and Muschel, R. J. 2007. Distribution of dendritic cell subtypes in primary oral squamous cell carcinoma is inconsistent with a functional response. Cancer Lett 255, 145–152.

Okunishi, K., Dohi, M., Nakagome, K., Tanaka, R., Mizuno, S., Matsumoto, K., Miyazaki, J., Nakamura, T., and Yamamoto, K. 2005. A novel role of hepatocyte growth factor as an immune regulator through suppressing dendritic cell function. J Immunol 175, 4745–4753.

Osada, T., Chong, G., Tansik, R., Hong, T., Spector, N., Kumar, R., Hurwitz, H. I., Dev, I., Nixon, A. B., Lyerly, H. K., et al. 2008. The effect of anti-VEGF therapy on immature myeloid cell and dendritic cells in cancer patients. Cancer Immunol Immunother 57, 1115–1124.

Peng, G., Wang, H. Y., Peng, W., Kiniwa, Y., Seo, K. H., and Wang, R. F. 2007. Tumor-infiltrating gammadelta T cells suppress T and dendritic cell function via mechanisms controlled by a unique toll-like receptor signaling pathway. Immunity 27, 334–348.

Perrot, I., Blanchard, D., Freymond, N., Isaac, S., Guibert, B., Pacheco, Y., and Lebecque, S. 2007. Dendritic cells infiltrating human non-small cell lung cancer are blocked at immature stage. J Immunol 178, 2763–2769.

Pockaj, B. A., Basu, G. D., Pathangey, L. B., Gray, R. J., Hernandez, J. L., Gendler, S. J., and Mukherjee, P. 2004. Reduced T-cell and dendritic cell function is related to cyclooxygenase-2 overexpression and prostaglandin E2 secretion in patients with breast cancer. Ann Surg Oncol 11, 328–339.

Poznansky, M. C., Olszak, I. T., Foxall, R., Evans, R. H., Luster, A. D., and Scadden, D. T. 2000. Active movement of T cells away from a chemokine. Nat Med 6, 543–548.

Preynat-Seauve, O., Schuler, P., Contassot, E., Beermann, F., Huard, B., and French, L. E. 2006. Tumor-infiltrating dendritic cells are potent antigen-presenting cells able to activate T cells and mediate tumor rejection. J Immunol 176, 61–67.

Sakakura, K., Chikamatsu, K., Takahashi, K., Whiteside, T. L., and Furuya, N. 2006. Maturation of circulating dendritic cells and imbalance of T-cell subsets in patients with squamous cell carcinoma of the head and neck. Cancer Immunol Immunother 55, 151–159.

Salio, M., Palmowski, M. J., Atzberger, A., Hermans, I. F., and Cerundolo, V. 2004. CpG-matured murine plasmacytoid dendritic cells are capable of in vivo priming of functional CD8 T cell responses to endogenous but not exogenous antigens. J Exp Med 199, 567–579.

Sharma, M. D., Baban, B., Chandler, P., Hou, D. Y., Singh, N., Yagita, H., Azuma, M., Blazar, B. R., Mellor, A. L., and Munn, D. H. 2007. Plasmacytoid dendritic cells from mouse tumor-draining lymph nodes directly activate mature Tregs via indoleamine 2,3-dioxygenase. J Clin Invest 117, 2570–2582.

Steinman, L. 2007. A brief history of T(H)17, the first major revision in the T(H)1/T(H)2 hypothesis of T cell-mediated tissue damage. Nat Med 13, 139–145.

Taieb, J., Chaput, N., Menard, C., Apetoh, L., Ullrich, E., Bonmort, M., Pequignot, M., Casares, N., Terme, M., Flament, C., et al. 2006. A novel dendritic cell subset involved in tumor immunosurveillance. Nat Med 12, 214–219.

Thomachot, M. C., Bendriss-Vermare, N., Massacrier, C., Biota, C., Treilleux, I., Goddard, S., Caux, C., Bachelot, T., Blay, J. Y., and Menetrier-Caux, C. 2004. Breast carcinoma cells promote the differentiation of CD34+ progenitors towards 2 different subpopulations of dendritic cells with CD1a(high)CD86(-)Langerin- and CD1a(+)CD86(+)Langerin+ phenotypes. Int J Cancer 110, 710–720.

Troy, A. J., Summers, K. L., Davidson, P. J., Atkinson, C. H., and Hart, D. N. 1998. Minimal recruitment and activation of dendritic cells within renal cell carcinoma. Clin Cancer Res 4, 585–593.

Valenti, R., Huber, V., Filipazzi, P., Pilla, L., Sovena, G., Villa, A., Corbelli, A., Fais, S., Parmiani, G., and Rivoltini, L. 2006. Human tumor-released microvesicles promote the differentiation of myeloid cells with transforming growth factor-beta-mediated suppressive activity on T lymphocytes. Cancer Res 66, 9290–9298.

van Cruijsen, H., Hoekman, K., Stam, A. G., van den Eertwegh, A. J., Kuenen, B. C., Scheper, R. J., Giaccone, G., and de Gruijl, T. D. 2007. Defective differentiation of myeloid and plasmacytoid dendritic cells in advanced cancer patients is not normalized by tyrosine kinase inhibition of the vascular endothelial growth factor receptor. Clin Dev Immunol 2007, 17315.

van der Vliet, H. J., Wang, R., Yue, S. C., Koon, H. B., Balk, S. P., and Exley, M. A. 2008. Circulating myeloid dendritic cells of advanced cancer patients result in reduced activation and a biased cytokine profile in invariant NKT cells. J Immunol 180, 7287–7293.

Waller, E. K. (2007). The role of sargramostim (rhGM-CSF) as immunotherapy. Oncologist 12 Suppl 2, 22–26.

Wei, S., Kryczek, I., Zou, L., Daniel, B., Cheng, P., Mottram, P., Curiel, T., Lange, A., and Zou, W. (2005). Plasmacytoid dendritic cells induce CD8+ regulatory T cells in human ovarian carcinoma. Cancer Res 65, 5020–5026.

Wolfl, M., Batten, W. Y., Posovszky, C., Bernhard, H., and Berthold, F. (2002). Gangliosides inhibit the development from monocytes to dendritic cells. Clin Exp Immunol 130, 441–448.

Xu, Y., Darcy, P. K., and Kershaw, M. H. (2007). Tumor-specific dendritic cells generated by genetic redirection of Toll-like receptor signaling against the tumor-associated antigen, erbB2. Cancer Gene Ther 14, 773–780.

Zou, W. (2005). Immunosuppressive networks in the tumour environment and their therapeutic relevance. Nat Rev Cancer 5, 263–274.

Zou, W. (2006). Regulatory T cells, tumour immunity and immunotherapy. Nat Rev Immunol 6, 295–307.

Zou, W., Machelon, V., Coulomb-L'Hermin, A., Borvak, J., Nome, F., Isaeva, T., Wei, S., Krzysiek, R., Durand-Gasselin, I., Gordon, A., et al. (2001). Stromal-derived factor-1 in human tumors recruits and alters the function of plasmacytoid precursor dendritic cells. Nat Med 7, 1339–1346.

Chapter 6
Elimination of Dendritic Cells in Cancer

Viktor Umansky and Dirk Schadendorf

Abstract Elimination of mature functional dendritic cells represents one of the most important mechanisms of tumor immune evasion. It includes inhibition of dendritic cell differentiation and maturation as well as a direct induction of apoptosis in dendritic cells or their precursors. Numerous experimental and clinical studies revealed that different factors produced by both tumor and stromal cells, such as VEGF, IL-10, TGF-β, gangliosides and other, could induce apoptotic death of dendritic cells and stimulate spontaneous apoptosis both in vitro and in vivo. Both mechanisms, i.e. suppression of dendritic cell differentiation and dendritic cell apoptosis, can contribute to the reduction of dendritic cell numbers observed in cancer, which was shown to be associated with the tumor progression. Therefore, neutralization of the suppressive tumor microenvironment will allow a proper dendritic cell differentiation from their precursors and protect functionally active dendritic cells from apoptotic death.

6.1 Introduction

Dendritic cells (DC) are regarded as the principal inducer of antitumor T-cell-mediated immunity (Palucka et al. 2007; Steinman 2007). These professional antigen-presenting cells (APC) have been characterized by a capacity to process and present tumor antigens, as well as by an excessive expression of MHC molecules and costimulatory molecules that enable them to prime and stimulate naïve T cells (Shurin 1996; Lanzavecchia and Sallusto 2001). In agreement with this function, DC are located in virtually all peripheral tissues where they display a first barrier against tumor cells. They are here the guardians for the integrity of the body tissues (Moser 2003). After antigen encounter, these cells can leave the periphery and migrate through the lymphatic system to secondary lymphoid organs. Given the key role of DC in connecting innate and adaptive immunity

V. Umansky (✉)
Clinical Cooperation Unit Dermato-Oncology (G300), German Cancer Research Center, Im Neuenheimer Feld 280, 69120 Heidelberg, Germany
e-mail: V.Umansky@dkfz-heidelberg.de

M.R. Shurin, R.D. Salter (eds.), *Dendritic Cells in Cancer*,
DOI 10.1007/978-0-387-88611-4_6, © Springer Science+Business Media, LLC 2009

and in up-regulation of T-cell reactions against cancer cells, tumors developed different mechanisms that modulate DC functions and/or numbers in order to escape immune recognition and elimination.

Various alterations in the DC system in the process of tumor progression have been described during the last decade, although tumor-derived factors and signaling pathways responsible for abnormal DC differentiation and elimination are still not completely clear. In general, the DC system is acting in cancer patients under multidirectional influences of multiple local and systemic factors derived from tumor cells as well as from other host cells in the tumor microenvironment. These factors include various cytokines, chemokines, growth factors, hormones, prostaglandins, gangliosides and many other soluble and cell surface molecules (Shurin et al. 2006). Physical and psychological stress associated with cancer diagnosis and effects of different antitumor therapies could result in additional changes in DC numbers and function. Since more than half of all tumors develop in patients older than 65 years, an aging immune system together with the increased incidence of infections and autoimmune pathologies might also modulate DC generation and functional activity and, thus, antitumor immune responses (Shurin et al. 2007). Most of the described molecular mechanisms of DC alterations in the immunosuppressive tumor milieu relate to (1) elimination of functional DC; (2) suppression of key DC functions; (3) driving toward tolerogenic and immunosuppressive DC subpopulations; and (4) prevention of direct contacts between DC and tumor cells by decreasing production of chemokines that attract DC.

In this chapter, we will focus on DC elimination during tumor progression, which could be due to the blockage of their generation and maturation or to the induction of apoptotic cell death in DC and/or their precursors. Numerous clinical and experimental studies reported a dramatic decrease in DC amounts in different secondary lymphoid organs and peripheral blood as well as in primary and metastatic tumor lesions. As early as in 1988, Stene et al. demonstrated a decline in Langerhans cells (one of skin DC subsets) under human melanoma progression (Stene et al. 1988). These observations were confirmed and extended by Alcalay et al. who demonstrated that a decrease in numbers of skin DC correlated with a down-regulation in antigen-presenting activity of draining lymph node cells during the evolution of mouse skin cancers induced by ultraviolet radiation (Alcalay et al. 1989; Alcalay and Kripke 1991). Later on, it was clearly shown in numerous publications that the number of DC could be dramatically reduced in the peripheral blood of patients with different types of tumors such as squamous cell carcinoma of the head and neck (Almand et al. 2000; Hoffmann et al. 2002; Sakakura et al. 2006), lung cancer (Almand et al. 2000; Tabarkiewicz et al. 2008), myeloma (Ratta et al. 2002), invasive breast cancer (Almand et al. 2000; Della Bella et al. 2003; Pinzon-Charry et al. 2007), hepatocellular carcinoma (Ormandy et al. 2006) and leukemia (Maecker et al. 2006). Importantly, the decline in blood DC counts correlated with disease progression indicating an incremental effect of the tumor on blood DC subsets (Pinzon-Charry et al. 2007). Moreover, the presence of visceral metastases resulted in more profound decrease in numbers of circulating mature blood DC in patients with colorectal, gastric, lung, breast and renal cell cancers (Lissoni et al. 1999; Bellik et al. 2006; Pinzon-Charry et al. 2007). While it might be suggested that this diminished number of mature DC could reflect increased migration into the tumor site, several groups reported that primary tumors and metastases contained rather immature DC and a very few or even no mature and activated DC (Bell et al. 1999; Chen et al. 2000; Coventry et al. 2002).

6.2 Blockage of Dendritic Cell Generation and Maturation

The decreased DC frequency and capability to stimulate T-cell-mediated anti-tumor responses in cancer patients were shown to be attributed to the inhibition of their generation and differentiation (Shurin and Gabrilovich 2001). Tumors developed at least two mechanisms, which are responsible for this inhibition. The first one is linked to the impairment of differentiation from hematopoietic progenitor cells during the early maturation stages induced by tumor-derived factors. It has been demonstrated that such growth factors and cytokines as VEGF, M-CSF, TGF-β, IL-6 and IL-10 produced by human and mouse tumor cells can suppress in vitro DC maturation from $CD34^+$ precursors (Gabrilovich et al. 1996; Menetrier-Caux et al. 1998; Gerlini et al. 2004; Zou 2005; Kim et al. 2006). Tumor-derived VEGF has been proposed as a key factor in vivo affecting the early stage of DC maturation in the bone marrow (Ishida et al. 1998; Gabrilovich et al. 1999; Almand et al. 2000) and recruiting immature DC from the bone marrow to the tumor microenvironment (Kim et al. 2006). Human lung cancer cells secreting bombesin-like peptides were reported to down-regulate co-stimulatory molecule expression and IL-12 production by DC and block their ability to activate T cells (Makarenkova et al. 2003). Gangliosides produced by various human tumor cells could also suppress DC maturation and function associated with reduced IL-12 and TNF-α secretion and abnormal MHC class I antigen-processing machinery in human DC (Shurin et al. 2001; Peguet-Navarro et al. 2003; Tourkova et al. 2005). In addition, Bellone et al. (2006) demonstrated that tumor-derived IL-10, IL-6 and TGF-β profoundly affected the phenotype and function of DC in patients with advanced pancreatic carcinoma in favor of immature DC that can hamper an effective antitumor T-cell response. It is important to note that most of the above-mentioned cytokines and growth factors were found to be produced also by some host cells like regulatory T cells, macrophages or even DC with a tolerogenic profile contributing thereby to further impairment of DC maturation in both tumor-bearing animals and cancer patients (Gabrilovich 2004; Zou 2005; Kim et al. 2006; Polak et al. 2007).

The second mechanism by which tumors can alter DC differentiation is the effect on their development from $CD14^+$ monocytes, which could provide another source for generating functional DC (Zhou and Tedder 1996). This mechanism, that was shown to involve also soluble tumor-derived factors, acts by promoting an early but dysfunctional maturation of DC lacking the capacity to produce IL-12, a critical cytokine for T-cell stimulation (Kiertscher et al. 2000). However, a number of factors such as VEGF, TGF-β, IL-10 and PgE_2 have been excluded to be implicated in this pathway (Kiertscher et al. 2000). Later, it has been demonstrated that elevated plasmatic levels of spermine, a member of polyamine family produced by tumor cells, could mediate suppression of IL-12 production and T-cell stimulating capacity of DC in the peripheral blood of breast cancer patients (Della Bella et al. 2003). The authors suggested that the described mechanism could affect later stages of DC differentiation resulting in the impairment of IL-12 production by blood DC. Thus both mechanisms lead to the accumulation of immature DC with a tolerogenic pattern at the primary tumor site as well as in the secondary lymphoid organs and peripheral vessels. These cells are able to induce T-cell anergy or even T-cell death by presenting tumor antigens in the absence of costimulatory molecules, to produce suppressor cytokines and to drive regulatory T-cell differentiation (Mahnke and Enk 2005; Fricke and Gabrilovich 2006; Lu and Finn 2008). Blocking normal DC differentiation and maturation can result also in the accumulation of cells with properties of myeloid-derived suppressor cells (MDSC) (Nefedova et al. 2004). This heterogeneous population of myeloid cells consisting of monocytes/macrophages, granulocytes and DC at different stages of differentiation and expressing various surface markers has been recently found to induce a dramatic suppression of T-cell functions in mouse tumor models and in cancer patients (Serafini et al. 2006; Marigo et al. 2008; Nagaraj and Gabrilovich 2008).

In summary, abnormal DC generation and differentiation significantly facilitates development of the immunosuppressive network that supports tumor growth and metastasis in cancer patients.

6.3 Induction of Apoptosis in Dendritic Cells and Their Precursors

Elimination of DC with the capacity to stimulate potent antitumor T-cell-mediated reactions during tumor progression may be due to the DC killing or acceleration of their turnover. Numerous studies have demonstrated that DC undergo apoptosis in vitro and in vivo after interacting with cancer cells or soluble tumor-derived factors (Esche et al. 1999; Kiertscher et al. 2000; Pirtskhalaishvili et al. 2000a,b; Esche et al. 2001; Peguet-Navarro et al. 2003; Pinzon-Charry et al. 2005). It is important to note that these factors can induce apoptotic cell death of DC precursors as shown by Katsnelson et al. (2007) using conditioned medium from human small-cell lung carcinoma and bronchial carcinoid tumor cells. Moreover, apoptosis of peripheral blood DC was found to develop even in cancer patients with early stages of disease

(Pinzon-Charry et al. 2006). After co-incubation with tumor cells in vitro, DC underwent fast apoptotic changes including cytoplasm shrinking, caspase-3 activation, up-regulation of pro-apoptotic protein Bax and down-regulation of anti-apoptotic proteins Bcl-x_L and Bcl-2 (Esche et al. 1999; Pirtskhalaishvili et al. 2000b; Balkir et al. 2004). Tumor-induced DC apoptosis has been reported to be mediated by cytochrome c release and not by CD95/CD95 ligand interactions (Esche et al. 2001; Balkir et al. 2004). Intratumoral injection of mouse DC transduced with an adenoviral vector encoding the murine Bcl-x_L gene led to significant inhibition of the growth of RM-1 prostate cancer cells as compared with the administration of non-transduced DC suggesting thereby that the protection of DC from tumor-induced apoptosis could markedly increase the efficacy of DC-based antitumor immunotherapies (Pirtskhalaishvili et al. 2000b). Further studies clarified that some anti-apoptotic molecules, which do not belong to the Bcl-2 family of proteins, were also able to inhibit tumor-mediated DC apoptosis (Balkir et al. 2004). Among these molecules are Fas-associated death domain (FADD)-like ICE inhibitory proteins (FLIP) that block procaspase-8 binding to FADD (Tschopp et al. 1998) and human inhibitors of apoptosis (IAP)-like protein (XIAP) that efficiently inhibits active caspases (Duckett et al. 1998).

Different human and mouse tumor cells may express and/or secrete various pro-apoptotic factors such as IL-10, nitric oxide (NO), TGF-β, gangliosides, ceramide and hyaluronan that induce DC-infiltrating primary tumors and metastatic lesions or DC in the peripheral blood and secondary lymphoid organs to undergo apoptosis (Shurin and Gabrilovich 2001). Thus, IL-10 treatment resulted in an earlier onset of DNA fragmentation in human mature Langerhans cells in vitro and, even at low concentrations, reverted the effects of TNF-α and CD40 ligand (CD40L) in inhibiting DC apoptosis (Ludewig et al. 1995).

NO is a highly reactive free radical produced by different cell types that induce apoptotic cell death (Brune et al. 1999; Umansky and Schirrmacher 2001). Increased levels of NO production have been demonstrated for different human and mouse tumors (Lala and Chakraborty 2001). Although it has been recently published that DC treated in vitro with NO donors could become resistant to tumor-mediated apoptosis (Perrotta et al. 2004), other data indicate an induction of apoptosis in DC exposed to NO via down-regulation of IAP and up-regulation of caspases activity (Bonham et al. 1996; Stanford et al. 2001). Moreover, hyaluronan, a major component of glioma extracellular matrix supporting tumor cell migration and metastasis, was shown to promote DC apoptosis by stimulating inducible NO synthase activity and NO production in DC (Yang et al. 2002). These opposite effects could be attributed to the different NO concentrations, so that limited amounts of NO in the tumor microenvironment might have a deleterious influence on DC survival in vivo.

Overexpression of TGF-β in the tumor and serum is known to be associated with a poor prognosis in cancer patients (Hasegawa et al. 2001). However, the effect of TGF-β1 on DC is interestingly equivocal. On the one hand, it is required for the generation of Langerhans cells from CD34$^+$ hematopoietic precursors, which express CD1a but not CD83 and are arrested at an immature differentiation stage (Strobl et al. 1997). On the other hand, TGF-β1 can suppress the differentiation of mouse bone marrow-derived DC and their capacity to secrete IL-12, to stimulate tumor-specific T cells and to migrate into tumor-draining lymph nodes, so-called sentinel lymph nodes (Kobie et al. 2003; Lyakh et al. 2005). Furthermore, it has been demonstrated that tumor-derived TGF-β1 can directly induce DC apoptotic cell death within sentinel lymph nodes without evidence of metastasis removed from patients with non-small cell lung cancer, thereby facilitating metastasis within those nodes (Cochran et al. 2001; Ito et al. 2006).

Similarly, gangliosides purified from human melanoma cells have been reported to inhibit the phenotypic and functional differentiation of monocyte-derived DC and induce their apoptosis (Peguet-Navarro et al. 2003). Notably, this effect was shown to occur at any step of DC differentiation in a dose-dependent manner. Gangliosides are ubiquitous membrane-associated sialic acid-containing glycosphingolipids, which are known to be involved in cell growth and to be shed by a number of tumors into their microenvironment (Ladisch et al. 1987; Spiegel and Merrill 1996). Moreover, it has been reported that supernatants from murine melanoma cell lines stimulated apoptosis in bone marrow-derived murine DC by increasing endogenous levels of ceramide and that blocking ceramide synthesis protected DC from tumor-induced apoptosis (Kanto et al. 2001). Given that ceramide is a constitutive component of ganglioside structure, the authors proposed that gangliosides might be metabolized into ceramide, which, in turn, would induce DC apoptotic cell death. Although there are limited data on the pro-apoptotic influence of both compounds on DC in vivo, they can constantly shed into tumor microenvironment, circulate in the body and impair DC viability locally in the tumor tissue as well as systemically in the peripheral blood and secondary lymphoid organs.

Investigating factors that protect blood DC from tumor-induced apoptosis in breast cancer patients, Pinzon-Charry et al. (2006) demonstrated that CD40 stimulation with the CD40 ligand (CD154) can inhibit apoptosis through secretion of IL-12, which induced a sustained Bcl-2 expression. In a mouse model, the beneficial effect of CD40 ligation has also been related to Bcl-2, which counter-balances the pro-apoptotic properties of various DC maturation stimuli (Hou and Van Parijs 2004). IL-12, a pleotropic proinflammatory cytokine, has been described during last years to enhance DC functions, promote their maturation and significantly enhance their survival (Esche et al. 1999; Pirtskhalaishvili et al. 2000a,b; Portielje et al. 2003; Trinchieri 2003). Occupation of IL-12 receptors expressed on DC initiates the nuclear translocation of the NF-κB family members, and a lack of IL-12 production by DC may result in their early death (Ouaaz et al. 2002). Interestingly, TRANCE/RANKL, another modulator of NF-κB activity, has been implicated in prolonged DC survival (Wong et al. 1997). Pirtskhalaishvili et al. (2000a) presented evidence that stimulation of human DC with CD154 elevated their resistance to prostate cancer-induced apoptosis via increased production not only of IL-12 but also of IL-15, a cytokine with similar IL-2 properties. This increased cytokine production resulted in the up-regulation of the expression of Bcl-x_L, another member of Bcl-2 family.

6.4 Conclusions

Taken together, tumor-induced elimination of functional DC due to apoptotic cell death or hampered differentiation and maturation has been demonstrated to severely impair antitumor T-cell-mediated responses in vivo. For cancer patients, this would result in a significant acceleration of tumor progression. Consequently, protection of DC longevity in patients with cancer as well as pre-treatment of DC vaccine to up-regulate their survival in appalling tumor milieu should provide novel approaches for cancer patient treatment, as has been already proven in pre-clinical studies.

References

Alcalay, J., Goldberg, L. H., Wolf, J. E., Jr. and Kripke, M. L. 1989. Variations in the number and morphology of Langerhans' cells in the epidermal component of squamous cell carcinomas. Arch Dermatol 125:917–920.

Alcalay, J. and Kripke, M. L. 1991. Antigen-presenting activity of draining lymph node cells from mice painted with a contact allergen during ultraviolet carcinogenesis. J Immunol 146:1717–1721.

Almand, B., Resser, J. R., Lindman, B., Nadaf, S., Clark, J. I., Kwon, E. D., Carbone, D. P. and Gabrilovich, D. I. 2000. Clinical significance of defective dendritic cell differentiation in cancer. Clin Cancer Res 6:1755–1766.

Balkir, L., Tourkova, I. L., Makarenkova, V. P., Shurin, G. V., Robbins, P. D., Yin, X. M., Chatta, G. and Shurin, M. R. 2004. Comparative analysis of dendritic cells transduced with different anti-apoptotic molecules: sensitivity to tumor-induced apoptosis. J Gene Med 6:537–544.

Bell, D., Chomarat, P., Broyles, D., Netto, G., Harb, G. M., Lebecque, S., Valladeau, J., Davoust, J., Palucka, K. A. and Banchereau, J. 1999. In breast carcinoma tissue, immature dendritic cells reside within the tumor, whereas mature dendritic cells are located in peritumoral areas. J Exp Med 190:1417–1426.

Bellik, L., Gerlini, G., Parenti, A., Ledda, F., Pimpinelli, N., Neri, B. and Pantalone, D. 2006. Role of conventional treatments on circulating and monocyte-derived dendritic cells in colorectal cancer. Clin Immunol 121:74–80.

Bellone, G., Carbone, A., Smirne, C., Scirelli, T., Buffolino, A., Novarino, A., Stacchini, A., Bertetto, O., Palestro, G., Sorio, C., Scarpa, A., Emanuelli, G. and Rodeck, U. 2006. Cooperative induction of a tolerogenic dendritic cell phenotype by cytokines secreted by pancreatic carcinoma cells. J Immunol 177:3448–3460.

Bonham, C. A., Lu, L., Li, Y., Hoffman, R. A., Simmons, R. L. and Thomson, A. W. 1996. Nitric oxide production by mouse bone marrow-derived dendritic cells: implications for the regulation of allogeneic T cell responses. Transplantation 62:1871–1877.

Brune, B., von Knethen, A. and Sandau, K. B. 1999. Nitric oxide (NO): an effector of apoptosis. Cell Death Differ 6:969–975.

Chen, S., Akbar, S. M., Tanimoto, K., Ninomiya, T., Iuchi, H., Michitaka, K., Horiike, N. and Onji, M. 2000. Absence of CD83-positive mature and activated dendritic cells at cancer nodules from patients with hepatocellular carcinoma: relevance to hepatocarcinogenesis. Cancer Lett 148:49–57.
Cochran, A. J., Morton, D. L., Stern, S., Lana, A. M., Essner, R. and Wen, D. R. 2001. Sentinel lymph nodes show profound downregulation of antigen-presenting cells of the paracortex: implications for tumor biology and treatment. Mod Pathol 14:604–608.
Coventry, B. J., Lee, P. L., Gibbs, D. and Hart, D. N. 2002. Dendritic cell density and activation status in human breast cancer – CD1a, CMRF-44, CMRF-56 and CD-83 expression. Br J Cancer 86:546–551.
Della Bella, S., Gennaro, M., Vaccari, M., Ferraris, C., Nicola, S., Riva, A., Clerici, M., Greco, M. and Villa, M. L. 2003. Altered maturation of peripheral blood dendritic cells in patients with breast cancer. Br J Cancer 89:1463–1472.
Duckett, C. S., Li, F., Wang, Y., Tomaselli, K. J., Thompson, C. B. and Armstrong, R. C. 1998. Human IAP-like protein regulates programmed cell death downstream of Bcl-xL and cytochrome c. Mol Cell Biol 18:608–615.
Esche, C., Lokshin, A., Shurin, G. V., Gastman, B. R., Rabinowich, H., Watkins, S. C., Lotze, M. T. and Shurin, M. R. 1999. Tumor's other immune targets: dendritic cells. J Leukoc Biol 66:336–344.
Esche, C., Shurin, G. V., Kirkwood, J. M., Wang, G. Q., Rabinowich, H., Pirtskhalaishvili, G. and Shurin, M. R. 2001. Tumor necrosis factor-alpha-promoted expression of Bcl-2 and inhibition of mitochondrial cytochrome c release mediate resistance of mature dendritic cells to melanoma-induced apoptosis. Clin Cancer Res 7:974s–979s.
Fricke, I. and Gabrilovich, D. I. 2006. Dendritic cells and tumor microenvironment: a dangerous liaison. Immunol Invest 35:459–483.
Gabrilovich, D. 2004. Mechanisms and functional significance of tumour-induced dendritic-cell defects. Nat Rev Immunol 4:941–952.
Gabrilovich, D. I., Chen, H. L., Girgis, K. R., Cunningham, H. T., Meny, G. M., Nadaf, S., Kavanaugh, D. and Carbone, D. P. 1996. Production of vascular endothelial growth factor by human tumors inhibits the functional maturation of dendritic cells. Nat Med 2:1096–1103.
Gabrilovich, D. I., Ishida, T., Nadaf, S., Ohm, J. E. and Carbone, D. P. 1999. Antibodies to vascular endothelial growth factor enhance the efficacy of cancer immunotherapy by improving endogenous dendritic cell function. Clin Cancer Res 5:2963–2970.
Gerlini, G., Tun-Kyi, A., Dudli, C., Burg, G., Pimpinelli, N. and Nestle, F. O. 2004. Metastatic melanoma secreted IL-10 down-regulates CD1 molecules on dendritic cells in metastatic tumor lesions. Am J Pathol 165:1853–1863.
Hasegawa, Y., Takanashi, S., Kanehira, Y., Tsushima, T., Imai, T. and Okumura, K. 2001. Transforming growth factor-beta1 level correlates with angiogenesis, tumor progression, and prognosis in patients with nonsmall cell lung carcinoma. Cancer 91:964–971.
Hoffmann, T. K., Muller-Berghaus, J., Ferris, R. L., Johnson, J. T., Storkus, W. J. and Whiteside, T. L. 2002. Alterations in the frequency of dendritic cell subsets in the peripheral circulation of patients with squamous cell carcinomas of the head and neck. Clin Cancer Res 8:1787–1793.
Hou, W. S. and Van Parijs, L. 2004. A Bcl-2-dependent molecular timer regulates the lifespan and immunogenicity of dendritic cells. Nat Immunol 5:583–589.
Ishida, T., Oyama, T., Carbone, D. P. and Gabrilovich, D. I. 1998. Defective function of Langerhans cells in tumor-bearing animals is the result of defective maturation from hemopoietic progenitors. J Immunol 161:4842–4851.
Ito, M., Minamiya, Y., Kawai, H., Saito, S., Saito, H., Nakagawa, T., Imai, K., Hirokawa, M. and Ogawa, J. 2006. Tumor-derived TGFbeta-1 induces dendritic cell apoptosis in the sentinel lymph node. J Immunol 176:5637–5643.

Kanto, T., Kalinski, P., Hunter, O. C., Lotze, M. T. and Amoscato, A. A. 2001. Ceramide mediates tumor-induced dendritic cell apoptosis. J Immunol 167:3773–3784.
Katsenelson, N. S., Shurin, G. V., Bykovskaia, S. N., Shogan, J. and Shurin, M. R. 2001. Human small cell lung carcinoma and carcinoid tumor regulate dendritic cell maturation and function. Mod Pathol 14:40–45.
Kiertscher, S. M., Luo, J., Dubinett, S. M. and Roth, M. D. 2000. Tumors promote altered maturation and early apoptosis of monocyte-derived dendritic cells. J Immunol 164:1269–1276.
Kim, R., Emi, M., Tanabe, K. and Arihiro, K. 2006. Tumor-driven evolution of immunosuppressive networks during malignant progression. Cancer Res 66:5527–5536.
Kobie, J. J., Wu, R. S., Kurt, R. A., Lou, S., Adelman, M. K., Whitesell, L. J., Ramanathapuram, L. V., Arteaga, C. L. and Akporiaye, E. T. 2003. Transforming growth factor beta inhibits the antigen-presenting functions and antitumor activity of dendritic cell vaccines. Cancer Res 63:1860–1864.
Ladisch, S., Kitada, S. and Hays, E. F. 1987. Gangliosides shed by tumor cells enhance tumor formation in mice. J Clin Invest 79:1879–1882.
Lala, P. K. and Chakraborty, C. 2001. Role of nitric oxide in carcinogenesis and tumour progression. Lancet Oncol 2:149–156.
Lanzavecchia, A. and Sallusto, F. 2001. Regulation of T cell immunity by dendritic cells. Cell 106:263–266.
Lissoni, P., Bolis, S., Mandala, M., Viviani, S., Pogliani, E. and Barni, S. 1999. Blood concentrations of tumor necrosis factor-alpha in malignant lymphomas and their decrease as a predictor of disease control in response to low-dose subcutaneous immunotherapy with interleukin-2. Int J Biol Markers 14:167–171.
Lu, B. and Finn, O. J. 2008. T-cell death and cancer immune tolerance. Cell Death Differ 15:70–79.
Ludewig, B., Graf, D., Gelderblom, H. R., Becker, Y., Kroczek, R. A. and Pauli, G. 1995. Spontaneous apoptosis of dendritic cells is efficiently inhibited by TRAP (CD40-ligand) and TNF-alpha, but strongly enhanced by interleukin-10. Eur J Immunol 25:1943–1950.
Lyakh, L. A., Sanford, M., Chekol, S., Young, H. A. and Roberts, A. B. 2005. TGF-beta and vitamin D3 utilize distinct pathways to suppress IL-12 production and modulate rapid differentiation of human monocytes into CD83+ dendritic cells. J Immunol 174:2061–2070.
Maecker, B., Mougiakakos, D., Zimmermann, M., Behrens, M., Hollander, S., Schrauder, A. , Schrappe, M., Welte, K. and Klein, C. 2006. Dendritic cell deficiencies in pediatric acute lymphoblastic leukemia patients. Leukemia 20:645–649.
Mahnke, K. and Enk, A. H. 2005. Dendritic cells: key cells for the induction of regulatory T cells? Curr Top Microbiol Immunol 293:133–150.
Makarenkova, V. P., Shurin, G. V., Tourkova, I. L., Balkir, L., Pirtskhalaishvili, G., Perez, L., Gerein, V., Siegfried, J. M. and Shurin, M. R. 2003. Lung cancer-derived bombesin-like peptides down-regulate the generation and function of human dendritic cells. J Neuroimmunol 145:55–67.
Marigo, I., Dolcetti, L., Serafini, P., Zanovello, P. and Bronte, V. 2008. Tumor-induced tolerance and immune suppression by myeloid derived suppressor cells. Immunol Rev 222:162–179.
Menetrier-Caux, C., Montmain, G., Dieu, M. C., Bain, C., Favrot, M. C., Caux, C. and Blay, J. Y. 1998. Inhibition of the differentiation of dendritic cells from CD34(+) progenitors by tumor cells: role of interleukin-6 and macrophage colony-stimulating factor. Blood 92:4778–4791.
Moser, M. 2003. Dendritic cells in immunity and tolerance-do they display opposite functions? Immunity 19:5–8.
Nagaraj, S. and Gabrilovich, D. I. 2008. Tumor escape mechanism governed by myeloid-derived suppressor cells. Cancer Res 68:2561–2563.

Nefedova, Y., Huang, M., Kusmartsev, S., Bhattacharya, R., Cheng, P., Salup, R., Jove, R. and Gabrilovich, D. 2004. Hyperactivation of STAT3 is involved in abnormal differentiation of dendritic cells in cancer. J Immunol 172:464–474.

Ormandy, L. A., Farber, A., Cantz, T., Petrykowska, S., Wedemeyer, H., Horning, M., Lehner, F., Manns, M. P., Korangy, F. and Greten, T. F. 2006. Direct ex vivo analysis of dendritic cells in patients with hepatocellular carcinoma. World J Gastroenterol 12:3275–3282.

Ouaaz, F., Arron, J., Zheng, Y., Choi, Y. and Beg, A. A. 2002. Dendritic cell development and survival require distinct NF-kappaB subunits. Immunity 16:257–270.

Palucka, A. K., Ueno, H., Fay, J. W. and Banchereau, J. 2007. Taming cancer by inducing immunity via dendritic cells. Immunol Rev 220:129–150.

Peguet-Navarro, J., Sportouch, M., Popa, I., Berthier, O., Schmitt, D. and Portoukalian, J. 2003. Gangliosides from human melanoma tumors impair dendritic cell differentiation from monocytes and induce their apoptosis. J Immunol 170:3488–3494.

Perrotta, C., Falcone, S., Capobianco, A., Camporeale, A., Sciorati, C., De Palma, C., Pisconti, A., Rovere-Querini, P., Bellone, M., Manfredi, A. A. and Clementi, E. 2004. Nitric oxide confers therapeutic activity to dendritic cells in a mouse model of melanoma. Cancer Res 64:3767–3771.

Pinzon-Charry, A., Ho, C. S., Maxwell, T., McGuckin, M. A., Schmidt, C., Furnival, C., Pyke, C. M. and Lopez, J. A. 2007. Numerical and functional defects of blood dendritic cells in early- and late-stage breast cancer. Br J Cancer 97:1251–1259.

Pinzon-Charry, A., Maxwell, T. and Lopez, J. A. 2005. Dendritic cell dysfunction in cancer: a mechanism for immunosuppression. Immunol Cell Biol 83:451–461.

Pinzon-Charry, A., Maxwell, T., McGuckin, M. A., Schmidt, C., Furnival, C. and Lopez, J. A. 2006. Spontaneous apoptosis of blood dendritic cells in patients with breast cancer. Breast Cancer Res 8:R5.

Pirtskhalaishvili, G., Shurin, G. V., Esche, C., Cai, Q., Salup, R. R., Bykovskaia, S. N., Lotze, M. T. and Shurin, M. R. 2000a. Cytokine-mediated protection of human dendritic cells from prostate cancer-induced apoptosis is regulated by the Bcl-2 family of proteins. Br J Cancer 83:506–513.

Pirtskhalaishvili, G., Shurin, G. V., Gambotto, A., Esche, C., Wahl, M., Yurkovetsky, Z. R., Robbins, P. D. and Shurin, M. R. 2000b. Transduction of dendritic cells with Bcl-xL increases their resistance to prostate cancer-induced apoptosis and antitumor effect in mice. J Immunol 165:1956–1964.

Polak, M. E., Borthwick, N. J., Gabriel, F. G., Johnson, P., Higgins, B., Hurren, J., McCormick, D., Jager, M. J. and Cree, I. A. 2007. Mechanisms of local immunosuppression in cutaneous melanoma. Br J Cancer 96:1879–1887.

Portielje, J. E., Gratama, J. W., van Ojik, H. H., Stoter, G. and Kruit, W. H. 2003. IL-12: a promising adjuvant for cancer vaccination. Cancer Immunol Immunother 52:133–144.

Ratta, M., Fagnoni, F., Curti, A., Vescovini, R., Sansoni, P., Oliviero, B., Fogli, M., Ferri, E., Della Cuna, G. R., Tura, S., Baccarani, M. and Lemoli, R. M. 2002. Dendritic cells are functionally defective in multiple myeloma: the role of interleukin-6. Blood 100:230–237.

Sakakura, K., Chikamatsu, K., Takahashi, K., Whiteside, T. L. and Furuya, N. 2006. Maturation of circulating dendritic cells and imbalance of T-cell subsets in patients with squamous cell carcinoma of the head and neck. Cancer Immunol Immunother 55:151–159.

Serafini, P., Borrello, I. and Bronte, V. 2006. Myeloid suppressor cells in cancer: recruitment, phenotype, properties, and mechanisms of immune suppression. Semin Cancer Biol 16:53–65.

Shurin, G. V., Shurin, M. R., Bykovskaia, S., Shogan, J., Lotze, M. T. and Barksdale, E. M., Jr. 2001. Neuroblastoma-derived gangliosides inhibit dendritic cell generation and function. Cancer Res 61:363–369.

Shurin, M. R. 1996. Dendritic cells presenting tumor antigen. Cancer Immunol Immunother 43:158–164.

Shurin, M. R. and Gabrilovich, D. I. 2001. Regulation of dendritic cell system by tumor. Cancer Res Ther Control 11:65–78.
Shurin, M. R., Shurin, G. V. and Chatta, G. S. 2007. Aging and the dendritic cell system: implications for cancer. Crit Rev Oncol Hematol 64:90–105.
Shurin, M. R., Shurin, G. V., Lokshin, A., Yurkovetsky, Z. R., Gutkin, D. W., Chatta, G., Zhong, H., Han, B. and Ferris, R. L. 2006. Intratumoral cytokines/chemokines/growth factors and tumor infiltrating dendritic cells: friends or enemies? Cancer Metastasis Rev 25:333–356.
Spiegel, S. and Merrill, A. H., Jr. 1996. Sphingolipid metabolism and cell growth regulation. Faseb J 10:1388–1397.
Stanford, A., Chen, Y., Zhang, X. R., Hoffman, R., Zamora, R. and Ford, H. R. 2001. Nitric oxide mediates dendritic cell apoptosis by downregulating inhibitors of apoptosis proteins and upregulating effector caspase activity. Surgery 130:326–332.
Steinman, R. M. 2007. Lasker Basic Medical Research Award. Dendritic cells: versatile controllers of the immune system. Nat Med 13:1155–1159.
Stene, M. A., Babajanians, M., Bhuta, S. and Cochran, A. J. 1988. Quantitative alterations in cutaneous Langerhans cells during the evolution of malignant melanoma of the skin. J Invest Dermatol 91:125–128.
Strobl, H., Bello-Fernandez, C., Riedl, E., Pickl, W. F., Majdic, O., Lyman, S. D. and Knapp, W. 1997. flt3 ligand in cooperation with transforming growth factor-beta1 potentiates in vitro development of Langerhans-type dendritic cells and allows single-cell dendritic cell cluster formation under serum-free conditions. Blood 90:1425–1434.
Tabarkiewicz, J., Rybojad, P., Jablonka, A. and Rolinski, J. 2008. CD1c+ and CD303+ dendritic cells in peripheral blood, lymph nodes and tumor tissue of patients with non-small cell lung cancer. Oncol Rep 19:237–243.
Tourkova, I. L., Shurin, G. V., Chatta, G. S., Perez, L., Finke, J., Whiteside, T. L., Ferrone, S. and Shurin, M. R. 2005. Restoration by IL-15 of MHC class I antigen-processing machinery in human dendritic cells inhibited by tumor-derived gangliosides. J Immunol 175:3045–3052.
Trinchieri, G. 2003. Interleukin-12 and the regulation of innate resistance and adaptive immunity. Nat Rev Immunol 3:133–146.
Tschopp, J., Irmler, M. and Thome, M. 1998. Inhibition of fas death signals by FLIPs. Curr Opin Immunol 10:552–558.
Umansky, V. and Schirrmacher, V. 2001. Nitric oxide-induced apoptosis in tumor cells. Adv Cancer Res 82:107–131.
Wong, B. R., Josien, R., Lee, S. Y., Sauter, B., Li, H. L., Steinman, R. M. and Choi, Y. 1997. TRANCE (tumor necrosis factor [TNF]-related activation-induced cytokine), a new TNF family member predominantly expressed in T cells, is a dendritic cell-specific survival factor. J Exp Med 186:2075–2080.
Yang, T., Witham, T. F., Villa, L., Erff, M., Attanucci, J., Watkins, S., Kondziolka, D., Okada, H., Pollack, I. F. and Chambers, W. H. 2002. Glioma-associated hyaluronan induces apoptosis in dendritic cells via inducible nitric oxide synthase: implications for the use of dendritic cells for therapy of gliomas. Cancer Res 62:2583–2591.
Zhou, L. J. and Tedder, T. F. 1996. CD14+ blood monocytes can differentiate into functionally mature CD83+ dendritic cells. Proc Natl Acad Sci USA 93:2588–2592.
Zou, W. 2005. Immunosuppressive networks in the tumour environment and their therapeutic relevance. Nat Rev Cancer 5:263–274.

Part II
Mechanisms of Dendritic Cell Dysfunction in Cancer

While the concept that dendritic cells function abnormally in cancer is generally accepted by tumor immunologists and clinicians interested in cancer immunotherapy, questions remain regarding the mechanisms and tumor-derived factors responsible for deficient or anomalous generation and function of dendritic cells in the tumor environment. In this section of the book, molecular mechanisms and agents important in inducing dendritic cell dysfunction in cancer are emphasized and described in detail. First, tumor-derived factors, including cytokines, neuropeptides, small molecules, prostaglandins, gangliosides and tumor-associated antigens that were proven to disrupt dendropoiesis and provoke dendritic cell malfunctioning are described in Chapter 7, with specific focus on tumor-derived exosomes in Chapter 8. Chapter 9 discusses signaling pathways known to mediate dendritic cell dysfunction in the tumor environment, such as Stat, JNK, MAPK, PI3K/Akt and NF-κB, while Chapter 10 focuses specifically on the role of STAT3 in this process, which is also involved in regulating activity of tumor cells and thus is clearly important for tumor escape. How inflammation in the tumor environment impacts dendritic cell function is reported in Chapter 11, which carefully analyzes inflammatory mechanisms during different stages of tumor progression from the pre-malignant lesions (proto-cancer) to the final steps of tumor evolution such as tumor eradication or tumor escape from immune recognition. How dendritic cell functions are altered during the aging process and by stress, which both are often accompanied malignant diseases, are discussed, respectively, in Chapters 12 and 13. Finally in Chapter 14, how current conventional therapies for cancer, such as surgery, radiation and chemotherapy, may impact dendritic cell differentiation, function and longevity in patients is described in detail, with an emphasis on novel data analyzing the effect of low-dose chemotherapy on dendritic cells.

Chapter 7
Tumor-Derived Factors Responsible for Dendritic Cell Dysfunction

Alberto Pinzon-Charry and J. Alejandro López

Abstract Perpetuation of immune deficiency throughout tumor development is, to a great degree, the result of impairment of dendritic cell function by products secreted by tumors. They include cytokines, non-tumor-specific molecules (gangliosides, prostanoids, nitric oxide, etc.) and tumor-(specific) antigens (MUC-1, PSA, Her-2 neu). They may engender a distortion of dendritic cell development, block dendritic cell maturation, induce dendritic cell apoptosis or interfere with antigen presentation. Identifying those molecules and their interaction with dendritic cells will accelerate the development of more efficient immunotherapies. In this chapter we review the current literature on these interactions and highlight the possible avenues of minimization of their deleterious effects.

7.1 Introduction

The induction of an effective immunological response against tumors is mainly dependent on the effector cells of the innate and adaptive immunity with DC playing a very important regulatory role. Therefore, the context in which tumor antigens are presented to the immune system dictates the efficacy of tumoricidal responses. The spontaneous remission observed in a number of human cancers suggests that the immune system has the potential to present antigens adequately and thus eliminate malignant cells (Bodey 2002). However, tumors usually evade immune clearance due to a number of mechanisms including recruitment of regulatory cell types (Pekarek et al. 1995; Vence et al. 2007), deletion of effector cells (Staveley-O'Carroll et al. 1998; Saito et al. 2000) or secretion of immunosuppressive factors (Gabrilovich 2004). Here, we examine evidence on the role of tumor-derived factors inducing DC dysfunction and particularly the alteration of DC differentiation, maturation and longevity as a mechanism for immune suppression.

A. Pinzon-Charry (✉)
Bancroft Centre, L floor, Queensland Institute of Medical Research, Brisbane,
QLD 4006, Australia
e-mail: AlbertoP@qimr.edu.au

M.R. Shurin, R.D. Salter (eds.), *Dendritic Cells in Cancer*,
DOI 10.1007/978-0-387-88611-4_7, © Springer Science+Business Media, LLC 2009

7.2 Cytokines and Growth Factors Affecting Dendritic Cell Differentiation

One of the first cytokines reported to have an inhibitory effect on DC function in cancer was IL-10. This is a suppressive cytokine which exerts primarily anti-inflammatory functions and antagonizes several functions of antigen-presenting cells (APC), including DC. It has been shown to inhibit the ability of DC to stimulate T cells, inducing antigen-specific anergy (Steinbrink et al. 1999). IL-10 release has been reported from tumors like melanoma, multiple myeloma and lung cancer (Kruger-Krasagakes et al. 1994; Smith et al. 1994; Gu et al. 1996) as well as tumor-infiltrating macrophages, lymphocytes and peripheral blood lymphocytes (Kim et al. 1995; Asselin-Paturel et al. 1998). Importantly, IL-10 has been shown to prevent the differentiation of monocytes to DC (Allavena et al. 1998) as well as inhibit the antigen-presenting function of DC (Enk et al. 1993; Buelens et al. 1997). In addition, increased serum levels of IL-10 correlate with numerical deficiency and immature phenotype of circulating DC subsets in patients with hepatocellular carcinoma (Beckebaum et al. 2004), indicating a clear association between tumor-related production of IL-10 and defects in DC differentiation .

Secretion of IL-6 and M-CSF from carcinoma cells has also been observed to inhibit the differentiation of DC from $CD34^+$ myeloid progenitors (Menetrier-Caux et al. 1998). The molecular mechanisms responsible for this effect involve the modulation of GM-CSF and M-CSF receptor expression by tumor-secreted IL-6 and M-CSF (Menetrier-Caux et al. 1998). In addition, high levels of IL-6 have been correlated with poor prognosis in patients with multiple myeloma, renal cell carcinoma, melanoma and colorectal cancer (Blay et al. 1992; Deehan et al. 1994; Tartour et al. 1996; Ratta et al. 2002). Tumor overproduction of IL-6 has been demonstrated to inhibit the colony growth of DC progenitors (Ratta et al. 2002) and sera from bone marrow of multiple myeloma patients (containing high levels of IL-6) have been shown to inhibit the induction of fully functional DC (Hayashi et al. 2003). Another study has demonstrated that IL-6 plays a crucial role in maintaining an immature phenotype on DC in vivo (Park et al. 2004), confirming the significant role of IL-6 in the inhibition of DC differentiation both in vitro and in vivo.

Granulocyte/monocyte-colony-stimulating factor (GM-CSF) has also been associated with tumor-induced dysfunction of myelopoiesis. Spontaneous production of GM-CSF has been reported for several types of human tumor cell lines (Bronte et al. 2000) and production of this cytokine has been associated with the ability of cancer cells to metastasize (Tsuchiya et al. 1988). Although cancer cells modified to produce GM-CSF elicit robust antitumor immune responses by recruiting DC, the aberrant secretion of GM-CSF by some tumors could be deleterious to the host immune response. In mice, chronic GM-CSF production by tumors has been reported to suppress tumor-specific CTL responses through the generation of a population of inhibitory immature

APC. Although these immature cells could be generated by the administration of GM-CSF alone, GM-CSF in combination with IL-4 induced their differentiation into mature APC (Bronte et al. 1999) underscoring the necessity of an appropriate cytokine milieu for adequate APC differentiation. Similarly, the production of large amounts of GM-CSF by some tumors could impair the immune response. This has been demonstrated in a study whereby vaccination with tumor cells producing large quantities of GM-CSF resulted in substantial accumulation of immature cells and immunosuppression in vivo (Serafini et al. 2004). The relevance of these findings in humans has been confirmed with reports indicating that immune suppression and increased recurrence and metastasis in patients with head and neck squamous cell carcinoma were related to the presence of immature cells that secreted GM-CSF (Pak et al. 1995; Young et al. 1997). Interestingly, these cells could be differentiated into fully functional DC in vitro, following culture with the pro-differentiating factor 1α,25-dihydroxyvitamin D3 (Garrity et al. 1997).

VEGF is another tumor-derived factor shown to affect differentiation of DC. VEGF is produced by different types of tumors and increased levels of this cytokine have been associated with poor prognosis in cancer (Toi et al. 1996). VEGF plays a role in inducing proliferation of endothelial cells and formation of neo-vasculature within the tumor. It has also been demonstrated that VEGF significantly affects the differentiation of multiple hematopoietic lineages in vivo, including DC (Gabrilovich et al. 1998). The inhibitory effect of VEGF on DC differentiation has been confirmed and increased VEGF levels have been reported to correlate with reduced number of infiltrating and circulating DC in patients with different types of cancer (Saito et al. 1998; Lissoni et al. 2001; Takahashi et al. 2004). In addition to altered differentiation of DC, elevated levels of VEGF have been associated with an increased number of immature cells with immunosuppressive function in the circulation of patients with cancer (Almand et al. 2000). Notably, these cells could be differentiated into mature DC in vitro, after culture with all-*trans* retinoic acid (Almand et al. 2001). More recently, however, inhibition of VEGF has shown to improve on DC maturation while yielding discreet improvement on immune responses (Fricke et al. 2007).

7.3 Other Immune Mediators Affecting Dendritic Cell Differentiation

Gangliosides (membrane-bound glycosphingolipids with a sialic acid moiety) could contribute to tumor-induced immune suppression by altering differentiation of several lineages and hematopoiesis (Sietsma et al. 1998) (Fig. 7.1). A number of tumors including medulloblastoma, lymphoma, melanoma, neuroblastoma, retinoblastoma and hepatoma are known to display an aberrant ganglioside composition (Birkle et al. 2003) also shedding some of these molecules into the tumor milieu and the circulation (Ladisch

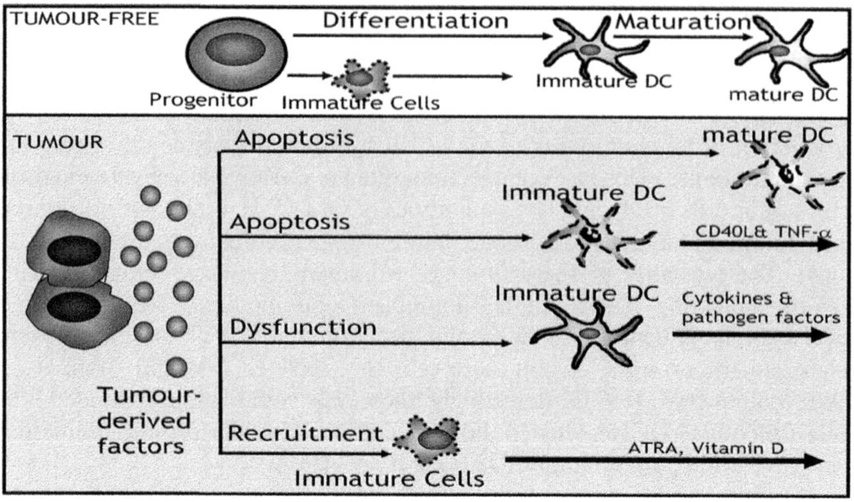

Fig. 7.1 Mechanisms for tumor-induced dendritic cell dysfunction. In a tumor-free environment (*upper panel*), hematopoietic precursors give rise to progenitors which differentiate into immature DC. Following antigen/'danger signal' encounter, immature DC undergo maturation and become specialized in antigen presentation. Under the influence of tumor-derived factors (*lower panel*), differentiation of DC is hampered resulting in the recruitment, accumulation and dysfunction of immature cells and immature DC. DC can also be induced to undergo apoptosis by tumor products. Immature cells can be differentiated into DC under the influence of factors like all-*trans* retinoic acid (ATRA) or vitamin D. Immature DC can be induced to mature with pro-inflammatory cytokines/pathogen-derived factors. Treatment of DC with TNF-α or CD40L confers protection to tumor-induced apoptosis promoting tumor clearance

et al. 1987; Portoukalian et al. 1993). More importantly, it has been shown that neuroblastoma and melanoma-derived gangliosides impair the phenotypic and functional differentiation of DC providing another mechanism for tumor-induced immunosuppression (Shurin et al. 2001; Peguet-Navarro et al. 2003) (Fig. 7.2). It appears that gangliosides interfere with the expression of costimulatory molecules and inhibit NF-κB (Caldwell et al. 2003). Interestingly, such DC dysfunction might be partly corrected with IL-15 (Tourkova et al. 2005).

There are other factors that could play redundant or synergistic roles on the inhibition of DC differentiation by tumors. Several reports now correlate alterations of arachidonic acid metabolism with carcinogenesis (Gately and Li 2004). Arachidonic acid metabolites (prostanoids) including prostaglandins and thromboxanes are synthesized by cyclooxygenase (COX)-1 and -2. In several types of cancer, including melanoma, colon, breast and lung carcinoma, alterations in the expression of these enzymes have been reported (Tsujii et al. 1997; Denkert et al. 2001). Most studies indicate that prostanoids have a role in tumorigenesis mostly through their pro-angiogenic effects (Gately and Li 2004).

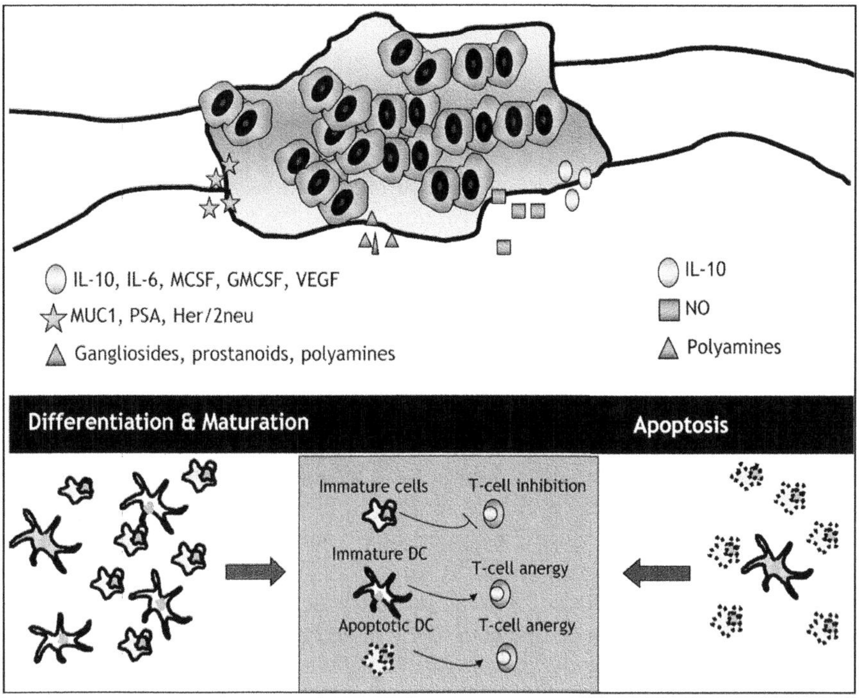

Fig. 7.2 Tumor-induced dendritic cell dysfunction generates ineffective immune responses. In the absence of tumor-derived factors, DC differentiate and mature appropriately. Mature DC increase their capacity to process antigens and express the cytokines and costimulatory molecules essential for initiating an effective immune response. Under the influence of tumor-derived factors, immature cells, immature DC and apoptotic DC are generated and accumulated. Although immature cells appear not to present antigens, they can suppress tumor-specific T-cell activation through the production of reactive oxygen species (ROS). Immature DC and apoptotic DC can present antigens but in the absence of adequate costimulation and cytokine secretion, they can induce anergy, abortive proliferation or induction of regulatory T cells, thus favouring tumor evasion

However, recent data indicate that prostanoids may also contribute to carcinogenesis through inhibition of DC differentiation. Indeed, it has recently been demonstrated that tumor COX-2 expression decreases host antitumor response by impairing DC maturation and activity (Sharma et al. 2003). This appears to be through specific signaling pathways. The EP2 receptor, one of the four receptors for prostaglandin E2 (PgE2), has been reported to mediate PgE2-induced inhibition of DC differentiation and function, playing an essential role in impaired antitumor responses (Yang et al. 2003). It has also been shown that COX-2-regulated prostanoids inhibit DC production of IL-12, increase secretion of IL-10 and contribute to tumor-induced inhibition of DC development (Sombroek et al. 2002).

7.4 Tumor Antigens and Metabolites Affecting Dendritic Cell Differentiation

Various reports indicate that other factors produced by malignant cells like tumor antigens (MUC1 and PSA) or polyamines (spermine) are also responsible for impaired DC maturation and function (Aalamian et al. 2003; Della Bella et al. 2003; Monti et al. 2004). Interestingly, it has been demonstrated that whereas DC readily take up tumor antigens from epithelial tumors such as MUC1 and HER-2/neu, these glycoproteins inhibit their own transport to late endosomal compartments for processing and binding to class II MHC molecules (Hiltbold et al. 2000).

Similarly, prostate-specific antigen clearly inhibits the DC differentiation and maturation of DC in vitro (Aalamian et al. 2003). Although inhibition of DC maturation by serum from prostate cancer patients correlate with PSA levels (Aalamian-Matheis et al. 2007), the physiological significance of these findings has not yet been proven in vivo. Polyamines, in contrast, appear to play a relevant role in vivo. Putrescine, spermidine and spermine are essential for mammalian cell proliferation and differentiation. They have been implicated in impaired DC maturation and, more importantly, a significant negative correlation has been found between plasma levels of spermine and DC production of IL-12 in patients with breast cancer (Della Bella et al. 2003). These molecules have also been implicated in the inhibition of T-cell activity by immature suppressive cells in vivo (Bronte et al. 2003). Given that polyamines are nutrients constantly produced in the tumor microenvironment, these molecules could assist tumor evasion by supporting tumor growth and affect DC differentiation and T-cell function. More recently, a product of the cleavage of surface death receptor 6 (DR6) overexpressed in tumors has been described to interfere with DC development (Derosa et al. 2008).

7.5 Implications of Altered Dendritic Cell Differentiation

DC precursors and immature DC gain access to peripheral blood en route to peripheral tissues where they play an essential role in immune surveillance. However, the accumulation of immature cells associated with decreased number of circulating DC (Young et al. 1997; Almand et al. 2000, 2001) has deleterious effects for tumor surveillance. In patients with head and neck carcinoma, immature cells play a role in suppression of T-cell responses to recall antigens (Pak et al. 1995) and elevated tumor infiltration with immature cells correlates with increased rate of recurrence and metastases (Young et al. 1997). Immature cells have the capacity to inhibit MHC-I as well as MHC-II restricted T-cell responses (Almand et al. 2001). Although the mechanism of suppression has not been elucidated, a strict cell-to-cell contact with T cells is required to exert their suppressive activity (Almand et al. 2001). In mice, immature cells obtained from tumor-bearing hosts have been reported to express higher levels of reactive oxygen species (ROS) and accumulation of H_2O_2 and arginase activity appears to contribute to this. Inhibition of ROS abrogates the inhibitory effect, indicating that immature cells suppress T-cell responses via production of ROS (Kusmartsev et al. 2004). These data suggest that short-range immune mediators (i.e., ROS, H_2O_2 or arginase) probably present at lymphoid organs or at the tumor site on immature cells could play a crucial role in T-cell suppression.

Accumulation of immature cells displacing competent APC populations (i.e., DC) could also account for some immune suppression. We have reported that in patients with solid tumors immature DC represent a significant proportion (up to 65%) of the circulating DC compartment, thus impairing immune activation (Pinzon-Charry et al. 2005). Interestingly and supporting a role for tumor products, a close correlation between accumulation of immature cells and tumor burden is evident. It is tempting to speculate that while the systemic accumulation of immature cells could facilitate generalized immune dysfunction as a late event, immature APC present at the tumor site or lymphoid organs could play a role at an earlier phase in tumor progression. In fact, in patients with head and neck squamous cell carcinoma, tumor infiltration with immature cells has long been correlated with increased rate of recurrence and metastases (Young et al. 1997)

Tumors may also interfere with DC maturation, the final differentiation step whereby DC specialized in antigen capture are transformed into DC specialized in stimulation of T cells. Because immature DC fail to provide an appropriate costimulatory signal to T cells and tolerance or anergy may develop. Indeed, antigen presentation by immature DC has been reported to result in induction of tolerance through abortive proliferation or anergy of antigen-specific T cells in vivo (Probst et al. 2003). Immature DC can also induce tolerance through the generation of regulatory T cells that suppress immune responses by producing IL-10 and TGF-β (Dhodapkar et al. 2001). Moreover, immature DC with deficient expression of costimulatory molecules and poor capacity to stimulate T-cell responses have been reported in patients with basal cell, colorectal and breast cancer (Chaux et al. 1996; Gabrilovich et al. 1997) and DC from melanoma patients have an immature phenotype, reduced T-cell stimulatory capacity and induce anergy in syngeneic $CD4^+$ T cells in vitro (Enk et al. 1997).

Finally, DC-expressing indoleamine 2,3-dioxygenase (IDO) inhibit T-cell proliferation and promote T-cell death as a result of prostaglandin induction (Braun et al. 2005). Notably, large numbers of IDO-expressing DC can be found in tumor-draining lymph nodes, suggesting that they may be involved in the immunological unresponsiveness seen in cancer patients (Munn et al. 2002). Therefore, it is likely that similar mechanisms for induction of anergy occur in vivo at certain stages of tumor growth. As detailed elsewhere (Munn and Mellor 2007).

7.6 Tumor-Derived Factors and Dendritic Cell Apoptosis

Additional studies revealing other mechanisms utilized by tumors to evade effective immune responses have also been described. Induction of programmed cell death impairing the function of the key elements of the immune response like DC is one of such mechanisms. It has been demonstrated that DC undergo

Table 7.1 Tumor-derived factors responsible for DC dysfunction

Factor	Effect on DC	References
Cytokines		
IL-10	Impairment of differentiation, maturation and function in vitro and in vivo	Enk et al. (1993), Buelens et al. (1997), Allavena et al. (1998), Beckebaum et al. (2004)
	Increased apoptosis in vitro	Ludewig et al. (1996)
IL-6	Impairment of differentiation and maturation in vitro and in vivo	Menetrier-Caux et al. (1998), Ratta et al. (2002), Hayashi et al. (2003), Park et al. (2004)
M-CSF	Inhibition of differentiation from $CD34^+$ progenitors in vitro	Menetrier-Caux et al. (1998)
GM-CSF	Generation of immature APC with inhibitory role in vitro and in vivo	Pak et al. (1995), Garrity et al. (1997), Young et al. (1997), Bronte et al. (2000), Serafini et al. (2004)
VEGF	Alteration of differentiation of multiple lineages including DC in vitro and in vivo	Gabrilovich et al. (1998), Saito et al. (1998), Lissoni et al. (2001), Takahashi et al. (2004)
	Accumulation of inhibitory immature cells in vitro and in vivo	Almand et al. (2000), Almand et al. (2001), Kusmartsev et al. (2004)
Other mediators		
Gangliosides	Impairment of phenotypic and functional differentiation in vitro	Shurin et al. (2001), Peguet-Navarro et al. (2003)
	Phenotypic alteration and apoptosis in vitro	Kanto et al. (2001), Peguet-Navarro et al. (2003)
Prostanoids	Impairment of maturation and activity in vitro	Sombroek et al. (2002), Sharma et al. (2003), Yang et al. (2003)
Nitric oxide	Induction of apoptosis in vitro	Bonham et al. (1996)
Hyaluronan	Induction of apoptosis through induction of NO in vitro	Yang et al. (2002)
Polyamines	Induction of altered maturation in vitro	Della Bella et al. (2003)
DR6	Induction of apoptosis in vitro	Derosa et al. (2008)
Tumor antigens		
MUC1	Impairment of maturation and function in vitro	Hiltbold et al. (2000), Monti et al. (2004)
PSA	Alteration of differentiation and maturation in vitro	Aalamian et al. (2003)
Her/*2neu*		Hiltbold et al. (2000)

apoptosis in vitro and in vivo after interacting with cancer cells or tumor-derived factors (Esche et al. 1999; Pirtskhalaishvili et al. 2000a,b; Balkir et al. 2004) confirming that malignancies use this mechanism to impair appropriate immune responses. Indeed, tumor cells are known to express or release numerous pro-apoptotic factors such as IL-10, nitric oxide, gangliosides or ceramides that induce DC to undergo apoptosis and may explain why T cells fail to become fully activated to eradicate adjacent tumor cells (Table 7.1).

Apoptotic DC are ineffective at inducing immunity. It has been demonstrated that DC undergoing apoptosis deliver unusual activation during antigen presentation, leading to cellular unresponsiveness rather than effective immunity (Kitajima et al. 1996). Similarly, pre-apoptotic DC loaded with antigens show a marked decrease in their ability to induce antigen-specific immune responses in vivo (Colino et al. 2002). Finally, in breast cancer, significant apoptosis of blood DC has been reported (Pinzon-Charry et al. 2005). This is relevant because circulating DC are essential for adequate immunity given that they continually replenish the pool of tissue-residing DC and play a critical role in shaping immune responses in vivo. Most circulating DC appear to be en route from the bone marrow to peripheral and lymphoid tissues or from non-lymphoid tissues to the regional lymph nodes and spleen (de la Rosa et al. 2003; Villadangos and Heath 2005). Given that apoptotic cells are rapidly cleared from the circulation, the observation of a higher fraction of blood DC undergoing apoptosis in patients with breast cancer suggests increased turnover of these cells in vivo. Thus, continual efforts to replace the pool of blood DC from bone marrow would impose chronic stress on the immune system of breast cancer patients resulting in (i) relative paucity of DC in the circulation (Lissoni et al. 1999; Coventry et al. 2002) as well as (ii) failure to effectively replenish DC that infiltrate breast tumor tissue (Bell et al. 1999; Satthaporn et al. 2004) or (iii) migrate to lymphoid organs (Gabrilovich et al. 1997) for the initiation of T-cell immunity. Accordingly, in patients with operable breast carcinoma blood DC numbers are significantly reduced over prolonged periods of time suggesting diminished availability of DC precursors in cancer (Pinzon-Charry et al. 2007) .

7.7 Concluding Remarks

Overall, the evidence presented here emphasize the importance of DC in the elicitation of effective antitumor responses as well as the tumor-induced DC defects as a mechanism to escape immune surveillance. Tumors have been demonstrated to release numerous immunosuppressive factors that exert systemic effects on immune cell function and in particular, affect DC. The resulting dysfunction or apoptosis of mature DC, accumulation of immature DC or other immature cells with inhibitory functions result in a significant deficiency in the induction of antitumor responses. The data examined also suggest that tumor-induced DC dysfunction represents one of the crucial mechanisms underlying tumor immune evasion and indicates that tumor-induced suppression of DC has to be avoided if attempts to improve immune response in cancer are sought.

There is active interest in using DC and exploiting their distinctive immune functions as vectors for immune therapy of cancer. Nevertheless, given the heterogenous nature of DC dysfunction in cancer, multiple strategies would have to be considered. One approach could be the induction of differentiation or optimization of immature DC or immature cell function with differentiation agents or growth factors in vivo (Garrity et al. 1997; Almand et al. 2001). Another strategy would be the blockade of tumor-released factors that impair differentiation/function of DC in vivo (Gabrilovich et al. 1999). Finally, treatment of DC preparations (vaccines) with factors that increase their survival or resistance to apoptosis would be beneficial in the generation of antitumor immunity (Pinzon-Charry et al. 2006). Clear understanding of DC-related tumor evasion mechanisms will harness the potential utilization of DC as natural adjuvants in immunological therapy and assist the development of new methods to overcome the ineffective immunity against cancer (Lopez and Hart 2002).

Acknowledgements APC had been supported by University of Queensland and Paul Mackay Bolton Research Scholarships and is currently recipient of a NHMRC Biomedical Postdoctoral Fellowship ID: 443041. JAL is supported by the National Breast Cancer Foundation and Griffith University.

References

Aalamian-Matheis, M., Chatta, G. S., Shurin, M. R., Huland, E., Huland, H. and Shurin, G. V. 2007. Inhibition of dendritic cell generation and function by serum from prostate cancer patients: correlation with serum-free PSA. Adv Exp Med Biol 601:173–182.

Aalamian, M., Tourkova, I. L., Chatta, G. S., Lilja, H., Huland, E., Huland, H., Shurin, G. V. and Shurin, M. R. 2003. Inhibition of dendropoiesis by tumor derived and purified prostate specific antigen. J Urol 170:2026– 2030.

Allavena, P., Piemonti, L., Longoni, D., Bernasconi, S., Stoppacciaro, A., Ruco, L. and Mantovani, A. 1998. IL-10 prevents the differentiation of monocytes to dendritic cells but promotes their maturation to macrophages. Eur J Immunol 28:359–369.

Almand, B., Clark, J. I., Nikitina, E., van Beynen, J., English, N. R., Knight, S. C., Carbone, D. P. and Gabrilovich, D. I. 2001. Increased production of immature myeloid cells in cancer patients: a mechanism of immunosuppression in cancer. J Immunol 166:678–689.

Almand, B., Resser, J. R., Lindman, B., Nadaf, S., Clark, J. I., Kwon, E. D., Carbone, D. P. and Gabrilovich, D. I. 2000. Clinical significance of defective dendritic cell differentiation in cancer. Clin Cancer Res 6:1755–1766.

Asselin-Paturel, C., Echchakir, H., Carayol, G., Gay, F., Opolon, P., Grunenwald, D., Chouaib, S. and Mami-Chouaib, F. 1998. Quantitative analysis of Th1, Th2 and TGF-beta1 cytokine expression in tumor, TIL and PBL of non-small cell lung cancer patients. Int J Cancer 77:7–12.

Balkir, L., Tourkova, I. L., Makarenkova, V. P., Shurin, G. V., Robbins, P. D., Yin, X. M., Chatta, G. and Shurin, M. R. 2004. Comparative analysis of dendritic cells transduced with different anti-apoptotic molecules: sensitivity to tumor-induced apoptosis. J Gene Med 6:537–544.

Beckebaum, S., Zhang, X., Chen, X., Yu, Z., Frilling, A., Dworacki, G., Grosse-Wilde, H., Broelsch, C. E., Gerken, G. and Cicinnati, V. R. 2004. Increased levels of interleukin-10 in serum from patients with hepatocellular carcinoma correlate with profound numerical deficiencies and immature phenotype of circulating dendritic cell subsets. Clin Cancer Res 10:7260–7269.

Bell, D., Chomarat, P., Broyles, D., Netto, G., Harb, G. M., Lebecque, S., Valladeau, J., Davoust, J., Palucka, K. A. and Banchereau, J. 1999. In breast carcinoma tissue, immature dendritic cells reside within the tumor, whereas mature dendritic cells are located in peritumoral areas. J Exp Med 190:1417–1426.

Birkle, S., Zeng, G., Gao, L., Yu, R. K. and Aubry, J. 2003. Role of tumor-associated gangliosides in cancer progression. Biochimie 85:455–463.

Blay, J. Y., Negrier, S., Combaret, V., Attali, S., Goillot, E., Merrouche, Y., Mercatello, A., Ravault, A., Tourani, J. M., Moskovtchenko, J. F. and et al. 1992. Serum level of interleukin 6 as a prognosis factor in metastatic renal cell carcinoma. Cancer Res 52:3317–3322.

Bodey, B. 2002. Spontaneous regression of neoplasms: new possibilities for immunotherapy. Expert Opin Biol Ther 2:459–476.

Bonham, C. A., Lu, L., Li, Y., Hoffman, R. A., Simmons, R. L. and Thomson, A. W. 1996. Nitric oxide production by mouse bone marrow-derived dendritic cells: implications for the regulation of allogeneic T cell responses. Transplantation 62:1871–1877.

Braun, D., Longman, R. S. and Albert, M. L. 2005. A two-step induction of indoleamine 2,3 dioxygenase (IDO) activity during dendritic-cell maturation. Blood 106:2375–2381.

Bronte, V., Apolloni, E., Cabrelle, A., Ronca, R., Serafini, P., Zamboni, P., Restifo, N. P. and Zanovello, P. 2000. Identification of a CD11b(+)/Gr-1(+)/CD31(+) myeloid progenitor capable of activating or suppressing CD8(+) T cells. Blood 96:3838–3846.

Bronte, V., Chappell, D. B., Apolloni, E., Cabrelle, A., Wang, M., Hwu, P. and Restifo, N. P. 1999. Unopposed production of granulocyte-macrophage colony-stimulating factor by tumors inhibits CD8+ T cell responses by dysregulating antigen-presenting cell maturation. J Immunol 162:5728–5737.

Bronte, V., Serafini, P., De Santo, C., Marigo, I., Tosello, V., Mazzoni, A., Segal, D. M., Staib, C., Lowel, M., Sutter, G., Colombo, M. P. and Zanovello, P. 2003. IL-4-induced arginase 1 suppresses alloreactive T cells in tumor-bearing mice. J Immunol 170:270–278.

Buelens, C., Verhasselt, V., De Groote, D., Thielemans, K., Goldman, M. and Willems, F. 1997. Interleukin-10 prevents the generation of dendritic cells from human peripheral blood mononuclear cells cultured with interleukin-4 and granulocyte/macrophage-colony-stimulating factor. Eur J Immunol 27:756–762.

Caldwell, S., Heitger, A., Shen, W., Liu, Y., Taylor, B. and Ladisch, S. 2003. Mechanisms of ganglioside inhibition of APC function. J Immunol 171:1676–1683.

Chaux, P., Moutet, M., Faivre, J., Martin, F. and Martin, M. 1996. Inflammatory cells infiltrating human colorectal carcinomas express HLA class II but not B7-1 and B7-2 costimulatory molecules of the T-cell activation. Lab Invest 74:975–983.

Colino, J., Shen, Y. and Snapper, C. M. 2002. Dendritic cells pulsed with intact Streptococcus pneumoniae elicit both protein- and polysaccharide-specific immunoglobulin isotype responses in vivo through distinct mechanisms. J Exp Med 195:1–13.

Coventry, B. J., Lee, P. L., Gibbs, D. and Hart, D. N. 2002. Dendritic cell density and activation status in human breast cancer – CD1a, CMRF-44, CMRF-56 and CD-83 expression. Br J Cancer 86:546–551.

de la Rosa, G., Longo, N., Rodriguez-Fernandez, J. L., Puig-Kroger, A., Pineda, A., Corbi, A. L. and Sanchez-Mateos, P. 2003. Migration of human blood dendritic cells across endothelial cell monolayers: adhesion molecules and chemokines involved in subset-specific transmigration. J Leukoc Biol 73:639–649.

Deehan, D. J., Heys, S. D., Simpson, W. G., Broom, J., Franks, C. and Eremin, O. 1994. In vivo cytokine production and recombinant interleukin 2 immunotherapy: an insight into the possible mechanisms underlying clinical responses. Br J Cancer 69:1130–1135.

Della Bella, S., Gennaro, M., Vaccari, M., Ferraris, C., Nicola, S., Riva, A., Clerici, M., Greco, M. and Villa, M. L. 2003. Altered maturation of peripheral blood dendritic cells in patients with breast cancer. Br J Cancer 89:1463–1472.

Denkert, C., Kobel, M., Berger, S., Siegert, A., Leclere, A., Trefzer, U. and Hauptmann, S. 2001. Expression of cyclooxygenase 2 in human malignant melanoma. Cancer Res 61:303–308.

Derosa, D. C., Ryan, P. J., Okragly, A., Witcher, D. R. and Benschop, R. J. 2008. Tumor-derived death receptor 6 modulates dendritic cell development. Cancer Immunol Immunother 57:777–787.

Dhodapkar, M. V., Steinman, R. M., Krasovsky, J., Munz, C. and Bhardwaj, N. 2001. Antigen-specific inhibition of effector T cell function in humans after injection of immature dendritic cells. J Exp Med 193:233–238.

Enk, A. H., Angeloni, V. L., Udey, M. C. and Katz, S. I. 1993. Inhibition of Langerhans cell antigen-presenting function by IL-10. A role for IL-10 in induction of tolerance. J Immunol 151:2390–2398.

Enk, A. H., Jonuleit, H., Saloga, J. and Knop, J. 1997. Dendritic cells as mediators of tumor-induced tolerance in metastatic melanoma. Int J Cancer 73:309–316.

Esche, C., Lokshin, A., Shurin, G. V., Gastman, B. R., Rabinowich, H., Watkins, S. C., Lotze, M. T. and Shurin, M. R. 1999. Tumor's other immune targets: dendritic cells. J Leukoc Biol 66:336–344.

Fricke, I., Mirza, N., Dupont, J., Lockhart, C., Jackson, A., Lee, J. H., Sosman, J. A. and Gabrilovich, D. I. 2007. Vascular endothelial growth factor-trap overcomes defects in dendritic cell differentiation but does not improve antigen-specific immune responses. Clin Cancer Res 13:4840–4848.

Gabrilovich, D. 2004. Mechanisms and functional significance of tumour-induced dendritic-cell defects. Nat Rev Immunol 4:941–952.

Gabrilovich, D., Ishida, T., Oyama, T., Ran, S., Kravtsov, V., Nadaf, S. and Carbone, D. P. 1998. Vascular endothelial growth factor inhibits the development of dendritic cells and dramatically affects the differentiation of multiple hematopoietic lineages in vivo. Blood 92:4150–4166.

Gabrilovich, D. I., Corak, J., Ciernik, I. F., Kavanaugh, D. and Carbone, D. P. 1997. Decreased antigen presentation by dendritic cells in patients with breast cancer. Clin Cancer Res 3:483–490.

Gabrilovich, D. I., Ishida, T., Nadaf, S., Ohm, J. E. and Carbone, D. P. 1999. Antibodies to vascular endothelial growth factor enhance the efficacy of cancer immunotherapy by improving endogenous dendritic cell function. Clin Cancer Res 5:2963–2970.

Garrity, T., Pandit, R., Wright, M. A., Benefield, J., Keni, S. and Young, M. R. 1997. Increased presence of CD34+ cells in the peripheral blood of head and neck cancer patients and their differentiation into dendritic cells. Int J Cancer 73:663–669.

Gately, S. and Li, W. W. 2004. Multiple roles of COX-2 in tumor angiogenesis: a target for antiangiogenic therapy. Semin Oncol 31:2–11.

Gu, Z. J., Costes, V., Lu, Z. Y., Zhang, X. G., Pitard, V., Moreau, J. F., Bataille, R., Wijdenes, J., Rossi, J. F. and Klein, B. 1996. Interleukin-10 is a growth factor for human myeloma cells by induction of an oncostatin M autocrine loop. Blood 88:3972–3986.

Hayashi, T., Hideshima, T., Akiyama, M., Raje, N., Richardson, P., Chauhan, D. and Anderson, K. C. 2003. Ex vivo induction of multiple myeloma-specific cytotoxic T lymphocytes. Blood 102:1435–1442.

Hiltbold, E. M., Vlad, A. M., Ciborowski, P., Watkins, S. C. and Finn, O. J. 2000. The mechanism of unresponsiveness to circulating tumor antigen MUC1 is a block in intracellular sorting and processing by dendritic cells. J Immunol 165:3730–3741.

Kanto, T., Kalinski, P., Hunter, O. C., Lotze, M. T. and Amoscato, A. A. 2001. Ceramide mediates tumor-induced dendritic cell apoptosis. J Immunol 167:3773–3784.

Kim, J., Modlin, R. L., Moy, R. L., Dubinett, S. M., McHugh, T., Nickoloff, B. J. and Uyemura, K. 1995. IL-10 production in cutaneous basal and squamous cell carcinomas. A mechanism for evading the local T cell immune response. J Immunol 155:2240–2247.

Kitajima, T., Ariizumi, K., Bergstresser, P. R. and Takashima, A. 1996. Ultraviolet B radiation sensitizes a murine epidermal dendritic cell line (XS52) to undergo apoptosis upon antigen presentation to T cells. J Immunol 157:3312–3316.

Kruger-Krasagakes, S., Krasagakis, K., Garbe, C., Schmitt, E., Huls, C., Blankenstein, T. and Diamantstein, T. 1994. Expression of interleukin 10 in human melanoma. Br J Cancer 70:1182–1185.

Kusmartsev, S., Nefedova, Y., Yoder, D. and Gabrilovich, D. I. 2004. Antigen-specific inhibition of CD8+ T cell response by immature myeloid cells in cancer is mediated by reactive oxygen species. J Immunol 172:989–999.

Ladisch, S., Wu, Z. L., Feig, S., Ulsh, L., Schwartz, E., Floutsis, G., Wiley, F., Lenarsky, C. and Seeger, R. 1987. Shedding of GD2 ganglioside by human neuroblastoma. Int J Cancer 39:73–76.

Lissoni, P., Malugani, F., Bonfanti, A., Bucovec, R., Secondino, S., Brivio, F., Ferrari-Bravo, A., Ferrante, R., Vigore, L., Rovelli, F., Mandala, M., Viviani, S., Fumagalli, L. and Gardani, G. S. 2001. Abnormally enhanced blood concentrations of vascular endothelial growth factor (VEGF) in metastatic cancer patients and their relation to circulating dendritic cells, IL-12 and endothelin-1. J Biol Regul Homeost Agents 15:140–144.

Lissoni, P., Vigore, L., Ferranti, R., Bukovec, R., Meregalli, S., Mandala, M., Barni, S., Tancini, G., Fumagalli, L. and Giani, L. 1999. Circulating dendritic cells in early and advanced cancer patients: diminished percent in the metastatic disease. J Biol Regul Homeost Agents 13:216–219.

Lopez, J. A. and Hart, D. N. 2002. Current issues in dendritic cell cancer immunotherapy. Curr Opin Mol Ther 4:54–63.

Ludewig, B., Henn, V., Schroder, J. M., Graf, D. and Kroczek, R. A. 1996. Induction, regulation, and function of soluble TRAP (CD40 ligand) during interaction of primary CD4+ CD45RA+ T cells with dendritic cells. Eur J Immunol 26:3137– 3143.

Menetrier-Caux, C., Montmain, G., Dieu, M. C., Bain, C., Favrot, M. C., Caux, C. and Blay, J. Y. 1998. Inhibition of the differentiation of dendritic cells from CD34(+) progenitors by tumor cells: role of interleukin-6 and macrophage colony-stimulating factor. Blood 92:4778–4791.

Monti, P., Leone, B. E., Zerbi, A., Balzano, G., Cainarca, S., Sordi, V., Pontillo, M., Mercalli, A., Di Carlo, V., Allavena, P. and Piemonti, L. 2004. Tumor-derived MUC1 mucins interact with differentiating monocytes and induce IL-10highIL-12low regulatory dendritic cell. J Immunol 172:7341–7349.

Munn, D. H. and Mellor, A. L. 2007. Indoleamine 2,3-dioxygenase and tumor-induced tolerance. J Clin Invest 117:1147–1154.

Munn, D. H., Sharma, M. D., Lee, J. R., Jhaver, K. G., Johnson, T. S., Keskin, D. B., Marshall, B., Chandler, P., Antonia, S. J., Burgess, R., Slingluff, C. L., Jr. and Mellor, A. L. 2002. Potential regulatory function of human dendritic cells expressing indoleamine 2,3-dioxygenase. Science 297:1867–1870.

Pak, A. S., Wright, M. A., Matthews, J. P., Collins, S. L., Petruzzelli, G. J. and Young, M. R. 1995. Mechanisms of immune suppression in patients with head and neck cancer: presence of CD34(+) cells which suppress immune functions within cancers that secrete granulocyte-macrophage colony-stimulating factor. Clin Cancer Res 1:95–103.

Park, S. J., Nakagawa, T., Kitamura, H., Atsumi, T., Kamon, H., Sawa, S., Kamimura, D., Ueda, N., Iwakura, Y., Ishihara, K., Murakami, M. and Hirano, T. 2004. IL-6 regulates in vivo dendritic cell differentiation through STAT3 activation. J Immunol 173:3844–3854.

Peguet-Navarro, J., Sportouch, M., Popa, I., Berthier, O., Schmitt, D. and Portoukalian, J. 2003. Gangliosides from human melanoma tumors impair dendritic cell differentiation from monocytes and induce their apoptosis. J Immunol 170:3488–3494.

Pekarek, L. A., Starr, B. A., Toledano, A. Y. and Schreiber, H. 1995. Inhibition of tumor growth by elimination of granulocytes. J Exp Med 181:435–440.
Pinzon-Charry, A., Ho, C. S., Laherty, R., Maxwell, T., Walker, D., Gardiner, R. A., O'Connor, L., Pyke, C., Schmidt, C., Furnival, C. and Lopez, J. A. 2005. A population of HLA-DR+ immature cells accumulates in the blood dendritic cell compartment of patients with different types of cancer. Neoplasia 7:1112–1122.
Pinzon-Charry, A., Ho, C. S., Maxwell, T., McGuckin, M. A., Schmidt, C., Furnival, C., Pyke, C. M. and Lopez, J. A. 2007. Numerical and functional defects of blood dendritic cells in early- and late-stage breast cancer. Br J Cancer 97:1251–1259.
Pinzon-Charry, A., Schmidt, C. and Lopez, J. A. 2006. Dendritic cell immunotherapy for breast cancer. Expert Opin Biol Ther 6:591–604.
Pirtskhalaishvili, G., Shurin, G. V., Esche, C., Cai, Q., Salup, R. R., Bykovskaia, S. N., Lotze, M. T. and Shurin, M. R. 2000a. Cytokine-mediated protection of human dendritic cells from prostate cancer-induced apoptosis is regulated by the Bcl-2 family of proteins. Br J Cancer 83:506–513.
Pirtskhalaishvili, G., Shurin, G. V., Gambotto, A., Esche, C., Wahl, M., Yurkovetsky, Z. R., Robbins, P. D. and Shurin, M. R. 2000b. Transduction of dendritic cells with Bcl-xL increases their resistance to prostate cancer-induced apoptosis and antitumor effect in mice. J Immunol 165:1956–1964.
Portoukalian, J., David, M. J., Gain, P. and Richard, M. 1993. Shedding of GD2 ganglioside in patients with retinoblastoma. Int J Cancer 53:948–951.
Probst, H. C., Lagnel, J., Kollias, G. and van den Broek, M. 2003. Inducible transgenic mice reveal resting dendritic cells as potent inducers of CD8+ T cell tolerance. Immunity 18:713–720.
Ratta, M., Fagnoni, F., Curti, A., Vescovini, R., Sansoni, P., Oliviero, B., Fogli, M., Ferri, E., Della Cuna, G. R., Tura, S., Baccarani, M. and Lemoli, R. M. 2002. Dendritic cells are functionally defective in multiple myeloma: the role of interleukin-6. Blood 100:230–237.
Saito, H., Tsujitani, S., Ikeguchi, M., Maeta, M. and Kaibara, N. 1998. Relationship between the expression of vascular endothelial growth factor and the density of dendritic cells in gastric adenocarcinoma tissue. Br J Cancer 78:1573–1577.
Saito, T., Dworacki, G., Gooding, W., Lotze, M. T. and Whiteside, T. L. 2000. Spontaneous apoptosis of CD8+ T lymphocytes in peripheral blood of patients with advanced melanoma. Clin Cancer Res 6:1351–1364.
Satthaporn, S., Robins, A., Vassanasiri, W., El-Sheemy, M., Jibril, J. A., Clark, D., Valerio, D. and Eremin, O. 2004. Dendritic cells are dysfunctional in patients with operable breast cancer. Cancer Immunol Immunother 53:510–518.
Serafini, P., Carbley, R., Noonan, K. A., Tan, G., Bronte, V. and Borrello, I. 2004. High-dose granulocyte-macrophage colony-stimulating factor-producing vaccines impair the immune response through the recruitment of myeloid suppressor cells. Cancer Res 64:6337–6343.
Sharma, S., Stolina, M., Yang, S. C., Baratelli, F., Lin, J. F., Atianzar, K., Luo, J., Zhu, L., Lin, Y., Huang, M., Dohadwala, M., Batra, R. K. and Dubinett, S. M. 2003. Tumor cyclooxygenase 2-dependent suppression of dendritic cell function. Clin Cancer Res 9:961–968.
Shurin, G. V., Shurin, M. R., Bykovskaia, S., Shogan, J., Lotze, M. T. and Barksdale, E. M., Jr. 2001. Neuroblastoma-derived gangliosides inhibit dendritic cell generation and function. Cancer Res 61:363–369.
Sietsma, H., Nijhof, W., Dontje, B., Vellenga, E., Kamps, W. A. and Kok, J. W. 1998. Inhibition of hemopoiesis in vitro by neuroblastoma-derived gangliosides. Cancer Res 58:4840–4844.
Smith, D. R., Kunkel, S. L., Burdick, M. D., Wilke, C. A., Orringer, M. B., Whyte, R. I. and Strieter, R. M. 1994. Production of interleukin-10 by human bronchogenic carcinoma. Am J Pathol 145:18–25.

Sombroek, C. C., Stam, A. G., Masterson, A. J., Lougheed, S. M., Schakel, M. J., Meijer, C. J., Pinedo, H. M., van den Eertwegh, A. J., Scheper, R. J. and de Gruijl, T. D. 2002. Prostanoids play a major role in the primary tumor-induced inhibition of dendritic cell differentiation. J Immunol 168:4333–4343.

Staveley-O'Carroll, K., Sotomayor, E., Montgomery, J., Borrello, I., Hwang, L., Fein, S., Pardoll, D. and Levitsky, H. 1998. Induction of antigen-specific T cell anergy: An early event in the course of tumor progression. Proc Natl Acad Sci USA 95:1178–1183.

Steinbrink, K., Jonuleit, H., Muller, G., Schuler, G., Knop, J. and Enk, A. H. 1999. Interleukin-10-treated human dendritic cells induce a melanoma-antigen-specific anergy in CD8(+) T cells resulting in a failure to lyse tumor cells. Blood 93:1634–1642.

Takahashi, A., Kono, K., Ichihara, F., Sugai, H., Fujii, H. and Matsumoto, Y. 2004. Vascular endothelial growth factor inhibits maturation of dendritic cells induced by lipopolysaccharide, but not by proinflammatory cytokines. Cancer Immunol Immunother 53:543–550.

Tartour, E., Blay, J. Y., Dorval, T., Escudier, B., Mosseri, V., Douillard, J. Y., Deneux, L., Gorin, I., Negrier, S., Mathiot, C., Pouillart, P. and Fridman, W. H. 1996. Predictors of clinical response to interleukin-2–based immunotherapy in melanoma patients: a French multiinstitutional study. J Clin Oncol 14:1697–1703.

Toi, M., Taniguchi, T., Yamamoto, Y., Kurisaki, T., Suzuki, H. and Tominaga, T. 1996. Clinical significance of the determination of angiogenic factors. Eur J Cancer 32A:2513–2519.

Tourkova, I. L., Shurin, G. V., Chatta, G. S., Perez, L., Finke, J., Whiteside, T. L., Ferrone, S. and Shurin, M. R. 2005. Restoration by IL-15 of MHC class I antigen-processing machinery in human dendritic cells inhibited by tumor-derived gangliosides. J Immunol 175:3045–3052.

Tsuchiya, Y., Igarashi, M., Suzuki, R. and Kumagai, K. 1988. Production of colony-stimulating factor by tumor cells and the factor-mediated induction of suppressor cells. J Immunol 141:699–708.

Tsujii, M., Kawano, S. and DuBois, R. N. 1997. Cyclooxygenase-2 expression in human colon cancer cells increases metastatic potential. Proc Natl Acad Sci USA 94:3336–3340.

Vence, L., Palucka, A. K., Fay, J. W., Ito, T., Liu, Y. J., Banchereau, J. and Ueno, H. 2007. Circulating tumor antigen-specific regulatory T cells in patients with metastatic melanoma. Proc Natl Acad Sci USA 104:20884–20889.

Villadangos, J. A. and Heath, W. R. 2005. Life cycle, migration and antigen presenting functions of spleen and lymph node dendritic cells: limitations of the Langerhans cells paradigm. Semin Immunol 17:262–272.

Yang, L., Yamagata, N., Yadav, R., Brandon, S., Courtney, R. L., Morrow, J. D., Shyr, Y., Boothby, M., Joyce, S., Carbone, D. P. and Breyer, R. M. 2003. Cancer-associated immunodeficiency and dendritic cell abnormalities mediated by the prostaglandin EP2 receptor. J Clin Invest 111:727–735.

Yang, T., Witham, T. F., Villa, L., Erff, M., Attanucci, J., Watkins, S., Kondziolka, D., Okada, H., Pollack, I. F. and Chambers, W. H. 2002. Glioma-associated hyaluronan induces apoptosis in dendritic cells via inducible nitric oxide synthase: implications for the use of dendritic cells for therapy of gliomas. Cancer Res 62:2583–2591.

Young, M. R., Wright, M. A., Lozano, Y., Prechel, M. M., Benefield, J., Leonetti, J. P., Collins, S. L. and Petruzzelli, G. J. 1997. Increased recurrence and metastasis in patients whose primary head and neck squamous cell carcinomas secreted granulocyte-macrophage colony-stimulating factor and contained CD34+ natural suppressor cells. Int J Cancer 74:69–74.

Chapter 8
Tumor-Derived Exosomes as Dendritic Cell Modulators

Roberta Valenti, Veronica Huber, Paola Filipazzi, Manuela Iero, Giorgio Parmiani, and Licia Rivoltini

Abstract Cancer cells constitutively release endosome-derived microvesicles, also called 'exosomes', carrying a broad array of molecular determinants involved in the remodeling of the peritumoral microenvironment. This recently defined alternative mechanism of intercellular communication is exploited by tumor cells to favor their own growth and survival through the delivery of detrimental signals to the host's innate and adaptive immune system. Initially described for their ability to transfer tumor antigens to dendritic cells in a protected and highly immunogenic membrane-embedded form, tumor-derived exosomes have been more recently hinted to exert immunosuppressive effects on the development of antitumor immune responses at different levels. In particular, due to the transport of FasL and TRAIL pro-apoptotic molecules, exosomes derived from different tumor histotypes proved to induce programmed cell death of activated antitumor-specific T cells. On the other hand, the same microvesicles seem to mine immune-mediated recognition and elimination of cancer cells since their initial stages, regarding antigen uptake and presentation by dendritic cells. As reported herein, cancer patients display several phenotypic and functional defects in this cell subset, together with a more generalized dysfunction of the myeloid cell compartment, due to the tumor-driven expansion and activation of the so-called 'myeloid suppressor cells'. A possible involvement of tumor-derived exosomes in the disruption of the homeostasis of the antigen-presenting cell compartment in cancer patients has been recently suggested by a series of experimental evidences, as it will be mainly discussed in this chapter.

L. Rivoltini (✉)
Unit of Immunotherapy of Human Tumors, Istituto Nazionale Tumori, Via Venezian 1, 20133 Milano, Italy
e-mail: Licia.Rivoltini@istitutotumori.mi.it

M.R. Shurin, R.D. Salter (eds.), *Dendritic Cells in Cancer*,
DOI 10.1007/978-0-387-88611-4_8,

8.1 Introduction

As dendritic cells (DC) exert a pivotal role in the induction of antitumor immune responses, they represent a preferential target of the mechanisms developed by cancer cells to escape immune recognition. In particular, tumor cells may affect DC differentiation, recruitment and function through many different strategies, including direct and indirect interactions. Besides the mechanisms requiring cell-to-cell contact or the release of soluble mediators, an alternative pathway of intercommunication between growing tumors and the antigen-presenting cell compartment has been recently described, consisting in the release of membrane vesicles, also known as 'exosomes' (Fevrier and Raposo 2004). First described for their supposed ability to facilitate antigen cross-presentation through the transport of HLA/peptide complexes (Wolfers et al. 2001), tumor-derived exosomes are now considered also as potential mediators of suppressive signals to the immune system (Valenti et al. 2007), and in particular to DC and their precursors (Valenti et al. 2006). In the following paragraphs, after a brief description of the expected immunostimulatory properties attributed to tumor-derived exosomes at their debut in cancer immunology, we will more extensively discuss the growing evidences indicating their involvement in the disruption of antigen-presenting cell function in tumor-bearing hosts.

8.2 Tumor-Derived Exosomes Immunophenotype

Microvesicles, better known as 'exosomes', are 50–200 nm-sized membrane vesicles originating from the endosomal compartment (as indicated by the transport of endo-lysosomal markers, such as CD63, CD81, CD82, LAMP-1/2) and eventually released in the extracellular milieu through active exocytosis (Thery et al. 2002). Exosome release is characteristic of virtually all cell types, which may apply to this specialized secretory system in both physiological and pathological conditions. In particular, tumor cells of different histotypes are known to exacerbate this mechanism, through the constitutive release of these subcellular particles, which are able to deliver their signals not only in the tumor microenvironment but also at systemic level, through their proven ability to circulate in the peripheral blood and other body fluids of cancer patients (Andre et al. 2002; Valenti et al. 2007).

Thanks to the possibility of purifying tumor-derived exosomes through serial centrifugations of tumor cell culture supernatants or body fluids (such as plasma, ascites or pleural effusions) from cancer patients, many proteomic analysis have been published so far (Hegmans et al. 2004; Mears et al. 2004), describing the protein content of these particles. In general, tumor-derived exosomes were described to carry a protein profile selectively reflecting that of the originating cell, including both membrane and transmembrane proteins,

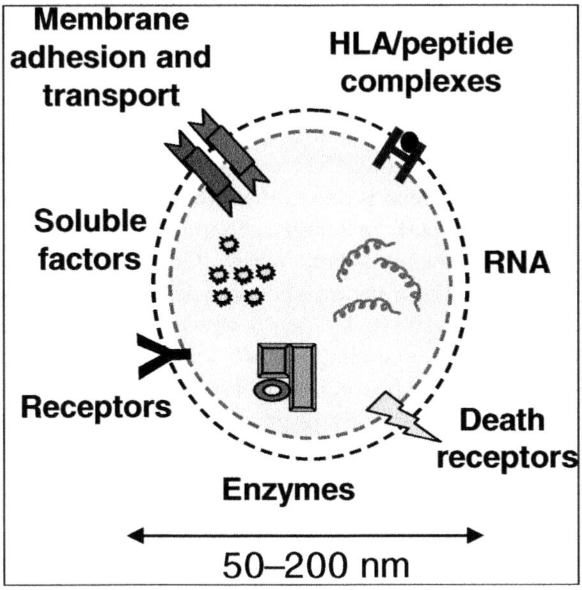

Fig. 8.1 Human tumor-derived exosome molecular content. Tumor-derived exosomes express a variety of molecular determinants involved in different biologic processes, such as adhesion, receptor discharge, signal transduction, antigen presentation, apoptosis and even transport nucleic acids

together with soluble molecules transported in their inner part, in a proteolytic enzyme-protected manner. As schematized in Fig. 8.1, the proteomic profiles of tumor-derived exosomes include molecular determinants involved in different cell functions. For instance, the presence of many adhesion molecules, such as integrins and annexins (Hegmans et al. 2004; Mears et al. 2004), and other cytoskeletal and membrane transport determinants (actin, ERM proteins and Rab proteins) underlies the strong fusogenic properties of exosomes, which is the basis for most of the functional features of these organelles.

The potential role of tumor exosomes in stimulating immune responses is actually rather controversial. The original works showed a potential stimulatory activity of tumor microvesicles on DC related to the transport of different tumor antigens, such as Mart-1, gp100, Her-2, CEA, together with HLA class I and II molecules, coupled with the transport of heat shock proteins, known to induce DC maturation and activation (Wolfers et al. 2001; Andre et al. 2002; Radons and Multhoff 2005). On the other hand, tumor-derived exosomes have been more recently described by several groups to consistently invalidate the ultimate outcomes of DC function, in terms of induction of apoptosis of antitumor-specific T cells, due to the transport on their surface of FasL and TRAIL pro-apoptotic molecules (Andreola et al. 2002; Huber et al. 2005; Kim et al. 2005; Taylor and Gercel-Taylor 2005).

Findings proving the ability of tumor-derived exosomes to transport bioactive cytokines and growth factors (such as TNF-α, IL-1β, TGF-β) and some functional receptors (such as TNFR1 and Tfr2) shed additional concerns and suggested an active role of these subcellular particles in modulating the tumor milieu (Calzolari et al. 2006; Clayton et al. 2007; Soderberg et al. 2007). Moreover, indirect evidences are suggesting the possible involvement of tumor-released microvesicles in neo-angiogenesis due to their ability to increase the secretion of metalloproteinase and VEGF in target cells, thanks to the transport of pro-angiogenic molecules of the tetraspanin family (Gesierich et al. 2006).

The involvement of these membrane particles in tumor progression and metastases is strongly hinted also by their newly described capacity to transport functional enzymes and even coding mRNA and microRNA (Hegmans et al. 2004; Valadi et al. 2007), advocating a novel scenario in which tumor cells may use exosome release to potentially re-program the whole cellular activity of autologous and heterologous target cells, turning their nearby and distant environment in a pro-tumorigenic surrounding.

8.3 Tumor-Derived Exosomes and the Induction of Dendritic Cell-Mediated Antigen Cross-Presentation

As mentioned above, the first publications on tumor-derived exosomes were mainly concerned with the study of their potential use in antigen presentation, as privileged carriers of antigens and immunostimulatory molecules to DC. When immunogenicity of tumor exosomes was tested in tumor transplanted animal models, they proved to prevent autologous cancer development in a CD4/CD8-dependent manner more efficiently than irradiated whole tumor cells (Zitvogel et al. 1998). In this setting, tumor-derived exosomes were able to induce antitumor responses leading to tumor rejection in both prophylaxis and therapy models, representing a potential source of shared tumor rejection antigens. These results could be easily justified by the rather artificial provision of tumor exosomes that were not in vivo produced by tumor cells, but were provided in a 'massive' quantity and intravenously. Subsequent work supporting the immunostimulatory role of tumor exosomes ex vivo studies was carried out using exosomes obtained from cancer patients and, in particular, isolated from ascitic fluids (Andre et al. 2002). We should mention that the exosome content of body fluids may be rather heterogeneous, including membrane vesicles released by immune cells themselves, which may account at least in part for the immunostimulatory activity of these preparations. Even if the first evidence collected in vitro were promising, the experimentation of this vaccination strategy in a human setting was rapidly supplanted by a safer and more controlled use of antigen-loaded DC-derived exosomes (Wieckowski and Whiteside 2006). In fact, while the vaccination of cancer patients with tumor-derived exosomes was discontinued, the feasibility of the use of DC-derived exosomes as a cancer vaccine was proved in Phase I clinical studies in melanoma and lung cancer patients and is currently under investigation (Escudier et al. 2005).

8.4 Tumor-Derived Exosomes and the Impairment of Dendritic Cell Differentiation

A critical finding suggesting a possible immunosuppressive role of tumor-derived exosomes consisted in demonstrating the presence of pro-apoptotic molecules, such as FasL and TRAIL, on their membranes (Andreola et al. 2002; Huber et al. 2005; Kim et al. 2005). As a consequence, exosomes obtained in vitro from many different tumor histotypes as well as from the plasma of advanced cancer patients proved to induce FasL- and TRAIL-mediated apoptosis of antitumor-specific effector T cells, thus undermining the effectiveness of antitumor immune responses at their final step.

However, FasL and TRAIL molecules were described to induce different effects on the antigen-presenting cell compartment: being immature, DC and their precursors intrinsically resistant to their pro-apoptotic effect, the same molecules may rather exert a pro-maturational effect on these cell subsets (Rescigno et al. 2000). In this view, FasL- and TRAIL-bearing tumor-derived exosomes were still expected to exert a stimulatory activity on the antigen-presenting cell compartment.

We initially tested this hypothesis by investigating the effect of melanoma-derived and colon carcinoma-derived exosomes on the in vitro differentiation of DC from their $CD14^+$ monocytes precursors (Valenti et al. 2006). Even if during the incubation of differentiating monocytes (i.e., cultured with IL-4 and GM-CSF) with tumor-derived exosomes no apoptosis was observed, as expected, the presence of these microvesicles exerted dramatic effects on DC differentiation. In particular, the resulting myeloid cell population was unable to differentiate into DC, instead remaining with a $CD14^+HLA\text{-}DR^{neg/low}$ phenotype. This was associated with a marked dysfunction in terms of lack of stimulatory effect on autologous T cells, which were inhibited in their ability to proliferate and release cytotoxic granules upon TCR activation. This was mainly ascribed to the significant changes induced by tumor-derived exosomes on the cytokine pattern released by differentiating monocytes. In fact, after few hours of incubation with the microvesicles, monocytes released relevant amount of inflammatory cytokine, such as TNF-α and IL-6, but also pro-angiogenic and immunosuppressive factors, such as IL-8, IL-10 and TGF-β. In blocking experiments, TGF-β further proved to be the main responsible of the suppressive activity exerted by exosome-treated monocytes on T cells.

Interestingly, myeloid cells with phenotype and functional characteristics reproducing those of monocyte-derived cells differentiated in vitro in the presence of tumor-derived exosomes were observed in the peripheral blood of melanoma patients (Filipazzi et al. 2007). While undetectable in healthy donors, a TGF-β-secreting $CD14^+HLA\text{-}DR^{neg}$ cell subset exerting inhibitory activity on autologous T-cell proliferation proved to be expanded in stage IV melanoma patients. Moreover, this newly identified myeloid suppressor cell subset seemed to be expanded upon GM-CSF-based vaccine treatment specifically in those patients who were unable to mount antitumor immune responses against the tumor, even upon vaccination. Even if the involvement of tumor-derived exosomes in this system can be demonstrated only indirectly, these data demonstrated for the first time the relevance of this immunosuppressive loop in melanoma patients and characterized a new immunosuppressive cell subset that, well defined in mouse models, has not been univocally described in the human setting.

While experiments were in progress with the aim of identifying the molecular determinants that in tumor-derived exosomes are responsible for this phenomenon, data on the role of these organelles in impairing DC differentiation were obtained in animal models (Yu et al. 2007). In fact, exosomes derived from a murine mammary cancer cell line induced an accumulation of myeloid precursors in the spleen which eventually proved to be unable to differentiate into DC ex vivo. In this setting, the activation of STAT-3 after a short-term incubation of DC precursors with the exosomes was observed and a possible role of IL-6 in mediating the inhibition of DC differentiation was suggested.

Tumor-released microvesicles may disrupt not only DC differentiation but also their maturation. At this purpose, we observed that monocyte-derived immature DC (obtained in vitro upon 5 days of culture with IL-4 and GM-CSF) exposed to melanoma-derived or colon carcinoma-derived exosomes acquired a mature phenotype, in terms of expression of HLA-DR, CD80, CD83 and CD86 surface molecules, but were unable to stimulate autologous T cells, probably due to their inability to secrete IL-12. In cross-presentation experiments, exosome-treated DC were unable to activate antigen-specific T cells in association with a marked down-regulation of the antigen-processing and presentation molecular machinery (Valenti R and Rivoltini L, unpublished observations).

Altogether these data imply a reconsideration of tumor-derived exosomes and their potential effect on antigen presentation in tumor-bearing hosts (Fig. 8.2). Indeed, exosomes still represent a privileged system for the delivery of membrane-embedded antigenic material to DC, but, due to the rather rich molecular milieu carried by exosomes in addition to tumor antigens, the outcome that this interaction can actually lead to in DC may be largely unpredictable. Furthermore, taking into account several studies that have demonstrated how tumor-derived exosomes further impair antitumor immune responses through the induction of apoptosis of effector T cells, the interaction of these subcellular entities with the host may result in a generalized immune suppression, through the undermining of tumor recognition from its initial to its final stages. In fact, the impairment of DC differentiation and function induced by tumor exosomes at different levels may result not only in the lack of activation of effector T cells but also in the indirect stimulation of regulatory T cells. In particular, the activation of this cell subset function seems to be at least in part associated with the interaction of regulatory T cells with myeloid suppressor cells through cell-to-cell contact or through the release of soluble

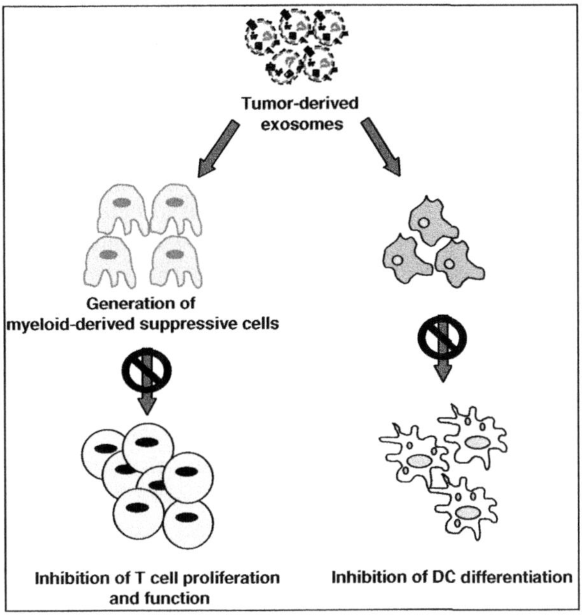

Fig. 8.2 Human tumor-derived exosomes effect on dendritic cell differentiation. Tumor-derived exosomes interfere with DC differentiation at different levels. In fact they not only impair monocyte differentiation into functional DC, but rather skew this process toward the generation of TGF-β-secreting myeloid cells exerting suppressive activity on effector T cells

mediators, such as TGF-β (Ghiringhelli et al. 2005). Direct evidence for the role of exosomes on this particular T-cell subset is not available yet, we can hypothesize that the immunosuppressive effect exerted by exosome-treated monocytes on T cells, may be at least in part due to the activation of regulatory T cells with the possible involvement of TGF-β released by differentiating monocytes in the presence of tumor-derived exosomes.

8.5 Driving the Interaction Between Tumor-Derived Exosomes and Dendritic Cells Toward Immune Stimulation: Is It Possible?

During the last few years intensive investigations were focused on the identification of pharmaceutical treatments able to down-regulate exosome release by tumor cells, mainly through a rather non-specific disruption of the cytoskeletal components that drive this pathway (Iero et al. 2008). However, since the ultimate goal of many DC-based immunotherapeutic strategies is the optimal stimulation of this cell subset in order to induce clinically relevant antitumor immune responses, an ideal approach should spare and exploit the immunogenic and pro-inflammatory characteristics of these vesicles simultaneously blunting their immunosuppressive potential, in order to drive the interaction between tumor-derived exosomes toward DC activation.

In this view, the molecular characterization of exosome protein content may lead to the identification of more specific strategies (such as the treatment with specific neutralizing antibodies) targeting the immunosuppressive baggage transported by tumor-derived exosomes, without affecting their immunogenic potential. Moreover, a deeper knowledge of the mechanisms regulating not only exosome release but also protein sorting to the exosome-mediated secretory pathway could lead to the possible quantitative but also qualitative control of this phenomenon. At this purpose, data about the possibility to influence exosome protein (and thus antigenic) content through the treatment of exosome-releasing cells with curcumin were recently published (Zhang et al. 2007).

Since the initial studies depicting exosomes as rather inert particles with a defined protein content randomly sorted from the cell of origin, evidence is now growing that these particles confer specific but versatile phenotype and functional characteristics to DC, and are able to affect many aspects of tumor/host interactions. Finally, only a deeper knowledge of these mechanisms will offer the opportunity to control and redirect exosome release to favor tumor recognition and rejection, with an optimal involvement of DC.

References

Andre, F., Schartz, N. E., Movassagh, M., Flament, C., Pautier, P., Morice, P., Pomel, C., Lhomme, C., Escudier, B., Le Chevalier, T., Tursz, T., Amigorena, S., Raposo, G., Angevin, E. and Zitvogel, L. 2002. Malignant effusions and immunogenic tumour-derived exosomes. Lancet 360:295–305.

Andreola, G., Rivoltini, L., Castelli, C., Huber, V., Perego, P., Deho, P., Squarcina, P., Accornero, P., Lozupone, F., Lugini, L., Stringaro, A., Molinari, A., Arancia, G., Gentile, M., Parmiani, G. and Fais, S. 2002. Induction of lymphocyte apoptosis by tumor cell secretion of FasL-bearing microvesicles. J Exp Med 195:1303–1316.

Calzolari, A., Raggi, C., Deaglio, S., Sposi, N. M., Stafsnes, M., Fecchi, K., Parolini, I., Malavasi, F., Peschle, C., Sargiacomo, M. and Testa, U. 2006. TfR2 localizes in lipid raft domains and is released in exosomes to activate signal transduction along the MAPK pathway. J Cell Sci 119:4486–4498.

Clayton, A., Mitchell, J. P., Court, J., Mason, M. D. and Tabi, Z. 2007. Human tumor-derived exosomes selectively impair lymphocyte responses to interleukin-2. Cancer Res 67:7458–7466.

Escudier, B., Dorval, T., Chaput, N., Andre, F., Caby, M. P., Novault, S., Flament, C., Leboulaire, C., Borg, C., Amigorena, S., Boccaccio, C., Bonnerot, C., Dhellin, O., Movassagh, M., Piperno, S., Robert, C., Serra, V., Valente, N., Le Pecq, J. B., Spatz, A., Lantz, O., Tursz, T., Angevin, E. and Zitvogel, L. 2005. Vaccination of metastatic melanoma patients with autologous dendritic cell (DC) derived-exosomes: results of thefirst phase I clinical trial. J Transl Med 3:10.

Fevrier, B. and Raposo, G. 2004. Exosomes: endosomal-derived vesicles shipping extracellular messages. Curr Opin Cell Biol 16:415–421.

Filipazzi, P., Valenti, R., Huber, V., Pilla, L., Canese, P., Iero, M., Castelli, C., Mariani, L., Parmiani, G. and Rivoltini, L. 2007. Identification of a new subset of myeloid suppressor

cells in peripheral blood of melanoma patients with modulation by a granulocyte-macrophage colony-stimulation factor-based antitumor vaccine. J Clin Oncol 25:2546–2553.
Gesierich, S., Berezovskiy, I., Ryschich, E. and Zoller, M. 2006. Systemic induction of the angiogenesis switch by the tetraspanin D6.1A/CO-029. Cancer Res 66:7083–7094.
Ghiringhelli, F., Puig, P. E., Roux, S., Parcellier, A., Schmitt, E., Solary, E., Kroemer, G., Martin, F., Chauffert, B. and Zitvogel, L. 2005. Tumor cells convert immature myeloid dendritic cells into TGF-beta-secreting cells inducing CD4+CD25+ regulatory T cell proliferation. J Exp Med 202:919–929.
Hegmans, J. P., Bard, M. P., Hemmes, A., Luider, T. M., Kleijmeer, M. J., Prins, J. B., Zitvogel, L., Burgers, S. A., Hoogsteden, H. C. and Lambrecht, B. N. 2004. Proteomic analysis of exosomes secreted by human mesothelioma cells. Am J Pathol 164:1807–1815.
Huber, V., Fais, S., Iero, M., Lugini, L., Canese, P., Squarcina, P., Zaccheddu, A., Colone, M., Arancia, G., Gentile, M., Seregni, E., Valenti, R., Ballabio, G., Belli, F., Leo, E., Parmiani, G. and Rivoltini, L. 2005. Human colorectal cancer cells induce T-cell death through release of proapoptotic microvesicles: role in immune escape. Gastroenterology 128:1796–1804.
Iero, M., Valenti, R., Huber, V., Filipazzi, P., Parmiani, G., Fais, S. and Rivoltini, L. 2008. Tumour-released exosomes and their implications in cancer immunity. Cell Death Differ 15:80–88.
Kim, J. W., Wieckowski, E., Taylor, D. D., Reichert, T. E., Watkins, S. and Whiteside, T. L. 2005. Fas ligand-positive membranous vesicles isolated from sera of patients with oral cancer induce apoptosis of activated T lymphocytes. Clin Cancer Res 11:1010–1020.
Mears, R., Craven, R. A., Hanrahan, S., Totty, N., Upton, C., Young, S. L., Patel, P., Selby, P. J. and Banks, R. E. 2004. Proteomic analysis of melanoma-derived exosomes by two-dimensional polyacrylamide gel electrophoresis and mass spectrometry. Proteomics 4:4019–4031.
Radons, J. and Multhoff, G. 2005. Immunostimulatory functions of membrane-bound and exported heat shock protein 70. Exerc Immunol Rev 11:17–33.
Rescigno, M., Piguet, V., Valzasina, B., Lens, S., Zubler, R., French, L., Kindler, V., Tschopp, J. and Ricciardi-Castagnoli, P. 2000. Fas engagement induces the maturation of dendritic cells (DCs), the release of interleukin (IL)-1beta, and the production of interferon gamma in the absence of IL-12 during DC-T cell cognate interaction: a new role for Fas ligand in inflammatory responses. J Exp Med 192:1661–1668.
Soderberg, A., Barral, A. M., Soderstrom, M., Sander, B. and Rosen, A. 2007. Redox-signaling transmitted in trans to neighboring cells by melanoma-derived TNF-containing exosomes. Free Radic Biol Med 43:90–99.
Taylor, D. D. and Gercel-Taylor, C. 2005. Tumour-derived exosomes and their role in cancer-associated T-cell signalling defects. Br J Cancer 92:305–311.
Thery, C., Zitvogel, L. and Amigorena, S. 2002. Exosomes: composition, biogenesis and function. Nat Rev Immunol 2:569–579.
Valadi, H., Ekstrom, K., Bossios, A., Sjostrand, M., Lee, J. J. and Lotvall, J. O. 2007. Exosome-mediated transfer of mRNAs and microRNAs is a novel mechanism of genetic exchange between cells. Nat Cell Biol 9:654–659.
Valenti, R., Huber, V., Filipazzi, P., Pilla, L., Sovena, G., Villa, A., Corbelli, A., Fais, S., Parmiani, G. and Rivoltini, L. 2006. Human tumor-released microvesicles promote the differentiation of myeloid cells with transforming growth factor-beta-mediated suppressive activity on T lymphocytes. Cancer Res 66:9290–9298.
Valenti, R., Huber, V., Iero, M., Filipazzi, P., Parmiani, G. and Rivoltini, L. 2007. Tumor-released microvesicles as vehicles of immunosuppression. Cancer Res 67:2912–2915.
Wieckowski, E. and Whiteside, T. L. 2006. Human tumor-derived vs dendritic cell-derived exosomes have distinct biologic roles and molecular profiles. Immunol Res 36:247–254.
Wolfers, J., Lozier, A., Raposo, G., Regnault, A., Thery, C., Masurier, C., Flament, C., Pouzieux, S., Faure, F., Tursz, T., Angevin, E., Amigorena, S. and Zitvogel, L. 2001.

Tumor-derived exosomes are a source of shared tumor rejection antigens for CTL cross-priming. Nat Med 7:297–303.

Yu, S., Liu, C., Su, K., Wang, J., Liu, Y., Zhang, L., Li, C., Cong, Y., Kimberly, R., Grizzle, W. E., Falkson, C. and Zhang, H. G. 2007. Tumor exosomes inhibit differentiation of bone marrow dendritic cells. J Immunol 178:6867–6875.

Zhang, H. G., Kim, H., Liu, C., Yu, S., Wang, J., Grizzle, W. E., Kimberly, R. P. and Barnes, S. 2007. Curcumin reverses breast tumor exosomes mediated immune suppression of NK cell tumor cytotoxicity. Biochim Biophys Acta 1773:1116–1123.

Zitvogel, L., Regnault, A., Lozier, A., Wolfers, J., Flament, C., Tenza, D., Ricciardi-Castagnoli, P., Raposo, G. and Amigorena, S. 1998. Eradication of established murine tumors using a novel cell-free vaccine: dendritic cell-derived exosomes. Nat Med 4:594–600.

Chapter 9
Signaling Pathways Mediating Dendritic Cell Dysfunction in Cancer

Jing Yang and Qing Yi

Abstract Dendritic cells are professional antigen-presenting cells that regulate the immune system. In cancers, they uptake tumor antigens, deliver them to T cells, and induce tumor-specific T-cell responses. However, tumor cells develop mechanisms to evade the immune system, partly by impairing dendritic cell differentiation, function and longevity. Accumulating evidence has indicated that intracellular signaling pathways, such as MAPK, JAK/STAT3, PI3K/Akt, and NF-κB, are critical to the regulation of dendritic cell differentiation, survival, and activity. However, these signaling pathways are found to be hyperactivated in both tumor cells and dendritic cells in malignancies. The constitutive activation of these pathways in cancer cells leads to tumor cell secretion of cytokines that activate intracellular signaling pathways, particularly p38 MAPK, in dendritic cells or their progenitor cells and impair dendritic cell differentiation and function. In this chapter we will discuss signaling pathways that mediate dendritic cell dysfunction. We will focus on the roles of MAPK, particularly p38 MAPK, in negatively regulating dendritic cell differentiation and function in cancers.

9.1 Introduction

Dendritic cells (DC) are populations of professional antigen-presenting cells that regulate the immune system (Santini and Belardelli 2003; Sheng et al. 2005; Evel-Kabler and Chen 2006). They originate from $CD34^+$ bone marrow stem cells and have high plasticity and common morphological and functional characteristics (Gabrilovich et al. 1996; Sheng et al. 2005). During their development, DC are classified as immature or mature cells. In the immature stage, DC are primarily localized in the peripheral tissues and perform specialized functions of antigen uptake and processing, in which they capture and carry

Q. Yi (✉)
Department of Lymphoma and Myeloma, M. D. Anderson Cancer Center, 1515 Holcombe Boulevard, Unit 0903, Houston, Texas 77030, USA
e-mail: qyi@mdanderson.org

M.R. Shurin, R.D. Salter (eds.), *Dendritic Cells in Cancer*,
DOI 10.1007/978-0-387-88611-4_9, © Springer Science+Business Media, LLC 2009

antigens to the lymph nodes. In the lymphoid organs, DC become mature and subsequently interact with antigen-specific T cells and initiate immune responses (Di Nicola and Lemoli 2000; Sinkovics and Horvath 2000; Aloysius et al. 2006).

DC are critically important for the stimulation of immune responses against tumor cells (Sinkovics and Horvath 2000; Gabrilovich 2002; Kusmartsev and Gabrilovich 2002). In this process, they are responsible for the uptake of tumor antigens for delivery to T cells and inducing tumor-specific T-cell responses. However, tumor cells release soluble factors to the microenvironment, impair DC differentiation and function, and induce immune tolerance to cancers (Di Nicola and Lemoli 2000; Kim et al. 2002; Yang et al. 2003). Thus far, the abnormalities of DC differentiation and function are considered one of the major factors limiting the success of cancer vaccines in clinical trials. Therefore, studies of the mechanisms of tumor-induced DC dysfunction may be a key point to improve antitumor immune responses in cancer patients (Santini and Belardelli 2003; Sheng et al. 2005; Rabinovich et al. 2007). Numerous recent studies have reported that tumor-induced activation of intracellular signaling pathways, such as mitogen-activated protein kinases (MAPK), JAK/STAT, and NF-κB, contributes to various defects of the immune system, particularly through compromising DC differentiation, function, and survival. Even though these signaling pathways are important for the development of normal hematopoietic cells, activation of these pathways is usually present in both tumor cells and abnormal DC to support tumor growth and survival (Nefedova et al. 2004; Philpott et al. 2004; Ade et al. 2007). In this chapter we will discuss signaling pathways mediating DC dysfunction. We will focus on the roles of MAPK, particularly p38 MAPK, in negatively regulating DC differentiation and function in cancers.

9.2 MAPK-Signaling Pathways

MAPK are proline-directed serine and threonine protein kinases, and are activated by dual-specificity kinases with phosphorylation of threonine and tyrosine in a Thr–Xaa–Tyr motif. Activation of MAPK-signaling pathways is through a MAPK-activating phosphorylation cascade, in which upstream kinases phosphorylate their downstream kinases on threonine and tyrosine residues, starting from MAPK kinase kinases (MAPKKK), to MAPK kinases (MAPKK), and finally to MAPK. The activated MAPK then interact with their cytoplasmic substrates and translocate into the nucleus, where they act as transcription factors and regulate target gene transcription (Nakamura et al. 1996; Ichijo 1999).

MAPK-signaling pathways are crucial for diverse cellular functions, including proliferation, differentiation, and apoptosis (Aplin et al. 2002; Budagian et al. 2003; Kawakami et al. 2003; Sigaud et al. 2005). There are three types of

MAPK, extracellular signal-regulated kinases (ERK), c-jun N-terminal kinases (JNK), and p38 MAPK, which are identified by the intervening amino acid. The ERK pathway, activated by polypeptide growth factors through their tyrosine kinase receptors, regulates cellular growth and survival. JNK- and p38-signaling pathways are activated by stress stimuli and inflammatory cytokines, and are involved in cellular differentiation, cytokine production, and apoptosis. MAPK-signaling pathways have been shown to be frequently activated in cancers, and may contribute to malignant phenotypes and uncontrolled cell growth. In addition, MAPK-signaling pathways are involved in the regulation of immune responses, including the initiation phase of innate immunity, activation of adaptive immunity, and cell death after completing immune function (Nakahara et al. 2004; Canesi et al. 2005; Kim et al. 2005; Zou and Hu 2005). Notably, recent studies have indicated that the MAPK-signaling pathways differentially regulate all aspects of DC phenotypic maturation, cytokine production, and DC functional development (Cruz et al. 1999; Nakahara et al. 2004; Xie et al. 2005; Wang et al. 2006b). Stimuli such as lipopolysaccharide (LPS), tumor-necrosis factor (TNF)-α, haptens, or ultraviolet-B (UVB) induce maturation of DC via MAPK-signaling pathways (Cruz et al. 1999; Nakahara et al. 2004; Tassiulas et al. 2007). On the other hand, tumor-induced abnormalities of DC differentiation and function are also associated with hyperactivation of MAPK-signaling pathways (Wang et al. 2006a,b).

9.3 p38 MAPK in Dendritic Cell Differentiation, Maturation, and Activity

There are four p38 MAPK: α and β, which are 75% homologous, and γ and δ, which are more distant relatives. All p38 MAPK can be activated by the same upstream MAP kinase kinases, such as MKK3 or MKK6, upon the stimulation of inflammatory cytokines or stress (Ichijo 1999; Lee et al. 2006). p38 MAPK signaling induces the activation of MAPK-activated protein kinase (MAPKAPK)-2 (Zaru et al. 2007), synthesis of TNF-α (Park et al. 1999; Lee et al. 2006), and phosphorylation of transcription factors such as activating-transcription-factor-2 (ATF-2), Elk-1, and SAP-1.

The p38 MAPK-signaling pathway is essential for normal DC maturation and activity (Ardeshna et al. 2000; Matos et al. 2005a,b; Xie et al. 2005; Osawa et al. 2006). LPS-induced maturation and up-regulation of surface antigens on DC such as CD40, CD80, CD83, CD86, and MHC class II molecules require p38 MAPK (West et al. 2004; Bharadwaj et al. 2005). The p38 MAPK inhibitor SB203580 abrogates the up-regulation of surface antigens in the process of DC maturation induced by LPS, $NiCl_2$, $NiSO_4$, and CD40L. Furthermore, LPS-induced DC secretion of cytokines such as TNF-α, IL-6, and IL-12 p40 and p70, also depends on the activation of p38 MAPK, because SB203580 has been shown to inhibit DC secretion of these cytokines (Geginat et al. 2003; Randolph et al. 2005; Lee et al. 2006; Saito et al. 2006). In addition, LPS-enhanced allostimulatory activity of DC is abrogated by SB203580 treatment, indicating that p38 MAPK are required for the endocytic and allostimulatory functions of DC (Kang et al. 2004).

However, we have shown that the importance of p38 MAPK-signaling pathways in DC is stage-dependent. While crucial for immature DC to mature and secrete cytokines, activation of p38 MAPK is detrimental to the generation and differentiation of DC from monocytes. During the differentiation of monocytes to immature DC, p38 MAPK activation induced by LPS impaired DC differentiation and p38 MAPK inhibitor SB203580 restored generation of functional DC in culture with LPS (Xie et al. 2003). Moreover, addition of SB203580 to cultures of normal monocytes accelerated the differentiation of the cells into immature DC (Xie et al. 2005). These results could be explained by the findings that inhibition of p38 MAPK enhances the phosphorylation of ERK and NF-κB activity and leads to enhanced up-regulation of expression of DC-related adhesion and costimulatory molecules and antigen-presentation capacity (Ardeshna et al. 2000; Xie et al. 2005; Lee et al. 2006; Osawa et al. 2006).

9.4 p38 MAPK in Tumor-Induced Dendritic Cell Dysfunction

DC from cancer patients are functionally defective; however, underlying molecular mechanisms are poorly understood at the present time. We have used the murine 5TGM1 myeloma model (Radl et al. 1988; Asosingh et al. 2000) to examine the effects and mechanism of tumor-derived factors on the differentiation and function of DC. Myeloma cells or tumor culture-conditioning medium (TCCM) were shown to inhibit differentiation and function of bone marrow-derived DC, as evidenced by the down-regulated expression of DC-related surface molecules, decreased IL-12 secretion, and compromised capacity of the cells to activate allospecific T cells. Moreover, TCCM-treated DC were inferior to normal DC at priming tumor-specific immune responses in vivo. Neutralizing antibodies against IL-6, IL-10, and TGF-β partially abrogated the effects. Our results showed that TCCM treatment activated p38 MAPK and JNK but inhibited ERK. Inhibiting p38 MAPK restored the phenotype, cytokine secretion, and function of TCCM-treated DC. DC from cultures with both TCCM and p38 inhibitor were as efficacious as normal DC at inducing tumor-specific antibody, type-1 T cell, and cytotoxic T lymphocyte responses, and prolonging mouse survival. Thus, our results suggest that tumor-induced p38 MAPK activation and ERK inhibition in DC may be a new mechanism for tumor evasion, and regulating these pathways during DC differentiation provides new strategies for generating potent DC vaccines for immunotherapy in cancer patients (Wang et al. 2006a, b).

Next, we examined whether the defects can be observed in DC from patients with myeloma. Previous studies have demonstrated that circulating DC in myeloma patients are functionally abnormal (Brown et al. 2001; Ratta et al. 2002). However, no study has been performed to examine monocyte-derived DC, which are commonly used for immunotherapy in patients. We found that patient-derived DC are phenotypically and functionally defective. Compared with their normal counterpart, patient-derived, mature DC expressed significantly lower levels of CD1a, CD40, CD80, and HLA-DR, and were poor at activating alloreactive T cells, presenting recall antigen, and activating autologous antigen-specific T cells. These abnormalities may be attributed to elevated production of autocrine cytokines such as IL-6, activated p38 MAPK and STAT3, and inhibited MEK/ERK-signaling pathways in the progenitor cells (Fig. 9.1). Treatment with neutralizing IL-6-specific antibody and more importantly, p38 MAPK inhibitor, or both, could correct these abnormalities. Treating patient-derived cells with these agents not only significantly increased cell yield but also produced DC that were as functional as their normal counterpart (Wang et al. 2006a). Thus, our studies have delineated the mechanistic defects of DC from myeloma patients, and identified ways for restoring the function of the cells to improve the efficacy of DC-based immunotherapy in this disease.

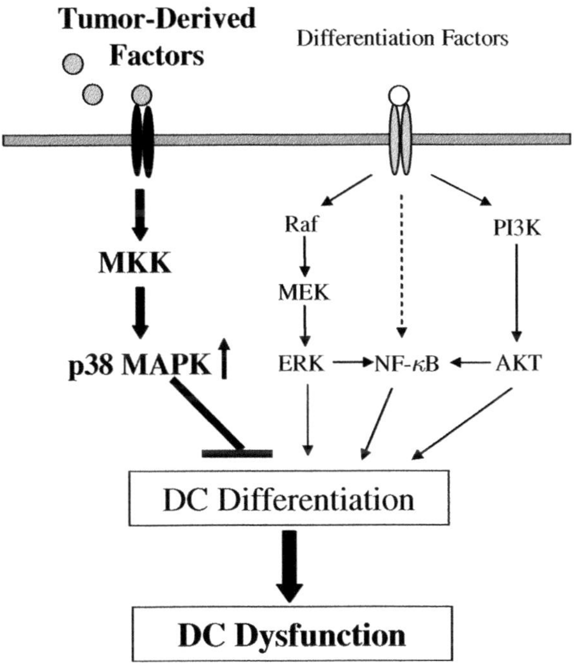

Fig. 9.1 Schematic representation of tumor-derived factor-mediated dendritic cell dysfunction by activating p38 MAPK-signaling pathway. During DC differentiation, cytokines bind to their receptors on DC precursors and activate Ras/Raf/ERK-, PI3K/Akt-, and NF-κB-signaling pathways. However, tumor-derived cytokines also activate p38 MAPK in the cells, which inhibit these pathways and impair DC differentiation

A recent study showed that inhibiting p38 MAPK signaling in DC attenuates regulatory T (Treg)-cell induction in response to Toll-like receptor (TLR) agonists and enhances their efficacy as vaccine adjuvants and cancer immunotherapeutics (Jarnicki et al. 2008). TLR ligands are commonly used adjuvants that promote type-1 T-cell responses against tumor antigens. However, TLR ligands also promoted the induction of IL-10-secreting Treg cells through p38 MAPK-induced IL-10 production by DC. Inhibition of p38 MAPK by SB203580 suppressed TLR-induced IL-10 and prostaglandin PgE_2 and enhanced IL-12 production in DC. Inhibition of p38 MAPK enhanced the antitumor therapeutic efficacy of DC pulsed with antigen and CpG, which was associated with an enhanced frequency of IFN-γ-secreting T cells and a reduction of $Foxp3^+$ Treg cells infiltrating the tumors. Taken together, these findings indicate that p38 is an important therapeutic target, and inhibiting p38 activity in DC obtained from cancer patients or DC pulsed with tumor antigens and TLR agonists will enhance the immunogenicity of the cells.

9.5 ERK and Dendritic Cell Dysfunction in Cancer

Recent studies have demonstrated that the ERK and p38 MAPK-signaling pathways differentially regulate DC maturation and modulate the initial commitment of naïve T helper cells toward Th1 or Th2 subsets (Aplin et al. 2002; Kandilci and Grosveld 2005; Lee et al. 2006). The p38 MAPK inhibitor SB203580 suppressed DC maturation, whereas the presence of ERK inhibitors PD98059 or U0126 enhanced LPS-induced phenotypic and functional maturation of DC, and increased the expression of MHC complex and costimulatory molecules. However, MAPK pathways, which are frequently activated in cancers, have active roles in immune evasion in cancer. By using ERK inhibitor U0126 and lentiviral $BRAF^{V600E}$ RNA interference, Sumimoto et al. (2006) demonstrated that the ERK-signaling pathway is essential for production of immunosuppressive factors by human melanoma cells that have constitutively activated ERK due to the $BRAF^{V600E}$ mutation, which can be detected in the majority of patients with melanomas (Sumimoto et al. 2004; Tanami et al. 2004). These findings indicate that ERK is another potential molecular target for overcoming cancer cell evasion of the immune system.

9.6 Role of JAK/STAT Signaling in Tumor-Induced Dendritic Cell Dysfunction

Over the past several years, investigators have been working on the JAK/STAT-signaling pathways in the context of cancer-mediated evasion of the immune system (Kortylewski et al. 2005; Nefedova et al. 2005b; Kim et al. 2006; Larmonier et al. 2006). JAK mutations and/or STAT abnormal activation are found in many types of cancers, such as myeloproliferative disorders with acquired JAK2 mutations (Taki and Taniwaki 2006; Jost et al. 2007; Mata et al. 2007), T-cell acute lymphoblastic leukemia (Taki and Taniwaki 2006), and leukemia or lymphoma with constitutive phosphorylation of JAK3, STAT1, STAT3, and STAT5 (Yu et al. 1997; Aboudola et al. 2007; Faderl et al. 2007). Among them, constitutive activation of STAT3 is common in a variety of lymphoid or myeloid malignancies and solid tumors, in human tumor cell lines and primary tumor cells from patients, and in virus-transformed cells (Yu et al. 1995; Campbell et al. 1997; Cheng et al. 2004; Park et al. 2005). Recent studies showed that hyperactivation of STAT3 is found in multiple myeloma, breast cancer, and prostate cancer (Wang et al. 2004). Interruption of STAT3 signaling in tumor cells reversed tumor-induced immunosuppression, resulting in enhanced ability of DC to present antigen, and consequently induce T-cell activation and break T-cell anergy (Wang et al. 2004). Hyperactivation of STAT3 in tumor cells has been shown to mediate immune evasion, particularly DC dysfunction.

Soluble factors released from tumor cells may prevent DC maturation. Tumor-derived proinflammatory factors such as IL-10 and VEGF inhibited DC maturation (Wang et al. 2004). Since JAK/STAT3-signaling pathway is a major signaling pathway that can be activated by cytokines binding to their membrane receptors (Li et al. 2007), tumor-derived factors inhibit DC differentiation and function via JAK/STAT3 activation. Accumulating evidence has shown that both IL-10 and VEGF released from tumor cells activate JAK/STAT3-signaling pathway in DC (Wang et al. 2004), and tumor-derived factors suppress myeloid cell differentiation by stimulating JAK/STAT3 signaling (Nefedova et al. 2004; Kusmartsev and Gabrilovich 2006). Nefedova (Nefedova et al. 2004; Nefedova et al. 2005a; Nefedova and Gabrilovich 2007) found that TCCM induced constitutive activation of JAK2 and STAT3, which prevented the differentiation of immature myeloid cells into mature DC. This negative regulation was abrogated after removal of TCCM. Inhibition of STAT3 abrogated the negative effects of the tumor-derived factors on myeloid cell differentiation. These observations suggest that activation of JAK/STAT-signaling pathways are negative regulators of DC differentiation and function in malignancies.

9.7 Other Signaling Pathways in Tumor-Induced Dendritic Cell Dysfunction

It is well known that activation of NF-κB plays an important role in DC maturation and function (Zou and Hu 2005; Osawa et al. 2006; Ade et al. 2007). JAK/STAT-, p38 MAPK-, and ERK-signaling pathways crosstalk with the NF-κB pathway, and factors activating STAT or MAPK also stimulate NF-κB, which includes members of p50, p52, RelA, RelB, and cRel. The proteins form active hetero- or homodimers, translocate to the nuclei, and initiate the transcription of target genes. NF-κB activity has been shown to be high in DC, and up-regulation of IL-12 expression requires activation of both p38 MAPK and NF-κB (Ade et al. 2007). Our previous studies showed that differentiation of immature DC is accompanied by increased NF-κB activity and that inhibiting p38 MAPK enhances the activity of NF-κB in immature DC (Wang et al. 2006a,b). Because high levels of NF-κB activity are frequently found in many types of cancers, NF-κB-signaling pathways may also contribute to tumor-induced DC dysfunction in cancer patients.

9.8 Conclusion

Hyperactivation of signaling pathways such as JAK/STAT, MAPK, and NF-κB, is not only important for tumor cell growth and survival but also critical for tumor-induced immunosuppression of tumor-bearing hosts, in which tumor cells escape from immune surveillance via deregulation of these signaling pathways. The activation of multiple signaling pathways in tumor cells mediates the expression and secretion of tumor-derived factors to the tumor microenvironment. Subsequently, these tumor-derived factors impair DC differentiation and impair their function, resulting in DC-mediated immune tolerance (Fig. 9.2).

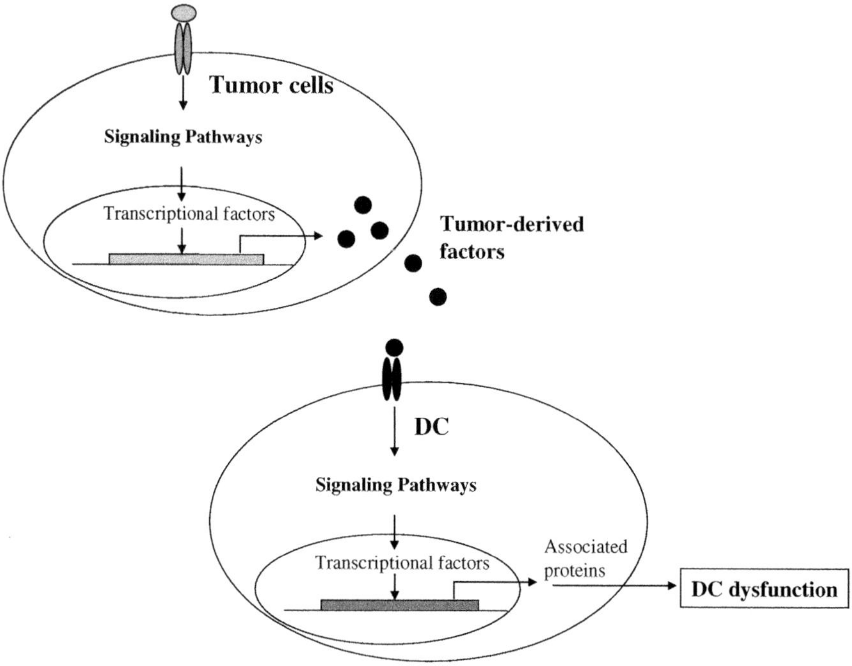

Fig. 9.2 A model for the roles of intracellular signaling pathways that mediate dendritic cell dysfunction in cancer. The *upper panel* shows activated intracellular signaling pathways, such as MAPK, JAK/STAT, and NF-κB, in tumor cells and secretion of tumor-derived factors into the tumor microenvironment. These tumor-derived factors activate intracellular signaling pathways in DC and impair their function, which partly contributes to immune suppression in cancer

Blockage of tumor-induced DC dysfunction by targeting signaling molecules or pathways may restore DC function.

Inhibitors to signaling molecules are already under investigation in clinical trials as therapeutic agents to treat cancers (McKay et al. 2000; Yoshikawa et al. 2001; Do et al. 2004; Chou et al. 2005; Jing et al. 2006; Barclay et al. 2007; Demuth et al. 2007; Jiang et al. 2007; Kirkwood et al. 2007). These antagonists as anti-cancer drugs are expected to not only improve DC function but also, and more importantly, may boost antitumor immunity in cancer patients. Even though these signaling pathways are pivotally important for normal cell proliferation and survival and blockade of them may possibly lead to toxicity in patients, some encouraging preliminary results have already been obtained from clinical trials testing the efficacy of the specific inhibitors. In future studies, it will be important to identify novel and specific targets in these signaling pathways for cancer therapy.

Acknowledgements This work was supported by institutional startup funds from the University of Texas M.D. Anderson Cancer Center, grants from the National Cancer Institute (R01 CA96569 and R01 CA103978), Multiple Myeloma Research Foundation, International Myeloma Foundation, and Commonwealth Foundation for Cancer Research. We thank Ms. Alison Woo for providing editorial assistance.

References

Aboudola, S., Murugesan, G., Szpurka, H., Ramsingh, G., Zhao, X., Prescott, N., Tubbs, R. R., Maciejewski, J. P. and Hsi, E. D. 2007. Bone marrow phospho-STAT5 expression in non-CML chronic myeloproliferative disorders correlates with JAK2 V617F mutation and provides evidence of in vivo JAK2 activation. American Journal of Surgical Pathology 31:233–239.

Ade, N., Antonios, D., Kerdine-Romer, S., Boisleve, F., Rousset, F. and Pallardy, M. 2007. NF-kappaB plays a major role in the maturation of human dendritic cells induced by NiSO(4) but not by DNCB. Toxicological Sciences 99:488–501.

Aloysius, M. M., Takhar, A., Robins, A. and Eremin, O. 2006. Dendritic cell biology, dysfunction and immunotherapy in gastrointestinal cancers. Surgeon Journal of the Royal Colleges of Surgeons of Edinburgh & Ireland 4:195–210.

Aplin, A. E., Hogan, B. P., Tomeu, J. and Juliano, R. L. 2002. Cell adhesion differentially regulates the nucleocytoplasmic distribution of active MAP kinases. Journal of Cell Science 115:2781–2790.

Ardeshna, K. M., Pizzey, A. R., Devereux, S. and Khwaja, A. 2000. The PI3 kinase, p38 SAP kinase, and NF-kappaB signal transduction pathways are involved in the survival and maturation of lipopolysaccharide-stimulated human monocyte-derived dendritic cells. Blood 96:1039–1046.

Asosingh, K., Radl, J., Van Riet, I., Van Camp, B. and Vanderkerken, K. 2000. The 5TMM series: a useful in vivo mouse model of human multiple myeloma. Hematol J 1:351–356.

Barclay, J. L., Anderson, S. T., Waters, M. J. and Curlewis, J. D. 2007. Characterization of the SOCS3 promoter response to prostaglandin E2 in T47D cells. Molecular Endocrinology 21:2516–2528.

Bharadwaj, U., Zhang, R., Yang, H., Li, M., Doan, L. X., Chen, C. and Yao, Q. 2005. Effects of cyclophilin A on myeloblastic cell line KG-1 derived dendritic like cells (DLC) through p38 MAP kinase activation. Journal of Surgical Research 127:29–38.

Brown, R. D., Pope, B., Murray, A., Esdale, W., Sze, D. M., Gibson, J., Ho, P. J., Hart, D. and Joshua, D. 2001. Dendritic cells from patients with myeloma are numerically normal but functionally defective as they fail to up-regulate CD80 (B7-1) expression after huCD40LT stimulation because of inhibition by transforming growth factor-beta1 and interleukin-10. Blood 98:2992–2998.

Budagian, V., Bulanova, E., Brovko, L., Orinska, Z., Fayad, R., Paus, R. and Bulfone-Paus, S. 2003. Signaling through P2X7 receptor in human T cells involves p56lck, MAP kinases, and transcription factors AP-1 and NF-kappa B. Journal of Biological Chemistry 278:1549–1560.

Campbell, G. S., Yu, C. L., Jove, R. and Carter-Su, C. 1997. Constitutive activation of JAK1 in Src-transformed cells. Journal of Biological Chemistry 272:2591–2594.

Canesi, L., Lorusso, L. C., Ciacci, C., Betti, M. and Gallo, G. 2005. Effects of the brominated flame retardant tetrabromobisphenol-A (TBBPA) on cell signaling and function of Mytilus hemocytes: involvement of MAP kinases and protein kinase C. Aquatic Toxicology 75:277–287.

Cheng, C. H., Yu, K. C., Chen, H. L., Chen, S. Y., Huang, C. H., Chan, P. C., Wung, C. W. and Chen, H. C. 2004. Blockade of v-Src-stimulated tumor formation by the Src homology 3 domain of Crk-associated substrate (Cas). FEBS Letters 557:221–227.

Chou, S. D., Khan, A. N., Magner, W. J. and Tomasi, T. B. 2005. Histone acetylation regulates the cell type specific CIITA promoters, MHC class II expression and antigen presentation in tumor cells. International Immunology 17:1483–1494.

Cruz, M. T., Duarte, C. B., Goncalo, M., Carvalho, A. P. and Lopes, M. C. 1999. Involvement of JAK2 and MAPK on type II nitric oxide synthase expression in skin-derived dendritic cells. American Journal of Physiology 277:C1050–C1057.

Demuth, T., Reavie, L. B., Rennert, J. L., Nakada, M., Nakada, S., Hoelzinger, D. B., Beaudry, C. E., Henrichs, A. N., Anderson, E. M. and Berens, M. E. 2007. MAP-ing glioma invasion: mitogen-activated protein kinase kinase 3 and p38 drive glioma invasion and progression and predict patient survival. Molecular Cancer Therapeutics 6:1212–1222.

Di Nicola, M. and Lemoli, R. M. 2000. Dendritic cells: specialized antigen presenting cells. Haematologica 85:202–207.

Do, Y., Hegde, V. L., Nagarkatti, P. S. and Nagarkatti, M. 2004. Bryostatin-1 enhances the maturation and antigen-presenting ability of murine and human dendritic cells. Cancer Research 64:6756–6765.

Evel-Kabler, K. and Chen, S. Y. 2006. Dendritic cell-based tumor vaccines and antigen presentation attenuators. Molecular Therapy: The Journal of the American Society of Gene Therapy 13:850–858.

Faderl, S., Ferrajoli, A., Harris, D., Van, Q., Kantarjian, H. M. and Estrov, Z. 2007. Atiprimod blocks phosphorylation of JAK-STAT and inhibits proliferation of acute myeloid leukemia (AML) cells. Leukemia Research 31:91–95.

Gabrilovich, D. I. 2002. Dendritic cell vaccines for cancer treatment. Current Opinion in Molecular Therapeutics 4:452–458.

Gabrilovich, D. I., Nadaf, S., Corak, J., Berzofsky, J. A. and Carbone, D. P. 1996. Dendritic cells in antitumor immune responses. II. Dendritic cells grown from bone marrow precursors, but not mature DC from tumor-bearing mice, are effective antigen carriers in the therapy of established tumors. Cellular Immunology 170:111–119.

Geginat, J., Sallusto, F. and Lanzavecchia, A. 2003. Cytokine-driven proliferation and differentiation of human naive, central memory and effector memory CD4+ T cells. Pathologie Biologie 51:64–66.

Ichijo, H. 1999. From receptors to stress-activated MAP kinases. Oncogene 18:6087–6093.

Jarnicki, A. G., Conroy, H., Brereton, C., Donnelly, G., Toomey, D., Walsh, K., Sweeney, C., Leavy, O., Fletcher, J., Lavelle, E. C., Dunne, P. and Mills, K. H. 2008. Attenuating regulatory T cell induction by TLR agonists through inhibition of p38 MAPK signaling in

dendritic cells enhances their efficacy as vaccine adjuvants and cancer immunotherapeutics. Journal of immunology (Baltimore, MD: 1950) 180:3797–3806.
Jiang, J. L., Wang, S., Li, N. S., Zhang, X. H., Deng, H. W. and Li, Y. J. 2007. The inhibitory effect of simvastatin on the ADMA-induced inflammatory reaction is mediated by MAPK pathways in endothelial cells. Biochemistry & Cell Biology 85:66–77.
Jing, N., Zhu, Q., Yuan, P., Li, Y., Mao, L. and Tweardy, D. J. 2006. Targeting signal transducer and activator of transcription 3 with G-quartet oligonucleotides: a potential novel therapy for head and neck cancer. Molecular Cancer Therapeutics 5:279–286.
Jost, E., do, O. N., Dahl, E., Maintz, C. E., Jousten, P., Habets, L., Wilop, S., Herman, J. G., Osieka, R. and Galm, O. 2007. Epigenetic alterations complement mutation of JAK2 tyrosine kinase in patients with BCR/ABL-negative myeloproliferative disorders. Leukemia 21:505–510.
Kandilci, A. and Grosveld, G. C. 2005. SET-induced calcium signaling and MAPK/ERK pathway activation mediate dendritic cell-like differentiation of U937 cells. Leukemia 19:1439–1445.
Kang, H. K., Lee, H. Y., Lee, Y. N., Jo, E. J., Kim, J. I., Aosai, F., Yano, A., Kwak, J. Y. and Bae, Y. S. 2004. Toxoplasma gondii-derived heat shock protein 70 stimulates the maturation of human monocyte-derived dendritic cells. Biochemical & Biophysical Research Communications 322:899–904.
Kawakami, Y., Rodriguez-Leon, J., Koth, C. M., Buscher, D., Itoh, T., Raya, A., Ng, J. K., Esteban, C. R., Takahashi, S., Henrique, D., Schwarz, M. F., Asahara, H. and Izpisua Belmonte, J. C. 2003. MKP3 mediates the cellular response to FGF8 signalling in the vertebrate limb.[see comment]. Nature Cell Biology 5:513–519.
Kim, C. J., Dessureault, S., Gabrilovich, D., Reintgen, D. S. and Slingluff, C. L., Jr. 2002. Immunotherapy for melanoma. Cancer Control 9:22–30.
Kim, G. Y., Kim, K. H., Lee, S. H., Yoon, M. S., Lee, H. J., Moon, D. O., Lee, C. M., Ahn, S. C., Park, Y. C. and Park, Y. M. 2005. Curcumin inhibits immunostimulatory function of dendritic cells: MAPKs and translocation of NF-kappa B as potential targets. Journal of Immunology 174:8116–8124.
Kim, R., Emi, M. and Tanabe, K. 2006. Cancer immunosuppression and autoimmune disease: beyond immunosuppressive networks for tumour immunity. Immunology 119:254–264.
Kirkwood, K. L., Li, F., Rogers, J. E., Otremba, J., Coatney, D. D., Kreider, J. M., D'Silva, N. J., Chakravarty, S., Dugar, S., Higgins, L. S., Protter, A. A. and Medicherla, S. 2007. A p38alpha selective mitogen-activated protein kinase inhibitor prevents periodontal bone loss. Journal of Pharmacology & Experimental Therapeutics 320:56–63.
Kortylewski, M., Jove, R. and Yu, H. 2005. Targeting STAT3 affects melanoma on multiple fronts. Cancer & Metastasis Reviews 24:315–327.
Kusmartsev, S. and Gabrilovich, D. I. 2002. Immature myeloid cells and cancer-associated immune suppression. Cancer Immunology, Immunotherapy 51:293–298.
Kusmartsev, S. and Gabrilovich, D. I. 2006. Effect of tumor-derived cytokines and growth factors on differentiation and immune suppressive features of myeloid cells in cancer. Cancer & Metastasis Reviews 25:323–331.
Larmonier, N., Merino, D., Nicolas, A., Cathelin, D., Besson, A., Bateman, A., Solary, E., Martin, F., Katsanis, E. and Bonnotte, B. 2006. Apoptotic, necrotic, or fused tumor cells: an equivalent source of antigen for dendritic cell loading. Apoptosis 11:1513–1524.
Lee, H. J., Cho, J. W., Kim, S. C., Kang, K. H., Lee, S. K., Pi, S. H. and Kim, E. C. 2006. Roles of p38 and ERK MAP kinases in IL-8 expression in TNF-alpha- and dexamethasone-stimulated human periodontal ligament cells. Cytokine 35:67–76.
Li, J., Yu, B., Song, L., Eschrich, S. and Haura, E. B. 2007. Effects of IFN-gamma and Stat1 on gene expression, growth, and survival in non-small cell lung cancer cells. Journal of Interferon & Cytokine Research 27:209–220.
Mata, R., Subira, D., Garcia-Raso, A. and Llamas, P. 2007. JAK2 as a molecular marker in myeloproliferative diseases. Cardiovascular & Hematological Agents in Medicinal Chemistry 5:198–203.

Matos, T. J., Duarte, C. B., Goncalo, M. and Lopes, M. C. 2005a. DNFB activates MAPKs and upregulates CD40 in skin-derived dendritic cells. Journal of Dermatological Science 39:113–123.

Matos, T. J., Duarte, C. B., Goncalo, M. and Lopes, M. C. 2005b. Role of oxidative stress in ERK and p38 MAPK activation induced by the chemical sensitizer DNFB in a fetal skin dendritic cell line. Immunology & Cell Biology 83:607–614.

McKay, D. M., Botelho, F., Ceponis, P. J. and Richards, C. D. 2000. Superantigen immune stimulation activates epithelial STAT-1 and PI 3-K: PI 3-K regulation of permeability. American Journal of Physiology – Gastrointestinal & Liver Physiology 279:G1094–G1103.

Nakahara, T., Uchi, H., Urabe, K., Chen, Q., Furue, M. and Moroi, Y. 2004. Role of c-Jun N-terminal kinase on lipopolysaccharide induced maturation of human monocyte-derived dendritic cells. International Immunology 16:1701–1709.

Nakamura, K., Zhou, C. J., Parente, J. and Chew, C. S. 1996. Parietal cell MAP kinases: multiple activation pathways. American Journal of Physiology 271:G640-G649.

Nefedova, Y., Cheng, P., Gilkes, D., Blaskovich, M., Beg, A. A., Sebti, S. M. and Gabrilovich, D. I. 2005a. Activation of dendritic cells via inhibition of Jak2/STAT3 signaling. Journal of Immunology 175:4338–4346.

Nefedova, Y. and Gabrilovich, D. I. 2007. Targeting of Jak/STAT pathway in antigen presenting cells in cancer. Current Cancer Drug Targets 7:71–77.

Nefedova, Y., Huang, M., Kusmartsev, S., Bhattacharya, R., Cheng, P., Salup, R., Jove, R. and Gabrilovich, D. 2004. Hyperactivation of STAT3 is involved in abnormal differentiation of dendritic cells in cancer. Journal of Immunology 172:464–474.

Nefedova, Y., Nagaraj, S., Rosenbauer, A., Muro-Cacho, C., Sebti, S. M. and Gabrilovich, D. I. 2005b. Regulation of dendritic cell differentiation and antitumor immune response in cancer by pharmacologic-selective inhibition of the janus-activated kinase 2/signal transducers and activators of transcription 3 pathway. Cancer Research 65:9525–9535.

Osawa, Y., Iho, S., Takauji, R., Takatsuka, H., Yamamoto, S., Takahashi, T., Horiguchi, S., Urasaki, Y., Matsuki, T. and Fujieda, S. 2006. Collaborative action of NF-kappaB and p38 MAPK is involved in CpG DNA-induced IFN-alpha and chemokine production in human plasmacytoid dendritic cells. Journal of Immunology 177:4841–4852.

Park, S., Kim, D., Kaneko, S., Szewczyk, K. M., Nicosia, S. V., Yu, H., Jove, R. and Cheng, J. Q. 2005. Molecular cloning and characterization of the human AKT1 promoter uncovers its up-regulation by the Src/Stat3 pathway. Journal of Biological Chemistry 280:38932–38941.

Park, S. M., Kim, H. S., Choe, J. and Lee, T. H. 1999. Differential induction of cytokine genes and activation of mitogen-activated protein kinase family by soluble CD40 ligand and TNF in a human follicular dendritic cell line. Journal of Immunology 163:631–638.

Philpott, N. J., Nociari, M., Elkon, K. B. and Falck-Pedersen, E. 2004. Adenovirus-induced maturation of dendritic cells through a PI3 kinase-mediated TNF-alpha induction pathway. Proceedings of the National Academy of Sciences of the United States of America 101:6200–6205.

Rabinovich, G. A., Gabrilovich, D. and Sotomayor, E. M. 2007. Immunosuppressive strategies that are mediated by tumor cells. Annual Review of Immunology 25:267–296.

Radl, J., Croese, J. W., Zurcher, C., Van den Enden-Vieveen, M. H. and de Leeuw, A. M. 1988. Animal model of human disease. Multiple myeloma. Am J Pathol 132:593–597.

Randolph, G. J., Sanchez-Schmitz, G. and Angeli, V. 2005. Factors and signals that govern the migration of dendritic cells via lymphatics: recent advances. Springer Seminars in Immunopathology 26:273–287.

Ratta, M., Fagnoni, F., Curti, A., Vescovini, R., Sansoni, P., Oliviero, B., Fogli, M., Ferri, E., Della Cuna, G. R., Tura, S., Baccarani, M. and Lemoli, R. M. 2002. Dendritic cells are functionally defective in multiple myeloma: the role of interleukin-6. Blood 100:230–237.

Saito, Y., Yanagawa, Y., Kikuchi, K., Iijima, N., Iwabuchi, K. and Onoe, K. 2006. Low-dose lipopolysaccharide modifies the production of IL-12 by dendritic cells in response to various cytokines. Journal of Clinical & Experimental Hematopathology 46:31–36.

Santini, S. M. and Belardelli, F. 2003. Advances in the use of dendritic cells and new adjuvants for the development of therapeutic vaccines. Stem Cells 21:495–505.

Sheng, K. C., Pietersz, G. A., Wright, M. D. and Apostolopoulos, V. 2005. Dendritic cells: activation and maturation–applications for cancer immunotherapy. Current Medicinal Chemistry 12:1783–1800.

Sigaud, S., Evelson, P. and Gonzalez-Flecha, B. 2005. H2O2-induced proliferation of primary alveolar epithelial cells is mediated by MAP kinases. Antioxidants & Redox Signaling 7:6–13.

Sinkovics, J. G. and Horvath, J. C. 2000. Vaccination against human cancers (review). International Journal of Oncology 16:81–96.

Sumimoto, H., Imabayashi, F., Iwata, T. and Kawakami, Y. 2006. The BRAF-MAPK signaling pathway is essential for cancer-immune evasion in human melanoma cells. Journal of Experimental Medicine 203:1651–1656.

Sumimoto, H., Miyagishi, M., Miyoshi, H., Yamagata, S., Shimizu, A., Taira, K. and Kawakami, Y. 2004. Inhibition of growth and invasive ability of melanoma by inactivation of mutated BRAF with lentivirus-mediated RNA interference. Oncogene 23:6031–6039.

Taki, T. and Taniwaki, M. 2006. Chromosomal translocations in cancer and their relevance for therapy. Current Opinion in Oncology 18:62–68.

Tanami, H., Imoto, I., Hirasawa, A., Yuki, Y., Sonoda, I., Inoue, J., Yasui, K., Misawa-Furihata, A., Kawakami, Y. and Inazawa, J. 2004. Involvement of overexpressed wild-type BRAF in the growth of malignant melanoma cell lines. Oncogene 23:8796–8804.

Tassiulas, I., Park-Min, K. H., Hu, Y., Kellerman, L., Mevorach, D. and Ivashkiv, L. B. 2007. Apoptotic cells inhibit LPS-induced cytokine and chemokine production and IFN responses in macrophages. Human Immunology 68:156–164.

Wang, S., Hong, S., Yang, J., Qian, J., Zhang, X., Shpall, E., Kwak, L. W. and Yi, Q. 2006a. Optimizing immunotherapy in multiple myeloma: Restoring the function of patients' monocyte-derived dendritic cells by inhibiting p38 or activating MEK/ERK MAPK and neutralizing interleukin-6 in progenitor cells. Blood 108:4071–4077.

Wang, S., Yang, J., Qian, J., Wezeman, M., Kwak, L. W. and Yi, Q. 2006b. Tumor evasion of the immune system: inhibiting p38 MAPK signaling restores the function of dendritic cells in multiple myeloma. Blood 107:2432–2439.

Wang, T., Niu, G., Kortylewski, M., Burdelya, L., Shain, K., Zhang, S., Bhattacharya, R., Gabrilovich, D., Heller, R., Coppola, D., Dalton, W., Jove, R., Pardoll, D. and Yu, H. 2004. Regulation of the innate and adaptive immune responses by Stat-3 signaling in tumor cells.[erratum appears in Nat Med. 2004 Feb;10(2):209]. Nature Medicine 10:48–54.

West, M. A., Wallin, R. P., Matthews, S. P., Svensson, H. G., Zaru, R., Ljunggren, H. G., Prescott, A. R. and Watts, C. 2004. Enhanced dendritic cell antigen capture via toll-like receptor-induced actin remodeling. Science 305:1153–1157.

Xie, J., Qian, J., Wang, S., Freeman, M. E., 3rd, Epstein, J. and Yi, Q. 2003. Novel and detrimental effects of lipopolysaccharide on in vitro generation of immature dendritic cells: involvement of mitogen-activated protein kinase p38. J Immunol 171:4792–4800.

Xie, J., Qian, J., Yang, J., Wang, S., Freeman, M. E., 3rd and Yi, Q. 2005. Critical roles of Raf/MEK/ERK and PI3K/AKT signaling and inactivation of p38 MAP kinase in the differentiation and survival of monocyte-derived immature dendritic cells. Experimental Hematology 33:564–572.

Yang, L., Ng, K. Y. and Lillehei, K. O. 2003. Cell-mediated immunotherapy: a new approach to the treatment of malignant glioma. Cancer Control 10:138–147.

Yoshikawa, H., Matsubara, K., Qian, G. S., Jackson, P., Groopman, J. D., Manning, J. E., Harris, C. C. and Herman, J. G. 2001. SOCS-1, a negative regulator of the JAK/STAT pathway, is silenced by methylation in human hepatocellular carcinoma and shows growth-suppression activity [see comment]. Nature Genetics 28:29–35.

Yu, C. L., Jove, R. and Burakoff, S. J. 1997. Constitutive activation of the Janus kinase-STAT pathway in T lymphoma overexpressing the Lck protein tyrosine kinase. Journal of Immunology 159:5206–5210.

Yu, C. L., Meyer, D. J., Campbell, G. S., Larner, A. C., Carter-Su, C., Schwartz, J. and Jove, R. 1995. Enhanced DNA-binding activity of a Stat3-related protein in cells transformed by the Src oncoprotein. Science 269:81–83.

Zaru, R., Ronkina, N., Gaestel, M., Arthur, J. S. and Watts, C. 2007. The MAPK-activated kinase Rsk controls an acute Toll-like receptor signaling response in dendritic cells and is activated through two distinct pathways. Nature Immunology 8:1227–1235.

Zou, G. M. and Hu, W. Y. 2005. LIGHT regulates CD86 expression on dendritic cells through NF-kappaB, but not JNK/AP-1 signal transduction pathway. Journal of Cellular Physiology 205:437–443.

Chapter 10
A Role for STAT3 in Dendritic Cell Regulation by Tumor-Derived Factors

Dmitry I. Gabrilovich

Abstract Ineffective dendritic cell differentiation and activation in cancer is a well-established phenomenon that is considered as one of the major mechanisms of tumor escape. The defects in dendritic cells are caused primarily by soluble tumor-derived factors. In recent years accumulated evidence suggested that the members of the family of signal transducers and activators of transcription (STAT), and more specifically STAT3, could be primarily responsible for dendritic cell defects in cancer. In this review we will discuss recent findings describing the role of STAT3 in dendritic cell differentiation and function.

10.1 Introduction

Dendritic cells (DC) are the most potent antigen-presenting cells (APC). Although their number is quite small their impact on generation and maintenance of immune responses is great (Albert et al. 1998; Inaba et al. 1998). DC play a critical role in an antitumor immune response (Rock et al. 1993; Huang et al. 1994). It is now well established that the induction of an effective antitumor immune response is seriously limited by the defects in the host immune system (Gabrilovich et al. 1996, 1997; Nestle et al. 1997; Gabrilovich 2004). The existence of host immune deficits also may be one of the explanations of the limited success of clinical trials reported by now. In recent years several groups described the decreased number and defective function of DC in tumor-bearing mice and in cancer patients (rev in (Gabrilovich 2004)). DC defects are now considered as one of the important mechanisms of tumor escape from immune system control.

Abnormal DC differentiation and function in cancer is shown to be mediated primarily by soluble tumor-derived factors. Several different tumor-derived factors including VEGF, GM-CSF, IL-10, IL-6, TGF-β, prostaglandins, and

D.I. Gabrilovich (✉)
H. Lee Moffitt Cancer Center, University of South Florida, MRC 2067,
12902 Magnolia Dr., Tampa, FL 33612, USA
e-mail: dgabril@moffitt.usf.edu

M.R. Shurin, R.D. Salter (eds.), *Dendritic Cells in Cancer*,
DOI 10.1007/978-0-387-88611-4_10, © Springer Science+Business Media, LLC 2009

gangliosides have been implicated in this phenomenon (Shurin and Gabrilovich 2001; Gabrilovich 2004). Importantly, these different tumor-derived factors affect myeloid cells at various stages of differentiation. The list of tumor-derived factors that affect myeloid cell differentiation is not complete and is constantly growing. Various tumors produce different levels of these factors and most of the studies failed to identify a single factor responsible for DC defects in cancer. Nevertheless, the abnormalities in DC differentiation induced by different tumors were rather similar. It may suggest that these factors may trigger some common signaling pathways specifically responsible for abnormal differentiation of myeloid cells. Recent evidence pointed out that Jak/STAT might play a critical role in this process.

10.2 Overview of Jak/STAT Pathway

The signal transducers and activators of transcription (STAT) were originally identified on the basis of their ability to transduce a signal downstream of interferon receptors (rev in (Darnell et al. 1994)). A number of studies have demonstrated that STAT are critical components of diverse signal transduction pathway involved in response to a wide range of cytokines and growth factors, the receptors for which are associated with Janus kinases (Jak) (Ihle 1995). Besides Jak, STAT can also be activated by growth factor receptors, such as epidermal growth factor (EGF), platelet-derived growth factor (PDGF), hepatocyte growth factor (HGF), and colony-stimulating factor-1 (CSF-1) receptors, which all possess an intrinsic tyrosine kinase activity (Calo et al. 2003). STAT proteins may also be activated by non-receptor tyrosine kinases (RTK) such as Src and Abl and through serpentine receptors, such as those for angiotensin II, serotonin, and α-melatonin-stimulating hormone (Williams 1999; Bromberg 2002; Rane and Reddy 2002).

The Jak family of tyrosine kinases includes four members: Jak1, Jak2, Jak3, and Tyk2 (Ihle 1995; Rane and Reddy 2000). The binding of a cytokine to its cell-surface receptor induces receptor dimerization or oligomerization and the subsequent activation of Jak tyrosine kinases, which are constitutively associated with the receptor (Ihle 1995). Activated Jak then trigger the phosphorylation of specific tyrosine residues on the receptor, which serve as a docking site for seven members of STAT family (Darnell 1997). STAT proteins consist of ~700–850 amino acids and vary from 90 to 115 kDa in size. All STAT share several conserved domains, which are necessary for their function. DNA binding domain has been suggested to regulate DNA binding specificity; SH2 (SRC homology 2) domain is required for STAT activation; α-helical linker domain connects the DNA binding and SH2 domains; the transcriptional activation domain located at the C-terminal region of STAT and involved in interaction with transcriptional complexes; NH2-terminal oligomerization domain is required for protein–protein interaction; and a four-stranded helical coiled coil domain is considered to be important for the association with a number of regulatory modifiers (Calo et al. 2003; Shuai and Liu 2003).

Recruitment of STAT proteins to the receptor occurs through the interaction of its SH2 domain with a phosphorylated tyrosine residue at the receptor. STAT are then phosphorylated by Jak at a single conserved tyrosine residue in the C-terminal domain (Tyr701 for Stat1, Tyr690 for Stat2, Tyr705 for Stat3, Tyr693 for Stat4, Tyr694 for Stat5, and Tyr641 for Stat6). At this point, STAT dissociate from the receptor and form homo- or heterodimers through interactions of phosphorylated tyrosine of one STAT and SH2 domain of another. Each STAT dimer is further phosphorylated in serine, which makes it an active transcription factor. The dimer translocates to the nucleus where it induces the expression of target genes (Imada and Leonard 2000). Recent studies identify another mechanism of STAT3 activation. STAT3 is acetylated by histone acetyltransferase p300 at Lys685 in the C-terminal transcriptional activation domain. STAT3 acetylation is reversible by type I histone deacetylase (HDAC). Furthermore, the acetylation of STAT3 can stimulate its sequence-specific DNA binding ability and transactivation activity (Wang et al. 2005; Yuan et al. 2005). This may play an important role in regulation of another transcription factor implicated in DC differentiation, NF-κB. Nadiminty et al. demonstrated that active STAT3 induced activation of non-canonical NF-κB involved in the proteolytic processing of p100 to p52. The STAT3-mediated p100 processing required activation of STAT3 by the acetyltransferase activity of cAMP response element-binding protein (CREB)-binding protein (CBP)/p300. A mutant of STAT3 defective in acetylation blocked STAT3-mediated p100 processing to p52 and acted as a dominant negative in blocking the production of p52 (Nadiminty et al. 2006).

As is the case for any signaling pathway there are several mechanisms of negative regulation of Jak/STAT activity. Suppressor of cytokine signaling (SOCS) proteins, protein inhibitor of activated STAT (PIAS) family, and tyrosine phosphatases are the main post-translational regulators of Jak and STAT (Chung et al. 1997; Shuai and Liu 2003).

Several tyrosine phosphatases have been shown to dephosphorylate Jak. They include SHP1, SHP2, CD45, PTP1B, and T-cell PTP (TCPTP). SHP1 and SHP2 are constitutively expressed in cells. Upon cytokine stimulation they bind and dephosphorylate receptor kinases and Jak (Levy et al. 2002). CD45 is a receptor tyrosine phosphatase, which can directly bind and dephosphorylate all Jak. Jak2 and Tyk2 are substrates for PTP1B, and Jak1 and Jak3 for TCPTP (Shuai and Liu 2003).

SOCS proteins were initially identified as negative regulators of cytokine signaling. SOCS proteins are expressed at low levels in unstimulated cells. A wide range of cytokines and growth factors, which induce Jak/STAT signaling, also rapidly induce SOCS expression. SOCS family consists of eight members which may inactivate Jaks by distinct mechanisms (Starr et al. 1997; Alexander et al. 1999). Those mechanisms include direct binding to tyrosine-phosphorylated Jak (SOCS1), binding to activated receptor (SOCS2 and SOCS3), competing with STAT for binding to the receptor docking site (CIS), or binding to components of ubiquitin E3 ligase complex and targeting proteins for degradation (Greenhalgh and Hilton 2001; Greenhalgh et al. 2005).

Unlike SOCS proteins, PIAS proteins are constitutively expressed in a number of cell lines. The PIAS family includes four members (Shuai and Liu 2005). Besides STAT, PIAS proteins can regulate other transcription factors, including NF-κB and SMAD. Each member of the PIAS family is involved in STAT regulation. It has been shown that PIAS proteins may interact only with the dimeric but not monomeric form of STAT (Liao et al. 2000; Shuai and Liu 2005). Thus, PIAS-3 binds to phosphorylated STAT3 dimers and inhibits DNA binding of STAT3.

10.3 Jak/STAT Signaling in Dendritic Cell Differentiation and Function

It is well established that Jak/STAT signaling is critically involved in normal myelopoiesis. Studies with dominant-negative mutants in a number of different experimental models suggested a necessary role for STAT3 in myeloid differentiation. In addition to promoting differentiation, STAT3 has been implicated in survival signaling from the IL-6/gp130 cytokine receptor system (rev. in (Coffer et al. 2000; Smithgall et al. 2000)).

Recent studies provided evidence that STAT3 is involved in the differentiation of DC. Using mice with conditional knockout of STAT3, Laouar and colleagues have demonstrated that the deletion of STAT3 caused profound deficiency in the DC differentiation and abrogated the effect of cytokine Flt3L on DC development. The pool of hematopoietic stem cells (HSC) was not changed and 2- to 3-fold accumulation of common lymphoid (CLP) and myeloid (CMP) progenitors was observed. Authors suggested that Flt3L-induced DC differentiation depends on STAT3 (Laouar et al. 2003). In contrast, DC differentiation induced by GM-CSF was independent of STAT3 signaling. Since it is known that GM-CSF activates STAT3 in myeloid cells it remains unclear whether observed effects of STAT3 on DC differentiation were indeed restricted to only one cytokine. Consistent with these findings, Onai et al. have demonstrated that overexpression of human Flt3 in hematopoietic progenitors rescues and enhances their DC differentiation potential. In Flt3-megakaryocyte/erythrocyte-restricted progenitors (MEP), enforced Flt3 signaling induces transcription of STAT3 and PU.1. Ectopic expression of STAT3 in Flt3-MEP induced Flt3 receptor expression and instructed differentiation into DC and myelomonocytic cells, whereas GATA-1 expression and consecutive megakaryocyte/erythrocyte development was suppressed (Onai et al. 2006).

An important role of STAT3 activation in DC development was demonstrated using Gfi1 knockout mice. Gfi1 is a zinc finger repressor molecule that interacts with PIAS3, a known inhibitor of STAT3. Lack of Gfi1 resulted in dramatic decrease in STAT3 activity in progenitor cells. This defect was associated with a global reduction of myeloid and lymphoid DC in all lymphoid organs whereas the number of epidermal Langerhans cells was increased. Gfi1$^{-/-}$ hematopoietic progenitor cells were unable to develop into DC. Instead, they differentiated into macrophages, suggesting that Gfi1 is a critical modulator of DC versus macrophage development (Rathinam et al. 2005).

Thus, STAT3 is required for DC development. However, its role in DC maturation/activation appears to be quite different. STAT3 is one of the major mediators of signaling via IL-6 receptor. IL-6, in turn, plays a major role in maintaining immature DC. IL-6 knockout mice had increased numbers of mature DC, indicating that IL-6 indeed blocks DC maturation in vivo. Park et al. (2004) demonstrated that STAT3 activation by IL-6 was required for the suppression of LPS-induced DC maturation. The follow-up study from this group identified one possible mechanism of this effect. STAT3 signaling reduced intracellular MHC class II $\alpha\beta$ dimer, Ii, and H2-DM levels in DC. IL-6-mediated STAT3 activation enhanced cathepsin S activities by decreasing the level of endogenous inhibitor cystatin C. Cathepsin S overexpression in DC resulted in the same effect on MHC class II expression as IL-6 treatment. Cathepsin S inhibitors blocked the effect of IL-6 on MHC class II in DC and the overexpression of cystatin C suppressed IL-6–STAT3-mediated increase of cathepsin S activity and the reduction of MHC class II $\alpha\beta$, Ii, and H2-DM levels in DC (Kitamura et al. 2005). The data obtained during the analysis of LPS-induced DC activation were consistent with these results. Constitutively active STAT3 blocked LPS-induced IL-12p40 gene expression and the recruitment to the IL-12p40 promoter NF-κB transcription factor (Hoentjen et al. 2005).

The role of several other members of the STAT family in DC maturation/activation was evaluated by Jackson et al. (2004). They have found that the STAT6 was constitutively activated in immature DC (iDC) and declines as iDC differentiate into mature DC (mDC). STAT6 downregulation was accompanied by dramatic induction of SOCS1, SOCS2, SOCS3, and cytokine-induced Src homology 2-containing protein expression, suggesting that inhibition of STAT6 signaling may be required for DC maturation. In contrast, STAT1 signaling was most robust in mDC and is not inhibited by the upregulated SOCS proteins, indicating that STAT1 and STAT6 pathways are distinctly regulated in maturing DC. Furthermore, optimal activation of STAT1 during DC maturation requires both IL-4 and GM-CSF. STAT1 was important for the expression of CD40 and CD11c in mDC. Since STAT1 and STAT3 often have antagonistic effects, these data indirectly support observations from other studies that DC activation is associated with downregulation of STAT3. These findings were further supported by a report that activation of macrophages and DC with CpG-ODN induces expression of SOCS1 and SOCS3 in vitro and in vivo. SOCS proteins were functional because they inhibited IFN-γ as well as IL-6- and GM-CSF-induced phosphorylation of STAT proteins (Dalpke et al. 2001). However, in a different study bone marrow-derived DC transduced with SOCS3 expressed a low level of MHC class II and CD86 on their surface, produced a high level of IL-10 but low levels of IL-12 and IFN-γ, and expressed a low level of IL-23 p19 mRNA. Functionally, SOCS3-transduced DC drove naïve myelin oligodendrocyte glycoprotein-specific T cells to a strong Th2 differentiation in vitro and in vivo (Li et al. 2006).

In the most recent study the role of SOCS1 in function of different DC subsets has been investigated. The $CD11c^{+}$ $CD8^{-}$ DC population in freshly isolated splenic DC from normal mice highly expressed SOCS1. However, in SOCS1-deficient environment, the proportion of $CD8\alpha^{+}$ DC noticeably increased without affecting the cell number of conventional and plasmacytoid DC populations. The $CD8^{+}$ DC secreted a large amount of IFN-γ, IL-12, and B lymphocyte stimulator/B cell activation factor of the tumor necrosis factor family in response to LPS and CpG stimulation. This was responsible for the development of DC-mediated systemic autoimmunity in the old age of SOCS1-deficient mice. Moreover, the $CD8^{+}$ DC subsets expressed more indoleamine 2,3-dioxygenase and IL-10, and hence inhibit the allogeneic proliferative T-cell response and antigen-induced Th1 responses (Tsukada et al. 2005).

Additional evidence that STAT3 is critically important for the function of macrophages and neutrophils was obtained in experiments with knockout mice. Takeda and colleagues have generated mice with a cell type-specific disruption of the STAT3 gene in macrophages and neutrophils. The mutant mice were highly susceptible to endotoxin shock with increased production of inflammatory cytokines such as TNF-α, IL-1, IFN-γ, and IL-6. Endotoxin-induced production of inflammatory cytokines was augmented because the suppressive effects of IL-10 on inflammatory cytokine production from macrophages and neutrophils were completely abolished. The mice showed a polarized immune response toward the Th1 type and developed chronic enterocolitis with age (Takeda et al. 1999). Using the same STAT3 knockout mice, Cheng and colleagues demonstrated that targeted disruption of STAT3 resulted in the priming of antigen-specific $CD4^{+}$ T cells in response to an otherwise tolerogenic stimulus in vivo. Furthermore, STAT3-deficient antigen-presenting cells effectively break antigen-specific T-cell anergy in vitro. Conversely, increased STAT3 activity in APC led to impaired antigen-specific T-cell responses (Cheng et al. 2003).

Thus, available data indicate that STAT3 is critically important for the development of myeloid cells, including DC, from HPC. However, activation of STAT3 results in the inhibition of DC maturation/activation in response to various stimuli. These cells were unable to effectively stimulate immune responses. These results turned out to be very relevant to the situation observed in tumor-bearing hosts.

10.4 Jak/STAT3 and Abnormal Dendritic Cell Differentiation and Maturation in Cancer

In an attempt to identify molecular mechanisms of abnormal DC differentiation in cancer we studied activation of different members of MAPK, PI3K, and Jak/STAT pathways. We found that tumor-derived factors induced substantial hyper-phosphorylation of Jak2 but not other members of the Jak family. This was associated with the activation of STAT3 (Nefedova et al. 2004). Importantly, HPC had high levels of STAT3 activity, which significantly decreased during DC differentiation in vitro. Conditioned medium from several different tumor cell lines prevents this downregulation and maintained high levels of STAT3 activity. Tumor cells produce a number of factors able to affect differentiation of DC: GM-CSF, M-CSF, IL-10, IL-6, and VEGF. Direct stimulation of STAT3 activity in hematopoietic cells has been demonstrated for all of them. Using neutralizing antibodies we have found that M-CSF, VEGF, and IL-10 were primarily responsible for STAT3 activation by tumor cell-conditioned medium (Nefedova et al. 2004). To confirm the critical role of STAT3 in observed abnormalities in myeloid cell differentiation, activated STAT3 was overexpressed in HPC. Constitutive activation of STAT3 in HPC resulted in the accumulation of immature myeloid cells and DC differentiation was significantly impaired (Nefedova et al. 2004).

The level of STAT3 activity was also evaluated in immature and mature DC generated in the presence of tumor-derived factors. Tumor-derived factors maintained increased STAT3 binding activity in iDC. These data suggested that tumor-derived factors might affect DC differentiation in cancer via constitutive activation of Jak2/STAT3. Similar effects were observed in another study by Wang et al. (2004). In addition, that study had demonstrated that constitutive activation of STAT3 in tumor cells suppressed expression of pro-inflammatory mediators. Blockade of STAT3 in tumor cells increased expression of pro-inflammatory cytokines and chemokines that activated innate immunity and DC, leading to tumor-specific T-cell responses. In addition, constitutive STAT3 activity induces production of pleiotropic factors by tumor cells that inhibit DC functional maturation. Authors suggested that STAT3 activity in tumor cells can mediate immune evasion by blocking both the production and sensing of inflammatory signals by multiple components of the immune system (Wang et al. 2004).

To further identify the role of STAT in tumor-associated abnormalities in tumor-bearing mice we compared splenic and tumor-associated macrophages (TAM) directly isolated from tumor-bearing mice. TAM but not macrophages freshly isolated from spleens of tumor-bearing or naïve mice were able to inhibit T-cell-mediated immune response in vitro via induction of T-cell apoptosis. Arginase and NO were both responsible for the apoptotic mechanism and were seen only in TAM. Using the analysis of STAT activity in combination with STAT knockout mice, we have determined that STAT1 but not STAT3 or

STAT6 was responsible for TAM-suppressive activity (Kusmartsev and Gabrilovich 2005).

Critical role of STAT3 in DC abnormalities was confirmed by Sumimoto et al. who showed that suppressive activity of the conditioned media from melanoma cells on the production of IL-12 and TNF-α by DC was markedly reduced after transduction with STAT3 RNAi (Sumimoto et al. 2006). Consistent with these findings, Bharadwaj et al. have demonstrated that media conditioned by a highly metastatic human pancreatic cancer cell line BxPC-3 inhibited DC differentiation from both $CD34^+$ and monocyte precursors. These resulted in inefficient allo- and antigen-specific T-cell stimulation by these cells. BxPC-3-conditioned media led to the activation of STAT3 in $CD14^+$ monocytes. Blocking of STAT3 activation reversed the inhibitory effect of conditioned media on DC differentiation (Bharadwaj et al. 2007).

10.5 Targeting Jak/STAT Pathway in Dendritic Cells

The data described above indicate that constitutive activation of Jak2 and Stat3 in antigen-presenting cells and especially DC in cancer can suppress their ability to induce immune response, which makes this pathway an attractive target for therapeutic intervention. Using a genetic approach, it has been shown that targeted disruption of STAT3 resulted in the activation of APC and their ability to break immune tolerance (Cheng et al. 2003). A more recent study has demonstrated that in $STAT3^{-/-}$ tumor-bearing mice function of DC, T cells, natural killer (NK) cells, and neutrophils were markedly upregulated, which resulted in tumor regression (Kortylewski et al. 2005).

However, this approach obviously cannot be used in clinical applications. Therefore, the possibility of pharmacological regulation of STAT3 activity has been explored. A new selective inhibitor of Jak2/STAT3 pathway, JSI-124 (cucurbitacin I), has been tested. JSI-124 is a member of the cucurbitacin family of compounds that are isolated from various plant families such as the Cucurbitaceae and Cruciferae. JSI-124 inhibits the cellular levels of phosphotyrosine-STAT3 and phospho-Jak2, but not phospho-ERK1/2, phospho-JNK, and phospho-Akt (Blaskovich et al. 2003). JSI-124 could downregulate phosphotyrosine-STAT3 levels by promoting the protein phosphatase activities of SHP1 and SHP2 (Schaper et al. 1998; Stofega et al. 1998). Alternatively, JSI-124 could also activate physiological inhibitors that are known to directly or indirectly downregulate STAT3 activation (Turkson and Jove 2000). Experiments in vitro have shown that JSI-124 overcomes the differentiation block induced by tumor-derived factors and promotes the differentiation of mature DC and macrophages. JSI-124 significantly reduced the presence of immature myeloid cells in vivo and promoted the accumulation of mature DC. In addition to a direct antitumor effect in several animal models, JSI-124 significantly enhanced the effect of cancer immunotherapy (Nefedova et al. 2005b). This indicates that pharmacologic inhibition of the JAK2/STAT3 pathway can be an important new therapeutic strategy to enhance antitumor activity of cancer immunotherapy. Inhibition of Jak2/STAT3 signaling resulted in a dramatic activation of immature DC generated in the presence of tumor-derived factors as well as in control medium. This activation manifested in upregulation of MHC class II, costimulatory molecules, and a dramatic increase in the ability to stimulate allogeneic or antigen-specific T cells. Inhibition of Jak2/STAT3 signaling resulted in the activation of the transcription factor NF-κB. This upregulation was not due to a conventional pathway involving IκBα, but was probably due to a block of the dominant-negative effect of STAT3 (Nefedova et al. 2005a).

In a different study, systemic administration of JSI-124 in glioma-bearing immunocompetent mice, but not athymic mice, resulted in prolonged survival, suggesting a role of adaptive immunity in the antitumor effect. Furthermore, JSI-124 promoted maturation of tumor-infiltrating DC and activation of tumor-conditioned cytotoxic T cells. When intraperitoneal JSI-124 administration was combined with intravenous transfer of Pmel-I mouse-derived type-1 CTL (Tc1), glioma-bearing mice exhibited prolonged survival compared with either therapy alone (Fujita et al. 2008). Combination of CpG with JSI-124 treatment resulted in synergistic antitumor effects compared to CpG or JSI-124 alone. Correlating with these findings, the combination therapy resulted in significantly higher intratumoral levels of several pro-inflammatory cytokines (IL-12, IFN-γ, TNF-α, and IL-2), increases in intratumoral CD8$^+$ and CD4$^+$ T cells, and increases in activated DC in the tumors and regional lymph nodes (Molavi et al. 2008).

A similar effect on antitumor immune response was reported recently for another compound – CPA-7. This is a platinum compound that blocks STAT3 activity at low micromolar concentrations. CPA-7 induces the regression of mouse CT26 colon tumor, which correlates with the abrogation of persistent STAT3 activity in tumors (Turkson et al. 2004). Targeting STAT3 with CPA-7 induces T-cell- and NK cell-dependent growth inhibition of established tumors otherwise resistant to direct killing by the inhibitor (Kortylewski et al. 2005).

Targeting of SOCS1 is another approach to regulate the Jak/STAT pathway in DC in cancer. Silencing SOCS1 in DC using siRNA strongly enhanced antigen-specific antitumor immunity, which was caused by hyper-production of IL-12 (Shen et al. 2004; Evel-Kabler et al. 2006). Unfortunately, these studies did not clarify the role of STAT1 and STAT3 in these effects of SOCS1 inhibition, and more studies are needed to understand the mechanism of the observed effects. Recently, Ling et al. reported that treatment of 4T1 breast tumor cells with novel triterpenoid C-28 methyl ester of 2-cyano-3, 12-dioxoolen-1, 9-dien-28-oic acid (CDDO-Me) resulted in inactivation of STAT3, Src, and Akt; reduction of c-Myc mRNA levels; and abrogation of invasive growth of 4T1 cells. CDDO-Me completely eliminated 4T1 breast cancer growth and lung metastases in vivo and restored the presence of mature DC in spleens (Ling et al. 2007).

10.6 Conclusions

Thus, recent data have demonstrated that STAT3 plays a critical role as a negative regulator of function of antigen-presenting cells, and DC in particular. Activation of STAT3 prevents maturation of DC and leads to the accumulation of immunosuppressive MDSC. Under physiological conditions this may protect organism from excessive immune reactivity and autoimmune abnormalities. However, persistent hyperactivation of STAT3 caused by tumor-derived factors results in the inability of DC to respond to different stimuli and to induce antigen-specific immune responses. Recent reports suggest that targeting this pathway may dramatically improve not only DC function but also, most importantly, antitumor immune reactivity. Coupled with the direct antitumor effect of STAT3 inhibitors, this approach can be very effective in the treatment of different types of cancer and can be potentially used to augment the effect of cancer vaccines.

References

Albert, M. L., Birthe, S. and Bhardwaj, N. 1998. Dendritic cells acquire antigen from apoptotic cells and induce class I-restricted CTLs. Nature 392:86–89.

Alexander, W. S., Starr, R., Metcalf, D., Nicholson, S. E., Farley, A., Elefanty, A. G., Brysha, M., Kile, B. T., Richardson, R., Baca, M., Zhang, J. G., Willson, T. A., Viney, E. M., Sprigg, N. S., Rakar, S., Corbin, J., Mifsud, S., DiRago, L., Cary, D., Nicola, N. A. and Hilton, D. J. 1999. Suppressors of cytokine signaling (SOCS): negative regulators of signal transduction. J Leukoc Biol 66:588–592.

Bharadwaj, U., Li, M., Zhang, R., Chen, C. and Yao, Q. 2007. Elevated interleukin-6 and G-CSF in human pancreatic cancer cell conditioned medium suppress dendritic cell differentiation and activation. Cancer Res 67:5479–5488.

Blaskovich, M. A., Sun, J., Cantor, A., Turkson, J., Jove, R. and Sebti, S. M. 2003. Discovery of JSI-124 (Cucurbitacin I), a selective janus kinase/signal transducer and activator of transcription 3 signaling pathway inhibitor with potent antitumor activity against human and murine cancer cells in mice. Cancer Res 63:1270–1279.

Bromberg, J. 2002. Stat proteins and oncogenesis. J Clin Invest 109:1139–1142.

Calo, V., Migliavacca, M., Bazan, V., Macaluso, M., Buscemi, M., Gebbia, N. and Russo, A. 2003. STAT proteins: from normal control of cellular events to tumorigenesis. J Cell Physiol 197:157–168.

Cheng, F., Wang, H. W., Cuenca, A., Huang, M., Ghansah, T., Brayer, J., Kerr, W. G., Takeda, K., Akira, S., Schoenberger, S. P., Yu, H., Jove, R. and Sotomayor, E. M. 2003. A critical role for Stat3 signaling in immune tolerance. Immunity 19:425–436.

Chung, J., Uchida, E., Grammer, T. C. and Blenis, J. 1997. STAT3 serine phosphorylation by ERK-dependent and -independent pathways negatively modulates its tyrosine phosphorylation. Mol Cell Biol 17:6508–6516.

Coffer, P. J., Koenderman, L. and de Groot, R. P. 2000. The role of STATs in myeloid differentiation and leukemia. Oncogene 19:2511–2522.

Dalpke, A. H., Opper, S., Zimmermann, S. and Heeg, K. 2001. Suppressors of cytokine signaling (SOCS)-1 and SOCS-3 are induced by CpG-DNA and modulate cytokine responses in APCs. J Immunol 166:7082–7089.

Darnell, J. E., Jr. 1997. STATs and gene regulation. Science 277:1630–1635.

Darnell, J. E., Jr., Kerr, I. M. and Stark, G. R. 1994. Jak-STAT pathways and transcriptional activation in response to IFNs and other extracellular signaling proteins. Science 264:1415–1421.

Evel-Kabler, K., Song, X. T., Aldrich, M., Huang, X. F. and Chen, S. Y. 2006. SOCS1 restricts dendritic cells' ability to break self tolerance and induce antitumor immunity by regulating IL-12 production and signaling. J Clin Invest 116:90–100.

Fujita, M., Zhu, X., Sasaki, K., Ueda, R., Low, K. L., Pollack, I. F. and Okada, H. 2008. Inhibition of STAT3 promotes the efficacy of adoptive transfer therapy using type-1 CTLs by modulation of the immunological microenvironment in a murine intracranial glioma. J Immunol 180:2089–2098.

Gabrilovich, D., Ciernik, F. and Carbone, D. P. 1996. Dendritic cells in anti-tumor immune responses. I. Defective antigen presentation in tumor-bearing hosts. Cell. Immunol 170:101–110.

Gabrilovich, D. I. 2004. The mechanisms and functional significance of tumour-induced dendritic-cell defects. Nat Rev Immunol 4:941–952.

Gabrilovich, D. I., Corak, J., Ciernik, I. F., Kavanaugh, D. and Carbone, D. P. 1997. Decreased antigen presentation by dendritic cells in patients with breast cancer. Clin Cancer Res. 3:483–490.

Greenhalgh, C. J. and Hilton, D. J. 2001. Negative regulation of cytokine signaling. J Leukoc Biol 70:348–356.

Greenhalgh, C. J., Rico-Bautista, E., Lorentzon, M., Thaus, A. L., Morgan, P. O., Willson, T. A., Zervoudakis, P., Metcalf, D., Street, I., Nicola, N. A., Nash, A. D., Fabri, L. J., Norstedt, G., Ohlsson, C., Flores-Morales, A., Alexander, W. S. and Hilton, D. J. 2005. SOCS2 negatively regulates growth hormone action in vitro and in vivo. J Clin Invest 115:397–406.

Hoentjen, F., Sartor, R. B., Ozaki, M. and Jobin, C. 2005. STAT3 regulates NF-{kappa}B recruitment to the IL-12p40 promoter in dendritic cells. Blood 105:689–696.

Huang, A. Y., Golumbek, P., Ahmadzadeh, M., Jaffee, E., Pardoll, D. and Levitsky, H. 1994. Role of bone marrow-derived cells in presenting MHC class I-restricted tumor antigens. Science 264:961–965.

Ihle, J. N. 1995. The Janus protein tyrosine kinases in hematopoietic cytokine signaling. Semin Immunol 7:247–254.

Imada, K. and Leonard, W. J. 2000. The JAK-STAT pathway. Mol Immunol 37:1–11.

Inaba, K., Turley, S., Yamaide, F., Iyoda, T., Mahnke, K., Inaba, M., Pack, M., Subklewe, M., Sauter, B., Sheff, D., Albert, M., Bhardwaj, N., Mellman, I. and Steinman, R. M. 1998. Efficient presentation of phagocytosed cellular fragments on the major histocompatibility complex class II products of dendritic cells. J Exp Med 188:2163–2173.

Jackson, S. H., Yu, C. R., Mahdi, R. M., Ebong, S. and Egwuagu, C. E. 2004. Dendritic cell maturation requires STAT1 and is under feedback regulation by suppressors of cytokine signaling. J Immunol 172:2307–2315.

Kitamura, H., Kamon, H., Sawa, S., Park, S. J., Katunuma, N., Ishihara, K., Murakami, M. and Hirano, T. 2005. IL-6-STAT3 controls intracellular MHC class II alphabeta dimer level through cathepsin S activity in dendritic cells. Immunity 23:491–502.

Kortylewski, M., Kujawski, M., Wang, T., Wei, S., Zhang, S., Pilon-Thomas, S., Niu, G., Kay, H., Mule, J., Kerr, W. G., Jove, R., Pardoll, D. and Yu, H. 2005. Inhibiting Stat3 signaling in the hematopoietic system elicits multicomponent antitumor immunity. Nat Med 11:1314–1321.

Kusmartsev, S. and Gabrilovich, D. 2005. STAT1 signaling regulates tumor-associated macrophage-mediated T cell deletion. J Immunol 174:4880–4891.

Laouar, Y., Welte, T., Fu, X.-Y. and Flavell, R. A. 2003. STAT3 Is required for Flt3L-dependent dendritic cell differentiation. Immunity 19:903–912.

Levy, C., Nechushtan, H. and Razin, E. 2002. A new role for the STAT3 inhibitor, PIAS3: a repressor of microphthalmia transcription factor. J Biol Chem 277:1962–1966.

Li, Y., Chu, N., Rostami, A. and Zhang, G. X. 2006. Dendritic cells transduced with SOCS-3 exhibit a tolerogenic/DC2 phenotype that directs type 2 Th cell differentiation in vitro and in vivo. J Immunol 177:1679–1688.

Liao, J., Fu, Y. and Shuai, K. 2000. Distinct roles of the NH2- and COOH-terminal domains of the protein inhibitor of activated signal transducer and activator of transcription (STAT) 1 (PIAS1) in cytokine-induced PIAS1-Stat1 interaction. Proc Natl Acad Sci USA 97:5267–5272.

Ling, X., Konopleva, M., Zeng, Z., Ruvolo, V., Stephens, L. C., Schober, W., McQueen, T., Dietrich, M., Madden, T. L. and Andreeff, M. 2007. The novel triterpenoid C-28 methyl ester of 2-cyano-3, 12-dioxoolen-1, 9-dien-28-oic acid inhibits metastatic murine breast tumor growth through inactivation of STAT3 signaling.
Cancer Res 67:4210–4218.

Molavi, O., Ma, Z., Hamdy, S., Lai, R., Lavasanifar, A. and Samuel, J. 2008 Synergistic antitumor effects of CpG oligodeoxynucleotide and STAT3 inhibitory agent JSI-124 in a mouse melanoma tumor model. Immunol Cell Biol Apr 8 [Epub ahead of print].

Nadiminty, N., Lou, W., Lee, S. O., Lin, X., Trump, D. L. and Gao, A. C. 2006. Stat3 activation of NF-{kappa}B p100 processing involves CBP/p300-mediated acetylation. Proc Natl Acad Sci USA 103:7264–7269.

Nefedova, Y., Cheng, P., Gilkes, D., Blaskovich, M., Beg, A. A., Sebti, S. M. and Gabrilovich, D. I. 2005a. Activation of dendritic cells via inhibition of Jak2/STAT3 signaling. J Immunol 175:4338–4346.

Nefedova, Y., Huang, M., Kusmartsev, S., Bhattacharya, R., Cheng, P., Salup, R., Jove, R. and Gabrilovich, D. 2004. Hyperactivation of STAT3 is involved in abnormal differentiation of dendritic cells in cancer. J Immunol 172:464–474.

Nefedova, Y., Nagaraj, S., Rosenbauer, A., Muro-Cacho, C., Sebti, S. M. and Gabrilovich, D. I. 2005b. Regulation of dendritic cell differentiation and antitumor immune response in cancer by pharmacologic-selective inhibition of the janus-activated kinase 2/signal transducers and activators of transcription 3 pathway. Cancer Res 65:9525–9535.

Nestle, F. O., Burg, G., Fah, J., Wrone-Smith, T. and Nickoloff, B. J. 1997. Human sunlight-induced basal-cell-carcinoma-associated dendritic cells are deficient in T cell co-stimulatory molecules and are impaired as antigen-presenting cells. Am J Pathol 150:641–651.

Onai, N., Obata-Onai, A., Tussiwand, R., Lanzavecchia, A. and Manz, M. G. 2006. Activation of the Flt3 signal transduction cascade rescues and enhances type I interferon-producing and dendritic cell development. J Exp Med 203:227–238.

Park, S. J., Nakagawa, T., Kitamura, H., Atsumi, T., Kamon, H., Sawa, S., Kamimura, D., Ueda, N., Iwakura, Y., Ishihara, K., Murakami, M. and Hirano, T. 2004. IL-6 regulates in vivo dendritic cell differentiation through STAT3 activation. J Immunol 173:3844–3854.

Rane, S. G. and Reddy, E. P. 2002. JAKs, STATs and Src kinases in hematopoiesis. Oncogene 21:3334–3358.

Rane, S. G. and Reddy, E. S. 2000. Janus kinases: components of multiple signaling pathways. Oncogene 19:5662–5679.

Rathinam, C., Geffers, R., Yucel, R., Buer, J., Welte, K., Moroy, T. and Klein, C. 2005. The transcriptional repressor Gfi1 controls STAT3-dependent dendritic cell development and function. Immunity 22:717–728.

Rock, K. L., Rothstein, L., Gamble, S. and Fleischacker, C. 1993. Characterization of antigen-presenting cells that present exogenous antigens in association with class I MHC molecules. J Immunol 150:438–446.

Schaper, F., Gendo, C., Eck, M., Schmitz, J., Grimm, C., Anhuf, D., Kerr, I. M. and Heinrich, P. C. 1998. Activation of the protein tyrosine phosphatase SHP2 via the interleukin-6 signal transducing receptor protein gp130 requires tyrosine kinase Jak1 and limits acute-phase protein expression. Biochem J 335(Pt 3):557–565.

Shen, L., Evel-Kabler, K., Strube, R. and Chen, S. Y. 2004. Silencing of SOCS1 enhances antigen presentation by dendritic cells and antigen-specific anti-tumor immunity. Nat Biotechnol 22:1546–1553.

Shuai, K. and Liu, B. 2003. Regulation of JAK-STAT signalling in the immune system. Nat Rev Immunol 3:900–911.

Shuai, K. and Liu, B. 2005. Regulation of gene-activation pathways by PIAS proteins in the immune system. Nat Rev Immunol 5:593–605.

Shurin, M. and Gabrilovich, D. I. 2001. Regulation of dendritic cell system by tumor. Cancer Res Ther Control 11:65–78.

Smithgall, T. E., Briggs, S. D., Schreiner, S., Lerner, E. C., Cheng, H. and Wilson, M. B. 2000. Control of myeloid differentiation and survival by Stats. Oncogene 19:2612–2618.

Starr, R., Willson, T. A., Viney, E. M., Murray, L. J., Rayner, J. R., Jenkins, B. J., Gonda, T. J., Alexander, W. S., Metcalf, D., Nicola, N. A. and Hilton, D. J. 1997. A family of cytokine-inducible inhibitors of signalling. Nature 387:917–921.

Stofega, M. R., Wang, H., Ullrich, A. and Carter-Su, C. 1998. Growth hormone regulation of SIRP and SHP-2 tyrosyl phosphorylation and association. J Biol Chem 273:7112–7117.

Sumimoto, H., Imabayashi, F., Iwata, T. and Kawakami, Y. 2006. The BRAF-MAPK signaling pathway is essential for cancer-immune evasion in human melanoma cells. J Exp Med. 203:1651–1656.

Takeda, K., Clausen, B. E., Kaisho, T., Tsujimura, T., Terada, N., Forster, I. and Akira, S. 1999. Enhanced Th1 activity and development of chronic enterocolitis in mice devoid of Stat3 in macrophages and neutrophils. Immunity 10:39–49.

Tsukada, J., Ozaki, A., Hanada, T., Chinen, T., Abe, R., Yoshimura, A. and Kubo, M. 2005. The role of suppressor of cytokine signaling 1 as a negative regulator for aberrant expansion of CD8alpha + dendritic cell subset. Int Immunol 17:1167–1178.

Turkson, J. and Jove, R. 2000. STAT proteins: novel molecular targets for cancer drug discovery. Oncogene 19:6613–6626.

Turkson, J., Zhang, S., Palmer, J., Kay, H., Stanko, J., Mora, L. B., Sebti, S., Yu, H. and Jove, R. 2004. Inhibition of constitutive signal transducer and activator of transcription 3 activation by novel platinum complexes with potent antitumor activity. Mol Cancer Ther 3:1533–1542.

Wang, R., Cherukuri, P. and Luo, J. 2005. Activation of Stat3 sequence-specific DNA binding and transcription by p300/CREB-binding protein-mediated acetylation. J Biol Chem 280:11528–11534.

Wang, T., Niu, G., Kortylewski, M., Burdelya, L., Shain, K., Zhang, S., Bhattacharya, R., Gabrilovich, D., Heller, R., Coppola, D., Dalton, W., Jove, R., Pardoll, D. and Yu, H. 2004. Regulation of the innate and adaptive immune responses by Stat-3 signaling in tumor cells. Nat Med 10:48–54.

Williams, J. G. 1999. Serpentine receptors and STAT activation: more than one way to twin a STAT. Trends Biochem Sci 24:333–334.

Yuan, Z. L., Guan, Y. J., Chatterjee, D. and Chin, Y. E. 2005. Stat3 dimerization regulated by reversible acetylation of a single lysine residue. Science 307:269–273.

Chapter 11
Tumor-Associated Inflammation and Impact on Dendritic Cell Function

Zoltan Pos, Marianna Sabatino, Tara Spivey, Hui Liu, Andras Falus, and Francesco M. Marincola

Abstract Tumor-associated inflammation is a frequently observed phenomenon considered to be one of the major hallmarks of neoplastic disease progression. Recent data show that inflammatory processes around malignancies are able to both support and suppress neoplastic progression depending on the phase of tumor progression or the cytokine context of ongoing immunological processes in the tumor microenvironment. Recent concepts of malignancy-associated inflammation and its impact on dendritic cell functions suggest that the net effect of inflammation on the balance between tumor growth and dendritic cell-controlled immunity is continuously changing over time, as inflammatory signals are frequently re-interpreted along with cancer progression. Initially, inflammation sustains malignant conversion and supports survival of tumor cells in cryptic cancers; however, it also allows their recognition by dendritic cells via damage-associated molecular patterns. In progressing cancers, inflammation contributes to malignant invasion, angiogenesis, and metastasis formation. In addition, via suppressing dendritic cell activation, maturation, and disrupting communication between dendritic cells and NK or T cells, it also corrupts antitumor immune responses launched by both the innate and adaptive immune systems. On the other hand, upon appropriate stimulation of specific toll-like receptors, tumor-infiltrating myeloid and plasmacytoid dendritic cells can be successfully activated, leading to harsh inflammatory reactions ultimately resulting in rapid rejection of tumor cells. Thus, ubiquitous presence of inflammation around tumors can be exploited not only for early detection of cryptic malignant lesions but also for inducing dendritic cell-mediated rejection of established malignancies.

F.M. Marincola (✉)
NIH CC, Bldg 10, R-1C711, Bethesda, MD 20892, USA
e-mail: FMarincola@cc.nih.gov

M.R. Shurin, R.D. Salter (eds.), *Dendritic Cells in Cancer*,
DOI 10.1007/978-0-387-88611-4_11, © Springer Science+Business Media, LLC 2009

11.1 Introduction: Tumor-Associated Inflammation

The idea that local inflammation plays a fundamental role in both cancer development and progression was originally suggested by Virchow in the nineteenth century. Virchow noted that tumors frequently arise at sites of sustained inflammation and that tumor development is often associated with the presence of inflammatory cells, i.e., a "lymphoreticular infiltrate". Soon, the term tumor-associated inflammation was coined to describe this phenomenon, and indeed, in recent years it has become increasingly clear that there is a complex interplay between tumors and inflammation during tumor progression. In this chapter, we will review our current knowledge about inflammation-mediated effects on tumor progression and describe the key factors affecting dendritic cell (DC) function and DC-controlled immune responses in the special microenvironment of a tumor-associated inflammation.

For a long time, one of the most important questions about tumor-associated inflammation has been if inflammation had detrimental or curative effects in cancer. For decades, heated debates have been conducted in this field, and indeed, there is a large body of evidence to support both concepts. In our opinion, however, these simplified paradigms are not appropriate to explain the true nature of inflammation in cancer, simply because the interaction between tumors and tumor-associated inflammation is continuously re-interpreted along with cancer progression. It seems that inflammation can have strikingly different effects on tumors depending on the phase of progression. We will demonstrate that inflammation has different impacts on (1) a proto-cancer consisting of a small group of cancer cells that have yet to adapt to their environment, survive the initial attacks of the immune system, and cope with numerous mutations acquired during malignant transformation that are actually incompatible with cell survival; (2) an established cancer being in transition to the phase of active growth, invasion, angiogenesis, and ultimately, formation of an altered self-"organ" in the host; (3) a clinically manifested, established tumor in the rare cases of spontaneous or treatment-induced tumor rejection, when a large solid tumor mass collides with a massive immune response, the latter somehow managing to break through the walls of a tumor-created immune tolerance; and (4) in the terminal phase of the disease, when cancer cells ultimately escape immune control. In other words, when interpreting the impact of inflammation on tumors, one major bias responsible for many apparently contradictory, but actually not necessarily confounding, results might have been the tumor type and the progression phase investigated, hence, the heterogeneity of cancer and the differences between different experimental models (Bui and Schreiber 2007). Therefore, in this chapter, we will make an attempt to separate these effects from each other and analyze the impact of tumor-associated inflammation on DC functions depending on the phase of tumor progression.

In addition, based on data available about the many interactions between inflammation, DC, and innate immune mechanisms in tumor rejection, we also intend to demonstrate that a major re-evaluation of the role of DC in antitumor responses, and immune-mediated tumor rejection in particular, is becoming more and more inevitable. Reviewing the lessons learned from T-cell-based immunotherapy so far, we will suggest that until recently, the overall importance of DC-induced T-cell response, and especially DC–T-cell communication in lymph nodes, has been somewhat overrated in the process of both spontaneous and therapy-induced rejection of established tumors, and we are just beginning to realize the actual importance of danger signals, local inflammation, and a successfully activated innate immune system in these processes. In our opinion, a growing body of evidence suggests that triggering a harsh inflammation and mounting an innate immune response are two major prerequisites of successful tumor elimination, as only they can provide the required support for sustained T-cell activation in the otherwise immunologically compromised environment of tumors. In our opinion, this might lead to a new, refined view of tumor rejection, where T-cell-mediated antigen recognition represents rather visualization of tumor cells for the innate immune system, than actual killing power, and where not a T-cell-mediated elimination of tumor cells but rather a strong T-cell-directed innate cellular burst and the associated tissue damage are the key mechanisms for rejection of established cancer (Shanker et al. 2007).

11.2 Inflammation and Dendritic Cell Functions in the Proto-cancer

In the natural history of cancer, inflammation is present from the very beginning of the malignant transformation. Persistent inflammatory processes, usually being consequences of unresolved bacterial (e.g., *Helicobacter pylori*) or viral infections (e.g., HCV), permanent chemical (e.g., asbestos, tobacco fume) or mechanical tissue stress, play a key role in the early steps of tumorigenesis (Coussens and Werb 2002; Mager 2006). It has been shown that such chronic inflammations, accompanied by constitutive activation of the tumor necrosis factor (TNF)-α – nuclear factor-κB (NF-κB) signaling pathway (Aggarwal et al. 2006; Karin 2006), create a supportive environment for the transformation of normal cells into neoplastic cancer cells via multiple ways. Most importantly, inflammation supports mutagenesis via enhanced production of reactive oxygen and nitrogen intermediates (Maeda and Akaike 1998) and also supports initial survival of transformed cells via abnormal signaling through TNF-α (Aggarwal et al. 2006) or by highjacking p53-mediated cell cycle control through persistent signaling by macrophage migration inhibitory factor (MIF) (Hudson et al. 1999). On the other hand, besides favoring the transition between normal and transformed neoplastic cells, inflammation associated to permanent tissue stress also creates a hostile environment for emerging tumor cells, as it supports their recognition by DC and other immune cells. As the overwhelming majority of proto-cancer cells are not yet well adapted to their environment, they initially succumb in large numbers to the challenges they face because of their intrinsic genomic instability, shortage of nutrients, hypoxia, lack of space and appropriate amounts of growth factors, or, ironically, even the inflammatory tissue stress that originally fostered their malignant transformation. Data show that DCs are able to recognize conserved molecular patterns appearing on the surface or in the local environment of cells dying due to intrinsic or extrinsic stress factors, also stress associated to inflammation, and it seems that they show particularly high affinity for dying tumor cells (Sauter et al. 2000; Viorritto et al. 2007; Dhodapkar et al. 2008). This recognition is based on the fact that cells affected by permanent stress exhibit so-called damage-associated molecular pattern molecules (DAMP). DAMP include several, otherwise intracellular, molecules translocated to the plasma membrane or to the extracellular compartment of stress-affected or dying cells, such as the heat shock proteins HSP70 and HSP90, the intracellular signaling molecule calreticulin, but also RNA, DNA, crystallized particles of uric acid, or the protein high-mobility group box 1 (HMGB1), originally serving as a non-histone chromatin-binding protein, but when released by dying cells, also heralding massive cellular stress (Tesniere et al. 2008). Studies show that DCs become rapidly alarmed when recognizing these signals using their impressive armamentarium of pattern recognition receptors, mainly via toll-like receptors (TLR) and nod-like receptors (NLR) (Rock and Kono 2008).

As for toll-like receptors, HSP70 and HSP90 were suggested to be recognized via TLR4; HMGB1 by TLR2, TLR4, or the receptor for advanced glycosylation end products (RAGE); uric acid through TLR2; DNA via TLR9; and RNA on the TLR3 receptor (Tesniere et al. 2008). Dying cells also release ample amounts of inflammatory cytokines such as tumor TNF-α, IL-6, or IL-8 and stimulate nearby immune cells such as macrophages or NK cells; consequently, DC activation, maturation, and release of further inflammatory cytokines take place. Interestingly, it was also shown that in DC, HSP not only allow recognition and uptake of debris from stress-affected cells, but also support processing and DC-mediated cross-presentation of rescued antigens to T cells (Tesniere et al. 2008).

Besides TLR, NLR also participate in recognition of dying cells and activation of DC (Carneiro et al. 2008). The cytoplasmatic nod-like receptor protein NALP3 is able to recognize several stress-related molecules such as uric acid, and ATP, and via binding the adaptor protein ASC, it recruits the inflammatory caspase-1. This protein complex, also called inflammasome, is able to convert the proproteins of the inflammatory cytokines IL-1β and IL-18, thereby allowing for their selection and further enhancement of the local immune response (Panelli et al. 2007). Interestingly, it has been shown that also self-DNA can activate the inflammasome in a NALP3-independent manner; thus, several DAMP released by dying cells can efficiently activate the inflammatory response of DC in a TLR-independent fashion too (Muruve et al. 2008).

Hence, it seems that the inflammatory environment surrounding a proto-cancer might provide the appropriate signals not only for the recognition and uptake of tumor antigens but also for DC maturation and subsequent activation of a T-cell-mediated immune response. However, as little is known about immune evolution of small cancer cell clusters (Dunn et al. 2002), and results are particularly contradictory in terms of cancer arising in chronically inflamed tissues (Bui and Schreiber 2007), information available about the actual relevance of DC-mediated T-cell induction in the elimination of cryptic cancers is scarce. As for indirect evidence, data show that in $RAG^{-/-}$ mice or HIV-infected humans with fully developed AIDS, thus in hosts with a compromised adaptive immune system, tumors arise at higher frequencies than in normal hosts having an intact adaptive antitumor immunity (Shankaran et al. 2001; Grulich et al. 2007). In addition, there is also direct evidence that T-cell-mediated immunity actively seeks and efficiently destroys cryptic cancer cells; e.g., in a methylcholanthrene-induced mouse sarcoma model, it has been shown that growth of sporadic cancer is controlled mainly by T cells and not the innate immune arm of the antitumor response (Koebel et al. 2007). On the other hand, in the same model, it was also shown that although sustained TNF-α signaling might support malignant transformation, at least for a proto-cancer, the same inflammatory pathway is also responsible for initial immune editing, thus suggesting that DC present in the inflamed environment of a cryptic tumor are fully functional in capturing and presenting antigens to specific T cells in locoregional lymph nodes. To sum up, in line with the cancer immunoediting theory (Dunn et al. 2002, 2004), the data available in the literature suggest that at least in this initial phase of tumor progression, spontaneously arising tumor cells, particularly in an inflammatory environment, are highly immunogenic, easy to recognize by DC, and thus very vulnerable to immune-mediated elimination.

11.3 Inflammation and Dendritic Cell Functions in Established Tumors

As cancer evolves from a proto-cancer state to an established tumor, its influence on inflammatory processes in its local microenvironment becomes more and more apparent. Cancer cells in established tumors typically secrete various cytokines and inflammatory mediators such as TNF-α, IL-1, and IL-6 in order to sustain atypical chronic inflammations in their local environment (Balkwill and Mantovani 2001; Aggarwal et al. 2006). Moreover, they actively recruit inflammatory cells via CC and CXC chemokines, creating an unusual inflammatory infiltrate consisting of selected subsets of macrophages, lymphocytes, DC, mast cells but usually low numbers of neutrophil granulocytes (Talmadge et al. 2007). This typical immune infiltrate, a mixture of type 2 (M2) macrophages, CD11b+ Gr-1+ myeloid-derived suppressor cells (MDSC), T cells having typical characteristics of the memory and regulatory phenotype, immature and tolerogenic DC, and immune complexes derived from B cell-secreted antibodies, actively contributes to tumor-associated abnormal inflammatory processes (Talmadge et al. 2007; Tan and Coussens 2007). On the other hand, this tumor-controlled inflammation never results in mobilization of an effective cellular response and elimination of the tumor cells.

The permanent presence of such smoldering tumor-associated inflammations for established tumors is of outstanding importance, for many reasons. First of all, inflammatory cytokines such as TNF-α, IL-1, and IL-6 often support growth and survival of tumor cells both via autocrine and paracrine signaling loops (Balkwill and Mantovani 2001; Aggarwal et al. 2006). Second, such tumor-associated inflammations were shown to enhance cancer invasion by fostering local matrix reorganization, corrupting the homeostasis of extracellular matrix components, e.g., by affecting the control of the production of collagen (Diegelmann and Evans 2004), and also forcing breakdown of the normal matrix structure by enhanced synthesis and release of matrix metalloproteinases like MMP-3 or MMP-9 (Arnott et al. 2002). Tumor-controlled inflammation can also facilitate motility of cancer cells invading neighboring tissues by preferentially upregulating diverse chemokines and chemokine receptors on tumor cells (Kedrin et al. 2007). Moreover, tumor-associated inflammation supports angiogenesis via stimulating the secretion of a limited array of CXC chemokines skewed toward the proangiogenic glutamic acid–leucine–arginine positive (ELR+) CXC family members, such as CXCL1, CXCL2, CXCL3, CXCL5, CXCL6, and CXCL8 (Strieter et al. 2006). These CXC ligands stimulate endothelial cell migration and angiogenic capillary vessel sprouting by activating the CXCR2 receptor on endothelial cells. In contrast, non-ELR+ CXC chemokines, many of them interferon inducible and having potent antiangiogenic properties, are typically absent in tumor-associated inflammation (Strieter et al. 2006). Finally, tumor-associated inflammation is definitively able to impede antitumor immune responses and create local immunosuppression, as well (Ben-Baruch 2006).

There is evidence that tumor-associated inflammation in established cancers can lead to severe impairment of DC functions. Several tumor-derived pro-inflammatory mediators, such as IL-6 (Chomarat et al. 2000), and also prostanoid products of the cyclooxygenases 1 and 2 (COX-1 and COX-2), such as prostaglandins and thromboxanes, massively interfere with DC maturation and their ability to mount Th1-type immune responses (Sombroek et al. 2002). Interestingly, these inflammatory mediators are present in the supernatant of cancer cells freshly isolated from clinically manifested tumors, but not established tumor cell lines, further emphasizing their importance in the context of the inflammatory tissue environment, but not in the growth of tumor cells per se. Prostanoids were shown to synergize with the inflammatory cytokine IL-6 in cancer-mediated suppression of DC maturation (Sombroek et al. 2002), and, recently it was shown that they not only compromise interactions between DC and T cells, but also impede communication between DC and NKT cells (Torres et al. 2008).

Besides inflammatory signals originating from the tumor itself, also tumor-attracted inflammatory cells hamper DC functions. First of all, several tumor-derived chemokines were shown to attract macrophages to the tumor environment, many of them members of the CC chemokine family, CCL2 having paramount importance in this phenomenon (Sica et al. 2006). The micromilieu of established tumors, rich in IL-10, prostaglandin E2 (PgE2), and immune complexes, lacking, however, IFN-α, LPS, or other danger signals, massively skews the phenotype of tumor-infiltrating macrophages (TAM) by inhibiting their maturation to DC and converting them into macrophages typically showing properties of M2 macrophages (Mantovani et al. 2008; Martinez et al. 2008). M2 cells display low levels or absent IL-12 production, diminished capacity to release NO or reactive oxygen intermediates (ROI), but on the other hand, actively secrete IL-10. Data show that TAM attracted to the tumor microenvironment are not only generally unable to exert antitumor functions, or support cellular immune response against a tumor, but also contribute to the suppression of DC-induced adaptive immunity, mainly by their altered cytokine profile containing large amounts of IL-10, VEGF, and IL-6, a combination not favoring DC antigen uptake, activation, and maturation (Sica et al. 2006; Mantovani et al. 2008).

Besides macrophages, also native CD4+ CD25+ regulatory T cells, expressing the forkhead box P3 (FOXP3) transcription factor, and the antigen-induced FOXP3-negative, IL-10, or TGF-β-secreting Treg cells are important components of the inflammatory infiltrate around tumors (Cools et al. 2007). These cells also heavily interfere with DC maturation both by direct cell–cell contact involving cross-binding of DC-displayed CD80 or CD86 with CTLA-4 on Treg, resulting in upregulation of the enzyme indoleamine 2,3-dioxygenase, a protein depleting the amino acid tryptophan and thereby inducing T-cell anergy (Fallarino et al. 2003), or by their suppressive cytokine milieu, as both Treg-derived IL-10 and TGF-β were repeatedly shown to suppress or inhibit upregulation of costimulatory molecules and MHC proteins on DC (Moore et al. 1993; Li et al. 2006).

11.4 Inflammation and Dendritic Cell Functions During Rejection of Established Tumors

Spontaneous rejection of established tumors by the host is a rare but consistently observed event. In the last decades, many attempts were made to induce rejection of established tumors based on the assumption that efficient, DC-mediated presentation of tumor antigens can result in T-cell-mediated tumor rejection. According to this concept, tumor rejection can be achieved by vaccinating cancer patients with purified tumor antigens, tumor antigens mixed with adjuvants, adoptively transferring DC, previously ex vivo loaded with tumor antigens and also stimulated by adjuvants and/or cytokines, or DC fused with tumor cells, hence, by creating activated mature DC migrating to the local lymph nodes, capable to present tumor antigens along with appropriate costimulatory signals to T cells, a process ultimately resulting in T-cell-mediated elimination of the tumor mass. After many disappointing results, however, it became clear that even under optimal conditions, the mere presence of mature, antigen-loaded DC in secondary lymphoid organs draining a tumor is usually not sufficient to elicit T-cell-mediated rejection (Rosenberg et al. 2004). We believe that the failure of active-specific immunization as an effective antitumor therapy was due to the fact that though generation of antigen-specific T cells might be relatively easily achievable using this approach, maintenance of effector T cells in the circulation, their direction to and re-activation at the site of the tumor were more difficult tasks than previously expected (Marincola et al. 2003; Mocellin et al. 2004). Results showed that shortly after leaving lymph nodes, T cells generated by DC-based vaccination usually acquire a quiescent phenotype, infiltrate tumors in numbers negligible compared to the size of the lesion, and usually do not show effector functions in the direct environment of the tumor (Monsurro et al. 2004; Mantovani et al. 2008). In fact, it was demonstrated that in order to achieve complete rejection of a solid tumor mass by T-cell-mediated adaptive immunity alone, tumor-infiltrating T cells must be isolated, transfected with TCR showing high affinity against dominant immunogenic tumor antigens, ex vivo expanded by IL-2, adoptively transferred in very large numbers to patients, and also supported by additional cytokines or antibodies neutralizing the immunosuppressive effects of the tumor or by chemo-/radiotherapy directly ablating the tumor-corrupted immune system of the host (Rosenberg et al. 2008).

Interestingly, however, it has also been shown that rapid rejection of established tumors can be induced by using a completely different approach, i.e., by converting smoldering tumor-associated inflammations into harsh inflammatory reactions. Even more interesting is the evidence suggesting that DC actively participate in these processes too. Investigation of the imidazoquinolines, a group of small molecular inducers of inflammation, showed that several members of this family, particularly imiquimod, resiquimod, or gardiquimod, are able to induce rejection of established tumors not only by activating tumor-specific T cells, but even more importantly, affecting tumor-associated inflammation (Schon and Schon 2008). Upon topical application on several tumors, these molecules activate TLR7 and/or TLR8 on several cells in the tumor stroma, inducing harsh inflammations inside and around the affected tumors, frequently resulting in tumor regression or even complete rejection of established cancer. Parallel to its pro-inflammatory effects, others have observed that imiquimod also induces apoptotic mechanisms in the tumor, an effect unrelated but probably also supporting immune-mediated tumor elimination (Schon and Schon 2008).

Based on the above observations, an imiquimod cream (Aldara) has been successfully evaluated in the clinical management of several unrelated tumors of the skin, and ultimately, this treatment approach also gained approval for clinical use by the FDA. Investigations conducted on imiquimod's mode of action revealed that imiquimod-induced TLR activation, resembling classical danger signals for the innate immune system, triggers a brisk infiltration of these tumors by many different DC subtypes, including not only skin-resident myeloid DC and Langerhans cells but also plasmacytoid DC (Schiller et al. 2006). In addition to DC, also macrophages, T cells, B cells, and large numbers of NK cells infiltrate the tumor; however, it was shown that the cell fraction most rapidly and sensitively reacting on imiquimod challenge is the DC compartment, particularly plasmacytoid DC (Schiller et al. 2006; Schon and Schon 2008). It seems that upon imiquimod treatment, pDC rapidly infiltrate the tissue affected, release IFN-α, thereby producing an initial inflammation, activating bystander macrophages and NK cells, which in turn release a vast array of cytokines and chemokines. These ligands include not only many inflammatory mediators (TNF-α, IL-6) and chemokines responsive for further recruitment of other inflammatory cells (CCL2, CCL3, CCL4, CXCL8), but also cytokines supporting DC survival and maturation (G-CSF, GM-CSF, IL-12) and mediators capable of triggering antitumor immune responses both by the native and adaptive arm of the immune system (IFN-α, IFN-γ, TNF-α, IL-2, IL-12) (Schon and Schon 2008). In fact, results show that these signals license both Langerhans cells and myeloid DC to migrate to local lymph nodes and activate a typically Th1-committed adaptive immune response (Schiller et al. 2006; Panelli et al. 2007; Papadavid et al. 2007).

Interestingly, however, the case of imiquimod also underlines the importance of local inflammation not only in the afferent, but in the effector phase of antitumor immune responses too. It was shown that imiquimod-induced tumor rejection happens in imiquimod-treated lesions of a given patient but not in others, even if they are in the close vicinity of the treatment-affected regressing tumor (Schon and Schon 2008). In other words, even if Th1-polarized T cells are successfully generated in lymph nodes draining a tumor, they are probably not capable to mount an effective immune response unless they infiltrate a massively inflamed tumor tissue. This observation is in line with the old concept suggesting that inflammation, inflammation-induced costimulatory molecules on activated accessory cells of the native immune system, such as DC, might be required not only to induce but also to direct and locally sustain an effective T-cell response (Kroczek et al. 2004; Marelli-Berg et al. 2007; Ward and Kaufman 2007; Mantovani et al. 2008). Apparently, for antitumoral T-cell responses, this secondary inflammatory support is crucial for T cells in the effector phase, and the case of imiquimod demonstrates that such harsh inflammations can be indeed strong enough to overcome the effects of the otherwise heavily T-cell suppressive milieu of an established, solid tumor.

Moreover, data suggest that imiquimod-induced tumor rejection relies heavily on innate immune mechanisms, most importantly NK cells, but also DC themselves. Recently, convincing data were presented showing that imiquimod treatment also triggers activation of NK cells, and these cells contribute to tumor rejection too. Imiquimod-induced NK cell activation may happen both directly (via TLR7) and indirectly, because engagement of TLR8 on DC or other accessory cells provides both appropriate cytokine and contact signals for NK cell activation, maturation, and also elicitation of their full killing activity (Gorski et al. 2006). Until recently, not much has been known about the exact mechanisms of DC–NK cell communication; however, recent studies shed some light in this process. It has been demonstrated that full activation of NK cells requires DC support at several levels, both inside and outside the secondary lymphoid organs, and of note, evidence is available that IFN-α released by TLR-stimulated bystander cells is critical for licensing DC for NK cell activation (Kijima et al. 2008). DC support for NK cell activation involves both secretion of NK-stimulatory cytokines, such as IFN-α, IL-12, and IL-15 (Arina et al. 2007), and direct cell-to-cell contact via IL-15 trans-presentation (Lucas et al. 2007), interaction between DC-presented Jagged2 and NK cell-displayed Notch (Kijima et al. 2008), or membrane-bound TNF-α on DC and TNF receptor type 2 on the surface of NK cells (Xu et al. 2007).

Finally, it has been shown that imiquimod-induced inflammation is also able to convert resident DC into so-called killer DC that have the potential to kill tumor cells directly. It was observed that both myeloid and plasmacytoid DC can acquire a killer phenotype upon TLR stimulation, the former typically producing perforin and granzyme B, while the latter expressing TRAIL (Stary et al. 2007). To what extent this pathway actually contributes to inflammation-mediated tumor regression remains to be clarified.

11.5 Inflammation and Dendritic Cell Function in Terminal Phase Cancer Ultimately Escaping Immune Control

Metastasis formation and complete escape from immune control are considered to be hallmarks of terminal phase disease progression in cancer. Alternating phases of tumor evasion and spontaneous and/or treatment-induced regression are supposed to happen repeatedly along with cancer progression; however, it is not yet clear if there are any, and if yes, what are the key steps leading to, and the basic criteria to be fulfilled for complete immunological escape. Actually, we do not even know what are the unique characteristics of tumors (if any) successfully escaping immune control.

Likewise, information about the role of inflammation in terminal phase cancer is scarce. There is some evidence that besides the obvious selective pressure represented by the immune system, i.e., step by step elimination of all immunogenic clones from the heterogenous tumor mass, paradoxically, also treatment- or regression-induced inflammation can contribute to immune escape and metastasis formation. First of all, it was suggested that even danger signals converting smoldering tumor-associated inflammation into harsh inflammations can support tumor survival. In fact, there is some evidence that tumor cells also frequently express receptors for danger signals, and stimulation of TLR on tumor cells can indeed provide survival advantage for some clones (Chen et al. 2008). Similarly, tumors can adapt astonishingly well even to aggressive inflammation-inducing immunotherapies, such as systemic IL-2 treatment, as there are reports suggesting that IL-2-dependent tumor cell growth and survival can happen in patients (Pawelec 2004). In addition, tumor-associated macrophages and mast cells were proposed to release proteolytic enzymes facilitating permeabilization of capillaries in the local circulation and thereby support metastatic emigration of tumor cells from the primary tumor. Finally, it has been suggested that inflammation can contribute to metastatic colony formation also by affecting chemokines, chemokine receptors expressed by cancer cells, and by directly supporting survival of cancer cells in freshly formed metastatic foci too (DeNardo et al. 2008).

However, the question to what extent these processes actually contribute to terminal phase disease progression remains to be further clarified, and so does the net effect of inflammation on dendritic cells in metastatic cancer.

11.6 Summary

To sum up, inflammation is one of the core components of the neoplastic phenotype. It has dramatic effects both on autonomous and immune-assisted tumor progression and also affects DC functions through many ways. In our opinion, chronic smoldering inflammation is a unique property of developing neoplasms that must be seriously considered as a potential target for future tumor therapies, not only as a factor discriminating between healthy tissues and cancer lesions but also as a phenomenon that can be exploited by converting it into a harsh inflammation fueling tumor rejection.

References

Aggarwal, B. B., Shishodia, S., Sandur, S. K., Pandey, M. K. and Sethi, G. 2006. Inflammation and cancer: how hot is the link? Biochem Pharmacol 72:1605–1621.

Arina, A., Murillo, O., Dubrot, J., Azpilikueta, A., Alfaro, C., Perez-Gracia, J. L., Bendandi, M., Palencia, B., Hervas-Stubbs, S. and Melero, I. 2007. Cellular liaisons of natural killer lymphocytes in immunology and immunotherapy of cancer. Expert Opin Biol Ther 7:599–615.

Arnott, C. H., Scott, K. A., Moore, R. J., Hewer, A., Phillips, D. H., Parker, P., Balkwill, F. R. and Owens, D. M. 2002. Tumour necrosis factor-alpha mediates tumour promotion via a PKC alpha- and AP-1-dependent pathway. Oncogene 21:4728–4738.

Balkwill, F. and Mantovani, A. 2001. Inflammation and cancer: back to Virchow? Lancet 357:539–545.

Ben-Baruch, A. 2006. Inflammation-associated immune suppression in cancer: the roles played by cytokines, chemokines and additional mediators. Semin Cancer Biol 16:38–52.

Bui, J. D. and Schreiber, R. D. 2007. Cancer immunosurveillance, immunoediting and inflammation: independent or interdependent processes? Curr Opin Immunol 19:203–208.

Carneiro, L. A., Magalhaes, J. G., Tattoli, I., Philpott, D. J. and Travassos, L. H. 2008. Nod-like proteins in inflammation and disease. J Pathol 214:136–148.

Chen, R., Alvero, A. B., Silasi, D. A., Steffensen, K. D. and Mor, G. 2008. Cancers take their Toll–the function and regulation of Toll-like receptors in cancer cells. Oncogene 27:225–233.
Chomarat, P., Banchereau, J., Davoust, J. and Palucka, A. K. 2000. IL-6 switches the differentiation of monocytes from dendritic cells to macrophages. Nat Immunol 1:510–514.
Cools, N., Ponsaerts, P., Van Tendeloo, V. F. and Berneman, Z. N. 2007. Regulatory T cells and human disease. Clin Dev Immunol 2007:89195.
Coussens, L. M. and Werb, Z. 2002. Inflammation and cancer. Nature 420:860–867.
DeNardo, D. G., Johansson, M. and Coussens, L. M. 2008. Immune cells as mediators of solid tumor metastasis. Cancer Metastasis Rev 27:11–18.
Dhodapkar, M. V., Dhodapkar, K. M. and Palucka, A. K. 2008. Interactions of tumor cells with dendritic cells: balancing immunity and tolerance. Cell Death Differ 15:39–50.
Diegelmann, R. F. and Evans, M. C. 2004. Wound healing: an overview of acute, fibrotic and delayed healing. Front Biosci 9:283–289.
Dunn, G. P., Bruce, A. T., Ikeda, H., Old, L. J. and Schreiber, R. D. 2002. Cancer immunoediting: from immunosurveillance to tumor escape. Nat Immunol 3:991–998.
Dunn, G. P., Old, L. J. and Schreiber, R. D. 2004. The immunobiology of cancer immunosurveillance and immunoediting. Immunity 21:137–148.
Fallarino, F., Grohmann, U., Hwang, K. W., Orabona, C., Vacca, C., Bianchi, R., Belladonna, M. L., Fioretti, M. C., Alegre, M. L. and Puccetti, P. 2003. Modulation of tryptophan catabolism by regulatory T cells. Nat Immunol 4:1206–1212.
Gorski, K. S., Waller, E. L., Bjornton-Severson, J., Hanten, J. A., Riter, C. L., Kieper, W. C., Gorden, K. B., Miller, J. S., Vasilakos, J. P., Tomai, M. A. and Alkan, S. S. 2006. Distinct indirect pathways govern human NK-cell activation by TLR-7 and TLR-8 agonists. Int Immunol 18:1115–1126.
Grulich, A. E., van Leeuwen, M. T., Falster, M. O. and Vajdic, C. M. 2007. Incidence of cancers in people with HIV/AIDS compared with immunosuppressed transplant recipients: a meta-analysis. Lancet 370:59–67.
Hudson, J. D., Shoaibi, M. A., Maestro, R., Carnero, A., Hannon, G. J. and Beach, D. H. 1999. A proinflammatory cytokine inhibits p53 tumor suppressor activity. J Exp Med 190:1375–1382.
Karin, M. 2006. Nuclear factor-kappaB in cancer development and progression. Nature 441:431–436.
Kedrin, D., van Rheenen, J., Hernandez, L., Condeelis, J. and Segall, J. E. 2007. Cell motility and cytoskeletal regulation in invasion and metastasis. J Mammary Gland Biol Neoplasia 12:143–152.
Kijima, M., Yamaguchi, T., Ishifune, C., Maekawa, Y., Koyanagi, A., Yagita, H., Chiba, S., Kishihara, K., Shimada, M. and Yasutomo, K. 2008. Dendritic cell-mediated NK cell activation is controlled by Jagged2-Notch interaction. Proc Natl Acad Sci USA 105:7010–7015.
Koebel, C. M., Vermi, W., Swann, J. B., Zerafa, N., Rodig, S. J., Old, L. J., Smyth, M. J. and Schreiber, R. D. 2007. Adaptive immunity maintains occult cancer in an equilibrium state. Nature 450:903–907.
Kroczek, R. A., Mages, H. W. and Hutloff, A. 2004. Emerging paradigms of T-cell co-stimulation. Curr Opin Immunol 16:321–327.
Li, M. O., Wan, Y. Y., Sanjabi, S., Robertson, A. K. and Flavell, R. A. 2006. Transforming growth factor-beta regulation of immune responses. Annu Rev Immunol 24:99–146.
Lucas, M., Schachterle, W., Oberle, K., Aichele, P. and Diefenbach, A. 2007. Dendritic cells prime natural killer cells by trans-presenting interleukin 15. Immunity 26:503–517.
Maeda, H. and Akaike, T. 1998. Nitric oxide and oxygen radicals in infection, inflammation, and cancer. Biochemistry (Mosc) 63:854–865.
Mager, D. L. 2006. Bacteria and cancer: cause, coincidence or cure? A review. J Transl Med 4:14.

Mantovani, A., Romero, P., Palucka, A. K. and Marincola, F. M. 2008. Tumour immunity: effector response to tumour and role of the microenvironment. Lancet 371:771–783.
Marelli-Berg, F. M., Okkenhaug, K. and Mirenda, V. 2007. A two-signal model for T cell trafficking. Trends Immunol 28:267–273.
Marincola, F. M., Wang, E., Herlyn, M., Seliger, B. and Ferrone, S. 2003. Tumors as elusive targets of T-cell-based active immunotherapy. Trends Immunol 24:335–342.
Martinez, F. O., Sica, A., Mantovani, A. and Locati, M. 2008. Macrophage activation and polarization. Front Biosci 13:453–461.
Mocellin, S., Mandruzzato, S., Bronte, V. and Marincola, F. M. 2004. Cancer vaccines: pessimism in check. Nat Med 10:1278–1279 (author reply 1279–1280).
Monsurro, V., Wang, E., Yamano, Y., Migueles, S. A., Panelli, M. C., Smith, K., Nagorsen, D., Connors, M., Jacobson, S. and Marincola, F. M. 2004. Quiescent phenotype of tumor-specific CD8 + T cells following immunization. Blood 104:1970–1978.
Moore, K. W., O'Garra, A., de Waal Malefyt, R., Vieira, P. and Mosmann, T. R. 1993. Interleukin-10. Annu Rev Immunol 11:165–190.
Muruve, D. A., Petrilli, V., Zaiss, A. K., White, L. R., Clark, S. A., Ross, P. J., Parks, R. J. and 1Tschopp, J. 2008. The inflammasome recognizes cytosolic microbial and host DNA and triggers an innate immune response. Nature 452:103–107.
Panelli, M. C., Stashower, M. E., Slade, H. B., Smith, K., Norwood, C., Abati, A., Fetsch, P., Filie, A., Walters, S. A., Astry, C., Arico, E., Zhao, Y., Selleri, S., Wang, E. and Marincola, F. M. 2007. Sequential gene profiling of basal cell carcinomas treated with imiquimod in a placebo-controlled study defines the requirements for tissue rejection. Genome Biol 8:R8.
Papadavid, E., Stratigos, A. J. and Falagas, M. E. 2007. Imiquimod: an immune response modifier in the treatment of precancerous skin lesions and skin cancer. Expert Opin Pharmacother 8:1743–1755.
Pawelec, G. 2004. Tumour escape: antitumour effectors too much of a good thing? Cancer Immunol Immunother 53:262–274.
Rock, K. L. and Kono, H. 2008. The inflammatory response to cell death. Annu Rev Pathol 3:99–126.
Rosenberg, S. A., Restifo, N. P., Yang, J. C., Morgan, R. A. and Dudley, M. E. 2008. Adoptive cell transfer: a clinical path to effective cancer immunotherapy. Nat Rev Cancer 8:299–308.
Rosenberg, S. A., Yang, J. C. and Restifo, N. P. 2004. Cancer immunotherapy: moving beyond current vaccines. Nat Med 10:909–915.
Sauter, B., Albert, M. L., Francisco, L., Larsson, M., Somersan, S. and Bhardwaj, N. 2000. Consequences of cell death: exposure to necrotic tumor cells, but not primary tissue cells or apoptotic cells, induces the maturation of immunostimulatory dendritic cells. J Exp Med 191:423–434.
Schiller, M., Metze, D., Luger, T. A., Grabbe, S. and Gunzer, M. 2006. Immune response modifiers–mode of action. Exp Dermatol 15:331–341.
Schon, M. P. and Schon, M. 2008. TLR7 and TLR8 as targets in cancer therapy. Oncogene 27:190–199.
Shankaran, V., Ikeda, H., Bruce, A. T., White, J. M., Swanson, P. E., Old, L. J. and Schreiber, R. D. 2001. IFNgamma and lymphocytes prevent primary tumour development and shape tumour immunogenicity. Nature 410:1107–1111.
Shanker, A., Verdeil, G., Buferne, M., Inderberg-Suso, E. M., Puthier, D., Joly, F., Nguyen, C., Leserman, L., Auphan-Anezin, N. and Schmitt-Verhulst, A. M. 2007. CD8 T cell help for innate antitumor immunity. J Immunol 179:6651–6662.
Sica, A., Schioppa, T., Mantovani, A. and Allavena, P. 2006. Tumour-associated macrophages are a distinct M2 polarised population promoting tumour progression: potential targets of anti-cancer therapy. Eur J Cancer 42:717–727.
Sombroek, C. C., Stam, A. G., Masterson, A. J., Lougheed, S. M., Schakel, M. J., Meijer, C. J., Pinedo, H. M., van den Eertwegh, A. J., Scheper, R. J. and de Gruijl, T. D. 2002.

Prostanoids play a major role in the primary tumor-induced inhibition of dendritic cell differentiation. J Immunol 168:4333–4343.
Stary, G., Bangert, C., Tauber, M., Strohal, R., Kopp, T. and Stingl, G. 2007. Tumoricidal activity of TLR7/8-activated inflammatory dendritic cells. J Exp Med 204:1441–1451.
Strieter, R. M., Burdick, M. D., Mestas, J., Gomperts, B., Keane, M. P. and Belperio, J. A. 2006. Cancer CXC chemokine networks and tumour angiogenesis. Eur J Cancer 42:768–778.
Talmadge, J. E., Donkor, M. and Scholar, E. 2007. Inflammatory cell infiltration of tumors: Jekyll or Hyde. Cancer Metastasis Rev 26:373–400.
Tan, T. T. and Coussens, L. M. 2007. Humoral immunity, inflammation and cancer. Curr Opin Immunol 19:209–216.
Tesniere, A., Panaretakis, T., Kepp, O., Apetoh, L., Ghiringhelli, F., Zitvogel, L. and Kroemer, G. 2008. Molecular characteristics of immunogenic cancer cell death. Cell Death Differ 15:3–12.
Torres, D., Paget, C., Fontaine, J., Mallevaey, T., Matsuoka, T., Maruyama, T., Narumiya, S., Capron, M., Gosset, P., Faveeuw, C. and Trottein, F. 2008. Prostaglandin D2 inhibits the production of IFN-gamma by invariant NK T cells: consequences in the control of B16 melanoma. J Immunol 180:783–792.
Viorritto, I. C., Nikolov, N. P. and Siegel, R. M. 2007. Autoimmunity versus tolerance: can dying cells tip the balance? Clin Immunol 122:125–134.
Ward, R. C. and Kaufman, H. L. 2007. Targeting costimulatory pathways for tumor immunotherapy. Int Rev Immunol 26:161–196.
Xu, J., Chakrabarti, A. K., Tan, J. L., Ge, L., Gambotto, A. and Vujanovic, N. L. 2007. Essential role of the TNF-TNFR2 cognate interaction in mouse dendritic cell-natural killer cell crosstalk. Blood 109:3333–3341.

Chapter 12
Changes in Dendritic Cells in Cancer and Aging

Annabelle Grolleau-Julius and Raymond L. Yung

Abstract The majority of common cancers preferentially affect the older population. However, despite the recognized role of dendritic cells in cancer immunotherapy, relatively little is known about the consequence of aging on these important cells. Recent studies have revealed that dendritic cells from old hosts exhibit significant functional defects at multiple levels. These changes may provide a mechanistic basis for the disappointing results of dendritic cell-based immunotherapy in older cancer patients. There is clearly a need to improve our knowledge on how aging modulates these parameters to optimally exploit DC in anticancer vaccines.

12.1 Introduction

Since their original description by Ralph Steinman in 1973, dendritic cells (DC) have received increasing scientific and clinical interest due to their key status as "directors of the immune system orchestra" (Banchereau and Steinman 1998). There are four innate specialized features of DC that contribute to their capacity to control T-cell recognition and responsiveness: (1) a specialized endocytic system for antigen capture and processing, (2) maturation in response to an array of stimuli, (3) location at body surfaces and in the T-cell areas of lymphoid organs, and (4) subsets with distinct pattern recognition receptors and functions. The discovery of in vitro culture systems yielding large amounts of mouse and human DC prompted their use in immunotherapy approaches, particularly in cancer (Banchereau and Palucka 2005). In the past decades, several in vitro and in vivo studies in rodents have demonstrated that immunizations with DC pulsed with tumor antigens (TA) result in protective immunity and rejection of established tumors in various malignancies (Flamand et al. 1994; Asavaroengchai et al. 2002; Grolleau et al. 2005). Trials in humans have shown the safety of TA-loaded DC as

R.L. Yung (✉)
University of Michigan, Room 3023 BSRB, 109 Zina Pitcher Place, Ann Arbor, Michigan 48109-2200, USA
e-mail: ryung@umich.edu

M.R. Shurin, R.D. Salter (eds.), *Dendritic Cells in Cancer*,
DOI 10.1007/978-0-387-88611-4_12, © Springer Science+Business Media, LLC 2009

well as beneficial clinical and immune responses. However, the objective responses obtained to date using DC cancer vaccines have been modest. A better understanding of how DC are able to induce and modulate immunity is necessary to optimally exploit DC in anticancer vaccines.

Aging has become an important field of research addressing medical problems in the understanding and treatment of aging-associated diseases and disorders. The immune system undergoes substantial changes with aging including a gradual decline in both cell-mediated and humoral immunity (Miller 1996). These changes are complex in nature and can include alterations in both the strength and quality of the immune responses. Evidence of dysfunctions of many immune parameters associated with aging and age-associated decline in T-cell-mediated immune responses is well characterized. Furthermore, proliferation and cytotoxic responses of T cells decrease with age (Miller 1991; Flood et al. 1998; Donnini et al. 2002). However, limited and partly controversial data are provided on the effect of aging on DC functions and on the role of DC in T-cell immunosenescence. In this chapter we will provide a short overview of our current knowledge on the effects of aging on both human and murine DC functions and discuss the potential implications of DC immunosenescence in cancer and cancer immunotherapy.

12.2 Age-Associated Alterations in Dendritic Cell Function

12.2.1 Dendritic Cell Numbers in Old Age

The most consistent observation is a reduction of Langerhans cell (LC) numbers with age. Many of the published studies have been summarized in detail in two excellent recent reviews (Shurin et al. 2007; Agrawal et al. 2008) and have cumulatively demonstrated that both natural aging and UV-induced skin aging markedly decreased LC density in aged animals and humans and altered their dendrite formation and length in the epidermis and in mucosal tissues. Reduction in the number of LC in human skin in normal aging is attenuated by UV exposure, causing a further reduction in LC number in the UV-exposed skin of elderly people. In mice, decreased LC density has been attributed to the deficiency in LC bone marrow progenitor cells (Sprecher et al. 1990). Whether or not the frequency of LC progenitors is altered in human aging and whether these cells have the same capability of producing peripheral DC as in the young is less clear. Other than skin-derived DC, little is known about the effect of aging on DC number, with however a general idea that the number varies with the subsets of DC and tissue of their residence. For instance, a decreased frequency of in situ DC in Peyer's patches, brain, and thymic DC has been reported (Agrawal et al. 2008). However, data on the plasmacytoid DC (pDC) subset are more controversial with some publications reporting a progressive loss of circulating pDC numbers during human aging (Shodell and Siegal 2002;

Teig et al. 2002; Narbutt et al. 2004; Vuckovic et al. 2004; Perez-Cabezas et al. 2007) and others showing no significant difference (Agrawal et al. 2007b; Della Bella et al. 2007). Pérez-Cabezas et al. (2007) provided a comparison of the different methods used to enumerate pDCs from blood samples in several published studies with the conclusion that the different protocols used might account for the controversial results. Della Bella et al. (2007) also showed a significant decrease of human myeloid DCs with age associated with a reduced frequency of CD34+ hematopoietic precursor cells.

12.2.2 Differentiation and Maturation of DC in Old Age

We and others have found that comparable numbers of bone marrow or monocyte-derived immature DC (BM-DC and MO-DC, respectively) from aged animals or aged persons could be generated in vitro following stimulation with IL-4 and GM-CSF as compared to their young counterparts (Lung et al. 2000; Grolleau-Julius et al. 2006; Tesar et al. 2006; Agrawal et al. 2007a). Several groups also tested the responsiveness of these cells to maturation-inducing stimuli. It was demonstrated that agents such as inactivated influenza virus, purified protein derivative, or *Mycobacterium tuberculosis* increased surface expression of MHC class II, CD54, CD80, and CD86 and triggered the secretion of IL-12 and TNF-α equally well in MO-DC from old and young persons (Saurwein-Teissl et al. 1998; Lung et al. 2000). Similar results were observed in human lin-/HLA-DR+ peripheral blood DC after short-period incubation with lipopolysaccharide (LPS), with the exception of IL-12 whose intracellular secretion was significantly lower in old than in young individuals. Additionally there was no difference in the percentage of DC expressing IL-10 (Della Bella et al. 2007). In mice, our group found that old LPS-stimulated BM-DC also have a normal phenotype but secrete higher levels of IL-10 and less IL-6 and TNF-α than their young counterparts (Grolleau-Julius et al. 2006). In contrast, Chiu et al. (2007) reported that myeloid DC from lymph nodes of aged mice displayed reduced CD40 and CD86 expression that could be restored to levels of young mice by inactivating T regulatory (Treg) cells. DC maturation is dependent on the binding of pathogen-associated molecular patterns derived from bacteria, fungi, parasites, and viruses to toll-like receptors (TLR) (Mazzoni and Segal 2004). In the past few years, there has been an exponential increase in our knowledge of TLR and the cell biological changes that occur in DC proximal to TLR activation (Watts et al. 2007). The signals that elicit TLR include both exogenous ligands, such as LPS, lipoteichoic acid, flagellin, CpG motifs, and dsRNA, and endogenous ligands such as heat-shock proteins, fibronectin, hyaluronic acid, and messenger RNA. Only one published study has examined aging and TLR expression and function in DC. It demonstrated that in vitro-derived BM-DC and freshly isolated splenic myeloid DC from aged C57BL/6 and CBA mice have preserved TLR expression as well as

TLR-mediated immune responses (Tesar et al. 2006). No data are currently available on whether DC from aged individuals have altered TLR function or expression.

12.2.3 Antigen Uptake in the Aged

Although an efficient uptake of antigen is essential for the generation of specific immune response and maintenance of peripheral self-tolerance, only limited data are available on the effect of aging on this DC function. Using flow cytometry, Agrawal et al. (2007a) have recently investigated micropinocytosis of Lucifer Yellow dye, endocytosis of dextran beads, and phagocytosis of apoptotic Jurkat T cells by in vitro MO-DC generated from young and old donors. They found that MO-DC from aged subjects were 50% less efficient in the phagocytosis of these foreign and self-antigens, suggesting an impairment of antigen uptake via both receptor-dependent and receptor-independent mechanisms in aging. Expression of mannose receptor, one of the receptors involved in the uptake of dextran, was not affected by aging. The authors proposed a model in which decreased phosphorylation of AKT, due to increased expression of phosphatase and tensin homolog (PTEN) in aged DC, results in impaired PI3K-signaling pathway and leads to reduced phagocytosis and increased secretion of inflammatory cytokines TNF-α and IL-6. Using CD11c+ bone-marrow-derived DC from young and old C57BL/6 mice we also found a decreased uptake of FITC-dextran in aged DC without any significant changes in the expression of mannose receptor, but no impairment in the uptake and processing of whole ovalbumin protein (Julius et al. 2008). This suggests a decrease in the expression and/or function of selective receptors associated with DC phagocytosis in aging.

12.2.4 DC Migration in Old Age

Evidence for impaired chemokine-dependent migration of DC with aging includes a recent report showing that despite similar levels of expression of the lymphoid homing chemokine receptor CCR7, in vitro migration of monocyte-derived LPS-stimulated DC from elderly subjects toward CCR7 ligand CCL19 was impaired as compared to young subjects (Agrawal et al. 2007a). Similar results were observed in response to SDF-1, a CXCR4 ligand. The investigators attributed their results to defects in the downstream signaling pathway, possibly in the PI3K pathway. However, the exact mechanism is unknown and further studies will be needed to unravel it. We have also found that aged BM-DC have impaired in vitro migratory response to CCL21, another CCR7 ligand (Julius et al. 2008). This was not due to a defect in CCR7 protein expression, but rather due to a defect in the signal transduction pathway. In vivo recruitment of airway DC to draining mediastinal lymph nodes was reported to be diminished in aged mice (Linton et al. 2005). Cumberbatch et al. (2002) also observed reduced LC migration in BALB/c mice in response to allergen. The same group additionally reported significantly impaired migration of LC in response to TNF-α in elderly patients (Bhushan et al. 2002) and later showed that exogenous interleukin-1β restores impaired LC migration in aged skin (Bhushan et al. 2004). Pietschmann et al. (2000) have assessed human DC transendothelial migration in non-inflammatory conditions. After exposure to an endothelial cell monolayer, no change in the percentage of peripheral blood myeloid-enriched DC recovered in the migrating fraction was observed with age. These findings suggest that aging has a selective effect on DC migration, depending on the specific cellular and molecular ligands involved.

12.2.5 T-Cell Stimulatory Properties in Old Age

There is some controversy regarding the capacity of aged DC to stimulate T cells. Two reports from the Grubeck-Loebenstein group have shown that MO-DC from old healthy individuals were equally effective as young control cells in presenting antigen to tetanus-specific T-cell clones as well as to influenza-specific resting CD4+ T cells (Steger et al. 1996; Lung et al. 2000). Agrawal et al. (2008) reported that young CD8+ T-cell responsiveness to aged MO-DC was comparable to young MO-DC whereas proliferation of CD4+ was decreased. In mice, Komatsubara et al. (1986a,b) found that the stimulatory capacity of splenic DC in both allogeneic and syngeneic mixed leukocyte reaction showed an age-dependent polymorphism with, depending on the mouse strain, an increase or decrease with age. The observed changes were not associated with the alteration in the expression of MHC molecules. Another study demonstrated a decreased capacity of epidermal DC from old mice to stimulate the proliferation of allogeneic T cells and to present ovalbumin to sensitized T cells (Sprecher et al. 1990). We found that old BM-DC were less effective than young DC in stimulating syngeneic ovalbumin-specific CD4+ and CD8+ proliferation, despite intact peptide presentation and no significant change in MHC and the classical co-stimulatory molecules CD80, CD86, CD40, and CD54 (Grolleau-Julius et al. 2006; Julius et al. 2008). Interestingly, we observed a decrease of DC-specific/intracellular adhesion molecule type 3-grabbing non-integrin (DC-SIGN) expression on the cell surface of aged DC compared to young DC. DC-SIGN has been recently identified as a novel adhesion receptor on DC that is essential in several key functions, including interactions between DC and T cells (Geijtenbeek et al. 2000). DC-SIGN binds ICAM-3 with high affinity and this cellular interaction establishes the first molecular interaction between DC and resting cells (van Kooyk and Geijtenbeek 2002). Others have demonstrated the relevance of DC-SIGN in DC-induced T-cell proliferation by showing that antibodies against human DC-SIGN inhibit DC-induced proliferation of resting T cells (Gijzen et al. 2007). It is clear from our data that the capacity to induce T-cell proliferation in vitro is reduced with aging and that this correlates with a selective reduced DC-SIGN expression.

12.3 Consequences of the Senescence of Dendritic Cells in Cancer Immunotherapy

Having outlined the broad spectrum of DC functional changes that accompany aging, it is quite unfortunate that although many studies have demonstrated the efficacy of autologous DC vaccines in stimulating the antitumor immune response in the young, almost none have considered the potential impact that aging may have on DC antitumor surveillance. Sharma et al. recently compared the efficacy of DC vaccines in young and old mice in the TRAMP-C2 prostate cancer model. Day 7 tumor-bearing young and old C57BL/6 mice were immunized with their respective DC pulsed with apoptotic TRAMP-C2 cells and the efficacy of autologous DC vaccines assessed by following tumor growth over time (Sharma et al. 2006a). The results showed that the effect of the DC vaccine against TRAMP-C2 tumor is dependent on the age of the host, with 60% tumor growth inhibition in young animals and only 5% inhibition observed in old animals. Interestingly, in vivo co-administration of anti-OX-40 or anti-4-1BB monoclonal antibodies with tumor-cell-pulsed DC resulted in normalization of antitumor response by aged DC vaccination not only with respect to tumor regression but also in the induction of antigen-specific CTL. These data suggest that with appropriate co-stimulating signals, in this case members of the TNF superfamily, DC-induced antitumor responses can be restored in aging mice. Shi et al. (2005) compared the ability of ovalbumin-pulsed young DC to induce protective immune responses against a subsequent ovalbumin-expressing melanoma challenge in young and old hosts. Aged mice displayed impaired DC vaccination response as measured by decreased percentage survival and reduced antigen-specific cytotoxicity of splenic CTL.

Our group investigated the consequences of DC immunosenescence in tumor-bearing young hosts and showed that mice bearing ovalbumin-expressing melanoma and treated with the ovalbumin peptide-pulsed old DC exhibited significantly reduced tumor regression than mice treated with ovalbumin peptide-pulsed young DC (Julius et al. 2008). This correlated with a reduced frequency of splenic antigen-specific CD8+ T cells in vivo and a significant decrease in the influx of CD8+ T cells into tumors 7 days after injection of DC vaccines. However, effector functions of these antigen-specific T cells, as determined by IFN-γ production and cytotoxic activity, were similar to those obtained from mice vaccinated with young OVA pulsed-DC on a per cell basis, suggesting that the quality of the DC-stimulated T cells remains intact. In our model, we also observed that old injected DC migrated to regional

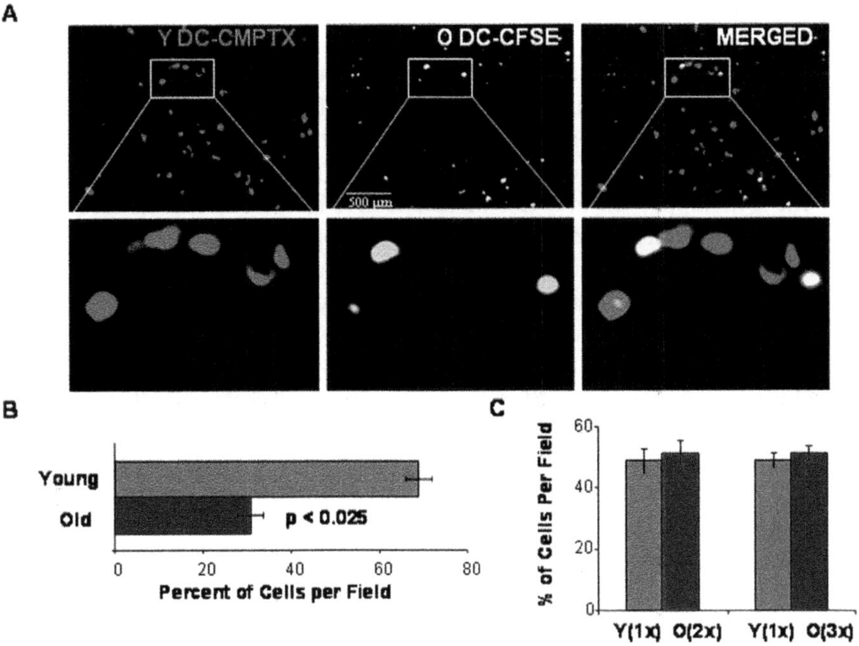

Fig. 12.1 In vivo lymph node dendritic cell migration in aging. Groups of young mice (n = 3) received 50 µl of equal numbers (2×10^6) of CMPTX-labeled young DC and CFSE-labeled old DC into their footpad hind. Popliteal draining lymph nodes were excised 24 h later and the frequency of DC/node was assessed by fluorescent microscopy. (**A**) Representative photos (200×) showing young DC in *light grey*, old DC in *white*, and the merge picture. (**B**) Close up (digital enlargement) of selected fields is also shown. Percent of red versus green DC per field was derived from microscopic examination data by counting at least 200 cells/group in randomly selected fields. (**C**) Similar experiments were done using two times (2×) and three times (3×) more old DC than young DC (Julius et al. 2008)

lymph nodes two times less than young DC (Fig. 12.1). Interestingly, we found that restoring old DC in vivo migration does not improve the age-associated defect in DC tumor surveillance, suggesting that DC migration through CCR7–CCL21 interaction is not the primary mechanism for the aging defect.

An interesting study demonstrated that antitumor responses in old mice could be restored by targeting antigen-presenting cells with the TLR ligand CpG-ODN (Sharma et al. 2008). This was associated with a drastic reduction of Treg number within the tumors. An earlier publication from the same group indicated that greater accumulation of CD25+Foxp3+ T cells in the spleen and lymph nodes of aged animals prevents the activation of immune responses in aged animals (Sharma et al. 2006b). Interestingly, using the B16-OVA melanoma model, another group reported that a small population of pDC can directly activate Treg in vivo in tumor-draining lymph nodes of tumor-bearing mice (Sharma et al. 2007). These results open new roads for DC vaccination whether by targeting DC in vivo or by improving their stimulatory capacity.

12.4 Conclusions

Overall, these studies demonstrate that aging affects several important features of DC. It is also clear that different subsets of DC are impaired at different levels. Experimental animal studies on the antitumor properties of DC suggest that immunotherapeutic intervention could be effective in young animals, but that the same approach may not be as effective in the older hosts. Promising studies indicate that it is possible to restore or improve DC immunosurveillance in old mice but significant gaps remain to be addressed. For instance, the mechanisms that contribute to the lack of an effective tumor immune response in DC immunotherapy are poorly defined. There is little work on DC in older cancer patients although DC-based vaccines rely on the patients' autologous cells. Likewise, a better understanding of the cross-talk between DC with other members of the immune system such as B cells, T regulatory cells, and NK cells in aging will be needed to allow investigators to devise improved therapeutic strategies for older cancer patients. There is clearly a need to improve our knowledge on how aging modulates these parameters to optimally exploit DC in anticancer vaccines.

References

Agrawal, A., Agrawal, S., Cao, J. N., Su, H., Osann, K. and Gupta, S. 2007a. Altered innate immune functioning of dendritic cells in elderly humans: a role of phosphoinositide 3-kinase-signaling pathway. J Immunol 178:6912–6922.

Agrawal, A., Agrawal, S. and Gupta, S. 2007b. Dendritic cells in human aging. Exp Gerontol 42:421–426.

Agrawal, A., Agrawal, S., Tay, J. and Gupta, S. 2008. Biology of dendritic cells in aging. J Clin Immunol 28:14–20.

Asavaroengchai, W., Kotera, Y. and Mule, J. J. 2002. Tumor lysate-pulsed dendritic cells can elicit an effective antitumor immune response during early lymphoid recovery. Proc Natl Acad Sci USA 99:931–936.

Banchereau, J. and Palucka, A. K. 2005. Dendritic cells as therapeutic vaccines against cancer. Nat Rev Immunol 5:296–306.

Banchereau, J. and Steinman, R. M. 1998. Dendritic cells and the control of immunity. Nature 392:245–252.

Bhushan, M., Cumberbatch, M., Dearman, R. J., Andrew, S. M., Kimber, I. and Griffiths, C. E. 2002. Tumour necrosis factor-alpha-induced migration of human Langerhans cells: the influence of ageing. Br J Dermatol 146:32–40.

Bhushan, M., Cumberbatch, M., Dearman, R. J., Kimber, I. and Griffiths, C. E. 2004. Exogenous interleukin-1beta restores impaired Langerhans cell migration in aged skin. Br J Dermatol 150:1217–1218.

Chiu, B. C., Stolberg, V. R., Zhang, H. and Chensue, S. W. 2007. Increased Foxp3(+) Treg cell activity reduces dendritic cell co-stimulatory molecule expression in aged mice. Mech Ageing Dev 128:618–627.

Cumberbatch, M., Dearman, R. J. and Kimber, I. 2002. Influence of ageing on Langerhans cell migration in mice: identification of a putative deficiency of epidermal interleukin-1beta. Immunology 105:466–477.

Della Bella, S., Bierti, L., Presicce, P., Arienti, R., Valenti, M., Saresella, M., Vergani, C. and Villa, M. L. 2007. Peripheral blood dendritic cells and monocytes are differently regulated in the elderly. Clin Immunol 122:220–228.

Donnini, A., Argentati, K., Mancini, R., Smorlesi, A., Bartozzi, B., Bernardini, G. and Provinciali, M. 2002. Phenotype, antigen-presenting capacity, and migration of antigen-presenting cells in young and old age. Exp Gerontol 37:1097–1112.

Flamand, V., Sornasse, T., Thielemans, K., Demanet, C., Bakkus, M., Bazin, H., Tielemans, F., Leo, O., Urbain, J. and Moser, M. 1994. Murine dendritic cells pulsed in vitro with tumor antigen induce tumor resistance in vivo. Eur J Immunol 24:605–610.

Flood, P. M., Liu, X., Alexander, R., Schreiber, H. and Haque, S. 1998. Loss of resistance to a highly immunogenic tumor with age corresponds to the decline of CD8 T cell activity. J Immunother 21:307–316.

Geijtenbeek, T. B., Torensma, R., van Vliet, S. J., van Duijnhoven, G. C., Adema, G. J., van Kooyk, Y. and Figdor, C. G. 2000. Identification of DC-SIGN, a novel dendritic cell-specific ICAM-3 receptor that supports primary immune responses. Cell 100:575–585.

Gijzen, K., Tacken, P. J., Zimmerman, A., Joosten, B., de Vries, I. J., Figdor, C. G. and Torensma, R. 2007. Relevance of DC-SIGN in DC-induced T cell proliferation. J Leukoc Biol 81:729–740.

Grolleau, A., Sloan, A. and Mule, J. J. 2005. Dendritic cell-based vaccines for cancer therapy. Cancer Treat Res 123:181–205.

Grolleau-Julius, A., Garg, M. R., Mo, R., Stoolman, L. L. and Yung, R. L. 2006. Effect of aging on bone marrow-derived murine CD11c + CD4-CD8alpha-dendritic cell function. J Gerontol A Biol Sci Med Sci 61:1039–1047.

Julius, A., Harning, E., Abernathy, L., and Yung, R. L. 2008. Impaired dendritic cell function in aging leads to defective antitumor immunity. Cancer Res 68:6341–6349.

Komatsubara, S., Cinader, B. and Muramatsu, S. 1986a. Functional competence of dendritic cells of ageing C57BL/6 mice. Scand J Immunol 24:517–525.

Komatsubara, S., Cinader, B. and Muramatsu, S. 1986b. Polymorphism of age-related changes in stimulatory capacity of murine dendritic cells. Mech Ageing Dev 37:163–173.

Linton, P. J., Li, S. P., Zhang, Y., Bautista, B., Huynh, Q. and Trinh, T. 2005. Intrinsic versus environmental influences on T-cell responses in aging. Immunol Rev 205:207–219.

Lung, T. L., Saurwein-Teissl, M., Parson, W., Schonitzer, D. and Grubeck-Loebenstein, B. 2000. Unimpaired dendritic cells can be derived from monocytes in old age and can mobilize residual function in senescent T cells. Vaccine 18:1606–1612.

Mazzoni, A. and Segal, D. M. 2004. Controlling the Toll road to dendritic cell polarization. J Leukoc Biol 75:721–730.

Miller, R. A. 1991. Aging and immune function, Int Rev Cytol 124:187–215. Abstract | View Record in Scopus | Cited By in Scopus (148)

Miller, R. A. 1996. The aging immune system: primer and prospectus. Science 273:70–74.

Narbutt, J., Lesiak, A., Zak-Prelich, M., Wozniacka, A., Sysa-Jedrzejowska, A., Tybura, M., Robak, T. and Smolewski, P. 2004. The distribution of peripheral blood dendritic cells assayed by a new panel of anti-BDCA monoclonal antibodies in healthy representatives of the polish population. Cell Mol Biol Lett 9:497–509.

Perez-Cabezas, B., Naranjo-Gomez, M., Fernandez, M. A., Grifols, J. R., Pujol-Borrell, R. and Borras, F. E. 2007. Reduced numbers of plasmacytoid dendritic cells in aged blood donors. Exp Gerontol 42:1033–1038.

Pietschmann, P., Hahn, P., Kudlacek, S., Thomas, R. and Peterlik, M. 2000. Surface markers and transendothelial migration of dendritic cells from elderly subjects. Exp Gerontol 35:213–224.

Saurwein-Teissl, M., Schonitzer, D. and Grubeck-Loebenstein, B. 1998. Dendritic cell responsiveness to stimulation with influenza vaccine is unimpaired in old age. Exp Gerontol 33:625–631.

Sharma, M. D., Baban, B., Chandler, P., Hou, D. Y., Singh, N., Yagita, H., Azuma, M., Blazar, B. R., Mellor, A. L. and Munn, D. H. 2007. Plasmacytoid dendritic cells from mouse tumor-draining lymph nodes directly activate mature Tregs via indoleamine 2,3-dioxygenase. J Clin Invest 117:2570–2582.

Sharma, S., Dominguez, A. L., Hoelzinger, D. B. and Lustgarten, J. 2008. CpG-ODN but not other TLR-ligands restore the antitumor responses in old mice: the implications for vaccinations in the aged. Cancer Immunol Immunother 57:549–561.

Sharma, S., Dominguez, A. L. and Lustgarten, J. 2006a. Aging affect the anti-tumor potential of dendritic cell vaccination, but it can be overcome by co-stimulation with anti-OX40 or anti-4-1BB. Exp Gerontol 41:78–84.

Sharma, S., Dominguez, A. L. and Lustgarten, J. 2006b. High accumulation of T regulatory cells prevents the activation of immune responses in aged animals. J Immunol 177:8348–8355.

Shi, M., Bi, X., Xu, S., He, Y., Guo, X. and Xiang, J. 2005. Increased susceptibility of tumorigenicity and decreased anti-tumor effect of DC vaccination in aged mice are potentially associated with increased number of NK1.1 + CD3 + NKT cells. Exp Oncol 27:125–129.

Shodell, M. and Siegal, F. P. 2002. Circulating, interferon-producing plasmacytoid dendritic cells decline during human ageing. Scand J Immunol 56:518–521.

Shurin, M. R., Shurin, G. V. and Chatta, G. S. 2007. Aging and the dendritic cell system: implications for cancer. Crit Rev Oncol Hematol 64:90–105.

Sprecher, E., Becker, Y., Kraal, G., Hall, E., Harrison, D. and Shultz, L. D. 1990. Effect of aging on epidermal dendritic cell populations in C57BL/6 J mice. J Invest Dermatol 94:247–253.

Steger, M. M., Maczek, C. and Grubeck-Loebenstein, B. 1996. Morphologically and functionally intact dendritic cells can be derived from the peripheral blood of aged individuals. Clin Exp Immunol 105:544–550.

Teig, N., Moses, D., Gieseler, S. and Schauer, U. 2002. Age-related changes in human blood dendritic cell subpopulations. Scand J Immunol 55:453–457.

Tesar, B. M., Walker, W. E., Unternaehrer, J., Joshi, N. S., Chandele, A., Haynes, L., Kaech, S. and Goldstein, D. R. 2006. Murine [corrected] myeloid dendritic cell-dependent toll-like receptor immunity is preserved with aging. Aging Cell 5:473–486.

van Kooyk, Y. and Geijtenbeek, T. B. 2002. A novel adhesion pathway that regulates dendritic cell trafficking and T cell interactions. Immunol Rev 186:47–56.

Vuckovic, S., Gardiner, D., Field, K., Chapman, G. V., Khalil, D., Gill, D., Marlton, P., Taylor, K., Wright, S., Pinzon-Charry, A., Pyke, C. M., Rodwell, R., Hockey, R. L., Gleeson, M., Tepes, S., True, D., Cotterill, A. and Hart, D. N. 2004. Monitoring dendritic cells in clinical practice using a new whole blood single-platform TruCOUNT assay. J Immunol Methods 284:73–87.

Watts, C., Zaru, R., Prescott, A. R., Wallin, R. P. and West, M. A. 2007. Proximal effects of Toll-like receptor activation in dendritic cells. Curr Opin Immunol 19:73–78.

Chapter 13
Stress, Immunity and Dendritic Cells in Cancer

Rachel Kohman and Alexander W. Kusnecov

Abstract The literature on stress and immunity clearly demonstrates that stress exposure modifies certain aspects of immune function. While stressful experiences and stress-associated molecules have well-documented effects on the activity of lymphocytes, stress exposure may disrupt processes that occur earlier in the immune response. Stress-induced alterations in immune function may, in part, result from changes in the development and/or function of cells involved in the initiation of the response, namely dendritic cells. Although the literature is relatively limited, the findings suggest that stress-associated hormones and neuropeptides can modulate maturation and migration of dendritic cells. The present chapter concentrates on stress-related alterations in dendritic cell function and the potential relevance of psychological factors to the progression of cancer and to the effectiveness of therapeutic treatments.

13.1 Psychosocial Factors and Health

The impact of psychosocial factors on health and disease has long been the subject of empirical investigation. This had initially emerged from psychoanalytic traditions in the first half of the twentieth century which argued that the unconscious emotional life of the individual was an important determinant in either the precipitation and/or perpetuation of certain, if not all, diseases (Weiss 1945). Inevitably, improvements in the tools of science and new theories of personality and behavior led to the abandonment of this perspective, but nonetheless continued to retain a psychosomatic orientation to illness (Wolf 1948). Indeed, the ascent of the empirical tradition in the behavioral sciences following the Second World War cemented the view that the behavior of an organism in some way contributed to disease onset and/or recovery from disease. This was supported by research in animals and humans showing that stress could contribute to illness, legitimizing what is now referred to as the biopsychosocial

A.W. Kusnecov (✉)
Department of Psychology, Rutgers University, Piscatway, NJ 08854, USA
e-mail: kusnecov@rci.rutgers.edu

M.R. Shurin, R.D. Salter (eds.), *Dendritic Cells in Cancer*,
DOI 10.1007/978-0-387-88611-4_13, © Springer Science+Business Media, LLC 2009

model of disease (Novack et al. 2007). The relevance of this model has been strongly demonstrated in the case of heart disease and obesity, which can form the basis of multiple diseases, including diabetes (Kop 1999; Golden 2007; Torres and Nowson 2007; Dimsdale 2008). However, in the case of cancer, the role of psychosocial factors has always been controversial, in spite of animal studies that have suggested that growth of tumors can be influenced by stress (Saul et al. 2005; Kerr et al. 2006; Thaker et al. 2006; Ben-Eliyahu et al. 2007; Thaker et al. 2007) and that behavioral factors can influence the rate of progression and outcome of cancer in humans (Chida et al. 2008; Gidron and Ronson 2008).

Cancer is one disease for which multiple theories of causation exist, largely because there are multiple carcinogens, and the emergence of cancer is not necessarily confined to specific tissues. Consequently, leaving aside the question of precisely how cellular mutations arise in different individuals with different primary tumor sites, the relevance of the psychosocial perspective is in addressing whether the prior and ongoing behavior of an individual is in any way relevant to the appearance and progression of the disease. In this regard, behaviors that are lifestyle characteristics, such as smoking, drug and alcohol use and diet, represent obvious biological modifiers that may help to promote, interact or even constitute the basis for carcinogenesis (e.g., the carcinogens in cigarette smoke). However, behavior can be further reduced to the affective and mood states and traits of an individual, and this may indeed be the basis for variations in neuroendocrine and autonomic nervous system (ANS) responses that modify the physiologic state of an individual either in the short term or on a chronic basis.

Such changes – which can be the basis of a stress response – may not be oncogenic, but they may be modifiers of opposing and/or facilitative processes in the development of a cancerous disease. From a clinical perspective, the emotional state of an individual with cancer, and the endocrine and autonomic nervous systems activated by the stress of cancer, may represent potential obstacles to therapeutic interventions. Nowhere is this likely to be more evident than in the use of dendritic cell (DC) therapy, where cells are introduced into a neurohormonal environment that is likely to be substantially different from that of normal, non-diseased individuals who do not face the adaptational and psychological coping demands of cancer patients (Miller et al. 2008). It is already well documented that chemotherapy is a particularly stressful procedure that produces substantial physical illness (e.g., nausea) and can, in fact, be behaviorally conditioned. That is, patients can develop learned aversions to the chemotherapeutic drug, procedure and hospital environment, a phenomenon that has been termed conditioned anticipatory nausea (Bovbjerg et al. 1992; Morrow and Rosenthal 1996; Bovbjerg 2006). Associated with such reactions are physiological changes indicative of stress, such as increased cortisol and reduced measures of lymphocyte mitogenic responses (Bovbjerg et al. 1990; Sabbioni et al. 1997). Conversely, it has been demonstrated that psychological interventions, such as moderated group discussions among people with cancer, have delayed mortality and increased measures of natural killer (NK) cell activity (Spiegel 1997; Spiegel and Sephton 2001). The basis for the clinical outcomes in these studies is not clear, but a very large literature exists on the impact of stress on the immune system function (see below). This literature is particularly instructive in designing immunotherapeutic interventions or studying the role of local neurohormonal factors that modify immunologic control of certain cancers.

With these comments in mind, the purpose of this chapter is to address the literature on how stress and/or stress-related factors affect DC function. In terms of stress-related factors, we will focus largely on the hypothalamic–pituitary–adrenal (HPA) axis and the sympathetic branch of the autonomic nervous system, which are exquisitely sensitive to stress. Due to space limitations, our treatment of stress and immune functions cannot be exhaustive, although the reader is urged to consider a number of reviews, both old and more recent, that have provided an in-depth treatment of the effects of stress on immune function (Kusnecov and Rabin 1994; Moynihan and Ader 1996; Glaser and Kiecolt-Glaser 2005). In the final analysis, it is hoped that the reader will appreciate the potential influence that efferent mechanisms under central nervous system control may have on the capability of modifying the therapeutic efficacy of exogenously introduced DC.

13.2 Stress, Homeostasis and Allostatic Load

The word "stress" is a much used and abused term, which has generated endless debates over how it should be defined, conceptualized and interpreted (Goldstein and Kopin 2007; Le Moal 2007). In many ways, it has been a useful idea that captures the reality of biological interactions with the environment, although it is for this very reason that it also fails as an explanation for what can go wrong in a biological system, since the stress response can be conceptualized as an adaptive response. The conventional view is that "stress" is an environmental challenge that impacts the organism negatively and disrupts normal biological processes. However, "stress" has also been referred to as a physiological or biological state, a perspective that has created confusion and the need for a clarifying nomenclature (Goldstein and Kopin 2007; Le Moal 2007). Therefore, we will opt for the term "stressor" to refer to any stimulus or event that causes a behavioral and/or physiological perturbation in the organism that is in fact a state of "stress". Finally, a "stress effect" is the actual behavioral and/or biological change from some normal, basal state as a function of stressor exposure, and in this sense is similar to a state of stress, but one that is empirically determined and may represent the final outcome of the dynamic sequence of events that are induced by a stressor. We note, however, that from a biological perspective, independent of the cognitive and emotional aspects that involve feelings of aversion and displeasure, the stress experience is perfectly normal and necessary. In fact, stress, as a physiological phenomenon, is a change in cellular function that is integrated into the normal flow of biological activity, but without necessarily disrupting the goal of those functions. For example, glucose utilization is a normal, ongoing process designed to maintain energy and biochemical synthetic processes within the cell. Acute and chronic stressors can increase circulating glucose levels (Flaherty et al. 1986; Hager et al. 2004) without necessarily affecting the utilization of glucose and the ultimate goal of supporting the biochemical machinery of the cell (e.g., increased consumption of glucose during neuronal arousal – a phenomenon capitalized on by modern neuroimaging tools). Therefore, while the rate of appearance of glucose and the synthesis and utilization of various proteins and peptides in the body may increase, the outcome is plastic and likely to return to a previous state of equilibrium. This is a cornerstone of all physiological systems and the basis of the concept of homeostasis. Consequently, it should not be the case that a rise in glucose in response to, say, public speaking, is representative of a threat to the individual; nor might for that matter would immune changes to an acute stressor necessarily result in serious repercussions (e.g., as reviewed below stressors are capable of enhancing immune responses). On the other hand, should glucose levels remain high upon termination of the real or perceived threat to an individual, then regular feedback mechanisms designed to restore equilibrium may be impaired. Were this to occur, it would be the basis of disease (e.g., diabetes) and the source of further challenges to the functioning of multiple biological systems, such as the brain, whose control of the neuroendocrine system would be forced to compensate with abnormal levels of activity (e.g., promote increased pancreatic insulin production).

This line of thought highlights a new perspective on stress research that places central importance on the concept of allostatic load, which refers to the cumulative burden exerted on biological systems as a function of persistent demands by the environment (McEwen 1998; Korte et al. 2005). It is a tenet of physiology that homeostasis requires the operation of various regulatory mechanisms that retain basal functioning within upper and lower boundaries. Maintenance of core body temperature is one example of a homeostatic state, since it remains stable under varying changes of the external environment. Allostasis, on the other hand, refers to an adaptational response that causes a migration of bodily functions outside the normal range set by homeostasis. For example, infections represent a challenge to the body that will result in an increase in a pyrogenic response, in which core body temperature has spiked to a new steady-state level as a function of increased production of proinflammatory cytokines, such as interleukin-1 (IL-1) (Roth et al. 2006; Blatteis 2007). Once an infection has been resolved, the core body temperature will return to its previous set-point level, until reset to a new level by the next infectious episode. In this situation, the fluctuations between abnormal and normal set-point levels of core body temperature occur on a different timescale and variable rates of onset, as determined by the appearance of infectious antigens that induce an immune response. However, the resetting of body temperature to a higher value that remains in place until infection or cytokine levels subside represents a single instance of an allostatic load. If this is repeated frequently across the life span of an organism, the allostatic load is considered severe, since it represents a cumulative index of the burden placed on the regulatory mechanisms that produce the bidirectional shifts in core temperature. Such shifts can exert a significant effect on brain function and may result in pathology, with the most telling example of allostatic load being aging (McEwen 2003). Throughout the life span of an organism, certain events are imperative, such as the hunger drive and its satisfaction through seeking and obtaining food. Secondary to this is reproduction, followed by the economic aspects of ensuring protection and shelter from the environment and obtaining adequate rest when the body experiences the drive to sleep. All of these are basic behavioral processes, with additional underlying biochemical and physiological processes of energy utilization and metabolism. Aging, therefore, is associated with a decline in the effectiveness of normal ongoing processes necessary for maintaining survival. There is an enormous literature demonstrating that aging is associated with a decline in a number of biological systems, including the nervous and immune systems. Insofar as the immune system has played a prominent role in the conceptualization of cancer, it is noteworthy that the incidence of cancer increases with age (Deng et al. 2008). Therefore, leaving aside the issue of "stress", it is noteworthy that time alone in a healthy, active individual ensures that regulatory systems become more fragile and subject to dysfunction when perturbed. As a case in point, research on telomeres has highlighted their use as an index of aging. Telomeres are chromosomal end-points of DNA that control the length and size of chromosomes subsequent to cellular division. Telomere length and the actions of telomerase, which normally maintains telomere stability, decline with normal aging (Deng et al. 2008; Shawi and Autexier 2008). However, a recent study demonstrated that the chronic stress of caregiving for children with compromised health or adults with Alzheimer's may shorten telomeres and, by inference, could accelerate aging (Epel et al. 2004; Damjanovic et al. 2007). Moreover, it has been proposed that chronic infections and inflammation can produce the same result (van Baarle et al. 2005). Given that stress can affect the immune system, it is likely that exposure of individuals to repeated stressors may impair their ability to mount an effective immune response against viral and/or bacterial pathogens. This in turn increases rates of infection, resulting in excess allostatic load and increased risk for chronic disease.

13.3 Stress and the Immune System: An Overview

As reviewed elsewhere (Kusnecov and Rabin 1994; Moynihan and Ader 1996; Segerstrom and Miller 2004; Glaser and Kiecolt-Glaser 2005), there is considerable evidence that exposure to various stressors produces significant alterations in a number of cellular and humoral parameters of immune function. In human studies, it is most notable that exposure to laboratory stressors, such as public speaking tasks or cognitive tasks designed to frustrate or increase sympathetic nervous activity (e.g., heart rate and blood pressure), produces a marked decline in the mitogenic activity of peripheral blood leukocytes and increased NK cell activity (Bachen et al. 1992; Herbert et al. 1994; Sgoutas-Emch et al. 1994; Marsland et al. 1995; Segerstrom and Miller 2004). The stressful nature of academic examinations has similarly produced reductions in lymphocyte proliferative capacity, as well as reactivation of latent viruses, such as EBV, as reflected by increased anti-EBV antibody titers in the serum of stressed individuals (Glaser and Kiecolt-Glaser, 2005). In recent years, there has been special attention directed toward chronic stressor exposure, and in the human literature, caregivers of chronically ill patients represent a particularly vulnerable group. Findings with caregivers have revealed altered cytokine production and telomere length erosion (Damjanovic et al. 2007), reduced antibody responses to influenza vaccines as well as reduced IL-1 and IL-2 production following exposure of cultured PBL to influenza antigens (Kiecolt-Glaser et al. 1996). Other studies by Cohen and colleagues have also shown that high levels of perceived stress, as well as prolonged experience of daily hassles, can result in increased symptomatic vulnerability to an inoculum of rhinovirus, which is the basis of the common cold (Cohen 2005). These in vivo studies, combined with evidence from other studies showing stress-related reductions in antibody titers following vaccinations (Marsland et al. 2001; Miller et al. 2004; Pressman et al. 2005; Marsland et al. 2006), provide strong support for the notion that the functioning of the immune system varies depending on the degree to which individuals are exposed to persistent stressors in their daily lives.

The mechanisms by which some of these effects occur remain to be determined. However, it is difficult to study this in human subjects, since manipulation of neuroendocrine and autonomic nervous system activity with antagonists is difficult to accomplish. For this purpose, animal studies have shown that stressor-induced modulation of the immune system can be modulated by increased glucocorticoids, noradrenaline and opioid peptides (Kusnecov and Rabin 1994; Elenkov and Chrousos 1999). However, the universal involvement of these factors is not always clear, since it may depend on the immune compartment examined (e.g., lymph nodes, spleen or blood) as well as the intensity, duration and frequency of stressor exposure. In some cases, alteration of NK cell cytotoxicity and/or antigen-specific T-cell cytotoxicity was shown to depend on a combination of glucocorticoid, noradrenergic and opioid influences (Dobbs et al. 1996; Tseng et al. 2005). Furthermore, different mechanisms may impact different immune compartments. For example, acute exposure of rats to a neurogenic stressor, such as footshock, produces suppression of mitogenic function in PBL and spleen cells (Lysle et al. 1987; Cunnick et al. 1988; Kusnecov and Rabin 1993). However, it was shown that whereas suppression of PBL was dependent on the major rodent glucocorticoid, corticosterone, the spleen cell suppression was dependent on sympathetic nervous system release of catecholamines (Cunnick et al. 1990).

Stressors may also differentially impact different immune compartments. For instance, it was shown that lymph nodes do not show suppression of lymphoproliferative activity under the same stressor exposure conditions that suppress the spleen and PBL (Shanks et al. 1997). Moreover, species differences may operate, in that mice exposed to a variety of different stressors do not routinely show suppressed mitogenic activity in the spleen (Cunnick et al. 1990). To extend this to within-species variations in genetic background, it was shown that BALB/c and C57BL/6 mice can differ in the modulation of their antibody response by exposure to a restraint stressor (Shanks and Kusnecov 1998). Specifically, a single 1 h restraint session was shown to enhance the antibody response to the antigen KLH in BALB/c mice, but this was not observed in C57BL/6 mice, which showed no major change in antibody titer or number of antibody-forming cells. Interestingly, it was demonstrated in this study that in vivo removal of macrophages abolished the stress effect in BALB/c mice. This suggested that macrophages are an important target for stressor-induced neurohormonal mechanisms. This is supported by other in vitro studies that showed the prominence of the macrophage as a potential mediator of the suppressive influence of stressor exposure on mitogen and antigen-induced lymphocyte proliferation (Coussons-Read et al. 1994; Fleshner et al. 1995).

13.4 Stress and Dendritic Cells

Much of the literature on stress and immunity has focused on studies of effector mechanisms (e.g., cytotoxicity, proliferation, antibody production), with relatively fewer studies addressing those factors involved in the inductive phase of the immune response. Given the importance of DC as a potential therapeutic tool for priming immune responses against tumor antigens, it is important to consider the impact on DC of hormonal, neuropeptide and neurotransmitter alterations due to various forms of stressor exposure. Other chapters in this volume describe in detail the immunobiology of DC and their role in T-cell activation. In addition, there are numerous reviews on various facets of DC function and therapeutic application. Therefore, the following will focus on those studies that have considered the impact of stress on DC function.

13.5 Effects of Psychological Stress on Dendritic Cell Function

Few studies have directly focused on DC function in animal or human subjects that have been subjected to a psychological stressor. As reviewed above, there are a number of animal studies that have applied stressors at the time of antigen presentation, or focused on antigen presentation per se, but only a few papers have directly focused on DC function. In an early study, it was observed that 8 h of a restraint stressor given to male BALB/c mice, followed immediately by sensitization with FITC, led to a depressed contact sensitivity response 5 days later to antigen-specific challenge in the ear (Kawaguchi et al. 1997). Interestingly, this phenomenon was associated with changes in the morphology of Ia^+ Langerhans cells (LC) observed in skin taken from a different group of animals that were sacrificed immediately after stressor exposure. Specifically, Ia^+ cells showed fewer dendrites and a more rounded appearance, and while the number of Ia^+ LC did not change, the area size occupied by LC was reduced by the stressor (ibid). Given that contact sensitivity, a T-cell-mediated response primarily involving $CD8^+$ T cells, is dependent on LC function, these observations suggest that stressor-induced disruption of epidermal DC at the time of antigen uptake may produce impaired development of an adaptive antigen-specific T-cell response.

The effects of stress on contact sensitivity or T-cell-mediated antigen-specific immune responses has been well documented (Kusnecov and Rabin 1993; Wood et al. 1993; Dhabhar and McEwen 1996; Dhabhar et al. 2000), with the study by Kawaguchi et al. being among the first to suggest a possible impairment of local DC antigen-presenting function (Kawaguchi et al. 1997). Interestingly, the effects reported by Kawaguchi et al. were related to stressor duration, since restraint prior to sensitization given for 1–2 h had no effect on the DTH response upon challenge or on LC morphology. However, in a different study, 2.5 h of restraint exposure prior to sensitization with DNFB produced an augmented response to antigen-specific ear challenge (Saint-Mezard et al. 2003). This was consistent with other studies demonstrating augmented sensitivity responses to a short period of stressor exposure in rats (Dhabhar and McEwen 1996; Dhabhar et al. 2000), but contrary to other evidence that acute stressor exposure of BALB/c mice prior to sensitization had a suppressive effect on the DTH challenge (Flint et al. 2001). These discrepancies may be related to differences in animal species and mouse strains, since such variability presents with different predictions for immunological outcome following stressor exposure (Lysle et al. 1990; Shanks and Kusnecov 1998; Flint and Tinkle 2001).

As noted, Saint-Mezard et al. observed enhanced contact sensitivity, which was further observed to be dependent on stressor effects on DC function (Saint-Mezard et al. 2003). This was demonstrated in two ways. First, it was found that bone marrow-derived DC that were antigen pulsed in vitro and then transferred subcutaneously to mice just prior to restraint exposure demonstrated enhanced ear swelling to a challenge 5 days later. Second, when DC were tagged with a fluorescent probe and injected into the footpads of recipient mice that were then restrained, there was an enhanced number of labeled DC in the draining popliteal lymph nodes. These observations led Saint-Mezard et al. to conclude that an acute stressor augments the T-cell-mediated contact sensitivity response by enhancing DC antigen presentation. Although this may involve increased maturation of DC – given their accelerated appearance in the lymph nodes where greater numbers are associated with induction of T-cell reactivity – DC from stressed animals did not show greater expression above that of control animals in the maturation markers CD80, CD86 and MHC class II (Saint-Mezard et al. 2003).

In a variation of these studies, Viswanathan et al. exposed male C57BL/6 mice to 2.5 h restraint, followed by injection of DNFB into the pinna of the ear (Viswanathan et al. 2005). Assessment of the pinna 6 and 24 h after sensitization revealed increased swelling. This was associated with increased leukocyte infiltration and mRNA for chemokines (MIP-3α and MCP-1) and proinflammatory cytokines (the α and β forms of IL-1, TNF-α, IL-6 and IFN-γ). Interestingly, both dermal and draining cervical lymph nodes from stressed mice contained greater numbers of mature DC (as defined by a CD11c and IAb^{hi} surface phenotype). However, this stressor-induced increased cellularity was not unique to DC, as there were also increased numbers of macrophages and both mature and naïve T cells in the lymph nodes (Viswanathan et al. 2005). Overall, therefore, an acute restraint stressor produced a coordinated convergence of antigen-presenting cells (DC and macrophages) and T cells that set the stage for augmented immunologic antigen-specific immunoreactivity. This increased recruitment of cells likely accounts for the greater stressor-induced reactivity to challenge with the sensitizing hapten observed in other studies.

Human studies examining stress effects on DC are as rare as those testing laboratory animals. However, two relatively recent studies suggest that as for other leukocyte subsets, stressor exposure can affect the number of DC in blood and epidermis. For example, Ho et al. used a new monoclonal antibody that specifically recognizes human circulating DC with a CD11c^{+}/CD14^{-}/CD19^{-} phenotype (Ho et al. 2001). In patients undergoing laparoscopic cholecystectomy, circulating DC were significantly elevated during surgery, but declined by the end of surgery to a level below the starting preoperative baseline numbers (Flint and Tinkle 2001). The effects of surgical stress on immune function have been well documented, elevating many of the classic indices of neuroendocrine and sympathetic nervous system activity (Ben-Eliyahu 2003). Therefore, these observations on DC circulation are likely related to stress-related physiological alterations. Interestingly, the initial rise and compensatory change may reflect a release of DC from sequestration sites with ready access to the vasculature, followed by redistribution of DC to tissues where antigen interactions may be required. A similar effect was observed in normal healthy individuals subjected to a physical exercise regimen, whereby DC increased significantly during the exercise (Ho et al. 2001). In another study, it was observed that healthy male and female volunteers subjected to a laboratory-based psychosocial stressor (a public speaking stressor) showed significantly reduced numbers of epidermal LC in skin biopsied 24 h after the stressor (Kleyn et al. 2008). This study was well controlled, in that biopsies were obtained prior to the stressor exposure, and a control group was used in which both 0 and 24 h biopsies were obtained without any intervening stressor exposure. The interpretation of this latter study is quite open, although the reduction in epidermal LC suggests possible egress from dermal tissue to draining lymph nodes, as suggested by the animal studies reviewed above.

These human studies demonstrate that stress can modify the numbers of circulating and dermal DC, although the functional properties of these effects are not known. Nevertheless, taking into account also the small number of animal studies, it is evident that as for macrophages and T cells, exposure of the organism to a stressor modifies the quantity and functional properties of DC. Additional studies are required in which DC antigen uptake and processing as well as cytokine production and cell–cell interactions with T cells are more closely assessed following stressor exposure. At present, however, it does appear that at least for acute stressors, there may be a positive effect on DC function.

13.6 Mechanisms of Stressor-Induced Modulation of Dendritic Cell Function

Two major stressor-activated mechanisms for influencing immune function have consistently received attention: the HPA axis and sympathetic nervous system (Kohm and Sanders 2000). The end-point products of each pathway consist of glucocorticoids and norepinephrine (NE, noradrenaline), respectively, for which receptors reside on most immune cells, including DC (Kohm and Sanders 2000; Delgado et al. 2005; Gonzalez-Rey et al. 2006; Truckenmiller et al. 2006). Traditional approaches to assessing the role of noradrenergic and glucocorticoid effects have involved surgical and pharmacological approaches. The role of the HPA axis has been assessed by removal of the adrenal gland (adrenalectomy) or antagonism of glucocorticoid receptors. Noradrenergic effects have largely been tested through the use of adrenergic receptor antagonists or catecholamine depletion from sympathetic nerve terminals. Given that there are very few studies investigating stressor effects on DC function, the opportunity to apply these tools has seldom occurred. In contrast, the potential impact of these mechanisms on DC function has been explored using in vitro systems.

Moreover, some additional evidence has been proposed for the influence of neuropeptides that are released from sensory nerve endings in the skin and other tissues. In this regard, recent evidence has focused on modulation of DC by vasoactive intestinal peptide (VIP) (Delgado et al. 2005; Gonzalez-Rey et al. 2006) and calcitonin gene-related peptide (CGRP), although other neuropeptides, such as alpha-melanocyte-stimulating hormone (α-MSH), pituitary adenylate cyclase-activating peptide (PACAP), substance P and opioid peptides, have also received some attention (Esche et al. 1999; Makarenkova et al. 2001; Seiffert and Granstein 2006). However, only CGRP has been investigated in relation to stressor effects on DC function. For example, as reported above, Kawaguchi et al. observed a restraint-induced suppression of contact sensitivity to FITC, which was associated with increased dermal expression of CGRP. Since CGRP has been shown to suppress the ability of LC to present antigen, the authors hypothesized that stressor-induced increases in this neuropeptide may have accounted for the reduced contact sensitivity in stressed animals (Kawaguchi et al. 1997). However, no manipulations of CGRP were performed to determine whether this was the case, and therefore, this idea still remains somewhat speculative. Similarly, alteration of epidermal LC numbers by psychosocial stress was associated with increased CGRP expression in human subjects (Kleyn et al. 2008), although the data remain correlative.

13.6.1 Glucocorticoids

The role of glucocorticoids in modifying DC function has been comprehensively reviewed by Truckenmiller et al. (2006). In general, glucocorticoids appear to inhibit differentiation of immature DC to mature T-cell-activating cells, whereas mature DC are resistant to glucocorticoid suppression (Truckenmiller et al. 2006). More recently, incubation of murine immature DC with 10^{-6} M corticosterone (CORT) blocked their progression to a maturation state following stimulation with LPS (Elftman et al. 2007). In addition, DC exposed to CORT failed to show a decrease in antigen uptake, as measured by uptake of soluble protein (FITC-OVA), and LPS-stimulated DC produced less IL-6, IL-12 and TNF-α (Elftman et al. 2007). However, pre-exposing DC to CORT decreased LPS-induced cytokine production. Furthermore, the authors showed that bone marrow-derived DC from male C57BL/6 mice exposed to CORT showed impaired in vivo antigen-specific (viz., HSV-1 $gB_{498-505}$ peptide) priming of naïve $CD8^+$T cells.

Similarly, Truckenmiller et al. found that physiologically relevant levels of CORT prior to infection decreased the number of antigen/MHC I complexes, with the effect occurring upstream of TAP, which transports the antigenic peptides into the endoplasmic reticulum (Truckenmiller et al. 2005). Specifically, mouse antigen-presenting cells were infected with recombinant vaccinia virus expressing ovalbumin (OVA), which results in H-2 K^b presentation of the OVA-derived SIINFEKL peptide. Addition of CORT (10^{-6} M) for 12 h suppressed the number of antigen/MHC complexes and reduced their ability (by 31–57%) to activate an antigen-specific T-cell line (Truckenmiller et al. 2005). Similarly, Pan et al. observed that the synthetic glucocorticoid, dexamethasone (DEX), blocked the surface expression of DC maturation markers CD86, CD40, CD54 and Ia (Pan et al. 2001). Moreover, DEX-treated DC had reduced ability to present antigens via the MHC class II, but not MHC class I, pathway (Pan et al. 2001). In addition, antigen uptake was assessed by measuring FITC-dextran internalization, although DEX exposure had no effect on mannose receptor-mediated endocytosis, which contrasted with findings in human DC by Piemonti et al. (Piemonti et al. 1999a). However, others also failed to find an effect on antigen uptake (Vieira et al. 1998; Dhabhar et al. 2000). Nonetheless, it appears that in the presence of normal antigen uptake, exposure of DC to DEX (10^{-7} to 10^{-8} M) still impairs their ability to activate T cells (Piemonti et al. 1999a, b; Woltman et al. 2000; Pan et al. 2001).

Fewer studies have assessed the effects of glucocorticoids in vivo on DC function in stressed animals. Interestingly, in the study reviewed above by Saint-Mezard et al., administration of the glucocorticoid receptor antagonist, RU486, failed to alter the augmenting effect of restraint on the contact sensitivity response to DNFB (Saint-Mezard et al. 2003). Alternatively, others have observed that the enhancing effect of stress on contact sensitivity reactions is glucocorticoid dependent (Viswanathan et al. 2005). Due to a lack of closer mechanistic assessment, it is not known whether in all cases of demonstrated stressor-induced immunoenhancement, DC function is responsible. However, at the very least, given the suppressive effects of glucocorticoids on DC in vitro, it is far more likely that glucocorticoids would be involved in phenomena where stressor exposure suppresses antigen-specific immune responses. However, this is merely a hypothesis and requires verification. Nonetheless, impaired immune function has been observed in mice that have chronically elevated corticosterone production due to overexpression of the brain HPA axis-stimulating neuropeptide, corticotrophin-releasing hormone (CRH) (Murray et al. 2004). These mice display a decrease in B-cell number, poor antibody isotope switching and secondary antigen-specific IgG responses. It was noted that both these transgenic mice and wild-type mice given CORT in their drinking water displayed reduced splenic germinal center formation that appeared to be related to impaired follicle DC, which trap antigen/antibody complexes and are important for germinal center reactions (Murray et al. 2004).

Overall, the available evidence provides a reasonable basis for hypothesizing a suppressive influence on DC of HPA axis activation following acute and chronic stressor exposure. In vitro studies, however, are somewhat misleading to the extent that incubation times between CORT and the glucocorticoid receptor are likely to be unrealistically prolonged, as opposed to the shorter period of hormonal increases that are seen following acute stressor exposure. Moreover, in vivo, glucocorticoids are likely to compete for cells with other peptides and neurotransmitters released by the peripheral autonomic nervous system. Therefore, more studies are required in which manipulations of endogenous levels of circulating glucocorticoids and glucocorticoid receptors may be manipulated to assess DC antigen processing, migration and antigen presentation in lymph nodes.

13.6.2 Role of the Sympathetic Nervous System: Norepinephrine

The effects of noradrenergic stimulation on T- and B-cell function have been well documented (Kohm and Sanders 2000), and similar effects have been observed in cultured DC. For example, short-term exposure (3 h) of DC to NE during stimulation with LPS or KLH decreased production of IL-12, but increased IL-10 synthesis (Maestroni 2000). Interestingly, these effects involved both α- and β-adrenergic receptors (AR) and was consistent with other evidence that DC express multiple AR subtypes, including $\beta1$, $\beta2$, $\alpha1b$, $\alpha2a$ and $\alpha2c$ adrenergic receptors (Seiffert et al. 2002; Maestroni and Mazzola 2003). The in vivo relevance of the β-AR was assessed in female BALB/c mice intradermally injected with peptidoglycan (PGN) or LPS in the presence or absence of the non-selective β-AR antagonist, propranolol (Manni and Maestroni 2008). Dermal tissue from PGN-injected animals showed increased IL-12, IL-23 and IFN-γ mRNA production after propranolol treatment, there being no effect when LPS was used as the stimulus (ibid). This effect on PGN-elevated cytokine production suggested that β-AR mediates the effects of local NE release, which serves to restrict the transcription of these cytokines. In a further experiment, it was shown that this propranolol effect contributed to enhanced antigen-specific delayed-type hypersensitivity as well as increased migration of DC to draining lymph nodes (Manni and Maestroni 2008).

Although these studies were not conducted utilizing a stress protocol, they do suggest that following stressor exposure, NE may serve to inhibit immature DC and their eventual migration and maturation to local, draining lymph nodes. Indeed, Saint-Mezard et al. observed that the increased recruitment of DC ostensibly responsible for increased stressor-induced contact sensitivity to DNFB could be attenuated by pharmacologic depletion of catecholamines (Saint-Mezard et al. 2003). This was not further investigated to more specifically assess adrenergic receptor involvement. However, the result does suggest that activation of the sympathetic nervous system may be an important influence on immature DC during coincidental exposure to a stressor and antigen. This finding is also consistent with a wealth of evidence that the noradrenergic system exerts a considerable influence on multiple aspects of the immune response both in vitro and in vivo (Kohm and Sanders 2000; Sanders and Straub 2002; Nance and Sanders 2007).

13.7 Concluding Comments

There is a considerable literature on the effects of stressor exposure on immune function in laboratory animals and human subjects. The activities of the immune system involve a complex, multi-faceted process that is largely appreciated in terms of its effector functions. However, the effectiveness of the immune response in neutralizing and eliminating antigens is predicated on successful induction by antigen-presenting cells. This chapter has not focused on DC biology per se but on the potential influence of local biological factors on DC function. These cells are conceptualized in terms of their ability to initiate the cascade of events that we call the immune response, and this conceptualization must consider what we call the local and systemic "neurohormonal" conditions created by factors emerging from the central nervous system. The latter fluctuates in levels of arousal, and under circumstances of stressor exposure, efferent outflow from the nervous system mobilizes musculature, metabolic activity and functions of visceral organs, and as we now recognize, the immune system. At present, the literature is relatively small, but it is already apparent that DC are susceptible to the neuroendocrine and autonomic sequelae of stressor exposure, which can be associated with disease diagnosis, therapy and progression in cancer patients. As more attention is placed on DC, and their role as clinical therapeutic tools, a better appreciation will be required of their interactions with neurally derived hormonal, neurotransmitter and neuropeptide ligands capable of modifying their antigen-presenting and cytokine-producing capabilities. Ultimately, as suggested by some of the findings reviewed in this chapter, it is possible that these neural mechanisms may serve actually to enhance DC function and render them more immunogenic. However, the conditions under which organismal levels of arousal favor such an effect remain to be determined, and if not clarified, may actually militate against the successful therapeutic use of exogenously administered DC in patients with cancer.

Acknowledgments Support contributed by grants MH60706, NIEHS P30 ES05022 and NIEHS Graduate Training grant 5T32 E507148.

References

Bachen, E. A., Manuck, S. B., Marsland, A. L., Cohen, S., Malkoff, S. B., Muldoon, M. F. and Rabin, B. S. 1992. Lymphocyte subset and cellular immune responses to a brief experimental stressor. Psychosom Med 54:673–679.

Ben-Eliyahu, S. 2003. The promotion of tumor metastasis by surgery and stress: immunological basis and implications for psychoneuroimmunology. Brain Behav Immun 17 Suppl 1:S27–36.

Ben-Eliyahu, S., Page, G. G. and Schleifer, S. J. 2007. Stress, NK cells, and cancer: still a promissory note. Brain Behav Immun 21:881–887.

Blatteis, C. M. 2007. The onset of fever: new insights into its mechanism. Prog Brain Res 162:3–14.

Bovbjerg, D. H. 2006. The continuing problem of post chemotherapy nausea and vomiting: contributions of classical conditioning. Auton Neurosci 129:92–98.

Bovbjerg, D. H., Redd, W. H., Jacobsen, P. B., Manne, S. L., Taylor, K. L., Surbone, A., Crown, J. P., Norton, L., Gilewski, T. A., Hudis, C. A. and et al. 1992. An experimental analysis of classically conditioned nausea during cancer chemotherapy. Psychosom Med 54:623–637.

Bovbjerg, D. H., Redd, W. H., Maier, L. A., Holland, J. C., Lesko, L. M., Niedzwiecki, D., Rubin, S. C. and Hakes, T. B. 1990. Anticipatory immune suppression and nausea in women receiving cyclic chemotherapy for ovarian cancer. J Consult Clin Psychol 58:153–157.

Chida, Y., Hamer, M., Wardle, J. and Steptoe, A. 2008. Do stress-related psychosocial factors contribute to cancer incidence and survival? Nat Clin Pract Oncol 5:466–475.

Cohen, S. 2005. Keynote presentation at the eight international congress of behavioral medicine: the Pittsburgh common cold studies: psychosocial predictors of susceptibility to respiratory infectious illness. Int J Behav Med 12:123–131.

Coussons-Read, M. E., Maslonek, K. A., Fecho, K., Perez, L. and Lysle, D. T. 1994. Evidence for the involvement of macrophage-derived nitric oxide in the modulation of immune status by a conditioned aversive stimulus. J Neuroimmunol 50:51–58.
Cunnick, J. E., Lysle, D. T., Armfield, A. and Rabin, B. S. 1988. Shock-induced modulation of lymphocyte responsiveness and natural killer activity: differential mechanisms of induction. Brain Behav Immun 2:102–113.
Cunnick, J. E., Lysle, D. T., Kucinski, B. J. and Rabin, B. S. 1990. Evidence that shock-induced immune suppression is mediated by adrenal hormones and peripheral beta-adrenergic receptors. Pharmacol Biochem Behav 36:645–651.
Damjanovic, A. K., Yang, Y., Glaser, R., Kiecolt-Glaser, J. K., Nguyen, H., Laskowski, B., Zou, Y., Beversdorf, D. Q. and Weng, N. P. 2007. Accelerated telomere erosion is associated with a declining immune function of caregivers of Alzheimer's disease patients. J Immunol 179:4249–4254.
Delgado, M., Gonzalez-Rey, E. and Ganea, D. 2005. The neuropeptide vasoactive intestinal peptide generates tolerogenic dendritic cells. J Immunol 175:7311–7324.
Deng, Y., Chan, S. S. and Chang, S. 2008. Telomere dysfunction and tumour suppression: the senescence connection. Nat Rev Cancer 8:450–458.
Dhabhar, F. S. and McEwen, B. S. 1996. Stress-induced enhancement of antigen-specific cell-mediated immunity. J Immunol 156:2608–2615.
Dhabhar, F. S., Satoskar, A. R., Bluethmann, H., David, J. R. and McEwen, B. S. 2000. Stress-induced enhancement of skin immune function: a role for gamma interferon. Proc Natl Acad Sci U S A 97:2846–2851.
Dimsdale, J. E. 2008. Psychological stress and cardiovascular disease. J Am Coll Cardiol 51:1237–1246.
Dobbs, C. M., Feng, N., Beck, F. M. and Sheridan, J. F. 1996. Neuroendocrine regulation of cytokine production during experimental influenza viral infection: effects of restraint stress-induced elevation in endogenous corticosterone. J Immunol 157:1870–1877.
Elenkov, I. J. and Chrousos, G. P. 1999. Stress hormones, Th1/Th2 patterns, pro/anti-inflammatory cytokines and susceptibility to disease. Trends Endocrinol Metab 10:359–368.
Elftman, M. D., Norbury, C. C., Bonneau, R. H. and Truckenmiller, M. E. 2007. Corticosterone impairs dendritic cell maturation and function. Immunology 122:279–290.
Epel, E. S., Blackburn, E. H., Lin, J., Dhabhar, F. S., Adler, N. E., Morrow, J. D. and Cawthon, R. M. 2004. Accelerated telomere shortening in response to life stress. Proc Natl Acad Sci U S A 101:17312–17315.
Esche, C., Makarenkova, V. P., Kost, N. V., Lotze, M. T., Zozulya, A. A. and Shurin, M. R. 1999. Murine dendritic cells express functional delta-type opioid receptors. Ann N Y Acad Sci 885:387–390.
Flaherty, C. F., Rowan, G. A. and Pohorecky, L. A. 1986. Corticosterone, novelty-induced hyperglycemia, and chlordiazepoxide. Physiol Behav 37:393–396.
Fleshner, M., Bellgrau, D., Watkins, L. R., Laudenslager, M. L. and Maier, S. F. 1995. Stress-induced reduction in the rat mixed lymphocyte reaction is due to macrophages and not to changes in T cell phenotypes. J Neuroimmunol 56:45–52.
Flint, M. S. and Tinkle, S. S. 2001. C57BL/6 mice are resistant to acute restraint modulation of cutaneous hypersensitivity. Toxicol Sci 62:250–256.
Flint, M. S., Valosen, J. M., Johnson, E. A., Miller, D. B. and Tinkle, S. S. 2001. Restraint stress applied prior to chemical sensitization modulates the development of allergic contact dermatitis differently than restraint prior to challenge. J Neuroimmunol 113:72–80.
Gidron, Y. and Ronson, A. 2008. Psychosocial factors, biological mediators, and cancer prognosis: a new look at an old story. Curr Opin Oncol 20:386–392.
Glaser, R. and Kiecolt-Glaser, J. K. 2005. Stress-induced immune dysfunction: implications for health. Nat Rev Immunol 5:243–251.

Golden, S. H. 2007. A review of the evidence for a neuroendocrine link between stress, depression and diabetes mellitus. Curr Diabetes Rev 3:252–259.

Goldstein, D. S. and Kopin, I. J. 2007. Evolution of concepts of stress. Stress 10:109–120.

Gonzalez-Rey, E., Chorny, A., Fernandez-Martin, A., Ganea, D. and Delgado, M. 2006. Vasoactive intestinal peptide generates human tolerogenic dendritic cells that induce CD4 and CD8 regulatory T cells. Blood 107:3632–3638.

Hager, P., Hagman, B., Wikstrom, A. C. and Strommer, L. 2004. CRF-receptor 1 blockade attenuates acute posttraumatic hyperglycemia in rats. J Surg Res 119:72–79.

Herbert, T. B., Cohen, S., Marsland, A. L., Bachen, E. A., Rabin, B. S., Muldoon, M. F. and Manuck, S. B. 1994. Cardiovascular reactivity and the course of immune response to an acute psychological stressor. Psychosom Med 56:337–344.

Ho, C. S., Lopez, J. A., Vuckovic, S., Pyke, C. M., Hockey, R. L. and Hart, D. N. 2001. Surgical and physical stress increases circulating blood dendritic cell counts independently of monocyte counts. Blood 98:140–145.

Kawaguchi, Y., Okada, T., Konishi, H., Fujino, M., Asai, J. and Ito, M. 1997. Reduction of the DTH response is related to morphological changes of Langerhans cells in mice exposed to acute immobilization stress. Clin Exp Immunol 109:397–401.

Kerr, L. R., Andrews, H. N., Strange, K. S., Emerman, J. T. and Weinberg, J. 2006. Temporal factors alter effects of social housing conditions on responses to chemotherapy and hormone levels in a Shionogi mammary tumor model. Psychosom Med 68:966–975.

Kiecolt-Glaser, J. K., Glaser, R., Gravenstein, S., Malarkey, W. B. and Sheridan, J. 1996. Chronic stress alters the immune response to influenza virus vaccine in older adults. Proc Natl Acad Sci U S A 93:3043–3047.

Kleyn, C. E., Schneider, L., Saraceno, R., Mantovani, C., Richards, H. L., Fortune, D. G., Cumberbatch, M., Dearman, R. J., Terenghi, G., Kimber, I. and Griffiths, C. E. 2008. The effects of acute social stress on epidermal Langerhans' cell frequency and expression of cutaneous neuropeptides. J Invest Dermatol 128:1273–1279.

Kohm, A. P. and Sanders, V. M. 2000. Norepinephrine: a messenger from the brain to the immune system. Immunol Today 21:539–542.

Kop, W. J. 1999. Chronic and acute psychological risk factors for clinical manifestations of coronary artery disease. Psychosom Med 61:476–487.

Korte, S. M., Koolhaas, J. M., Wingfield, J. C. and McEwen, B. S. 2005. The Darwinian concept of stress: benefits of allostasis and costs of allostatic load and the trade-offs in health and disease. Neurosci Biobehav Rev 29:3–38.

Kusnecov, A. W. and Rabin, B. S. 1993. Inescapable footshock exposure differentially alters antigen- and mitogen-stimulated spleen cell proliferation in rats. J Neuroimmunol 44:33–42.

Kusnecov, A. W. and Rabin, B. S. 1994. Stressor-induced alterations of immune function: mechanisms and issues. Int Arch Allergy Immunol 105:107–121.

Le Moal, M. 2007. Historical approach and evolution of the stress concept: a personal account. Psychoneuroendocrinology 32 Suppl 1:S3–9.

Lysle, D. T., Cunnick, J. E. and Rabin, B. S. 1990. Stressor-induced alteration of lymphocyte proliferation in mice: evidence for enhancement of mitogenic responsiveness. Brain Behav Immun 4:269–277.

Lysle, D. T., Lyte, M., Fowler, H. and Rabin, B. S. 1987. Shock-induced modulation of lymphocyte reactivity: suppression, habituation, and recovery. Life Sci 41:1805–1814.

Maestroni, G. J. 2000. Dendritic cell migration controlled by alpha 1b-adrenergic receptors. J Immunol 165:6743–6747.

Maestroni, G. J. and Mazzola, P. 2003. Langerhans cells beta 2-adrenoceptors: role in migration, cytokine production, Th priming and contact hypersensitivity. J Neuroimmunol 144:91–99.

Makarenkova, V. P., Esche, C., Kost, N. V., Shurin, G. V., Rabin, B. S., Zozulya, A. A. and Shurin, M. R. 2001. Identification of delta- and mu-type opioid receptors on human and murine dendritic cells. J Neuroimmunol 117:68–77.

Manni, M. and Maestroni, G. J. 2008. Sympathetic nervous modulation of the skin innate and adaptive immune response to peptidoglycan but not lipopolysaccharide: involvement of beta-adrenoceptors and relevance in inflammatory diseases. Brain Behav Immun 22:80–88.

Marsland, A. L., Cohen, S., Rabin, B. S. and Manuck, S. B. 2001. Associations between stress, trait negative affect, acute immune reactivity, and antibody response to hepatitis B injection in healthy young adults. Health Psychol 20:4–11.

Marsland, A. L., Cohen, S., Rabin, B. S. and Manuck, S. B. 2006. Trait positive affect and antibody response to hepatitis B vaccination. Brain Behav Immun 20:261–269.

Marsland, A. L., Manuck, S. B., Fazzari, T. V., Stewart, C. J. and Rabin, B. S. 1995. Stability of individual differences in cellular immune responses to acute psychological stress. Psychosom Med 57:295–298.

McEwen, B. S. 1998. Protective and damaging effects of stress mediators. N Engl J Med 338:171–179.

McEwen, B. S. 2003. Interacting mediators of allostasis and allostatic load: towards an understanding of resilience in aging. Metabolism 52:10–16.

Miller, A. H., Ancoli-Israel, S., Bower, J. E., Capuron, L. and Irwin, M. R. 2008. Neuroendocrine-immune mechanisms of behavioral comorbidities in patients with cancer. J Clin Oncol 26:971–982.

Miller, G. E., Cohen, S., Pressman, S., Barkin, A., Rabin, B. S. and Treanor, J. J. 2004. Psychological stress and antibody response to influenza vaccination: when is the critical period for stress, and how does it get inside the body? Psychosom Med 66:215–223.

Morrow, G. R. and Rosenthal, S. N. 1996. Models, mechanisms and management of anticipatory nausea and emesis. Oncology 53 Suppl 1:4–7.

Moynihan, J. A. and Ader, R. 1996. Psychoneuroimmunology: animal models of disease. Psychosom Med 58:546–558.

Murray, S. E., Rosenzweig, H. L., Johnson, M., Huising, M. O., Sawicki, K. and Stenzel-Poore, M. P. 2004. Overproduction of corticotropin-releasing hormone blocks germinal center formation: role of corticosterone and impaired follicular dendritic cell networks. J Neuroimmunol 156:31–41.

Nance, D. M. and Sanders, V. M. 2007. Autonomic innervation and regulation of the immune system (1987–2007). Brain Behav Immun 21:736–745.

Novack, D. H., Cameron, O., Epel, E., Ader, R., Waldstein, S. R., Levenstein, S., Antoni, M. H. and Wainer, A. R. 2007. Psychosomatic medicine: the scientific foundation of the biopsychosocial model. Acad Psychiatry 31:388–401.

Pan, J., Ju, D., Wang, Q., Zhang, M., Xia, D., Zhang, L., Yu, H. and Cao, X. 2001. Dexamethasone inhibits the antigen presentation of dendritic cells in MHC class II pathway. Immunol Lett 76:153–161.

Piemonti, L., Monti, P., Allavena, P., Leone, B. E., Caputo, A. and Di Carlo, V. 1999a. Glucocorticoids increase the endocytic activity of human dendritic cells. Int Immunol 11:1519–1526.

Piemonti, L., Monti, P., Allavena, P., Sironi, M., Soldini, L., Leone, B. E., Socci, C. and Di Carlo, V. 1999b. Glucocorticoids affect human dendritic cell differentiation and maturation. J Immunol 162:6473–6481.

Pressman, S. D., Cohen, S., Miller, G. E., Barkin, A., Rabin, B. S. and Treanor, J. J. 2005. Loneliness, social network size, and immune response to influenza vaccination in college freshmen. Health Psychol 24:297–306.

Roth, J., Rummel, C., Barth, S. W., Gerstberger, R. and Hubschle, T. 2006. Molecular aspects of fever and hyperthermia. Neurol Clin 24:421–439, v.

Sabbioni, M. E., Bovbjerg, D. H., Mathew, S., Sikes, C., Lasley, B. and Stokes, P. E. 1997. Classically conditioned changes in plasma cortisol levels induced by dexamethasone in healthy men. Faseb J 11:1291–1296.
Saint-Mezard, P., Chavagnac, C., Bosset, S., Ionescu, M., Peyron, E., Kaiserlian, D., Nicolas, J. F. and Berard, F. 2003. Psychological stress exerts an adjuvant effect on skin dendritic cell functions in vivo. J Immunol 171:4073–4080.
Sanders, V. M. and Straub, R. H. 2002. Norepinephrine, the beta-adrenergic receptor, and immunity. Brain Behav Immun 16:290–332.
Saul, A. N., Oberyszyn, T. M., Daugherty, C., Kusewitt, D., Jones, S., Jewell, S., Malarkey, W. B., Lehman, A., Lemeshow, S. and Dhabhar, F. S. 2005. Chronic stress and susceptibility to skin cancer. J Natl Cancer Inst 97:1760–1767.
Segerstrom, S. C. and Miller, G. E. 2004. Psychological stress and the human immune system: a meta-analytic study of 30 years of inquiry. Psychol Bull 130:601–630.
Seiffert, K. and Granstein, R. D. 2006. Neuroendocrine regulation of skin dendritic cells. Ann N Y Acad Sci 1088:195–206.
Seiffert, K., Hosoi, J., Torii, H., Ozawa, H., Ding, W., Campton, K., Wagner, J. A. and Granstein, R. D. 2002. Catecholamines inhibit the antigen-presenting capability of epidermal Langerhans cells. J Immunol 168:6128–6135.
Sgoutas-Emch, S. A., Cacioppo, J. T., Uchino, B. N., Malarkey, W., Pearl, D., Kiecolt-Glaser, J. K. and Glaser, R. 1994. The effects of an acute psychological stressor on cardiovascular, endocrine, and cellular immune response: a prospective study of individuals high and low in heart rate reactivity. Psychophysiology 31:264–271.
Shanks, N. and Kusnecov, A. W. 1998. Differential immune reactivity to stress in BALB/cByJ and C57BL/6 J mice: in vivo dependence on macrophages. Physiol Behav 65:95–103.
Shanks, N., Kusnecov, A., Pezzone, M., Berkun, J. and Rabin, B. S. 1997. Lactation alters the effects of conditioned stress on immune function. Am J Physiol 272:R16–25.
Shawi, M. and Autexier, C. 2008. Telomerase, senescence and ageing. Mech Ageing Dev 129:3–10.
Spiegel, D. 1997. Psychosocial aspects of breast cancer treatment. Semin Oncol 24:S1-36–S31-47.
Spiegel, D. and Sephton, S. E. 2001. Psychoneuroimmune and endocrine pathways in cancer: effects of stress and support. Semin Clin Neuropsychiatry 6:252–265.
Thaker, P. H., Han, L. Y., Kamat, A. A., Arevalo, J. M., Takahashi, R., Lu, C., Jennings, N. B., Armaiz-Pena, G., Bankson, J. A., Ravoori, M., Merritt, W. M., Lin, Y. G., Mangala, L. S., Kim, T. J., Coleman, R. L., Landen, C. N., Li, Y., Felix, E., Sanguino, A. M., Newman, R. A., Lloyd, M., Gershenson, D. M., Kundra, V., Lopez-Berestein, G., Lutgendorf, S. K., Cole, S. W. and Sood, A. K. 2006. Chronic stress promotes tumor growth and angiogenesis in a mouse model of ovarian carcinoma. Nat Med 12:939–944.
Thaker, P. H., Lutgendorf, S. K. and Sood, A. K. 2007. The neuroendocrine impact of chronic stress on cancer. Cell Cycle 6:430–433.
Torres, S. J. and Nowson, C. A. 2007. Relationship between stress, eating behavior, and obesity. Nutrition 23:887–894.
Truckenmiller, M. E., Bonneau, R. H. and Norbury, C. C. 2006. Stress presents a problem for dendritic cells: corticosterone and the fate of MHC class I antigen processing and presentation. Brain Behav Immun 20:210–218.
Truckenmiller, M. E., Princiotta, M. F., Norbury, C. C. and Bonneau, R. H. 2005. Corticosterone impairs MHC class I antigen presentation by dendritic cells via reduction of peptide generation. J Neuroimmunol 160:48–60.
Tseng, R. J., Padgett, D. A., Dhabhar, F. S., Engler, H. and Sheridan, J. F. 2005. Stress-induced modulation of NK activity during influenza viral infection: role of glucocorticoids and opioids. Brain Behav Immun 19:153–164.

van Baarle, D., Tsegaye, A., Miedema, F. and Akbar, A. 2005. Significance of senescence for virus-specific memory T cell responses: rapid ageing during chronic stimulation of the immune system. Immunol Lett 97:19–29.

Vieira, P. L., Kalinski, P., Wierenga, E. A., Kapsenberg, M. L. and de Jong, E. C. 1998. Glucocorticoids inhibit bioactive IL-12p70 production by in vitro-generated human dendritic cells without affecting their T cell stimulatory potential. J Immunol 161:5245–5251.

Viswanathan, K., Daugherty, C. and Dhabhar, F. S. 2005. Stress as an endogenous adjuvant: augmentation of the immunization phase of cell-mediated immunity. Int Immunol 17:1059–1069.

Weiss, E. 1945. Psychosomatic Medicine. The American Journal of Nursing 45:189–193.

Wolf, S. 1948. Experimental Research Into Psychosomatic Phenomena in Medicine. Science 107:637–639.

Woltman, A. M., de Fijter, J. W., Kamerling, S. W., Paul, L. C., Daha, M. R. and van Kooten, C. 2000. The effect of calcineurin inhibitors and corticosteroids on the differentiation of human dendritic cells. Eur J Immunol 30:1807–1812.

Wood, P. G., Karol, M. H., Kusnecov, A. W. and Rabin, B. S. 1993. Enhancement of antigen-specific humoral and cell-mediated immunity by electric footshock stress in rats. Brain Behav Immun 7:121–134.

Chapter 14
Cancer Therapy and Dendritic Cell Immunomodulation

Galina V. Shurin, Neil Amina, and Michael R. Shurin

Abstract Conventional cancer treatment still uses three important modalities of the last four decades: surgery, radiotherapy, and cytotoxic chemotherapy. For treatment of metastatic disease, cytotoxic chemotherapy is the mainstay of treatment, although it is broadly targeted and results in toxicity to normal tissues with limited expectation of curing metastatic tumors. Immunotherapies have also been explored over several decades. Among immunotherapies, approaches based on dendritic cell vaccines are particularly promising, since dendritic cells, as professional antigen-presenting cells, can utilize apoptosis/necrosis-induced therapy of tumors to elicit improved antitumor immunity through the acquisition of tumor antigens from dying tumor cells. The combination of conventional therapy with dendritic cell vaccine is one of the approaches to induce protective antitumor immunity and therapeutic efficacy against cancer. However, conventional therapy is impacting endogenous and exogenous dendritic cell activities and is commonly associated with myelosuppression. New strategies are necessary to develop feasible and effective combinatorial therapeutic approaches for cancer treatment. We have recently shown that short-term non-toxic low-dose chemotherapy, so-called chemomodulation, prior to intralesional injection of dendritic cell vaccine targets multiple immunological and stromal elements in the tumor environment, opening a new opportunity for cancer treatment.

14.1 Modulation of Dendritic Cell Activity by Surgical Procedures

Surgical resection of the primary tumor is a necessary and effective treatment for a number of cancers including breast, lung, gastric, kidney, colorectal, etc. (Shakhar and Ben-Eliyahu 2003; Page 2005; Goldfarb and Ben-Eliyahu 2006; Brancato and Miner 2008). However, immunosuppression induced by

G.V. Shurin (✉)
Departments of Pathology and Immunology, University of Pittsburgh Medical Center, S730 Scaife Hall, 3550 Terrace Street, Pittsburgh, PA, 15261, USA
e-mail: shuringv@upmc.edu

M.R. Shurin, R.D. Salter (eds.), *Dendritic Cells in Cancer*,
DOI 10.1007/978-0-387-88611-4_14, © Springer Science+Business Media, LLC 2009

preoperative and postoperative stress is a common clinical complication that has been widely studied in both animals and humans. It was observed that the bigger the surgical incision is, the more pronounced immune suppression would be, including impairment in the dendritic cell (DC) system (Utoh et al. 1988; Ogata et al. 2000; Ng et al. 2005; Kawasaki et al. 2007; Liao et al. 2007). Decline in the number of circulating DCs has been recently demonstrated in cancer patients who underwent surgery (Brivio et al. 2000a, b; Bellik et al. 2006; Takahashi et al. 2007). For example, surgically treated colorectal cancer patients had nearly one-half circulating DC subset as well as drastically decreased number of DCs obtained from monocytes compared to unoperated and healthy subjects (Bellik et al. 2006).

Several mechanisms acting in synergy might contribute to the surgery-induced immunosuppression. Many groups reported that physiological and psychological stress responses to surgery involve activation of sympathetic nervous system and the hypothalamic–pituitary–adrenal (HPA) axis, which results in the release of stress-related factors such as catecholamines, glucocorticoids, and opioids, known as mediators of immunosuppression (Ben-Eliyahu 2003; Shakhar and Ben-Eliyahu 2003; Shakhar and Blumenfeld 2003). For instance, plasma cortisol concentrations may be significantly increased during upper abdominal surgery, such as gastrectomy or hepatectomy. Cortisol, as has been shown, inhibits TNF-α production and potentiates IL-10 production and this was correlated with decreased level of circulating CD14+ monocytes as well as decreased HLA-DR expression by CD14+ monocytes (Steer et al. 1997; Kawasaki et al. 2007). Recently, several studies have indicated the role of DC in glucocorticoid-mediated immunosuppression. Rozkova et al. observed that DC differentiation in the presence of glucocorticoids was skewed to a qualitatively distinct population incapable of inducing an efficient immune response, with reduced IL-12 production and impaired T-cell stimulatory capacity of DCs (Rozkova et al. 2006). The development of DCs isolated from human bronchoalveolar lavage was inhibited in the presence of glucocorticoids. These DCs expressed low levels of costimulatory CD80 and CD86 molecules and had weak allostimulatory ability (Verhoeven et al. 2000). Interestingly, human DC treated in vitro with glucocorticoids produced the glucocorticoid-induced leucine zipper (GILZ), which oriented DC maturation to induce tolerance and generate regulatory T cells (Treg) expressing $CD25^{high}Foxp3^{+}CTLA\text{-}4/CD152^{+}$ and producing IL-10. GILZ prevented expression of costimulatory and CD83 molecules on DC and upregulated expression of co-inhibitory B7-H1/CD274 and ILT3/CD85κ, as well as production of IL-10 by DCs. Production of pro-inflammatory chemokines by these DCs was also impaired (Rea et al. 2000; Berrebi et al. 2003; Hamdi et al. 2007). Furthermore, DCs obtained from patients treated with glucocorticoids and producing GILZ were able to generate IL-10-secreting Treg ex vivo (Hamdi et al. 2007).

In addition to the release of glucocorticoids, surgical stress results in the secretion of higher levels of catecholamines, which contribute to the rising levels of pro-angiogenic factors, such as vascular endothelial growth factor (VEGF)

(Lutgendorf et al. 2003; Thaker et al. 2006). VEGF as a potent inducer of angiogenesis is necessary for wound healing, and elevation of plasma VEGF has been noted early after resection of different kind of tumors. Because VEGF is a potent promoter of angiogenesis, which is critical for tumor progression, a sustained increase in blood VEGF levels after surgery may stimulate the growth of residual metastases (Belizon et al. 2006; Kirman et al. 2007; Belizon et al. 2008). In addition to its angiogenic activity, VEGF may also play an immunosuppressant role by inhibiting DC maturation. It has been reported that patients with abnormally elevated blood concentrations of VEGF showed significantly lower values of immature and mature DCs, IL-12 expression, and significantly higher levels of endothelin-1 than patients with normal concentrations of VEGF (Lissoni et al. 2001).

Taken together, the surgery-induced release of stress hormones and rising levels of pro-angiogeneic factors contribute to the decline in the circulating DC in operable cancer patients and to the switch of DC differentiation from immune stimulating toward tolerogenic/regulatory DC with lower levels of costimulatory molecules, lower production of inflammatory cytokines and chemokines, and high ability to induce Treg, which might lead to downregulation of antitumor immunity and decrease the efficiency of cancer therapy. Thus, postoperative immunomodulation contributes to the overall response of cancer patients to the treatment and should be controlled for therapeutic purposes (Fig. 14.1).

Different approaches are used to overcome postoperative immunosuppression. One approach includes development of minimally invasive techniques, such as laparoscopy, which is associated with better-preserved systemic immune function (Allendorf et al. 1997; Whelan et al. 2003; Liao et al. 2007). Our own data revealed that laparoscopic invasion in the liver in mice caused significantly lower suppression of functional activity of DC when compared to open surgery (Shurin, unpublished). The use of prophylactic immunostimulatory therapy during the preoperative period is another common approach to decrease postoperative immunosuppression (Cerea et al. 2001; Goldfarb and Ben-Eliyahu 2006). Preoperative IL-2 immunotherapy has been proven to downregulate surgery-induced stimulation of angiogenesis, by either reduction of VEGF level or abrogation of decline of anti-angiogenic IL-12 (Brivio et al. 2002). Brivio et al. also demonstrated that presurgical administration of IL-2 abrogated surgery-induced decline in immature DC amount in operable cancer patients (Brivio et al. 2000b). Perioperatively administered recombinant human granulocyte-macrophage colony-stimulating factor (GM-CSF) induced increase in the level of soluble VEGF receptor 1, an endogenous inhibitor of VEGF, in patients with colon cancer undergoing minimally invasive surgery (Kirman et al. 2007). The other immunostimulatory regimen involves the use of poly I-C, which is a synthetic double-stranded RNA that elicits an immune response resembling that induced by viruses through the engagement of toll-like receptor-3 (TLR3) on immunocytes, including DC (Kunzmann et al. 2004; Werling et al. 2004). Poly I-C regimen significantly reduced, and in some experiments abolished, the establishment of MADB106 metastasis following

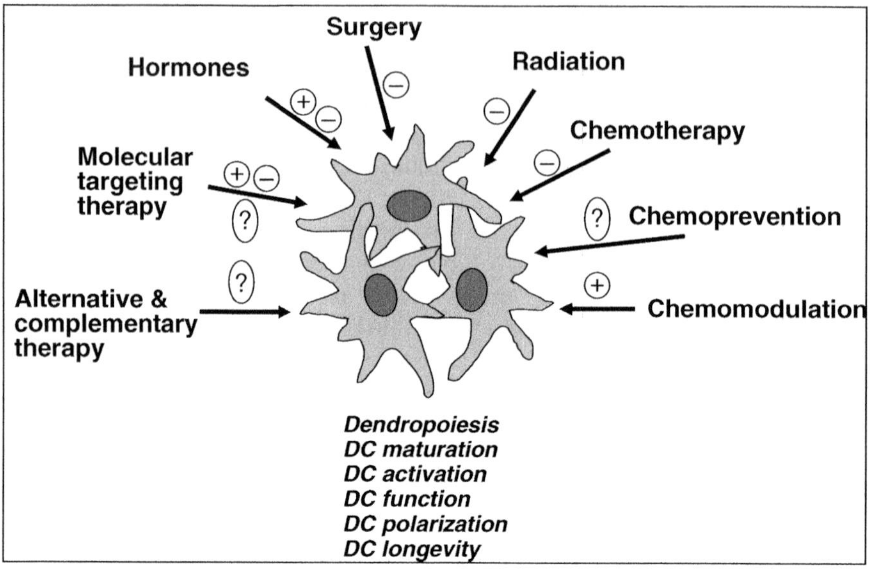

Fig. 14.1 Conventional and novel anti-cancer treatment approaches might markedly alter dendritic cell generation and function in tumor-bearing hosts. Mechanical stress associated with surgery and mediated by release of stress hormones might significantly downregulate dendritic cell (DC) generation and differentiation (dendropoiesis), as well as DC maturation, activation, and functioning. Similar anti-dendropoietic and DC-inhibiting properties were described for conventional radiation and cytotoxic chemotherapy. Although not systemically investigated, hormonal therapy and molecular targeting therapy might display both stimulatory and inhibitory influences on DC, which would depend on specific receptor and signaling molecule expression in a particular DC subtype. The effect of chemoprevention and alternative therapeutic strategies on the DC system in cancer patients has not been yet investigated. Newly appeared chemomodulatory strategy, i.e., treatment with very low doses of chemotherapeutic agents, in addition to suppressing tumor-mediated inhibition of DC activity, displays a direct stimulatory effect on DC maturation and function

surgery (Shakhar et al. 2007). Despite the positive effect of this immunostimulating strategy, clinically, however, postoperative immune suppression is often so profound that it can obliterate preoperative immunostimulation. Thus, immunostimulating approaches should be further tested in preclinical studies. Specifically, when using the stimulatory therapy, the cytokines induced by this therapy might interact with the malignant tissue in a way which supports its continued growth (Okazaki et al. 2005, 2006).

14.2 Impact of Radiation Therapy on Dendritic Cells

Sixty percent of cancer patients get radiation therapy. Exposure to ionizing radiation often leads to immunosuppression traditionally focused on lymphocytes as the primary cellular target. Radiation-induced immunosuppression is mainly attributed to significant lymphocytopenia. In general, memory T and NK cells are relatively radioresistant when compared to B cells and naïve T cells (De Ruysscher et al. 1992; Belka et al. 1999). DCs are not particularly sensitive to the cytotoxic effects of radiation since they are commonly non-proliferative cells. However, accumulating evidence indicates that radiation therapy might functionally affect DCs (Liao et al. 2007). For example, expression of MHC class II and CD86 on DCs decreased marginally over a 24-h culture period after irradiation, although their viability was not affected. The same authors investigated the effect of ionizing radiation on processing and presentation by the endogenous pathway and an exogenous loading of human MART-1 peptide in DC. The authors demonstrated that irradiation of DC in vitro can inhibit their ability to endogenously process and present MART-1 tumor-associated antigen to T cells and to generate protective antitumor immunity. It has been suggested that this effect might be mediated by blocking proteasome-mediated degradation essential for the production of antigenic peptides for loading onto MHC class I molecules, since proteasome in DC was affected following radiation (Liao et al. 2004). Reduction in skin DCs, particularly Langerhans cell density, has been reported in humans and mice after radiation therapy (Cole 1986; Kawase et al. 1990).

Traditionally radiation has been considered as a "silent" killer, since it was not translated into tumor immunity. Recent studies demonstrated that radiation has much more to offer than being a powerful cytotoxic agent. Radiation has definite immunomodulatory properties and is a potential adjuvant for cancer immunotherapy (Liao et al. 2004, 2007). Radiation therapy might broaden the spectrum of tumor-associated antigen epitopes that are recognized by the immune system. Irradiated tumor cells have been shown to serve effectively as a source of tumor antigens to elicit T-cell responses when processed and presented by DCs (Huang et al. 2007). Several studies with tumor irradiation followed by DC administration showed enhanced antitumor effects in mouse models. DC vaccines might efficiently recruit tumor antigens and endogenous infiltrating DC might augment their abilities for antigen processing and presentation following radiation therapy (Friedman 2002; Teitz-Tennenbaum et al. 2003, 2008). However, acute and chronic bone marrow toxicities (including inhibition of myelopoiesis) and intestinal damage limit the efficacy of radiation therapy and immunotherapy. When DC vaccineis given in high doses, it decreases a number of hematopoietic stem cells and progenitors and reduces self-renewal capacity of stem cells (Tubiana et al. 1993; Kyoizumi et al. 1994; Blumenthal et al. 1998; Liu et al. 2006). In addition, early and late toxicities limit the deliverable intensity of radiotherapy and might affect the long-term health-related quality of life of the patients. As a result, patients may experience symptoms associated with damage to normal tissue during the course of therapy, for a few weeks after therapy, or months or years later. Symptoms may be due to cell death or wound healing initiated within irradiated tissue and may be precipitated by exposure to further injury or trauma (Stone et al. 2003; Bentzen 2006). Further research is necessary to assess new strategies for preventing or reducing the side effects of radiation therapy.

14.3 Regulation of Dendritic Cell Function by Chemotherapy: Impact of High- Versus Low-Dose Chemotherapy on Dendritic Cell Activity

For almost half a century, systemic therapy of cancer has been dominated by the use of cytotoxic chemotherapeutics. Antiproliferative and cytotoxic agents are widely used for the treatment of cancer patients alone or in combination with surgery, radiation therapy, and biological therapy. Chemotherapy is the most actively used modality in the treatment of advanced stage cancer. In most cases chemotherapeutic drugs are administered at or near the maximum tolerated dose (MTD) for the purpose of maximal eradication of neoplastic cells. Historically, oncologists have used the highest possible dose that the body can tolerate in order to kill as many cancer cells as possible (Mihich 1967, 2000a, b). Since the available drugs are neither specific nor selective enough in their anticancer action, the main problems in cancer chemotherapy are related to acute and cumulative toxicities to normal tissues, including the cells of the immune system (Mihich 2000a; Perrotta et al. 2007). Conventional chemotherapy is commonly associated with myelosuppression and causes decreased longevity and number of DCs in vivo (Fig. 14.1) (Shin et al. 2003; Bellik et al. 2006; Laane et al. 2007; Perrotta et al. 2007). The suppression of the immune system by conventional chemotherapy might support tumor-escape mechanisms and promote the propagation of chemoresistant clones by maintaining a cytokine milieu that favors proliferation of tumor cells (Shurin et al. 2006).

One promising strategy to reduce the toxicity of chemotherapy maintaining its efficacy is the combination of chemotherapy with immunotherapy (Liao et al. 2007). Such a strategy could exploit the mutual effects of therapy to treat cancers due to the different mechanisms of action of chemotherapy and immunotherapy. Recent findings suggest that some chemotherapeutic agents can improve the efficacy of immunotherapy by different means (Lake and Robinson 2005; Nowak et al. 2006). Cell death induced by chemotherapeutic drugs is often accompanied by the release of a number of danger signals that lead to the activation of immune cells and may facilitate tumor antigen engulfing and processing (Nowak et al. 2003). Among immunotherapies, approaches based on DC vaccines are particularly promising (O'Neill and Bhardwaj 2007; Shu et al. 2007; Zhong et al. 2007b). DC therapy has been considered to be one of the emerging strategies for the treatment of patients with advanced cancer, especially for patients that resist conventional therapies, such as surgery, irradiation, and chemotherapy (Choi et al. 2005). A number of published clinical trials and preclinical studies investigated the ability of immune system to overcome tumor tolerance using DCs generated and loaded with tumor antigens ex vivo and injected into cancer patients. To avoid the requirement of ex vivo antigen-loading process, some researchers have suggested the administration of DC after systemic chemotherapy, because tumor cell death induced by chemotherapeutic drugs should liberate potential antigenic peptides for cross-presentation (Tong et al. 2001; Shin et al. 2003;

Heath et al. 2004; Choi et al. 2005). Various chemotherapeutic drugs are currently being used for combination therapy studies of DC and chemotherapy (Nestle et al. 1998; Gong et al. 2000; Tong et al. 2001; Le Poole et al. 2003; Shin et al. 2003). Improved tumor antigen presentation and T-cell activation due to drug-induced expression of MHC and costimulatory molecules on tumor and DC were demonstrated by a number of investigators (Galetto et al. 2003; Ohtsukasa et al. 2003; Do et al. 2004). The antitumor effect after the combination of chemotherapy with DC vaccine has been found to be significantly more profound. These experiments established proof-of-concept that the combination of the intratumoral placement of ex vivo-generated autologous DC in concert with apoptosis-inducing chemotherapy might lead to a generation of systemic antitumor immunity (Yu et al. 2003).

However, conventional chemotherapy is given in several cycles over relatively long periods of time. The prolonged exposure to the drugs induces the apoptotic death of immune cells, including both DC in a tumor-bearing host and DC from vaccine administered after chemotherapy (Yu et al. 2003; Cavanagh et al. 2007). For instance, Shin et al. and Perrotta et al. reported that cisplatin and vincristine triggered DC apoptosis both in vitro and in vivo (Shin et al. 2003; Perrotta et al. 2007). Mitomycin C could downregulate expression of CD80 and CD86 on DC (Jiga et al. 2004). In vivo, the levels of conventional DC subset were markedly decreased in the bone marrow of cancer patients receiving cytostatic treatment (Laane et al. 2007). DC function and blood counts were decreased in patients receiving various types of chemotherapy (Markowicz et al. 2002). In this case therapeutic efficacy of DC vaccines might be ascribed to immune-induced unbalancing of the intratumoral environment and to insufficient in vivo function of DC. Thus, in spite of noticeable antitumor potential of combination chemo/DC therapy, toxic effects of conventional doses of chemotherapeutic agents limit clinical utilization of this regime.

Another strategy to limit toxic effects of chemotherapeutic agents has been recently reported. Chemotherapy given in low, frequent doses – a novel strategy called "metronomic" delivery – achieved partial shrinkage of tumor due to its anti-angiogenic activity (Laquente et al. 2007). Metronomic (=steady) chemotherapy means the patient receives daily low doses of the drug over a certain period of time as opposed to standard therapy where one-time maximum dose is administered at about every 3 weeks. Main targets of metronomic treatment are the proliferating endothelial cells of tumor vasculature, which are more sensitive to low-dose chemotherapy-induced toxicity than tumor cells. Although this is an intriguing concept, there is already a suggestion that vascular parameters alone may not be predictive of survival or local or distant recurrences in cancer patients.

We have recently initiated a novel therapeutic approach on incorporation of very low-dose chemotherapy, so-called chemomodulation, into vaccine therapy, where low-dose chemotherapeutic agent modifies immunogenicity of the tumor microenvironment and the immunotherapeutic component targets tumor and induces protective antitumor immunity. We demonstrated, for instance, that a chemotherapeutic agent paclitaxel at low non-toxic (for tumor, DC, and bone marrow cells) doses stimulated maturation and function of DC and reduced tumor-mediated immunosuppression (Fig. 14.1). A single injection of a low non-toxic dose of paclitaxel in mice bearing lung carcinoma significantly upregulated the antitumor potential of DC vaccines in preclinical models (Zhong et al. 2007a). We suggested that the tumor cell modification by low-dose chemotherapeutic agents changes the local tumor microenvironment in a way that increases the source of tumor antigens for immune cells, upregulates their functional activity, and thus accelerates and sustains antitumor immunity and overall therapeutic effect.

The tumor microenvironment is now recognized as a major factor influencing not only malignant progression and metastatic potential but also treatment resistance to conventional anti-cancer therapies. Contradicting forces, such as cytokines operating in the tumor microenvironment, create a balance of pro- and anti-malignancy effects, and preclinical and clinical data support the notion that a conducive microenvironment is necessary for cancer immune rejection to take place (Shurin et al. 2006). Control of host–tumor interaction, activity of regulatory cells, and cytokine network at the tumor site with low-dose chemotherapy, i.e., chemomodulation, have the potential to reverse the malignant phenotype and reestablish normal control mechanism. We have recently developed a unique approach that allows evaluating secretion of intratumoral cytokines for several days in live freely moving mice. The method utilizes intratumoral insertion of unique microdialysis microprobes, and samples (80 μl/7–8 h) are analyzed by Luminex multiplex-based technique for the presence of more than 50 cytokines (Shurin et al. 2006). We have used this approach to evaluate intratumoral and peritumoral (non-malignant tissue) cytokines, chemokines, and growth factors in lung cancer-bearing mice receiving low-dose paclitaxel followed by intratumoral injection of DC vaccine. We demonstrated that a single injection of paclitaxel resulted in increased expression of macrophage inflammatory protein (MCP)-1 at the tumor site, while the application of low-dose paclitaxel followed by DC vaccine induced an increase in intratumoral MCP-1, γ-interferon-inducible protein-10 (IP-10), and decrease in intratumoral IL-1α release in vivo that was correlated with significant inhibition of tumor growth (Zhong et al. 2007a). These data suggest that low-dose chemotherapy might cause specific alterations in the intratumoral microenvironment, which can be beneficial for boosting antitumor responses and inhibiting tumor growth in vivo. The pathways of antitumor activity of chemotherapeutic agents used in therapeutic versus low concentration are different and unknown yet. For instance, paclitaxel is known to irreversibly polymerize α/β tubulin, thereby suppressing microtubule dynamics and arresting the cell cycle at G_2/M phase at high (>200 nM) concentration, but cannot form multipolar spindles in low concentration (Zhao et al. 2005). The studies to reveal the molecular mechanisms underlying immunomodulating effects of low-dose chemomodulators are under progress in our laboratory.

14.4 Chemomodulation of Small Rho GTPase Activity in Dendritic Cells

Motility, endocytosis, and antigen processing/presentation are the most important DC functions, which depend on the reorganization of actin cytoskeleton. Regulation of a wide variety of cellular activities, from endocytosis and cell motility to cytokinesis and cellular trafficking, falls largely to the small Rho GTPase family. It is well established that Rho GTPases control DC adherence, antigen processing and presentation, migration, chemotaxis, and endocytosis (Kobayashi et al. 2001; Benvenuti et al. 2004; Shurin et al. 2005; Tourkova et al. 2007). Our data suggest that chemotherapeutic agents at low concentrations might directly influence the key DC functions in vitro and in vivo. We hypothesized that Rho GTPases might be involved in the effects of low-dose chemotherapeutics on DC. Using Rac G-LISA, RhoA G-LISA activation assay kits and Western blot technology, we studied putative regulatory roles for various chemotherapeutic agents in modulating small Rho GTPases in DCs. Our data revealed that different classes of chemotherapeutic drugs, including paclitaxel, methotrexate, and doxorubicin, at low non-toxic concentrations regulated the activity of classic Rho family members Rac and RhoA, as well as non-classic Rho GTPase RhoE in murine DC, suggesting that small Rho GTPases might serve as new targets for modulating functional activity of DC vaccines or endogenous DC in various immunotherapeutic or chemoimmunotherapeutic strategies (Shurin et al. 2008). Thus, these results support the view that chemotherapeutic agents used in low concentrations might activate unusual signaling pathways in immune cells and thus display unusual immunomodulating properties.

14.5 Chemomodulation of Cytokine Expression in Dendritic Cells

Growing interest in combining chemotherapy and immunotherapy prompted evaluation of the ability of cytotoxic agents to modulate DC function. Our recent data demonstrated that a number of chemotherapeutic drugs in low concentrations (chemomodulators) induced antigen uptake, processing and presentation, motility, and expression of costimulatory molecules in DCs (Shurin et al. submitted).

We have tested whether low-dose non-toxic chemotherapy (chemomodulation) might alter cytokine profiling in DC resulting in downregulation of immunosuppressive cytokines, while immunostimulatory cytokines might be upregulated. Using eight cytotoxic agents, selected according to their non-toxic and immunostimulating effects on tumor cells and DCs, we have determined and compared the pattern of cytokines/chemokines/growth factor expression in control and chemomodulator-treated DCs. Murine bone marrow precursors were treated with chemotherapeutic agents (0.1–50 nM) for 72 h. On day 7, DCs were lysed and 30 biological markers were measured in DC lysates by Luminex assay based on xMAP technology in collaboration with Dr. Anna Lokshin,

Director of Luminex Core Facility, University of Pittsburgh Cancer Institute. Analysis of cytokine profiles in control and treated DCs demonstrated that several chemotherapeutic agents significantly upregulated production of IL-12p70 in DCs. For instance, the level of IL-12p70 was 363.0 ± 24.0 pg/ml for control (non-treated) DC, 590.0 ± 28.0 pg/ml for 1 nM vincristine-treated DC, and 617.0 ± 36.0 0 pg/ml for 5 nM paclitaxel-treated DC. Moderate upregulation of IL-12p70 expression was observed in DCs treated with 5 nM methotrexate, 50 nM mitomycin, and 10 nM doxorubicin, while 5-azacytidine at 10 nM caused inhibition of IL-12p70 production in DCs. Bleomycin (1 nM) did not alter the expression of IL-12p70 in DCs (Fig. 14.2A). As was shown, the production of immunosuppressive IL-10 cytokine was decreased in DCs by

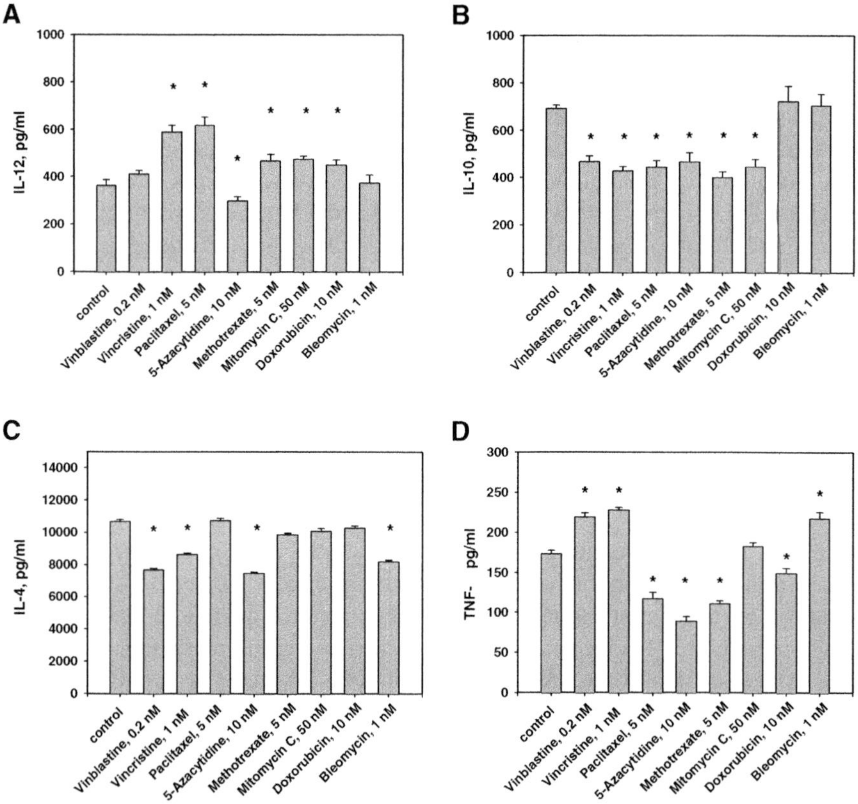

Fig. 14.2 Effect of chemomodulation, i.e., low non-toxic chemotherapy, on cytokine profiling in dendritic cells. Murine dendritic cells (DCs) were generated from the bone marrow hematopoietic precursors in cultures supplemented with GM-CSF and IL-4. Chemotherapeutic agents were added on day 1 and DC were harvested, lysed, and analyzed on day 7. Cytokines, chemokines, and growth factors were assessed in DC lysates by Luminex multiplex-based analysis. Control, non-treated DC; $p < 0.01$ versus control. Two independent experiments provided similar data

30–40% after their treatment with low concentrations of vinblastine, vincristine, paclitaxel, 5-azacytidine, methotrexate, and mitomycin C (Fig. 14.2B), while IL-4 production in DCs treated with chemotherapeutic drugs did not change with exception of slight (25–30%) but significant inhibition by 0.2 nM vinblastine, 1 nM vincristine, 10 nM 5-azacytidine, and 1 nM bleomycin (Fig. 14.2C). TNF-1α production was decreased in DCs treated with low-dose paclitaxel, 5-azacytidine, methotrexate, and doxorubicin by 20–50%, but increased in the presence of vinblastine, vincristine, and bleomycin up to 25% (Fig. 14.2D). The levels of other cytokines, chemokines, and growth factors were similar in control and chemotherapeutic-treated DC.

Based on these studies we conclude that at least several chemotherapeutic drugs at low concentration increased the production of pro-inflammatory cytokine IL-12 and decreased anti-inflammatory cytokine IL-10 in DCs, suggesting that low-dose chemomodulation might change the cytokine profiling in DCs toward their immunostimulating potential.

14.6 Conclusions

Our data justify the importance of systematic evaluation of chemotherapeutic agents for their ability to enhance the antitumor immune response when used in low non-toxic doses in tumor-bearing hosts. Modulation of the tumor environment by low-dose chemotherapy, i.e., chemomodulation, prior to cancer vaccine administration may provide a unique opportunity for therapeutic intervention by shaping the repertoire toward reactivity to tumor antigens. This might be due to the upregulation of antigen-processing machinery in tumor cells, downregulation of immunosuppressive potential of tumor cells, and/or enhancing activity of immune effector cells. An understanding of the underlying cellular and immunological events in both animal models and patients undergoing chemotherapy will guide decisions about which immunomodulatory approaches may be effective with different cytostatic drugs and hence to develop appropriate scheduling for integration of the treatment modalities.

References

Allendorf, J. D., Bessler, M., Whelan, R. L., Trokel, M., Laird, D. A., Terry, M. B. and Treat, M. R. 1997. Postoperative immune function varies inversely with the degree of surgical trauma in a murine model. Surg Endosc 11:427–430.

Belizon, A., Balik, E., Feingold, D. L., Bessler, M., Arnell, T. D., Forde, K. A., Horst, P. K., Jain, S., Cekic, V., Kirman, I. and Whelan, R. L. 2006. Major abdominal surgery increases plasma levels of vascular endothelial growth factor: open more so than minimally invasive methods. Ann Surg 244:792–798.

Belizon, A., Balik, E., Horst, P., Feingold, D., Arnell, T., Azarani, T., Cekic, V., Skitt, R., Kumara, S. and Whelan, R. L. 2008. Persistent elevation of plasma vascular endothelial growth factor levels during the first month after minimally invasive colorectal resection. Surg Endosc 22:287–297.

Belka, C., Ottinger, H., Kreuzfelder, E., Weinmann, M., Lindemann, M., Lepple-Wienhues, A., Budach, W., Grosse-Wilde, H. and Bamberg, M. 1999. Impact of localized radiotherapy on blood immune cells counts and function in humans. Radiother Oncol 50:199–204.

Bellik, L., Gerlini, G., Parenti, A., Ledda, F., Pimpinelli, N., Neri, B. and Pantalone, D. 2006. Role of conventional treatments on circulating and monocyte-derived dendritic cells in colorectal cancer. Clin Immunol 121:74–80.

Ben-Eliyahu, S. 2003. The promotion of tumor metastasis by surgery and stress: immunological basis and implications for psychoneuroimmunology. Brain Behav Immun 17 Suppl 1:S27–36.

Bentzen, S. M. 2006. Preventing or reducing late side effects of radiation therapy: radiobiology meets molecular pathology. Nat Rev Cancer 6:702–713.

Benvenuti, F., Hugues, S., Walmsley, M., Ruf, S., Fetler, L., Popoff, M., Tybulewicz, V. L. and Amigorena, S. 2004. Requirement of Rac1 and Rac2 expression by mature dendritic cells for T cell priming. Science 305:1150–1153.

Berrebi, D., Bruscoli, S., Cohen, N., Foussat, A., Migliorati, G., Bouchet-Delbos, L., Maillot, M. C., Portier, A., Couderc, J., Galanaud, P., Peuchmaur, M., Riccardi, C. and Emilie, D. 2003. Synthesis of glucocorticoid-induced leucine zipper (GILZ) by macrophages: an anti-inflammatory and immunosuppressive mechanism shared by glucocorticoids and IL-10. Blood 101:729–738.

Blumenthal, R. D., Alisauskas, R., Lew, W., Sharkey, R. M. and Goldenberg, D. M. 1998. Myelosuppressive changes from single or repeated doses of radioantibody therapy: effect of bone marrow transplantation, cytokines, and hematopoietic suppression. Exp Hematol 26:859–868.

Brancato, S. and Miner, T. J. 2008. Surgical management of gastric cancer: review and consideration for total care of the gastric cancer patient. Curr Treat Options Gastroenterol 11:109–118.

Brivio, F., Gilardi, R., Bucocev, R., Ferrante, R., Rescaldani, R., Vigore, L., Fumagalli, L., Nespoli, A. and Lissoni, P. 2000a. Surgery-induced decline in circulating dendritic cells in operable cancer patients: a possible explanation of postoperative immunosuppression. Hepatogastroenterology 47:1337–1339.

Brivio, F., Lissoni, P., Gilardi, R., Ferrante, R., Vigore, L., Curzi, L., Uggeri, F., Nespoli, A. and Fumagalli, L. 2000b. Abrogation of surgery-induced decline in circulating dendritic cells by subcutaneous preoperative administration of IL-2 in operable cancer patients. J Biol Regul Homeost Agents 14:200–203.

Brivio, F., Lissoni, P., Rovelli, F., Nespoli, A., Uggeri, F., Fumagalli, L. and Gardani, G. 2002. Effects of IL-2 preoperative immunotherapy on surgery-induced changes in angiogenic regulation and its prevention of VEGF increase and IL-12 decline. Hepatogastroenterology 49:385–387.

Cavanagh, W. A., Tjoa, B. A. and Ragde, H. 2007. Chemotherapy followed by syngeneic dendritic cell injection in the mouse: findings and implications for human treatment. Urology 70:36–41.

Cerea, K., Romano, F., Bravo, A. F., Motta, V., Uggeri, F., Brivio, F., Fumagalli, L. A. and Uggeri, F. 2001. Phase IB study on prevention of surgery-induced immunodeficiency with preoperative administration of low-dose subcutaneous interleukin-2 in gastric cancer patients. J Surg Oncol 78:32–37.

Choi, G. S., Lee, M. H., Kim, S. K., Kim, C. S., Lee, H. S., Im, M. W., Kil, H. Y., Seong, D. H. , Lee, J. R., Kim, W. C., Lee, M. G. and Song, S. U. 2005. Combined treatment of an intratumoral injection of dendritic cells and systemic chemotherapy (Paclitaxel) for murine fibrosarcoma. Yonsei Med J 46:835–842.

Cole, S. 1986. Long-term effects of local ionizing radiation treatment on Langerhans cells in mouse footpad epidermis. J Invest Dermatol 87:608–612.

De Ruysscher, D., Waer, M., Vandeputte, M., Aerts, R., Vantongelen, K. and van der Schueren, E. 1992. Changes of lymphocyte subsets after local irradiation for early stage breast cancer and seminoma testis: long-term increase of activated (HLA-DR+) T cells and decrease of "naive" (CD4–CD45R) T lymphocytes. Eur J Cancer 28A:1729–1734.

Do, Y., Hegde, V. L., Nagarkatti, P. S. and Nagarkatti, M. 2004. Bryostatin-1 enhances the maturation and antigen-presenting ability of murine and human dendritic cells. Cancer Res 64:6756–6765.

Friedman, E. J. 2002. Immune modulation by ionizing radiation and its implications for cancer immunotherapy. Curr Pharm Des 8:1765–1780.

Galetto, A., Buttiglieri, S., Forno, S., Moro, F., Mussa, A. and Matera, L. 2003. Drug- and cell-mediated antitumor cytotoxicities modulate cross-presentation of tumor antigens by myeloid dendritic cells. Anticancer Drugs 14:833–843.

Goldfarb, Y. and Ben-Eliyahu, S. 2006. Surgery as a risk factor for breast cancer recurrence and metastasis: mediating mechanisms and clinical prophylactic approaches. Breast Dis 26:99–114.

Gong, J., Nikrui, N., Chen, D., Koido, S., Wu, Z., Tanaka, Y., Cannistra, S., Avigan, D. and Kufe, D. 2000. Fusions of human ovarian carcinoma cells with autologous or allogeneic dendritic cells induce antitumor immunity. J Immunol 165:1705–1711.

Hamdi, H., Godot, V., Maillot, M. C., Prejean, M. V., Cohen, N., Krzysiek, R., Lemoine, F. M., Zou, W. and Emilie, D. 2007. Induction of antigen-specific regulatory T lymphocytes by human dendritic cells expressing the glucocorticoid-induced leucine zipper. Blood 110:211–219.

Heath, W. R., Belz, G. T., Behrens, G. M., Smith, C. M., Forehan, S. P., Parish, I. A., Davey, G. M., Wilson, N. S., Carbone, F. R. and Villadangos, J. A. 2004. Cross-presentation, dendritic cell subsets, and the generation of immunity to cellular antigens. Immunol Rev 199:9–26.

Huang, J., Wang, Y., Guo, J., Lu, H., Lin, X., Ma, L., Teitz-Tennenbaum, S., Chang, A. E. and Li, Q. 2007. Radiation-induced apoptosis along with local and systemic cytokine elaboration is associated with DC plus radiotherapy-mediated renal cell tumor regression. Clin Immunol 123:298–310.

Jiga, L. P., Bauer, T. M., Chuang, J. J., Opelz, G. and Terness, P. 2004. Generation of tolerogenic dendritic cells by treatment with mitomycin C: inhibition of allogeneic T-cell response is mediated by downregulation of ICAM-1, CD80, and CD86. Transplantation 77:1761–1764.

Kawasaki, T., Ogata, M., Kawasaki, C., Okamoto, K. and Sata, T. 2007. Effects of epidural anaesthesia on surgical stress-induced immunosuppression during upper abdominal surgery. Br J Anaesth 98:196–203.

Kawase, Y., Naito, S., Ito, M., Sekine, I. and Fujii, H. 1990. The effect of ionizing radiation on epidermal Langerhans cells – a quantitative analysis of autopsy cases with radiation therapy. J Radiat Res (Tokyo) 31:246–255.

Kirman, I., Belizon, A., Balik, E., Feingold, D., Arnell, T., Horst, P., Kumara, S., Cekic, V., Jain, S., Nasar, A. and Whelan, R. L. 2007. Perioperative sargramostim (recombinant human GM-CSF) induces an increase in the level of soluble VEGFR1 in colon cancer patients undergoing minimally invasive surgery. Eur J Surg Oncol 33:1169–1176.

Kobayashi, M., Azuma, E., Ido, M., Hirayama, M., Jiang, Q., Iwamoto, S., Kumamoto, T., Yamamoto, H., Sakurai, M. and Komada, Y. 2001. A pivotal role of Rho GTPase in the regulation of morphology and function of dendritic cells. J Immunol 167:3585–3591.

Kunzmann, V., Kretzschmar, E., Herrmann, T. and Wilhelm, M. 2004. Polyinosinic-polycytidylic acid-mediated stimulation of human gammadelta T cells via CD11c dendritic cell-derived type I interferons. Immunology 112:369–377.

Kyoizumi, S., McCune, J. M. and Namikawa, R. 1994. Direct evaluation of radiation damage in human hematopoietic progenitor cells in vivo. Radiat Res 137:76–83.

Laane, E., Bjorklund, E., Mazur, J., Lonnerholm, G., Soderhall, S. and Porwit, A. 2007. Dendritic cell regeneration in the bone marrow of children treated for acute lymphoblastic leukaemia. Scand J Immunol 66:572–583.

Lake, R. A. and Robinson, B. W. 2005. Immunotherapy and chemotherapy – a practical partnership. Nat Rev Cancer 5:397–405.
Laquente, B., Vinals, F. and Germa, J. R. 2007. Metronomic chemotherapy: an antiangiogenic scheduling. Clin Transl Oncol 9:93–98.
Le Poole, I. C., Bommiasamy, H. and Kast, W. M. 2003. Recent progress in tumour vaccine development. Expert Opin Investig Drugs 12:971–981.
Liao, Y. P., Schaue, D. and McBride, W. H. 2007. Modification of the tumor microenvironment to enhance immunity. Front Biosci 12:3576–3600.
Liao, Y. P., Wang, C. C., Butterfield, L. H., Economou, J. S., Ribas, A., Meng, W. S., Iwamoto, K. S. and McBride, W. H. 2004. Ionizing radiation affects human MART-1 melanoma antigen processing and presentation by dendritic cells. J Immunol 173:2462–2469.
Lissoni, P., Malugani, F., Bonfanti, A., Bucovec, R., Secondino, S., Brivio, F., Ferrari-Bravo, A., Ferrante, R., Vigore, L., Rovelli, F., Mandala, M., Viviani, S., Fumagalli, L. and Gardani, G. S. 2001. Abnormally enhanced blood concentrations of vascular endothelial growth factor (VEGF) in metastatic cancer patients and their relation to circulating dendritic cells, IL-12 and endothelin-1. J Biol Regul Homeost Agents 15:140–144.
Liu, W. C., Wang, S. C., Tsai, M. L., Chen, M. C., Wang, Y. C., Hong, J. H., McBride, W. H. and Chiang, C. S. 2006. Protection against radiation-induced bone marrow and intestinal injuries by Cordyceps sinensis, a Chinese herbal medicine. Radiat Res 166:900–907.
Lutgendorf, S. K., Cole, S., Costanzo, E., Bradley, S., Coffin, J., Jabbari, S., Rainwater, K., Ritchie, J. M., Yang, M. and Sood, A. K. 2003. Stress-related mediators stimulate vascular endothelial growth factor secretion by two ovarian cancer cell lines. Clin Cancer Res 9:4514–4521.
Markowicz, S., Walewski, J., Zajda, K., Wiechno, P. J., Skurzak, H. M., Giermek, J., Gajkowska, A., Krzakowski, M. and Pienkowski, T. 2002. Recovery of dendritic cell counts and function in peripheral blood of cancer patients after chemotherapy. Cytokines Cell Mol Ther 7:15–24.
Mihich, E. 1967. [Antitumor chemotherapy. Considerations on the potential role of immunity]. Pathol Biol 15:209–214.
Mihich, E. 2000a. New leads in cancer therapeutics: a keynote address. Medicina (B Aires) 60 Suppl 2:4–8.
Mihich, E. 2000b. On the immunomodulating effects of anti-cancer drugs and their therapeutic exploitation. Jpn J Clin Oncol 30:469–471.
Nestle, F. O., Alijagic, S., Gilliet, M., Sun, Y., Grabbe, S., Dummer, R., Burg, G. and Schadendorf, D. 1998. Vaccination of melanoma patients with peptide- or tumor lysate-pulsed dendritic cells. Nat Med 4:328–332.
Ng, C. S., Lee, T. W., Wan, S., Wan, I. Y., Sihoe, A. D., Arifi, A. A. and Yim, A. P. 2005. Thoracotomy is associated with significantly more profound suppression in lymphocytes and natural killer cells than video-assisted thoracic surgery following major lung resections for cancer. J Invest Surg 18:81–88.
Nowak, A. K., Lake, R. A., Marzo, A. L., Scott, B., Heath, W. R., Collins, E. J., Frelinger, J. A. and Robinson, B. W. 2003. Induction of tumor cell apoptosis in vivo increases tumor antigen cross-presentation, cross-priming rather than cross-tolerizing host tumor-specific CD8 T cells. J Immunol 170:4905–4913.
Nowak, A. K., Lake, R. A. and Robinson, B. W. 2006. Combined chemoimmunotherapy of solid tumours: improving vaccines? Adv Drug Deliv Rev 58:975–990.
O'Neill, D. W. and Bhardwaj, N. 2007. Exploiting dendritic cells for active immunotherapy of cancer and chronic infections. Mol Biotechnol 36:131–141.
Ogata, M., Okamoto, K., Kohriyama, K., Kawasaki, T., Itoh, H. and Shigematsu, A. 2000. Role of interleukin-10 on hyporesponsiveness of endotoxin during surgery. Crit Care Med 28:3166–3170.
Ohtsukasa, S., Okabe, S., Yamashita, H., Iwai, T. and Sugihara, K. 2003. Increased expression of CEA and MHC class I in colorectal cancer cell lines exposed to chemotherapy drugs. J Cancer Res Clin Oncol 129:719–726.

Okazaki, T., Ebihara, S., Asada, M., Kanda, A., Sasaki, H. and Yamaya, M. 2006. Granulocyte colony-stimulating factor promotes tumor angiogenesis via increasing circulating endothelial progenitor cells and Gr1 + CD11b + cells in cancer animal models. Int Immunol 18:1–9.

Okazaki, T., Ebihara, S., Takahashi, H., Asada, M., Kanda, A. and Sasaki, H. 2005. Macrophage colony-stimulating factor induces vascular endothelial growth factor production in skeletal muscle and promotes tumor angiogenesis. J Immunol 174:7531–7538.

Page, G. G. 2005. Surgery-induced immunosuppression and postoperative pain management. AACN Clin Issues 16:302–309; quiz 416–308.

Perrotta, C., Bizzozero, L., Falcone, S., Rovere-Querini, P., Prinetti, A., Schuchman, E. H., Sonnino, S., Manfredi, A. A. and Clementi, E. 2007. Nitric oxide boosts chemoimmunotherapy via inhibition of acid sphingomyelinase in a mouse model of melanoma. Cancer Res 67:7559–7564.

Rea, D., van Kooten, C., van Meijgaarden, K. E., Ottenhoff, T. H., Melief, C. J. and Offringa, R. 2000. Glucocorticoids transform CD40-triggering of dendritic cells into an alternative activation pathway resulting in antigen-presenting cells that secrete IL-10. Blood 95:3162–3167.

Rozkova, D., Horvath, R., Bartunkova, J. and Spisek, R. 2006. Glucocorticoids severely impair differentiation and antigen presenting function of dendritic cells despite upregulation of Toll-like receptors. Clin Immunol 120:260–271.

Shakhar, G., Abudarham, N., Melamed, R., Schwartz, Y., Rosenne, E. and Ben-Eliyahu, S. 2007. Amelioration of operation-induced suppression of marginating pulmonary NK activity using poly I-C: a potential approach to reduce postoperative metastasis. Ann Surg Oncol 14:841–852.

Shakhar, G. and Ben-Eliyahu, S. 2003. Potential prophylactic measures against postoperative immunosuppression: could they reduce recurrence rates in oncological patients? Ann Surg Oncol 10:972–992.

Shakhar, G. and Blumenfeld, B. 2003. Glucocorticoid involvement in suppression of NK activity following surgery in rats. J Neuroimmunol 138:83–91.

Shin, J. Y., Lee, S. K., Kang, C. D., Chung, J. S., Lee, E. Y., Seo, S. Y., Lee, S. Y., Baek, S. Y., Kim, B. S., Kim, J. B. and Yoon, S. 2003. Antitumor effect of intratumoral administration of dendritic cell combination with vincristine chemotherapy in a murine fibrosarcoma model. Histol Histopathol 18:435–447.

Shu, S., Zheng, R., Lee, W. T. and Cohen, P. A. 2007. Immunogenicity of dendritic-tumor fusion hybrids and their utility in cancer immunotherapy. Crit Rev Immunol 27:463–483.

Shurin, G. V., Tourkova, I. L., Chatta, G. S., Schmidt, G., Wei, S., Djeu, J. Y. and Shurin, M. R. 2005. Small rho GTPases regulate antigen presentation in dendritic cells. J Immunol 174:3394–3400.

Shurin, G. V., Tourkova, I. L. and Shurin, M. R. 2008. Low-dose chemotherapeutic agents regulate small Rho GTPase activity in dendritic cells. J Immunother 31:491–499.

Shurin, M. R., Shurin, G. V., Lokshin, A., Yurkovetsky, Z. R., Gutkin, D. W., Chatta, G., Zhong, H., Han, B. and Ferris, R. L. 2006. Intratumoral cytokines/chemokines/growth factors and tumor infiltrating dendritic cells: friends or enemies? Cancer Metastasis Rev 25:333–356.

Steer, J. H., Vuong, Q. and Joyce, D. A. 1997. Suppression of human monocyte tumour necrosis factor-alpha release by glucocorticoid therapy: relationship to systemic monocytopaenia and cortisol suppression. Br J Clin Pharmacol 43:383–389.

Stone, H. B., Coleman, C. N., Anscher, M. S. and McBride, W. H. 2003. Effects of radiation on normal tissue: consequences and mechanisms. Lancet Oncol 4:529–536.

Takahashi, K., Satoi, S., Yanagimoto, H., Terakawa, N., Toyokawa, H., Yamamoto, T., Matsui, Y., Takai, S., Kwon, A. H. and Kamiyama, Y. 2007. Circulating dendritic cells and development of septic complications after pancreatectomy for pancreatic cancer. Arch Surg 142:1151–1157; discussion 1157.

Teitz-Tennenbaum, S., Li, Q., Okuyama, R., Davis, M. A., Sun, R., Whitfield, J., Knibbs, R. N., Stoolman, L. M. and Chang, A. E. 2008. Mechanisms involved in radiation enhancement of intratumoral dendritic cell therapy. J Immunother 31:345–358.

Teitz-Tennenbaum, S., Li, Q., Rynkiewicz, S., Ito, F., Davis, M. A., McGinn, C. J. and Chang, A. E. 2003. Radiotherapy potentiates the therapeutic efficacy of intratumoral dendritic cell administration. Cancer Res 63:8466–8475.

Thaker, P. H., Han, L. Y., Kamat, A. A., Arevalo, J. M., Takahashi, R., Lu, C., Jennings, N. B., Armaiz-Pena, G., Bankson, J. A., Ravoori, M., Merritt, W. M., Lin, Y. G., Mangala, L. S., Kim, T. J., Coleman, R. L., Landen, C. N., Li, Y., Felix, E., Sanguino, A. M., Newman, R. A., Lloyd, M., Gershenson, D. M., Kundra, V., Lopez-Berestein, G., Lutgendorf, S. K., Cole, S. W. and Sood, A. K. 2006. Chronic stress promotes tumor growth and angiogenesis in a mouse model of ovarian carcinoma. Nat Med 12:939–944.

Tong, Y., Song, W. and Crystal, R. G. 2001. Combined intratumoral injection of bone marrow-derived dendritic cells and systemic chemotherapy to treat existing murine tumors. Cancer Res 61:7530–7535.

Tourkova, I. L., Shurin, G. V., Wei, S. and Shurin, M. R. 2007. Small rho GTPases mediate tumor-induced inhibition of endocytic activity of dendritic cells. J Immunol 178:7787–7793.

Tubiana, M., Carde, P. and Frindel, E. 1993. Ways of minimising hematopoietic damage induced by radiation and cytostatic drugs – the possible role of inhibitors. Radiother Oncol 29:1–17.

Utoh, J., Yamamoto, T., Utsunomiya, T., Kambara, T., Goto, H. and Miyauchi, Y. 1988. Effect of surgery on neutrophil functions, superoxide and leukotriene production. Br J Surg 75:682–685.

Verhoeven, G. T., Van Haarst, J. M., De Wit, H. J., Simons, P. J., Hoogsteden, H. C. and Drexhage, H. A. 2000. Glucocorticoids hamper the ex vivo maturation of lung dendritic cells from their low autofluorescent precursors in the human bronchoalveolar lavage: decreases in allostimulatory capacity and expression of CD80 and CD86. Clin Exp Immunol 122:232–240.

Werling, D., Hope, J. C., Howard, C. J. and Jungi, T. W. 2004. Differential production of cytokines, reactive oxygen and nitrogen by bovine macrophages and dendritic cells stimulated with Toll-like receptor agonists. Immunology 111:41–52.

Whelan, R. L., Franklin, M., Holubar, S. D., Donahue, J., Fowler, R., Munger, C., Doorman, J., Balli, J. E., Glass, J., Gonzalez, J. J., Bessler, M., Xie, H. and Treat, M. 2003. Postoperative cell mediated immune response is better preserved after laparoscopic vs open colorectal resection in humans. Surg Endosc 17:972–978.

Yu, B., Kusmartsev, S., Cheng, F., Paolini, M., Nefedova, Y., Sotomayor, E. and Gabrilovich, D. 2003. Effective combination of chemotherapy and dendritic cell administration for the treatment of advanced-stage experimental breast cancer. Clin Cancer Res 9:285–294.

Zhao, J., Kim, J. E., Reed, E. and Li, Q. Q. 2005. Molecular mechanism of antitumor activity of taxanes in lung cancer (Review). Int J Oncol 27:247–256.

Zhong, H., Han, B., Tourkova, I. L., Lokshin, A., Rosenbloom, A., Shurin, M. R. and Shurin, G. V. 2007a. Low-dose paclitaxel prior to intratumoral dendritic cell vaccine modulates intratumoral cytokine network and lung cancer growth. Clin Cancer Res 13:5455–5462.

Zhong, H., Shurin, M. R. and Han, B. 2007b. Optimizing dendritic cell-based immunotherapy for cancer. Expert Rev Vaccines 6:333–345.

Part III
Role of Dendritic Cells in Tumor Escape

Identification of functional and molecular abnormalities in dendritic cells in the tumor microenvironment allowed formulating the concept of insufficient antigen presentation and consequent deficient induction of antitumor immune response in cancer. However, soon it became clear that the biological and clinical significance of dendritic cells in controlling tumor dispensation is not limited by a deficient performance: emerging evidence demonstrates that under certain circumstances and specific conditions, e.g., those observed at the site of primary or metastatic tumors, regulatory dendritic cells might directly, e.g., through the cytokine and growth factor release, and indirectly, e.g. through regulatory T cells and probably macrophages, support tumor growth and protect tumor cells from immunological recognition and elimination. Chapters 15–18 in this section of the book describe diverse characteristics of subsets of dendritic cells that facilitate tumor escape. Chapter 15 discusses how the expression of tryptophan-catabolizing enzyme indoleamine-pyrrole 2,3-dioxygenase (IDO) is associated with the regulatory tumor-promoting activity of tumor-associated dendritic cells, along with other tolerogenic mechanisms, including production of immunosuppressing cytokines and expression of immune-inhibitory receptors. Chapter 16 focuses on alternatively activated dendritic cells, which in the tumor microenvironment can promote angiogenesis and sustain tumor escape, while Chapter 17 spotlights other direct mechanisms by which different dendritic cell subtypes might promote tumor survival, including transdifferentiation of dendritic cells. Thus with the apprehension that the dendritic cell lineage consists of several distinct cell populations, a question has occurred as to whether specific subtypes of dendritic cells are dysregulated in the cancer environment or whether the state of dendritic cell maturation characterizes challenging features of immune response vs. unresponsiveness in the tumor milieu. This important question is discussed in Chapter 18.

Chapter 15
Role of IDO in Dendritic Cell Differentiation and Function in Cancer

Alexey Popov and Joachim L. Schultze

Abstract Expression of tryptophan-catabolizing enzyme indoleamine-pyrrole 2,3-dioxygenase (IDO) has been associated with the regulatory phenotype of tumor-associated dendritic cells, along with other tolerogenic mechanisms, including production of immunomodulatory cytokines and expression of immune-inhibitory receptors. IDO activation in dendritic cells leads to tryptophan depletion and accumulation of its toxic downstream metabolites which in concert directly suppress proliferation of T cells and induce T-cell apoptosis. Furthermore, IDO-positive dendritic cells promote the induction of regulatory T cells, which further impair protective immunity against tumors. In the context of cancer, IDO induction in dendritic cells can be triggered by receptors expressed by regulatory T cells, such as CTLA4 and GITR, and by soluble tumor-associated factors, such as prostaglandin E2, in close alliance with TNF signaling. Immunosuppressive effects of enzymatically active IDO can be overcome by specific inhibitors such as 1-methyl-tryptophan which can be used for therapeutic purposes.

15.1 IDO – A Tryptophan-Catabolizing Enzyme with Immunoregulatory Properties

Indoleamine 2,3-dioxygenase (IDO) – a first and rate-limiting enzyme in the kynurenine pathway of tryptophan catabolism – was initially described by the group of Osamu Hayaishi almost 45 years ago (Taylor and Feng 1991). Tryptophan is one of the essential amino acids required for protein synthesis which cannot be synthesized by mammals, making the latter fully dependent on their dietary uptake (Moffett and Namboodiri 2003). Once ingested, tryptophan is distributed throughout the blood stream and can be acquired by the cells with

A. Popov (✉)
Laboratory for Genomics and Immunoregulation, Program Unit Molecular Immune & Cell Biology, LIMES (Life and Medical Sciences), University of Bonn, Karlrobert-Kreiten Street 13, 53115 Bonn, Germany
e-mail: alexey.popov@uni-bonn.de

M.R. Shurin, R.D. Salter (eds.), *Dendritic Cells in Cancer*,
DOI 10.1007/978-0-387-88611-4_15, © Springer Science+Business Media, LLC 2009

the help of a system L heteromeric amino acid transporter, which consists of the catalytic subunit/light chain (large amino acid transporter 1, LAT1) associated covalently with a surface glycoprotein 4F2hc/heavy chain (Verrey 2003). In mammalian cells, tryptophan can be catabolized by IDO and tryptophan 2,3-dioxygenase (TDO); in contrast to TDO, which is liver-specific, IDO can be expressed by various non-immune and immune cells, including dendritic cells (DC) (Mellor and Munn 2004). Depletion of tryptophan by IDO-expressing cells results in the accumulation of its first main metabolite, kynurenine, which is further catabolized by various enzymes along the so-called kynurenine pathway (Fig. 15.1).

In the first decades after discovery, IDO was characterized to be involved mostly in the repression of viral, parasitic and bacterial infections and in the

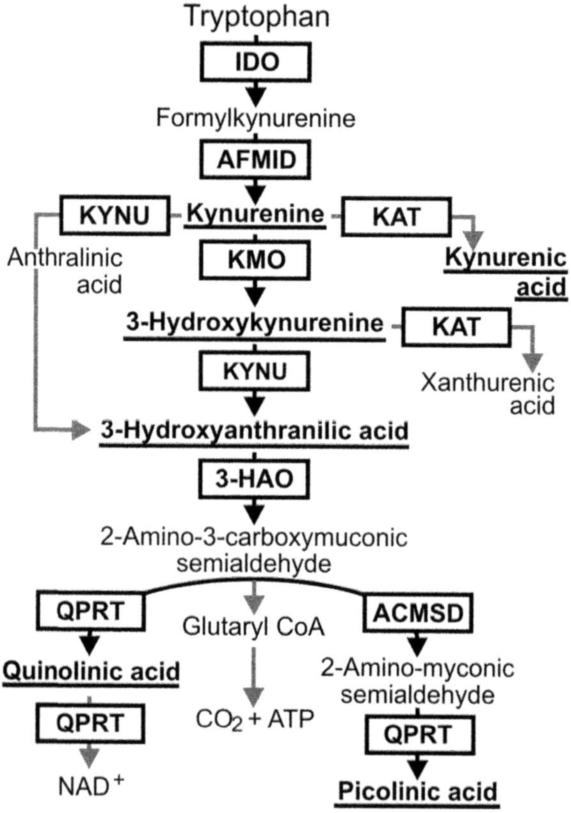

Fig. 15.1 Overview of the kynurenine pathway of tryptophan catabolism. Immunotoxic tryptophan metabolites are highlighted in bold and are underscored. Kynurenine pathway is highlighted by black arrows. IDO, indoleamine 2,3-dioxygenase; AFMID, arylformamidase; KYNU, kynureninase; KMO, kynurenine 3-monooxygenase; KAT, kynurenine aminotransferase; 3-HAO, 3-hydroxyanthranilic acid oxidase; QPRT, quinolinic acid phosphoribosyl transferase; ACSMD, aminocarboxymuconate semialdehyde decarboxylase

inhibition of tumor cell proliferation (Taylor and Feng 1991). Later on, evidence emerged that IDO-induced tryptophan depletion can promote an immune privileged environment in pregnancy (Munn et al. 1998) and during tumor progression (Uyttenhove et al. 2003) and can inhibit organ-specific autoimmunity (Hayashi et al. 2004). Furthermore, IDO-expressing human and murine DC (Hwu et al. 2000; Grohmann et al. 2001), as well as macrophages (Munn et al. 1999), were reported to inhibit antigen-specific T-cell proliferation and to induce T-cell apoptosis (Fallarino et al. 2002a; Mellor et al. 2003). Very recently a new tryptophan-catabolizing enzyme, closely related to IDO and named therefore indoleamine 2,3-dioxygenase-like protein (INDOL1 or IDO2), was described; in contrast to IDO (also referred to as IDO1), IDO2 protein is expressed mostly in the murine kidney, liver and organs of the reproductive system (Ball et al. 2007). Although concomitant expression of both IDO1 and IDO2 has been described in human DC (Lob et al. 2008a), there is experimental evidence that tryptophan degradation in human DC and tumor cells is entirely provided by IDO1 (Lob et al. 2008a, b).

15.2 Mechanisms of IDO-Mediated Suppression: Tryptophan Starvation and Downstream Metabolite Toxicity

Different mechanisms contribute to the IDO-mediated immune suppression. Due to the necessity of tryptophan for cell growth, the effects mediated by IDO^+ DC have been associated with tryptophan starvation. In the absence of tryptophan, T-cell cycle progression stops at a mid-G1 arrest point (Munn et al. 1999). Notably, plain addition of tryptophan to the arrested T cells is not sufficient to permit further cell cycle progression, but requires additional T-cell receptor (TCR) triggering to release the cell cycle arrest (Munn et al. 1999). Activation of a GCN2 kinase pathway by IDO^+ DC was proposed to act as a downstream molecular sensor of IDO-mediated effects in T cells (Munn et al. 2005). GCN2 (general control of non-derepressible-2) kinase is a stress-response kinase, which responds to elevation of uncharged tRNA due to amino acid withdrawal; its induction in T cells leads to the activation of an integrated stress-response (ISR) pathway, resulting in cell cycle arrest and creation of anergy (Munn et al. 2005).

Furthermore, most of the tryptophan metabolites are reported to be immunotoxic (Fig. 15.1). Such tryptophan catabolites as kynurenine, 3-hydroxyanthranilic and quinolinic acids (3-HAA and QA, respectively) were reported to induce T-cell apoptosis (Fallarino et al. 2002a; Frumento et al. 2002; Terness et al. 2002). Induction of T-cell apoptosis by tryptophan catabolites is associated with activation of caspase-8 and release of cytochrome c from mitochondria and affects mostly Th1 cells (Fallarino et al. 2002a). However, 3-HAA was reported to induce apoptosis of both Th1 and Th2 cells by specifically targeting PDK1 and inhibiting NF-κB activation upon T-cell antigen receptor triggering (Hayashi et al. 2007). In addition to apoptosis induction, several tryptophan catabolites (kynurenine, 3-hydroxykynurenine and 3-HAA) were reported to exert cytotoxic action on T cells, B cells and natural killer (NK) cells, whereas DC were not affected (Terness et al. 2002). Moreover, suppressive effects of tryptophan catabolites affect mostly activated T cells; in contrast, resting cells are not affected and can be activated normally thereafter (Frumento et al. 2002; Terness et al. 2002).

The suppressive effect of tryptophan catabolites is amplified in the absence of tryptophan and also the combination of several catabolites results in additional suppression (Frumento et al. 2002; Terness et al. 2002; Fallarino et al. 2006). In almost the same manner, combinatory effects of tryptophan deprivation and tryptophan catabolites result in GCN2-dependent downregulation of the TCR ξ-chain in murine $CD8^+$ T cells in vitro and in vivo (Fallarino et al. 2006).

In addition to suppression of conventional T cells, IDO^+ DC were reported to induce regulatory T cells in vitro and in vivo. Tryptophan starvation and tryptophan catabolites induce a regulatory phenotype in naive $CD4^+CD25^+$ T cells via induction of the forkhead transcription factor Foxp3 in a TGF-β- and GCN2-dependent manner (Fallarino et al. 2006). IDO^+ plasmacytoid DC (pDC) in tumor-draining lymph nodes (TDLN) of B16F10 melanoma-bearing mice can trigger the suppressive activity of resting $CD4^+CD25^+Foxp3^+$ regulatory T cells (Treg); Treg activation by IDO^+ pDC is MHC restricted and requires the GCN2 pathway and CTLA4 (cytotoxic T lymphocyte-associated antigen 4) signaling (Sharma et al. 2007). Moreover, IDO^+ DC were shown to induce $Foxp3^+$ regulatory T cells and even to promote tumor growth in vivo in advanced human melanoma (Wobser et al. 2007).

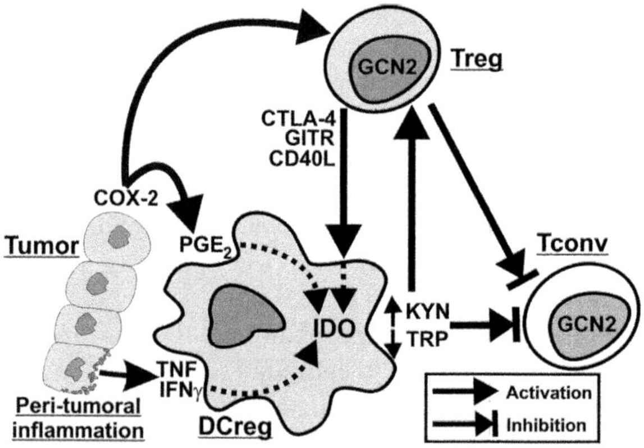

Fig. 15.2 Interaction of IDO^+ regulatory dendritic cells with T cells in the tumor microenvironment. CD40L, CD40 ligand; COX-2, cyclooxygenase-2; CTLA-4, cytotoxic T lymphocyte-associated antigen 4; DCreg, regulatory dendritic cells; GCN2, general control non-depressing 2 protein kinase; GITR, glucocorticoid-induced tumor necrosis factor receptor family-related protein; IDO, indoleamine 2,3-dioxygenase; IFNγ, interferon gamma; KYN, kynurenine; PgE_2, prostaglandin E2; Tconv, conventional T cells; TNF, tumor necrosis factor α; Treg, regulatory T cells; TRP, tryptophan

Besides the direct IDO-mediated suppressive mechanisms, several other molecules associated with IDO^+ DC, such as HLA-G and CD25, were described to be involved in the promotion of IDO-associated immune suppression (Lopez et al. 2006; von Bergwelt-Baildon et al. 2006). Expression of the non-classical HLA class I molecule HLA-G on the surface of DC can be induced by IDO and was shown to be complementary to IDO in inducing and maintaining immune tolerance (Lopez et al. 2006). Concomitant induction of IDO and CD25 provides DC with two potent inhibitory mechanisms: (i) IDO-mediated tryptophan starvation and accumulation of toxic downstream tryptophan metabolites are assisted by (ii) IL-2 scavenging through DC-derived surface and soluble CD25 and in concert with the latter inhibit T-cell proliferation (von Bergwelt-Baildon et al. 2006).

Taken together, IDO^+ regulatory DC possess several potent inhibitory mechanisms for inhibiting $CD4^+$ as well as $CD8^+$ T-cell function: suppression of T-cell proliferation is elicited mainly by tryptophan depletion and accumulation of toxic tryptophan metabolites and is further maintained by induction of Treg (Fig. 15.2).

15.3 Induction of IDO in Dendritic Cells

Distinct human and murine DC subsets have been described to express enzymatically active IDO; some of these DC express IDO constitutively, either in vivo or in vitro, in the others IDO expression can be induced by diverse stimuli (Popov and Schultze 2008). Constitutive expression of IDO has been described for human monocyte-derived DC, which co-express CD123 and CCR6 (Munn et al. 2002); later on the existence of $CD123^+CCR6^+IDO^+$ DC was put in question by other groups (Terness et al. 2002; Lob et al. 2007). In mice, plasmacytoid $CD19^+$ DC derived from TDLN of melanoma-bearing animals (Munn et al. 2004) and spleen-derived $CD8\alpha^+B220^-$ DC (Fallarino et al. 2002b) were reported to express IDO constitutively under steady-state conditions, whereas in spleen plasmacytoid $B220^+$ DC and in vitro-generated bone marrow-derived DC IDO expression can be induced (Popov and Schultze 2008). Although various factors including inflammatory cytokines and toll-like receptor (TLR) ligands can induce IDO activity, only some of them may play a role in the tumor context.

In general, ligation of B7-1/B7-2 (CD80/CD86) on DC by Treg-expressed CTLA4 and glucocorticoid-induced tumor necrosis factor receptor (GITR) was suggested to be a part of the interaction between IDO^+ DC and Treg in vivo (Grohmann et al. 2002, 2007) (Fig. 15.2). Induction of IDO by CTLA4 triggering on a subset of $CD19^+$ murine DC depends on expression of type I interferon receptors (Baban et al. 2005); moreover, secretion of IFN-α, which is indispensable for high-level expression of IDO following B7-1/B7-2 ligation, is itself controlled by IDO activity (Manlapat et al. 2007). On the other hand, induction of IDO by GITR–GITR ligand interaction requires both IFN-α and non-canonical NF-κB signaling (Grohmann et al. 2007).

Prostaglandin E_2 (PgE_2), a soluble factor associated with tumor-induced suppression, was reported to induce enzymatically active IDO in human DC (Braun et al. 2005; von Bergwelt-Baildon et al. 2006). In this context, PgE_2 signaling is executed over PgE_2 receptor 2 (EP2) and results in the activation of adenylate cyclase (Braun et al. 2005) but strongly depends on TNF signaling (Popov et al. 2006) (Fig. 15.2). Induction of IDO by TNF-α and PgE_2 is not restricted to in vitro-generated DC but can also be seen in circulating BDCA-1-positive myeloid DC derived from peripheral blood, as well (von Bergwelt-Baildon et al. 2006).

Taken together, IDO expression and functional activity in DC can be triggered by various soluble tumor-associated factors (interferons, TNF-α and PgE_2) and by reverse signaling through membrane-bound receptors expressed on regulatory T cells (CTLA4 and GITR) (Fig. 15.2).

15.4 IDO-Mediated Dendritic Cell-Driven Immune Suppression in the Tumor Environment

IDO^+ DC were revealed in TDLN in invasive cutaneous melanoma in humans (Lee et al. 2003), B16 melanoma-bearing mice (Munn et al. 2004) and different carcinomas associated with COX-2 overexpression and elevated levels of PgE_2 (von Bergwelt-Baildon et al. 2006). Of interest, in the latter study DC appeared to be of myeloid origin, whereas in the former two they were either plasmacytoid or of mixed origin, respectively, which points out that all DC compartments can promote IDO-mediated tolerance (Popov and Schultze 2008). Furthermore, IDO^+ DC described in peritumoral regions of gastric, breast and colon carcinomas were reported to co-express CD25, providing the latter with a supplementary inhibitory mechanism, which further impairs protective immunity (von Bergwelt-Baildon et al. 2006).

In all studies describing IDO^+ DC in vivo, they were found to represent a relatively minor subset, which raises the question if this small population can efficiently regulate host immune function. In vivo, even few IDO^+ DC were reported to suppress the immune responses elicited by higher amounts of stimulatory IDO^- DC (Grohmann et al. 2001). Moreover, IDO^+ DC can confer their suppressive ability to IDO^- DC in a kynurenine-dependent manner even in the absence of functional IDO (Belladonna et al. 2006), thereby extending the pool of inhibitory DC. However, IDO is expressed not only by tumor-associated DC, but by other tumor-infiltrating cells, including macrophages (Nakamura et al. 2007) and eosinophils (Astigiano et al. 2005), and by the neoplastic cells themselves (Uyttenhove et al. 2003). All these IDO^+ cells might contribute to DC-mediated immune suppression, in concert generating a significant gradient of tryptophan and its toxic metabolites. What portion of the regulatory effect is mediated by IDO^+ DC in the tumor environment needs to be further elucidated.

15.5 IDO Inhibitors and Their Role in Dendritic Cell-Targeted Cancer Immunotherapy

Immunosuppressive effects promoted by IDO$^+$ DC can be surmounted by specific competitive inhibitors, like 1-methyl-tryptophan (1-MT) (Hwu et al. 2000). In murine tumor models, 1-MT significantly inhibited tumor growth, enhanced survival of mice with breast cancer and melanoma treated with chemotherapeutic agents (Hou et al. 2007) and delayed tumor growth, when applied in combination with a DC/Lewis lung carcinoma fusion vaccine in the lung cancer model (Ou et al. 2008). Due to the poor solubility of 1-MT, its clinical application remains principally restricted to the murine models, where it can be applied orally with food and drinking water or applied subcutaneously as a "sustain–release form" to enhance the bioavailability (Hou et al. 2007; Romani et al. 2008). Of note, 1-methyl-tryptophan exists in two stereo-isoforms with potentially different biological properties: the levo-isoform (L-1-MT) and dextro-isoform (D-1-MT), respectively (Hou et al. 2007). Both isoforms are reported to inhibit IDO activity; however, recent data point out that 1-MT stereo-isoforms might exhibit species- and cell type-specific activity, while the D-isoform is described to be very potent in reversing the IDO-mediated T-cell suppression in murine DC in vitro and in vivo (Hou et al. 2007); the L-isoform, which targets IDO1, is the inhibitor of choice in human DC. Interestingly, D-1-MT is inefficient in human cells (Lob et al. 2008a, b).

The diversity and poor bioavailability of 1-MT isoforms encouraged the search for IDO inhibitors suitable for the treatment of human disease in clinical setting. Until now, several alternative medications have been proposed, i.e., synthetic IDO inhibitors with higher oral bioavailability such as methyl-thiohydantoin-tryptophan or more potent 1-MT isomers with higher IDO specificity (Muller et al. 2005; Hou et al. 2007) with other potent IDO inhibitors under current intensive investigation (Kumar et al. 2008).

15.6 Role of IDO in Dendritic Cell Differentiation and Function

Given that IDO-expressing DC potently suppress T cells and NK cells (Terness et al. 2002), one might expect that tryptophan deprivation and accumulation of toxic tryptophan catabolites can affect DC themselves. However, current data point out that DC might not be sensible to the downstream effects of IDO enzymatic activation. For instance, DC were shown to be insensitive to downstream catabolites of tryptophan: HAA, a strong suppressor of CD4$^+$ T cells, cannot inhibit NF-κB and cannot induce apoptosis in TLR4-activated DC (Hayashi et al. 2007). Moreover, DC are not affected by the cytotoxic effects promoted by various tryptophan catabolites (Terness et al. 2002).

Using fluorescence resonance energy transfer nanosensors, Kaper et al. have recently shown that LAT1 is implicated in tryptophan/kynurenine exchange in human cells, where tryptophan influx is strictly linked to kynurenine efflux, coupling tryptophan starvation to elevation of extracellular kynurenine levels (Kaper et al. 2007). According to the authors, this strict coupling protects IDO-competent cells from kynurenine accumulation and related inhibitory effects. Recently, LAT1-mediated kynurenine uptake was described for IDO^+ bone marrow-derived DC, therefore the same exchange mechanism may apply for DC, as well (Hara et al. 2008). In the light of these recent data, which point out that IDO-competent antigen-presenting cells can utilize tryptophan transporters with much higher affinity than LAT1 (Seymour et al. 2006), one might expect that the higher turnover of tryptophan in IDO^+ DC protects them from starvation. Besides, DC lack molecular sensors for IDO-mediated downstream events, such as GCN2 kinase, which is responsible for tryptophan- and kynurenine-mediated inhibitory effects described for T cells (Munn et al. 2005; Sharma et al. 2007).

Recently, pharmacological inhibition of IDO function during DC maturation in vitro has been reported to impair DC phenotype and inhibit DC responsiveness to chemokines (Hwang et al. 2005). Of interest, suppression of DC function, reported in this study, was mostly pronounced when DC were matured with proinflammatory stimuli, such as LPS and TNF-α, whereas DC stimulated by CD40 ligation were less sensible to IDO inhibition. It was therefore hypothesized that tryptophan might be essential for DC maturation and may play a role in the regulation of DC migratory function (Hwang et al. 2005). However, despite a possible role for IDO expression in regulating DC differentiation from bone marrow precursors in vitro, in vivo $IDO1^{-/-}$ mice display a normal DC compartment, pointing out that IDO is dispensable for DC development and function (de Faudeur et al. 2008).

Taken together, although suppression of T-cell responses by IDO^+ DC is well documented (Munn and Mellor 2007), the role of DC-expressed IDO on their own function further remains to be elucidated.

References

Astigiano, S., Morandi, B., Costa, R., Mastracci, L., D'Agostino, A., Ratto, G. B., Melioli, G. and Frumento, G. 2005. Eosinophil granulocytes account for indoleamine 2,3-dioxygenase-mediated immune escape in human non-small cell lung cancer. Neoplasia 7:390–396.

Baban, B., Hansen, A. M., Chandler, P. R., Manlapat, A., Bingaman, A., Kahler, D. J., Munn, D. H. and Mellor, A. L. 2005. A minor population of splenic dendritic cells expressing CD19 mediates IDO-dependent T cell suppression via type I IFN signaling following B7 ligation. Int Immunol 17:909–919.

Ball, H. J., Sanchez-Perez, A., Weiser, S., Austin, C. J., Astelbauer, F., Miu, J., McQuillan, J. A., Stocker, R., Jermiin, L. S. and Hunt, N. H. 2007. Characterization of an indoleamine 2,3-dioxygenase-like protein found in humans and mice. Gene 396:203–213.

Belladonna, M. L., Grohmann, U., Guidetti, P., Volpi, C., Bianchi, R., Fioretti, M. C., Schwarcz, R., Fallarino, F. and Puccetti, P. 2006. Kynurenine pathway enzymes in dendritic cells initiate tolerogenesis in the absence of functional IDO. J Immunol 177:130–137.

Braun, D., Longman, R. S. and Albert, M. L. 2005. A two-step induction of indoleamine 2,3 dioxygenase (IDO) activity during dendritic-cell maturation. Blood 106:2375–2381.

de Faudeur, G., de Trez, C., Muraille, E. and Leo, O. 2008. Normal development and function of dendritic cells in mice lacking IDO-1 expression. Immunol Lett 118:21–29.

Fallarino, F., Grohmann, U., Vacca, C., Bianchi, R., Orabona, C., Spreca, A., Fioretti, M. C. and Puccetti, P. 2002a. T cell apoptosis by tryptophan catabolism. Cell Death Differ 9:1069–1077.

Fallarino, F., Grohmann, U., You, S., McGrath, B. C., Cavener, D. R., Vacca, C., Orabona, C., Bianchi, R., Belladonna, M. L., Volpi, C., Santamaria, P., Fioretti, M. C. and Puccetti, P. 2006. The combined effects of tryptophan starvation and tryptophan catabolites down-regulate T cell receptor zeta-chain and induce a regulatory phenotype in naive T cells. J Immunol 176:6752–6761.

Fallarino, F., Vacca, C., Orabona, C., Belladonna, M. L., Bianchi, R., Marshall, B., Keskin, D. B., Mellor, A. L., Fioretti, M. C., Grohmann, U. and Puccetti, P. 2002b. Functional expression of indoleamine 2,3-dioxygenase by murine CD8 alpha(+) dendritic cells. Int Immunol 14:65–68.

Frumento, G., Rotondo, R., Tonetti, M., Damonte, G., Benatti, U. and Ferrara, G. B. 2002. Tryptophan-derived catabolites are responsible for inhibition of T and natural killer cell proliferation induced by indoleamine 2,3-dioxygenase. J Exp Med 196:459–468.

Grohmann, U., Fallarino, F., Bianchi, R., Belladonna, M. L., Vacca, C., Orabona, C., Uyttenhove, C., Fioretti, M. C. and Puccetti, P. 2001. IL-6 inhibits the tolerogenic function of CD8 alpha + dendritic cells expressing indoleamine 2,3-dioxygenase. J Immunol 167:708–714.

Grohmann, U., Orabona, C., Fallarino, F., Vacca, C., Calcinaro, F., Falorni, A., Candeloro, P., Belladonna, M. L., Bianchi, R., Fioretti, M. C. and Puccetti, P. 2002. CTLA-4-Ig regulates tryptophan catabolism in vivo. Nat Immunol 3:1097–1101.

Grohmann, U., Volpi, C., Fallarino, F., Bozza, S., Bianchi, R., Vacca, C., Orabona, C., Belladonna, M. L., Ayroldi, E., Nocentini, G., Boon, L., Bistoni, F., Fioretti, M. C., Romani, L., Riccardi, C. and Puccetti, P. 2007. Reverse signaling through GITR ligand enables dexamethasone to activate IDO in allergy. Nat Med 13:579–586.

Hara, T., Ogasawara, N., Akimoto, H., Takikawa, O., Hiramatsu, R., Kawabe, T., Isobe, K. and Nagase, F. 2008. High-affinity uptake of kynurenine and nitric oxide-mediated inhibition of indoleamine 2,3-dioxygenase in bone marrow-derived myeloid dendritic cells. Immunol Lett 116:95–102.

Hayashi, T., Beck, L., Rossetto, C., Gong, X., Takikawa, O., Takabayashi, K., Broide, D. H., Carson, D. A. and Raz, E. 2004. Inhibition of experimental asthma by indoleamine 2,3-dioxygenase. J Clin Invest 114:270–279.

Hayashi, T., Mo, J. H., Gong, X., Rossetto, C., Jang, A., Beck, L., Elliott, G. I., Kufareva, I., Abagyan, R., Broide, D. H., Lee, J. and Raz, E. 2007. 3-Hydroxyanthranilic acid inhibits PDK1 activation and suppresses experimental asthma by inducing T cell apoptosis. Proc Natl Acad Sci U S A 104:18619–18624.

Hou, D. Y., Muller, A. J., Sharma, M. D., DuHadaway, J., Banerjee, T., Johnson, M., Mellor, A. L., Prendergast, G. C. and Munn, D. H. 2007. Inhibition of indoleamine 2,3-dioxygenase in dendritic cells by stereoisomers of 1-methyl-tryptophan correlates with antitumor responses. Cancer Res 67:792–801.

Hwang, S. L., Chung, N. P., Chan, J. K. and Lin, C. L. 2005. Indoleamine 2,3-dioxygenase (IDO) is essential for dendritic cell activation and chemotactic responsiveness to chemokines. Cell Res 15:167–175.

Hwu, P., Du, M. X., Lapointe, R., Do, M., Taylor, M. W. and Young, H. A. 2000. Indoleamine 2,3-dioxygenase production by human dendritic cells results in the inhibition of T cell proliferation. J Immunol 164:3596–3599.

Kaper, T., Looger, L. L., Takanaga, H., Platten, M., Steinman, L. and Frommer, W. B. 2007. Nanosensor detection of an immunoregulatory tryptophan influx/kynurenine efflux cycle. PLoS Biol 5:e257.

Kumar, S., Malachowski, W. P., Duhadaway, J. B., Lalonde, J. M., Carroll, P. J., Jaller, D., Metz, R., Prendergast, G. C. and Muller, A. J. 2008. Indoleamine 2,3-dioxygenase is the anticancer target for a novel series of potent naphthoquinone-based inhibitors. J Med Chem 51:1706–1718.

Lee, J. R., Dalton, R. R., Messina, J. L., Sharma, M. D., Smith, D. M., Burgess, R. E., Mazzella, F., Antonia, S. J., Mellor, A. L. and Munn, D. H. 2003. Pattern of recruitment of immunoregulatory antigen-presenting cells in malignant melanoma. Lab Invest 83:1457–1466.

Lob, S., Ebner, S., Wagner, S., Weinreich, J., Schafer, R. and Konigsrainer, A. 2007. Are indoleamine-2,3-dioxygenase producing human dendritic cells a tool for suppression of allogeneic t-cell responses? Transplantation 83:468–473.

Lob, S., Konigsrainer, A., Schafer, R., Rammensee, H. G., Opelz, G. and Terness, P. 2008a. Levo- but not dextro-1-methyl tryptophan abrogates the IDO activity of human dendritic cells. Blood 111:2152–2154.

Lob, S., Konigsrainer, A., Zieker, D., Brucher, B. L., Rammensee, H. G., Opelz, G. and Terness, P. 2008b. IDO1 and IDO2 are expressed in human tumors: levo- but not dextro-1-methyl tryptophan inhibits tryptophan catabolism. Cancer Immunol Immunother 58:153–157.

Lopez, A. S., Alegre, E., LeMaoult, J., Carosella, E. and Gonzalez, A. 2006. Regulatory role of tryptophan degradation pathway in HLA-G expression by human monocyte-derived dendritic cells. Mol Immunol 43:2151–2160.

Manlapat, A. K., Kahler, D. J., Chandler, P. R., Munn, D. H. and Mellor, A. L. 2007. Cell-autonomous control of interferon type I expression by indoleamine 2,3-dioxygenase in regulatory CD19(+) dendritic cells. Eur J Immunol 37:1064–1071.

Mellor, A. L., Baban, B., Chandler, P., Marshall, B., Jhaver, K., Hansen, A., Koni, P. A., Iwashima, M. and Munn, D. H. 2003. Cutting edge: induced indoleamine 2,3 dioxygenase expression in dendritic cell subsets suppresses T cell clonal expansion. J Immunol 171:1652–1655.

Mellor, A. L. and Munn, D. H. 2004. IDO expression by dendritic cells: tolerance and tryptophan catabolism. Nat Rev Immuno l 4:762–774.

Moffett, J. R. and Namboodiri, M. A. 2003. Tryptophan and the immune response. Immunol Cell Biol 81:247–265.

Muller, A. J., DuHadaway, J. B., Donover, P. S., Sutanto-Ward, E. and Prendergast, G. C. 2005. Inhibition of indoleamine 2,3-dioxygenase, an immunoregulatory target of the cancer suppression gene Bin1, potentiates cancer chemotherapy. Nat Med 11:312–319.

Munn, D. H. and Mellor, A. L. 2007. Indoleamine 2,3-dioxygenase and tumor-induced tolerance. J Clin Invest 117:1147–1154.

Munn, D. H., Shafizadeh, E., Attwood, J. T., Bondarev, I., Pashine, A. and Mellor, A. L. 1999. Inhibition of T cell proliferation by macrophage tryptophan catabolism. J Exp Med 189:1363–1372.

Munn, D. H., Sharma, M. D., Baban, B., Harding, H. P., Zhang, Y., Ron, D. and Mellor, A. L. 2005. GCN2 kinase in T cells mediates proliferative arrest and anergy induction in response to indoleamine 2,3-dioxygenase. Immunity 22:633–642.

Munn, D. H., Sharma, M. D., Hou, D., Baban, B., Lee, J. R., Antonia, S. J., Messina, J. L., Chandler, P., Koni, P. A. and Mellor, A. L. 2004. Expression of indoleamine 2,3-dioxygenase by plasmacytoid dendritic cells in tumor-draining lymph nodes. J Clin Invest 114:280–290.

Munn, D. H., Sharma, M. D., Lee, J. R., Jhaver, K. G., Johnson, T. S., Keskin, D. B., Marshall, B., Chandler, P., Antonia, S. J., Burgess, R., Slingluff, C. L., Jr. and Mellor, A. L. 2002. Potential regulatory function of human dendritic cells expressing indoleamine 2,3-dioxygenase. Science 297:1867–1870.

Munn, D. H., Zhou, M., Attwood, J. T., Bondarev, I., Conway, S. J., Marshall, B., Brown, C. and Mellor, A. L. 1998. Prevention of allogeneic fetal rejection by tryptophan catabolism. Science 281:1191–1193.

Nakamura, T., Shima, T., Saeki, A., Hidaka, T., Nakashima, A., Takikawa, O. and Saito, S. 2007. Expression of indoleamine 2,3-dioxygenase and the recruitment of Foxp3-expressing regulatory T cells in the development and progression of uterine cervical cancer. Cancer Sci 98:874–881.

Ou, X., Cai, S., Liu, P., Zeng, J., He, Y., Wu, X. and Du, J. 2008. Enhancement of dendritic cell-tumor fusion vaccine potency by indoleamine-pyrrole 2,3-dioxygenase inhibitor, 1-MT. J Cancer Res Clin Oncol 134:525–533.

Popov, A., Abdullah, Z., Wickenhauser, C., Saric, T., Driesen, J., Hanisch, F. G., Domann, E., Raven, E. L., Dehus, O., Hermann, C., Eggle, D., Debey, S., Chakraborty, T., Kronke, M., Utermohlen, O. and Schultze, J. L. 2006. Indoleamine 2,3-dioxygenase-expressing dendritic cells form suppurative granulomas following Listeria monocytogenes infection. J Clin Invest 116:3160–3170.

Popov, A. and Schultze, J. L. 2008. IDO-expressing regulatory dendritic cells in cancer and chronic infection. J Mol Med 86:145–160.

Romani, L., Fallarino, F., De Luca, A., Montagnoli, C., D'Angelo, C., Zelante, T., Vacca, C., Bistoni, F., Fioretti, M. C., Grohmann, U., Segal, B. H. and Puccetti, P. 2008. Defective tryptophan catabolism underlies inflammation in mouse chronic granulomatous disease. Nature 451:211–215.

Seymour, R. L., Ganapathy, V., Mellor, A. L. and Munn, D. H. 2006. A high-affinity, tryptophan-selective amino acid transport system in human macrophages. J Leukoc Biol 80:1320–1327.

Sharma, M. D., Baban, B., Chandler, P., Hou, D. Y., Singh, N., Yagita, H., Azuma, M., Blazar, B. R., Mellor, A. L. and Munn, D. H. 2007. Plasmacytoid dendritic cells from mouse tumor-draining lymph nodes directly activate mature Tregs via indoleamine 2,3-dioxygenase. J Clin Invest 117:2570–2582.

Taylor, M. W. and Feng, G. S. 1991. Relationship between interferon-gamma, indoleamine 2,3-dioxygenase, and tryptophan catabolism. Faseb J 5:2516–2522.

Terness, P., Bauer, T. M., Rose, L., Dufter, C., Watzlik, A., Simon, H. and Opelz, G. 2002. Inhibition of allogeneic T cell proliferation by indoleamine 2,3-dioxygenase-expressing dendritic cells: mediation of suppression by tryptophan metabolites. J Exp Med 196:447–457.

Uyttenhove, C., Pilotte, L., Theate, I., Stroobant, V., Colau, D., Parmentier, N., Boon, T. and Van den Eynde, B. J. 2003. Evidence for a tumoral immune resistance mechanism based on tryptophan degradation by indoleamine 2,3-dioxygenase. Nat Med 9:1269–1274.

Verrey, F. 2003. System L: heteromeric exchangers of large, neutral amino acids involved in directional transport. Pflugers Arch 445:529–533.

von Bergwelt-Baildon, M. S., Popov, A., Saric, T., Chemnitz, J., Classen, S., Stoffel, M. S., Fiore, F., Roth, U., Beyer, M., Debey, S., Wickenhauser, C., Hanisch, F. G. and Schultze, J. L. 2006. CD25 and indoleamine 2,3-dioxygenase are up-regulated by prostaglandin E2 and expressed by tumor-associated dendritic cells in vivo: additional mechanisms of T-cell inhibition. Blood 108:228–237.

Wobser, M., Voigt, H., Houben, R., Eggert, A. O., Freiwald, M., Kaemmerer, U., Kaempgen, E., Schrama, D. and Becker, J. C. 2007. Dendritic cell based antitumor vaccination: impact of functional indoleamine 2,3-dioxygenase expression. Cancer Immunol Immunother 56:1017–1024.

Chapter 16
Dendritic Cells: From Inducers of Specific T-Cell Responses to Promoters of Angiogenesis

George Coukos and Fabian Benencia

Abstract Dendritic cells are the most efficient antigen-presenting cells. They capture, process and present antigens to T cells, thus initiating specific immune responses. Taking into account this capability, dendritic cells have been proposed as therapeutic agents against tumors. In recent years other properties of dendritic cells have surfaced. In particular, dendritic cells have been shown to have immunosuppressive properties in some settings and were also capable of inducing proliferation of regulatory T cells. Moreover, it has been shown that dendritic cells are able to generate angiogenic factors and might be able to participate in the angiogenic process. Thus, for tumor therapeutic purposes, further studies on the biology of dendritic cells are necessary in order to generate cells with optimized immunogenic properties, but avoiding a pro-angiogenic profile.

16.1 Dendritic Cells in Mouse and Humans

Dendritic cells (DC) are professional antigen-presenting cells (APC) found in peripheral tissues and in immunological organs such as thymus, bone marrow, spleen, lymph nodes and Peyer's patches (Banchereau et al. 2000; Lanzavecchia and Sallusto 2001; Bonasio and von Andrian 2006). DC present in peripheral tissues sample the organism for the presence of antigens, which they take up, process and present in the context of major histocompatibility molecules (MHC). Then, antigen-loaded DC migrate to lymphoid organs where they present the processed antigen to T lymphocytes triggering specific immune responses. Different DC subsets have been identified both in mouse and humans based on their differential expression of surface antigens (Liu 2001; Shortman and Liu 2002; Ardavin 2003; Hieronymus et al. 2005). In the mouse,

G. Coukos (✉)
Division of Gynecologic Oncology, Center for Research on Ovarian Cancer, University of Pennsylvania, 1209 Biomedical Research Building II/III, 421 Curie Boulevard, Philadelphia, PA, 19104-6142, USA
e-mail: gcks@mail.med.upenn.edu

M.R. Shurin, R.D. Salter (eds.), *Dendritic Cells in Cancer*,
DOI 10.1007/978-0-387-88611-4_16, © Springer Science+Business Media, LLC 2009

CD11c is a highly restricted marker of DC. Bone marrow-derived myeloid DC are present in most tissues and are characterized by coexpression of CD11c and CD11b. Three different subsets of splenic DC have been reported in the mouse characterized by differential expression of CD4 and CD8 markers: CD8α+CD4−, CD8α−CD4+ and CD8α−CD4− DC (Liu et al. 2001; Hieronymus et al. 2005). Another subset of DC is comprised of plasmacytoid DC (pDC). In the mouse, these cells are characterized by the expression of B220 and CD45RB, but no CD11b (Liu 2005). Plasmacytoid DC are characterized by producing large amounts of type 1 interferons (IFN) in response to viral infections (Penna et al. 2002; Liu 2005). Adding to the heterogeneity of mouse DC subsets, two novel subsets of DC have been described in the last years. One of these subsets, named killer DC, is characterized by the expression of B220 and NK1.1 receptors and is able to kill tumor cells, thus preventing tumor growth when used in adoptive therapies (Taieb et al. 2006). The second subset of DC is restricted to the spleen, expresses CD19 and suppresses T-cell responses via indoleamine 2,3-dioxygenase (IDO) expression (Baban et al. 2005). Adding to the complexity, it has been shown that DC subsets can change their phenotype. For example, CD8α−CD4− can give raise to the other two splenic DC subsets (Shortman and Liu 2002), while other reports showed that plasmacytoid DC could acquire myeloid DC characteristics under the influence of viral infection (Diebold et al. 2003). Different protocols have been developed for generating mouse DC in vitro to use in vaccination studies. Usually, these cells are obtained in vitro from bone marrow using GM-CSF alone or in combination with IL-4 (Inaba et al. 1992; Lutz et al. 1999; Masurier et al. 1999). Alternatively, myeloid and plasmacytoid DC can be obtained from bone marrow progenitors by using fms-related tyrosine kinase 3 ligand (Flt3) and different cytokines such as IL-6, stem cell factor, IL-3 or insulin-like growth factor 1 (Maraskovsky et al. 1996; Hieronymus et al. 2005). Flt3 can be also used for expansion of murine DC in vivo (Maraskovsky et al. 1996; Shurin et al. 1997).

In humans, circulating DC are characterized based on the expression of CD11c and CD123 antigens as myeloid ($CD11c^{+}CD123^{-}$) or plasmacytoid ($CD11c^{-}CD123^{+}$). Subsets of myeloid DC are further characterized based on expression of BDCA1 or BDCA3 molecules (Bonasio and von Andrian 2006). Human DC can be prepared in vitro for vaccination purposes from monocytes or $CD34^{+}$ precursors (Gluckman et al. 1997).

16.2 Dendritic Cell Maturation Process

Immature DC have a high phagocytic capability. These cells can take up antigen by phagocytosis, macropinocytosis or receptor-mediated endocytosis (Lanzavecchia and Sallusto 2001). The antigen is then processed into peptides which are presented on the DC surface associated with MHC molecules. Then, under the influence of pro-inflammatory cytokines or products released by damaged tissues ("danger signals"), immature DC turn into mature DC. The maturation process involves upregulation of MHC class II molecules and costimulatory molecules such as CD40, CD80 and CD86 and the chemokine receptor CCR7 as well as downregulation of the chemokine receptor CCR6. Upon maturation, DC show decreased phagocytic capability, increased efficacy to present processed antigens in the context of MHC molecules and consequently improved capability to activate T cells. Chemokines CCL19 (ELC/MIP-3α) and CCL21 (SLC/6Ckine), ligands for CCR7, are constitutively expressed at high levels in lymph nodes (Sanchez-Sanchez et al. 2004). Thus, mature DC migrate from the sites of antigen capture to T-cell regions of draining lymph nodes, where they contact naïve or memory T cells and initiate a specific immune response (Banchereau et al. 2000). In the absence of inflammatory signals, DC can remain phenotypically immature upon antigen capture, expressing low levels of MHC class II and costimulatory molecules. Importantly, antigen presentation in the absence of effective positive costimulation can lead to T-cell anergy and tolerance (Lutz and Schuler 2002). Thus, these DC are considered as "tolerogenic" in comparison with "immunogenic" DC capable of inducing potent specific immune responses. Interestingly, DC can switch from immunogenic to tolerogenic depending on the microenvironment conditions (Heath and Carbone 2001). For example, viral infections can differentiate pDC into Th1-inducing DC (Diebold et al. 2003), while IL-3 can induce Th1-inducing DC to differentiate into Th2-inducing DC (Liu 2001).

16.3 Dendritic Cells with Immunosuppressive Properties in the Tumor Microenvironment

DC are conspicuous members of the microenvironment of several types of cancer (Curiel et al. 2004; Mantovani et al. 2004; Shurin et al. 2006; Whiteside 2006; Baleeiro et al. 2008). Interestingly, tumor-associated DC can show immunosuppressive properties, may be incapable of inducing specific immune responses or can induce regulatory T-cell expansion. Tumor-associated cytokines such as vascular endothelial growth factor (VEGF), IL-10 and prostaglandin E_2 (PgE_2) can profoundly affect the nature of APC. In particular, DC showing low levels of costimulatory molecules have been detected in tumors expressing high levels of VEGF (Gabrilovich et al. 1996, 1999). VEGF is a central angiogenic molecule which has been proposed as a target for antitumor therapies (Ferrara 2004, 2005; Kenny et al. 2007). Highlighting the importance of VEGF in this process, patients treated with anti-VEGF antibody showed a decrease in immunosuppressive DC (Osada et al. 2008). Besides "paralysis" of their immune function, we and others have recently shown that tumor-associated DC or tumor monocyte subsets expressing DC markers are able to produce angiogenic factors and can promote angiogenic processes in the tumor microenvironment (Curiel et al. 2004; Mantovani et al. 2004; Coukos et al.

2005). This phenotype may be however reversible. For example, treatment of tumor-associated DC with inflammatory molecules induces them to acquire a typical APC phenotype with restored capability to activate T cells (Vicari et al. 2002). Interestingly, it has been recently shown that the endothelial cell-produced anti-angiogenic cytokine vascular endothelial growth inhibitor also induces DC maturation (Tian et al. 2007).

16.4 Dendritic Cells Showing Endothelial Cell Properties

Tumors require blood supply for expansive growth. With increasing distance from vessels, hypoxic tumor cells produce angiogenic factors that induce the formation of neovessels (Patan 2000). These are different from vessels of normal tissue at the morphologic and molecular levels (Papetti and Herman 2002; Djonov et al. 2003). Until recently, angiogenesis, or sprouting of endothelial cells (EC) from existing vessels, was the only accepted mechanism of tumor vascularization. Recent studies have suggested that vasculogenesis, or recruitment of endothelial progenitors that differentiate into endothelial cells, might contribute to the formation of tumor neovessels (Bailey and Fleming 2003). Endothelial cell progenitors were first identified by expression of the hematopoietic stem cell antigens, CD34 and flk-1, by other hematopoietic stem cell antigens, such as CD133 (AC133) (Bailey and Fleming 2003). Several populations of hematopoietic cells assume an endothelial phenotype when cultured under pro-angiogenic conditions. These include $CD34^+$, $Sca1^+$, $CD133^+$ and $CD14^+$ cells. Emerging studies have reported the capability of a $CD34^-$ monocyte/macrophage-containing mononuclear cell population to differentiate into endothelial-like cells in vitro (Fernandez Pujol et al. 2000; Schmeisser et al. 2001). More recently, different studies have demonstrated that monocytes or monocyte-like cells can also function as endothelial cell progenitors and incorporate into growing vasculature in experimental models (Rehman et al. 2003; Lewis et al. 2007). We and others have shown that mouse bone marrow-derived DC cultured in the presence of tumor factors can undergo an endothelization process characterized by the loss of CD14 and CD45 and the upregulation of endothelial markers CD31, CD34, von Willebrand factor, VEGF receptor-2 (VEGFR-2) and VE-cadherin (Fernandez Pujol et al. 2001; Conejo-Garcia et al. 2004; Gottfried et al. 2007). Furthermore, these "vascularized" DC displayed other characteristics of vascular cells such as LDL uptake, lectin binding and formation of cord-like structures in 3D gels. Several reports have shown that these cells are able to assemble into vascular structures in vitro and in vivo (Fernandez Pujol et al. 2001; Schmeisser et al. 2001; Conejo-Garcia et al. 2004; Gottfried et al. 2007). Furthermore, it has been recently reported that these cells even acquire functional properties similar to brain microvascular endothelial cells under the appropriate stimuli (Glod et al. 2006).

We have reported, in human ovarian carcinoma, the presence of a subset of CD45+ CD11c+ DC expressing endothelial markers and low levels of costimulatory molecules CD80 and CD86 (Conejo-Garcia et al. 2005). By means of cell sorting we were able to obtain a highly purified population of these cells and we showed that they have the capacity to participate in vascular structures in vivo (Conejo-Garcia et al. 2005). Recent studies have supported these data, showing the capability of human tumor-associated DC to express endothelial markers and assemble into endothelial-like structures in vitro (Gottfried et al. 2007). The capability of tumor-associated DC to behave as functional vascular cells in vivo is a matter of current debate, but DC can also contribute to angiogenic process by producing factors that can promote growth of bona fide endothelial cells (Sozzani et al. 2007).

16.5 Dendritic Cells as a Source of Angiogenic Factors

We have previously shown that DC precursors participate in tumor progression and angiogenesis in a mouse model of ovarian cancer (Conejo-Garcia et al. 2004). For these studies, we used the ID8-Vegf mouse model of ovarian carcinoma. ID8, a cell line derived from spontaneous in vitro malignant transformation of C57BL/6 mouse ovarian surface epithelial cells, was engineered to overexpress $VEGF_{164}$. Our published data support that our engineered model expresses levels of VEGF similar to that found in human ovarian tumor cells. ID8-Vegf-A tumor cells are able to generate solid tumor or ascites when injected into syngeneic C57BL/6 mice subcutaneously or via the intraperitoneal route, respectively. In particular, we reported that admixing immature DC with tumor cells increases tumor progression and angiogenesis. A close analysis of DC isolated from the tumor microenvironment showed that these cells expressed low levels of costimulatory molecules and also expressed VEGF receptor 2 (VEGFR-2) at the level of RNA and protein, thus being susceptible to the immunosuppressive effects of VEGF. Interestingly, these cells were able to produce VEGF, thus indicating not only a possible autocrine mechanism, but also their capability to contribute to endothelial cell growth (Conejo-Garcia et al. 2004). In the same direction, recent data from Dr. J. Folkman's lab (Fainaru et al. 2008) highlight the contribution of DC to angiogenesis in a murine model of endometriosis and in the peritoneal Lewis lung carcinoma tumor model. Similar to what we observed in our model, they showed that these pro-angiogenic DC have an immature phenotype and express VEGFR-2. Also, the same group showed that immature DC participate in angiogenesis processes in non-tumor settings (Nakai et al. 2008).

Further, we have isolated myeloid DC from the tumor microenvironment of human ovarian cancer and analyzed their transcriptome profile by means of low-density arrays (LDA). As shown in Fig. 16.1, these cells produce several angiogenic factors such as VEGF, IL-8 and matrix metalloproteases which can

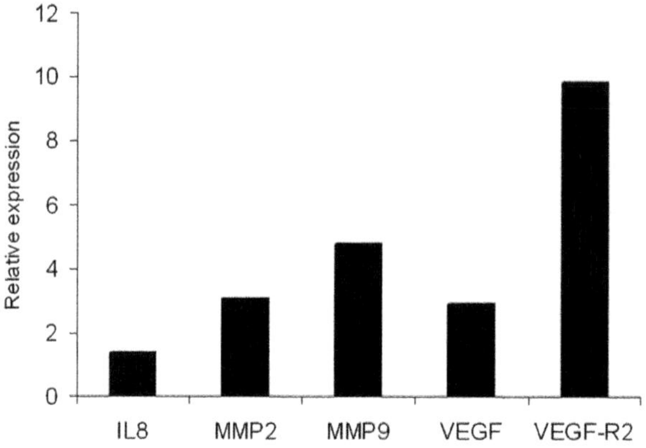

Fig. 16.1 Ovarian cancer-associated immature dendritic cells express pro-angiogenic factors at the level of RNA. Human myeloid CD83− dendritic cells were isolated from the ascites of different patients by means of fluorescence-activated cell sorting. RNA was isolated, reverse transcribed and pooled. Then, samples were analyzed by means of customized low-density arrays. Values were normalized to human immature dendritic cells generated in vitro from peripheral blood mononuclear cells. IL8, interleukin 8; MMP, matrix metalloproteinase; VEGF, vascular endothelial growth factor; VEGF-R2, VEGF receptor 2

contribute to endothelial cell proliferation and also express VEGFR-2. The production of pro-angiogenic factors by tumor-associated immature DC was also observed at the protein level as shown in Fig. 16.2. Interestingly, it has also been reported that tumor-associated pDC are capable of contributing to angiogenic process (Curiel et al. 2004), thus indicating that this capability may not be restricted to a specific DC subtype. All these data indicate that tumor-associated DC, showing an immature profile, are capable of contributing to

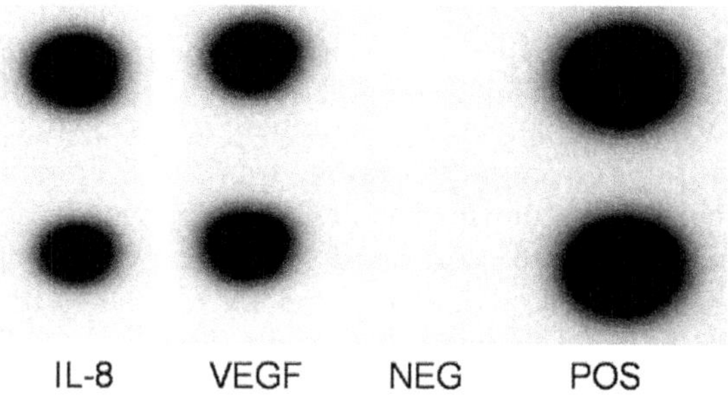

Fig. 16.2 Ovarian cancer-associated immature dendritic cells express pro-angiogenic factors at the level of protein. Human myeloid CD83− dendritic cells were isolated from the ascites of different patients by means of fluorescence-activated cell sorting. Proteins were isolated and analyzed by means of antibody arrays. IL-8, interleukin 8; VEGF, vascular endothelial growth factor; NEG, negative control; POS, positive control

angiogenesis, but recent reports consider that also mature DC might have angiogenic potential.

16.6 Angiogenic Properties of Mature Dendritic Cells

An array of maturation stimuli have been used in order to generate clinical-grade DC for antitumor therapies (Lee et al. 2002; Gigante et al. 2008). Interestingly, it has been shown that although some of these cocktails, specifically those containing PgE_2, generate highly efficient DC in terms of antigen presentation, they can also induce production of angiogenic factors such as VEGF. In particular, Riboldi et al. have reported that human blood myeloid DC can secrete VEGF in a dose-dependent manner when stimulated with LPS in the presence of calcitriol (1,25 dihydroxyvitamin D3), PgE_2 or IL-10, while no VEGF production was observed in in vitro cultures of immature myeloid blood DC (Riboldi et al. 2005). These cells also express other molecules that can act on endothelial cells such as fibroblast growth factor 2 (FGF-2). Moreover, mature DC can secrete chemokines such as IL-8 which can contribute to angiogenic processes (Ghanekar et al. 2007). These mature DC with pro-angiogenic capabilities have been named alternatively activated DC in opposition to the classical activated DC (Riboldi et al. 2005). Furthermore, human myeloid DC matured in the presence of PgE_2 have shown to increase regulatory T-cell levels when used in cancer patients for vaccination purposes (Banerjee et al. 2006).

16.7 Final Comments

The relevance of DC has been growing in the last 20 years, with regard to their exquisite capability to induce specific T-cell responses and their use as vaccines. In recent years, new DC properties have been unraveled, in particular their capability to participate in angiogenic process. In order to further enhance the vaccination capabilities of DC in a tumor setting, without compromising the patients' health, further studies regarding the biology of tumor-associated DC and the use of novel maturation cocktails that avoid enhancing their pro-angiogenic capabilities must be assessed.

References

Ardavin, C. 2003. Origin, precursors and differentiation of mouse dendritic cells. Nat Rev Immunol 3:582–590.

Baban, B., Hansen, A. M., Chandler, P. R., Manlapat, A., Bingaman, A., Kahler, D. J., Munn, D. H. and Mellor, A. L. 2005. A minor population of splenic dendritic cells

expressing CD19 mediates IDO-dependent T cell suppression via type I IFN signaling following B7 ligation. Int Immunol 17:909–919.
Bailey, A. S. and Fleming, W. H. 2003. Converging roads: evidence for an adult hemangioblast. Exp Hematol 31:987–993.
Baleeiro, R. B., Anselmo, L. B., Soares, F. A., Pinto, C. A., Ramos, O., Gross, J. L., Haddad, F., Younes, R. N., Tomiyoshi, M. Y., Bergami-Santos, P. C. and Barbuto, J. A. 2008. High frequency of immature dendritic cells and altered in situ production of interleukin-4 and tumor necrosis factor-alpha in lung cancer. Cancer Immunol Immunother 57:1335–1345.
Banchereau, J., Briere, F., Caux, C., Davoust, J., Lebecque, S., Liu, Y. J., Pulendran, B. and Palucka, K. 2000. Immunobiology of dendritic cells. Annu Rev Immunol 18:767–811.
Banerjee, D. K., Dhodapkar, M. V., Matayeva, E., Steinman, R. M. and Dhodapkar, K. M. 2006. Expansion of FOXP3high regulatory T cells by human dendritic cells (DCs) in vitro and after injection of cytokine-matured DCs in myeloma patients. Blood 108:2655–2661.
Bonasio, R. and von Andrian, U. H. 2006. Generation, migration and function of circulating dendritic cells. Curr Opin Immunol 18:503–511.
Conejo-Garcia, J. R., Benencia, F., Courreges, M. C., Kang, E., Mohamed-Hadley, A., Buckanovich, R. J., Holtz, D. O., Jenkins, A., Na, H., Zhang, L., Wagner, D. S., Katsaros, D., Caroll, R. and Coukos, G. 2004. Tumor-infiltrating dendritic cell precursors recruited by a beta-defensin contribute to vasculogenesis under the influence of Vegf-A. Nat Med 10:950–958.
Conejo-Garcia, J. R., Buckanovich, R. J., Benencia, F., Courreges, M. C., Rubin, S. C., Carroll, R. G. and Coukos, G. 2005. Vascular leukocytes contribute to tumor vascularization. Blood 105:679–681.
Coukos, G., Benencia, F., Buckanovich, R. J. and Conejo-Garcia, J. R. 2005. The role of dendritic cell precursors in tumour vasculogenesis. Br J Cancer 92:1182– 1187.
Curiel, T. J., Cheng, P., Mottram, P., Alvarez, X., Moons, L., Evdemon-Hogan, M., Wei, S., Zou, L., Kryczek, I., Hoyle, G., Lackner, A., Carmeliet, P. and Zou, W. 2004. Dendritic cell subsets differentially regulate angiogenesis in human ovarian cancer. Cancer Res 64:5535–5538.
Diebold, S. S., Montoya, M., Unger, H., Alexopoulou, L., Roy, P., Haswell, L. E., Al-Shamkhani, A., Flavell, R., Borrow, P. and Reis e Sousa, C. 2003. Viral infection switches non-plasmacytoid dendritic cells into high interferon producers. Nature 424:324–328.
Djonov, V., Baum, O. and Burri, P. H. 2003. Vascular remodeling by intussusceptive angiogenesis. Cell Tissue Res 314:107–117.
Fainaru, O., Adini, A., Benny, O., Adini, I., Short, S., Bazinet, L., Nakai, K., Pravda, E., Hornstein, M. D., D'Amato, R. J. and Folkman, J. 2008. Dendritic cells support angiogenesis and promote lesion growth in a murine model of endometriosis. Faseb J 22:522–529.
Fernandez Pujol, B., Lucibello, F. C., Gehling, U. M., Lindemann, K., Weidner, N., Zuzarte, M. L., Adamkiewicz, J., Elsasser, H. P., Muller, R. and Havemann, K. 2000. Endothelial-like cells derived from human CD14 positive monocytes. Differentiation 65:287–300.
Fernandez Pujol, B., Lucibello, F. C., Zuzarte, M., Lutjens, P., Muller, R. and Havemann, K. 2001. Dendritic cells derived from peripheral monocytes express endothelial markers and in the presence of angiogenic growth factors differentiate into endothelial-like cells. Eur J Cell Biol 80:99–110.
Ferrara, N. 2004. Vascular endothelial growth factor: basic science and clinical progress. Endocr Rev 25:581–611.
Ferrara, N. 2005. VEGF as a therapeutic target in cancer. Oncology 69 Suppl 3:11–16.
Gabrilovich, D. I., Chen, H. L., Girgis, K. R., Cunningham, H. T., Meny, G. M., Nadaf, S., Kavanaugh, D. and Carbone, D. P. 1996. Production of vascular endothelial growth factor by human tumors inhibits the functional maturation of dendritic cells. Nat Med 2:1096–1103.

Gabrilovich, D. I., Ishida, T., Nadaf, S., Ohm, J. E. and Carbone, D. P. 1999. Antibodies to vascular endothelial growth factor enhance the efficacy of cancer immunotherapy by improving endogenous dendritic cell function. Clin Cancer Res 5:2963–2970.

Ghanekar, S. A., Bhatia, S., Ruitenberg, J. J., DeLa Rosa, C., Disis, M. L., Maino, V. C., Maecker, H. T. and Waters, C. A. 2007. Phenotype and in vitro function of mature MDDC generated from cryopreserved PBMC of cancer patients are equivalent to those from healthy donors. J Immune Based Ther Vaccines 5:7.

Gigante, M., Mandic, M., Wesa, A. K., Cavalcanti, E., Dambrosio, M., Mancini, V., Battaglia, M., Gesualdo, L., Storkus, W. J. and Ranieri, E. 2008. Interferon-alpha (IFN-alpha)-conditioned DC preferentially stimulate type-1 and limit Treg-type in vitro T-cell responses from RCC patients. J Immunother 31:254–262.

Glod, J., Kobiler, D., Noel, M., Koneru, R., Lehrer, S., Medina, D., Maric, D. and Fine, H. A. 2006. Monocytes form a vascular barrier and participate in vessel repair after brain injury. Blood 107:940–946.

Gluckman, J. C., Canque, B., Chapuis, F. and Rosenzwajg, M. 1997. In vitro generation of human dendritic cells and cell therapy. Cytokines Cell Mol Ther 3:187–196.

Gottfried, E., Kreutz, M., Haffner, S., Holler, E., Iacobelli, M., Andreesen, R. and Eissner, G. 2007. Differentiation of human tumour-associated dendritic cells into endothelial-like cells: an alternative pathway of tumour angiogenesis. Scand J Immunol 65:329–335.

Heath, W. R. and Carbone, F. R. 2001. Cross-presentation, dendritic cells, tolerance and immunity. Annu Rev Immunol 19:47–64.

Hieronymus, T., Gust, T. C., Kirsch, R. D., Jorgas, T., Blendinger, G., Goncharenko, M., Supplitt, K., Rose-John, S., Muller, A. M. and Zenke, M. 2005. Progressive and controlled development of mouse dendritic cells from Flt3+CD11b+ progenitors in vitro. J Immunol 174:2552–2562.

Inaba, K., Inaba, M., Romani, N., Aya, H., Deguchi, M., Ikehara, S., Muramatsu, S. and Steinman, R. M. 1992. Generation of large numbers of dendritic cells from mouse bone marrow cultures supplemented with granulocyte/macrophage colony-stimulating factor. J Exp Med 176:1693–1702.

Kenny, P. A., Lee, G. Y. and Bissell, M. J. 2007. Targeting the tumor microenvironment. Front Biosci 12:3468–3474.

Lanzavecchia, A. and Sallusto, F. 2001. The instructive role of dendritic cells on T cell responses: lineages, plasticity and kinetics. Curr Opin Immunol 13:291–298.

Lee, A. W., Truong, T., Bickham, K., Fonteneau, J. F., Larsson, M., Da Silva, I., Somersan, S., Thomas, E. K. and Bhardwaj, N. 2002. A clinical grade cocktail of cytokines and PGE2 results in uniform maturation of human monocyte-derived dendritic cells: implications for immunotherapy. Vaccine 20 Suppl 4:A8–22.

Lewis, C. E., De Palma, M. and Naldini, L. 2007. Tie2-expressing monocytes and tumor angiogenesis: regulation by hypoxia and angiopoietin-2. Cancer Res 67:8429–8432.

Liu, Y. J. 2001. Dendritic cell subsets and lineages, and their functions in innate and adaptive immunity. Cell 106:259–262.

Liu, Y. J. 2005. IPC: professional type 1 interferon-producing cells and plasmacytoid dendritic cell precursors. Annu Rev Immunol 23:275–306.

Liu, Y. J., Kanzler, H., Soumelis, V. and Gilliet, M. 2001. Dendritic cell lineage, plasticity and cross-regulation. Nat Immunol 2:585–589.

Lutz, M. B., Kukutsch, N., Ogilvie, A. L., Rossner, S., Koch, F., Romani, N. and Schuler, G. 1999. An advanced culture method for generating large quantities of highly pure dendritic cells from mouse bone marrow. J Immunol Methods 223:77–92.

Lutz, M. B. and Schuler, G. 2002. Immature, semi-mature and fully mature dendritic cells: which signals induce tolerance or immunity? Trends Immunol 23:445–449.

Mantovani, A., Sozzani, S., Locati, M., Schioppa, T., Saccani, A., Allavena, P. and Sica, A. 2004. Infiltration of tumours by macrophages and dendritic cells: tumour-associated

macrophages as a paradigm for polarized M2 mononuclear phagocytes. Novartis Found Symp 256:137–145; discussion 146–148, 259–269.
Maraskovsky, E., Brasel, K., Teepe, M., Roux, E. R., Lyman, S. D., Shortman, K. and McKenna, H. J. 1996. Dramatic increase in the numbers of functionally mature dendritic cells in Flt3 ligand-treated mice: multiple dendritic cell subpopulations identified. J Exp Med 184:1953–1962.
Masurier, C., Pioche-Durieu, C., Colombo, B. M., Lacave, R., Lemoine, F. M., Klatzmann, D. and Guigon, M. 1999. Immunophenotypical and functional heterogeneity of dendritic cells generated from murine bone marrow cultured with different cytokine combinations: implications for anti-tumoral cell therapy. Immunology 96:569–577.
Nakai, K., Fainaru, O., Bazinet, L., Pakneshan, P., Benny, O., Pravda, E., Folkman, J. and D'Amato, R. 2008. Dendritic cells augment choroidal neovascularization. Invest Ophthalmol Vis Sci 49:3666–3670.
Osada, T., Chong, G., Tansik, R., Hong, T., Spector, N., Kumar, R., Hurwitz, H. I., Dev, I., Nixon, A. B., Lyerly, H. K., Clay, T. and Morse, M. A. 2008. The effect of anti-VEGF therapy on immature myeloid cell and dendritic cells in cancer patients. Cancer Immunol Immunother 57:1115–1124.
Papetti, M. and Herman, I. M. 2002. Mechanisms of normal and tumor-derived angiogenesis. Am J Physiol Cell Physiol 282:C947–970.
Patan, S. 2000. Vasculogenesis and angiogenesis as mechanisms of vascular network formation, growth and remodeling. J Neurooncol 50:1–15.
Penna, G., Vulcano, M., Sozzani, S. and Adorini, L. 2002. Differential migration behavior and chemokine production by myeloid and plasmacytoid dendritic cells. Hum Immunol 63:1164–1171.
Rehman, J., Li, J., Orschell, C. M. and March, K. L. 2003. Peripheral blood "endothelial progenitor cells" are derived from monocyte/macrophages and secrete angiogenic growth factors. Circulation 107:1164–1169.
Riboldi, E., Musso, T., Moroni, E., Urbinati, C., Bernasconi, S., Rusnati, M., Adorini, L., Presta, M. and Sozzani, S. 2005. Cutting edge: proangiogenic properties of alternatively activated dendritic cells. J Immunol 175:2788–2792.
Sanchez-Sanchez, N., Riol-Blanco, L., de la Rosa, G., Puig-Kroger, A., Garcia Bordas, J., Martin, D., Longo, N., Cuadrado, A., Cabanas, C., Corbi, A. L., Sanchez-Mateos, P. and Rodriguez-Fernandez, J. L. 2004. Chemokine receptor CCR7 induces intracellular signaling that inhibits apoptosis of mature dendritic cells. Blood 104:619–625.
Schmeisser, A., Garlichs, C. D., Zhang, H., Eskafi, S., Graffy, C., Ludwig, J., Strasser, R. H. and Daniel, W. G. 2001. Monocytes coexpress endothelial and macrophagocytic lineage markers and form cord-like structures in Matrigel under angiogenic conditions. Cardiovasc Res 49:671–680.
Shortman, K. and Liu, Y. J. 2002. Mouse and human dendritic cell subtypes. Nat Rev Immunol 2:151–161.
Shurin, M. R., Pandharipande, P. P., Zorina, T. D., Haluszczak, C., Subbotin, V. M., Hunter, O., Brumfield, A., Storkus, W. J., Maraskovsky, E. and Lotze, M. T. 1997. FLT3 ligand induces the generation of functionally active dendritic cells in mice. Cell Immunol 179:174–184.
Shurin, M. R., Shurin, G. V., Lokshin, A., Yurkovetsky, Z. R., Gutkin, D. W., Chatta, G., Zhong, H., Han, B. and Ferris, R. L. 2006. Intratumoral cytokines/chemokines/growth factors and tumor infiltrating dendritic cells: friends or enemies? Cancer Metastasis Rev 25:333–356.
Sozzani, S., Rusnati, M., Riboldi, E., Mitola, S. and Presta, M. 2007. Dendritic cell-endothelial cell cross-talk in angiogenesis. Trends Immunol 28:385–392.
Taieb, J., Chaput, N., Menard, C., Apetoh, L., Ullrich, E., Bonmort, M., Pequignot, M., Casares, N., Terme, M., Flament, C., Opolon, P., Lecluse, Y., Metivier, D., Tomasello, E. , Vivier, E., Ghiringhelli, F., Martin, F., Klatzmann, D., Poynard, T., Tursz, T., Raposo, G.,

Yagita, H., Ryffel, B., Kroemer, G. and Zitvogel, L. 2006. A novel dendritic cell subset involved in tumor immunosurveillance. Nat Med 12:214–219.

Tian, F., Grimaldo, S., Fujita, M., Cutts, J., Vujanovic, N. L. and Li, L. Y. 2007. The endothelial cell-produced antiangiogenic cytokine vascular endothelial growth inhibitor induces dendritic cell maturation. J Immunol 179:3742–3751.

Vicari, A. P., Chiodoni, C., Vaure, C., Ait-Yahia, S., Dercamp, C., Matsos, F., Reynard, O., Taverne, C., Merle, P., Colombo, M. P., O'Garra, A., Trinchieri, G. and Caux, C. 2002. Reversal of tumor-induced dendritic cell paralysis by CpG immunostimulatory oligonucleotide and anti-interleukin 10 receptor antibody. J Exp Med 196:541–549.

Whiteside, T. L. 2006. The role of immune cells in the tumor microenvironment. Cancer Treat Res 130:103–124.

Chapter 17
Protumorigenic Function of Dendritic Cells

Anjli Kukreja

Abstract A growing body of evidence suggests a strong role of the tumor microenvironment in shaping up the final outcome of tumor progression. The focus is now on how various stromal components interact with cancer cells to promote tumor growth and metastasis. Infiltration of dendritic cells is a common feature of most human tumors and until recently, these tumor-infiltrating dendritic cells have largely been studied for their immunologic properties. Current studies reveal a more active role of dendritic cells in the microenvironment leading to tumor enhancement, angiogenesis and bone resorption. These observations have tremendous applications for tumor therapy and suggest a need to better understand the ways by which dendritic cells regulate tumor biology in terms of their antigen-presenting as well as tumor-supporting functions including transdifferentiation into other cell types. This review focuses on the protumorigenic effects of dendritic cells in tumor progression that can be targeted to improve cancer therapy.

17.1 Introduction

Tumor microenvironment/stroma plays an important role in shaping up the tumor outcome and is therefore an important target for therapeutic intervention. There is a dynamic, bidirectional interplay between tumor and its stroma where stroma has both tumor-supporting and tumor-inhibiting roles. In spite of the fact that tumor stroma plays an important role in tumor progression, very little is known about the tumor microenvironment, its constituents, how they are recruited, altered and how they reciprocally affect tumor progression. A complex interplay of growth and differentiation factors orchestrated by tumor cells, stromal cells or tumor-infiltrating immune cells promote tumor cell

A. Kukreja (✉)
Section of Hematology, Yale University, 333 Cedar Street, Box 208021, New Haven, CT 06520-8021, USA
e-mail: Kukreja@rockefeller.edu

M.R. Shurin, R.D. Salter (eds.), *Dendritic Cells in Cancer*,
DOI 10.1007/978-0-387-88611-4_17,

survival, proliferation and metastasis (Krtolica et al. 2001; Coussens and Werb 2002; Iyengar et al. 2003).

Tumor cells, like normal cells, live in a complex microenvironment that includes the extracellular matrix (ECM), diffusible growth factors, cytokines and a variety of non-epithelial cell types, including those comprising the vasculature, cells of the immune system (lymphocytes, macrophages, dendritic cells (DC) and mast cells) and fibroblasts. Of the various cells that form a part of reticuloendothelial/phagocytic system including macrophages, DC and neutrophils, DC are unique. DC are professional antigen-presenting cells (APC) specialized to initiate and regulate adaptive immune responses. Immature DC present in most tissues are specialized to capture and process antigens in vivo (Trombetta and Mellman 2005) to be presented by MHC molecules on the cell surface and recognized by T cells. DC also upregulate co-stimulatory molecules and migrate to secondary lymphoid organs, e.g., spleen and lymph nodes, where they activate antigen-specific T cells. DC are also important in activating innate immune cells, e.g., natural killer (NK) cells (Lucas et al. 2007) and NKT cells (Fujii et al. 2002). Besides their ability to recognize and eliminate microbial/foreign antigens, DC have the inherent capacity to induce tolerance to self-antigens both in the thymus and periphery (Steinman et al. 2003). Just like lymphocytes, there are different subsets of DC with distinct functional properties that exist not only in skin (Langerhans cells and interstitial DC) but also in the peripheral blood (myeloid and plasmacytoid DC). The localization and migration of different subsets of DC in the tissues is controlled by a variety of cell-trafficking molecules, predominantly chemokines (Caux et al. 2000).

17.2 Role of Dendritic Cells in the Tumor Microenvironment

Infiltration of DC is a common feature of most human tumors (Rettig et al. 1997; Banchereau et al. 2000) and yet tumors evade immune responses, progress, metastasize and eventually result in death of the host. The relationship between the presence of DC and the clinical prognosis is however complex and depends not only on the numbers of DC but also on their maturation and functional status. DC have largely been studied in the context of their antigen presentation and induction and maintenance of antitumor immune responses. Both animal models and patient studies have provided compelling evidence suggesting that the immune system plays a major role in surveillance of spontaneous tumors (Dunn et al. 2002; Buell et al. 2005). DC play a major role in prevention and treatment of tumors and therefore are an attractive target for therapeutic manipulation of the immune system to enhance insufficient immune response to tumor antigens. On the contrary, very little is known about the ability of DC to support tumor pathogenesis in situ and whether this could play a role in tumor progression. Studies are now underway to establish the non-immunologic effects of DC on tumors. In this chapter, we will discuss the recent insights unraveling tumor/DC interactions favoring tumor progression, sustenance and survival that can be harnessed to improve cancer immunotherapy.

17.3 Protumorigenic Effects of Dendritic Cells

Although the presence of DC signature in some types of cancers has been shown to be associated with improved prognosis, other studies failed to show a correlation between DC infiltration and overall and/or metastasis-free survival (Vicari et al. 2004).

Tumors use several approaches to subvert immune responses and use DC in various ways to prevent immune response, induce immune tolerance and suppressive T cells. Recently, a more active role of DC in cancer pathogenesis has been suggested as discussed below by directly altering tumor growth. Tumors therefore have evolved ways to evade and exploit DC for their own benefit. Listed below are the various mechanisms that tumors use to subvert DC that are at the center stage by not only supporting tumor growth but also transdifferentiating into other cell types, leading to cancer progression and its complications.

17.3.1 DC and Enhanced Tumor Growth

Presence of DC in the tumor bed has largely been interpreted in terms of their ability to invoke an immune response. It is now increasingly clear that DC may interact more directly with the tumor cells. Furthermore, several studies have correlated the presence of DC signature in the tumor microenvironment with an adverse prognosis (Bell et al. 1999; Lin and Pollard 2004; Sandel et al. 2005). Both myeloma tumors in patients and mouse plasmacytomas are extensively infiltrated by DC (Said et al. 1997; Dembic et al. 2000; Bahlis et al. 2007). Our recent data suggest the possibility that DC may provide a niche to support the clonogenic growth of human myeloma (Kukreja et al. 2006) that was later corroborated by another study by Bahlis et al. (2007). The clonogenic growth of myeloma was supported not only by monocyte-derived immature and mature DC but also by blood-derived myeloid and plasmacytoid DC (Kukreja et al. 2006, 2007). DC not only supported the growth of myeloma but also Mantle cell lymphoma and breast cancer in in vitro assays. Furthermore, DC–tumor interactions led to upregulation of B-cell lymphoma 6 (Bcl-6) expression in myeloma cells.

DC express several molecules implicated in B-cell differentiation as well as co-stimulatory molecules and integrins (MacLennan and Vinuesa 2002). Two of these pathways involving the tumor necrosis factor (TNF) superfamily members, B-cell activating factor (BAFF)/a proliferation-inducing ligand (APRIL) (Moreaux et al. 2004; Novak et al. 2004; Moreaux et al. 2005) and receptor activator of nuclear factor κB ligand (RANK-L) pathways (Pearse et al. 2001) have been previously implicated in survival of myeloma cells. Blockade of BAFF/APRIL-mediated interactions with transmembrane activator and calcium modulator and cyclophilin ligand interactor-Fc (TACI-Fc) or blockade of RANK/RANK-L-mediated interactions with osteoprotegerin (OPG) (Giuliani et al. 2001) led to inhibition of DC-mediated enhancement of tumor clonogenicity in myeloma. This suggested that DC-mediated enhancement of myeloma clonogenicity was mediated in part by RANK-L and BAFF/APRIL-mediated interactions (Kukreja et al. 2006). However, this does not rule out additional mechanisms of DC-mediated regulation of myeloma growth such as integrins and co-stimulatory molecules expressed by DC.

Tumor-supporting role of DC was also shown in breast cancer where immature DC primarily localized in the tumor bed, whereas mature DC confined mainly to the peritumoral areas (Bell et al. 1999). DC further skewed CD4+ T-cell response to a Th2 type that promoted early tumor progression (Asprod et al. 2007). These observations suggest that DC have the ability to directly impact the biology of human tumors and may be a target of therapeutic intervention. In vivo experimental models are needed to address these questions to decipher protumorigenic roles of DC subsets and whether these subsets would be appropriate targets for cancer immunotherapy.

17.3.2 DC and Immunosuppression

Various studies during the past few years have shown that tumor-induced DC defects lead to defects in tumor-specific T-cell responses resulting in failure of immune system to respond to tumors (Enk et al. 1997; Nestle et al. 1997; Della Bella et al. 2003). Several cytokines including IL-6 (Chomarat et al. 2000), tumor-derived IL-10 (Steinbrink et al. 1999) and vascular endothelial growth factor (VEGF) (Gabrilovich et al. 1996; Ohm and Carbone 2001) have been implicated in the suppression of DC functions in the tumor microenvironment.

IL-6 is secreted by large numbers of tumors and is also known to be involved in tumor-mediated regulation of DC differentiation (Menetrier-Caux et al. 1998). IL-6 has been shown to be a potent inhibitor of DC maturation in vivo that required *stat3* activation in DC (Park et al. 2004). A study by Hayashi et al. (2003) showed that bone marrow sera from myeloma patients inhibited DC generation that could be reversed by anti-VEGF and/or anti-IL-6 antibodies. Another study further showed an important role of IL-6 in abnormal DC differentiation in multiple myeloma (Ratta et al. 2002).

IL-10 plays a major role in DC defects in cancer. IL-10 has been reported to convert immature DC into tolerogenic APC through decreased expression of co-stimulatory molecules (Steinbrink et al. 1997). Furthermore, it has been shown that IL-10-treated DC mediated suppression of antigen-specific proliferation of CD4+ and CD8+ T cells (Steinbrink et al. 2002). Tumor-induced

IL-10 has been shown to be specifically responsible for DC dysfunction in response to antigen-driven maturation in a mouse tumor model. This was confirmed by restoration of DC defect in IL-10 knockout mice (Yang and Lattime 2003).

VEGF was the first tumor-derived factor shown to inhibit DC differentiation. It plays an important role in neoangiogenesis and in hematopoiesis during embryogenesis. Sustained angiogenesis is a hallmark of most, if not all, tumors (Hanahan and Weinberg 2000). A number of studies in cancer patients have shown inhibitory effects of VEGF on DC differentiation (Saito et al. 1998; Lissoni et al. 2001; Takahashi et al. 2004). VEGF-treated DC are less potent stimulating antigen-specific T cells, suggesting a unique role of VEGF in down-regulating host immunity to cancer (Laxmanan et al. 2005).

It therefore appears intriguing that tumors secrete a variety of distinct cytokines that block or induce defective DC differentiation.

Another mechanism by which tumors evade immune responses is through the induction of suppressive cells thereby linking tumor expansion to immunosuppression. A study by Ghiringhelli et al. (2005) showed that a subset of DC exposed to tumor cells acquired the ability to secrete transforming growth factor-β (TGF-β) and to stimulate the expansion of regulatory T cells. Another study by Aspord et al. (2007) showed that breast cancer targets and instructs DC to prime IL-13+ CD4+ T cells that facilitated tumor development.

Tumors also have the ability to subvert DC functions by altering DC antigen capture and antigen-presenting pathway. Tumor antigen mucin 1 (MUC-1), overexpressed by breast cancer cells, interacts with DC-specific intercellular adhesion molecule 3 (ICAM-3)-grabbing nonintegrin (DC-SIGN) on DC and is inefficiently endocytosed by DC, leading to defect in antigen processing and presentation and causes fewer frequency of antigen-specific T cells (Finn et al. 1995; Hiltbold et al. 2000). Another study showed that MUC-1 antigen prevents development of efficient Th1-type antitumor immunity and favors a Th2-type response (Carlos et al. 2005).

17.3.3 DC and Angiogenesis

Angiogenesis plays an important role in tissue repair and remodeling. Similarly angiogenesis is important for both primary and metastatic tumor growth. DC are increasingly being recognized as a highly plastic cell population. Recent observations have shown that different DC subsets produce and release various pro- and anti-angiogenic mediators depending on their activation status and cytokine milieu. DC mediate vasculogenesis through the cooperation of β-defensins and VEGF-A in breast cancer (Conejo-Garcia et al. 2004). Similarly other studies have shown that DC in the presence of specific cytokines have the ability to transdifferentiate into endothelial-like cells (Fernandez Pujol et al. 2001; Gottfried et al. 2007). Another report by Curiel et al. in ovarian cancer further suggested that plasmacytoid DC induced angiogenesis in vivo through the production of tumor necrosis factor-α (TNF-α) and IL-8. Myeloid DC on the other hand inhibited angiogenesis through the production of IL-12 (Curiel et al. 2004). DC therefore are not only relevant in tumor immunopathogenesis but also might support angiogenesis in the tumor microenvironment.

17.3.4 Transdifferentiation of DC

Osteoimmunology is an emerging field that deals with the role of the immune system in bone remodeling (Arron and Choi 2000). The mass and shape of healthy skeleton is maintained by a delicate balance between bone-forming osteoblasts and bone-resorbing osteoclasts. Osteoclasts (OC), the multinucleated cells, are located in the vicinity of bones and play an important role in bone remodeling and homeostasis (Boyle et al. 2003; Teitelbaum and Ross 2003). Similar to DC, OC are derived from the monocytic–macrophage lineage of the hematopoietic precursors and are formed as a result of fusion of mononuclear cells. OC are also derived from blood monocytes, but in the presence of osteoclastic cytokines, RANK-L and macrophage colony-stimulating factor (M-CSF). There is excessive osteoclastic activity in many human diseases such as rheumatoid arthritis, Langerhans cell histiocytosis, Paget's disease and multiple myeloma leading to bone resorption or osteoporosis. Recent studies have shown alternate pathways of OC formation from DC in inflammatory states, suggesting highly plastic nature of these APC (Rivollier et al. 2004; Speziani et al. 2007). Langerhans cell histiocytosis is another rare disorder of unknown etiology characterized by aggressive chronic granuloma formation, bone resorption and soft tissue lesions (Grois et al. 2005). There is characteristic accumulation of CD1a+ Langerhans-like cells, myeloid cells and multinucleated giant cells (MGC) in one or multiple organs particularly in bone, skin and lymph nodes (da Costa et al. 2005). It was recently shown that IL-17 produced by DC led to IL-17-dependent DC fusion and formation of MGC, seen typically in Langerhans cell histiocytosis (Coury et al. 2008).

Our preliminary data suggest that DC, essential contributors to normal human immune responses, also have the ability to fuse and become MGC when co-cultured with myeloma cells (Kukreja et al. in preparation). Osteolytic lesions are a major source of morbidity in myeloma, and our data provide evidence that DC are not just immune-modulating specialized APC, but play an important role in not only supporting myeloma growth (Kukreja et al. 2006) but also serving as precursors to OC leading to characteristic osteoporosis in myeloma. Further, it will be interesting to see if DC are part of the malignant clone, forming OC by fusing with the tumor cells and may contribute to drug resistance.

RANK/RANK-L pathway is known to be essential for the osteoclast formation and has been studied extensively (Boyle et al. 2003; Teitelbaum and

Ross 2003). Besides RANK/RANK-L pathway, macrophage inflammatory protein-1α (MIP-1α) from tumor cells (Han et al. 2001), colony-stimulating factor (CSF)/macrophage colony-stimulating factor receptor (M-CSFR) (Teitelbaum 2000) and immunoreceptor tyrosine-based activation motif (ITAM) (Koga et al. 2004) pathways have been shown to promote osteoclastogenesis. The molecular mechanism of cell fusion leading to the formation of MGC is however not known. To date several molecules have been identified as candidates for cell fusion. Signal regulatory protein-α (SIRP-α)/CD47 has been shown to support macrophage fusion to form giant cell (Han et al. 2000). Kukita et al. showed that the seven-transmembrane receptor-like molecule, DC-specific transmembrane protein (DC-STAMP), expressed on human monocyte-derived DC is involved in osteoclastogenesis (Kukita et al. 2004). DC-STAMP-deficient mice showed complete abrogation of multinucleation of OC (Yagi et al. 2005). Taken together, these findings provide new insights into molecular pathogenesis of osteoclastogenesis and identify new potential markers for disease progression and therapeutic targeting.

17.3.5 DC and Metastasis

The significance of active recruitment and presence of DC subsets in tumors is beginning to be realized. Extensive data exist which suggests that tumors exploit the normal matrix remodeling capacities of macrophages enabling them to egress into and migrate through the surrounding stroma (Condeelis and Pollard 2006), enhancing tumor cell migration and invasion (Condeelis and Segall 2003).

The migratory pattern of DC is integral to their function as APC. The migration of DC to reach the inflammatory site requires chemokines (Caux et al. 2000), adhesion molecules (D'Amico et al. 1998; Pendl et al. 2002) and proteinases (matrix metalloproteinases (MMP)-2 and -9) (Osman et al. 2002; Vermaelen et al. 2003) involved in degradation of extracellular matrix. Several studies have reported that monocyte-derived DC and Langerhans cells in humans produce MMP-9 and low levels of MMP-2 (Hollender et al. 2002; Kouwenhoven et al. 2002; Osman et al. 2002). We speculate that tumors might also exploit DC for proteolysis of extracellular matrix in their migration through basement membrane resulting in tumor metastasis (Chabot et al. 2006).

17.4 Dendritic Cells Versus Macrophages

As already mentioned above, reticuloendothelial/phagocytic system is a complex network of specialized cells consisting of DC, macrophages as well as osteoclasts that infiltrate tumors. While the role of DC in tumor enhancement is beginning to be explored, the role of tumor-associated macrophages is extensively characterized. Various studies have shown that macrophages support tumor growth and infiltration of tumors with macrophages has been correlated with adverse prognosis (Lissbrant et al. 2000; Koide et al. 2004; Tsutsui et al. 2005). Macrophages have been shown to secrete various factors (epidermal growth factor (EGF), fibroblast growth factor (FGF), TGF-β, VEGF, chemokines and cytokines) that support tumor cell growth, migration and metastasis (Sica et al. 2008). Furthermore, macrophages are actively involved in tumor progression by producing pro-angiogenic factors such as VEGF, TNF-α, IL-8 and bFGF (Lewis et al. 1995). Additionally they express a broad range of angiogenesis-modulating enzymes, including MMP-2, MMP-7, MMP-9, MMP-12 and cyclooxygenase-2 (COX-2) (Sunderkotter et al. 1991; Klimp et al. 2001).

DC are, however, functionally and biologically distinct from macrophages. Both DC and macrophages recognize and uptake dying cells, but the immunological consequences of antigen processing and presentation are distinctly different between the two cell types (Trombetta and Mellman 2005). This may be due to marked differences in the lysosomal protease content between DC and macrophages. Unlike macrophages where proteins are rapidly degraded, limited lysosomal protease content in DC favors antigen persistence and efficient class II presentation (Delamarre et al. 2005). Furthermore, DC can efficiently present tumor antigens not only on class II but also on class I, termed cross-presentation (Albert et al. 1998; Heath et al. 2004). Cross-presentation of antigens from apoptotic tumor cells has been shown to be critical for induction of antitumor immunity (Heath et al. 2004). Our studies in myeloma have shown that DC and not macrophages promote tumor clonogenicity (Kukreja et al. 2006) and that DC undergo cell fusion to form tartrate-resistant acid phosphatase (TRAP) + bone-resorbing MGC/OC in the presence of myeloma (Kukreja et al. in preparation).

17.5 Summary

The importance of tumor microenvironment can be highlighted by the recent success of drugs specifically targeting tumor stroma. Proteasome inhibitor bortezomib that targets not only tumor cells but also DC (Kukreja et al. 2007), that have been shown to be present in tumor microenvironment in myeloma patients as already discussed, is being increasing used to treat refractory myeloma and its associated bone disease (Richardson and Anderson 2003; Richardson et al. 2005; Uy et al. 2007). Furthermore, the emergence of anti-angiogenic drugs (bevacizumab, anti-VEGF antibody; lenalidomide; thalidomide) targeting endothelial cells in the tumor stroma strengthens the progress toward this goal (Ferrara and Kerbel 2005).

The role of immune cells in human neoplasia is not clear (de Visser et al. 2006). On the one hand, immunocytes by releasing inflammatory mediators support angiogenesis and metastasis; however, they also inhibit tumor growth and have been shown to be associated with improved prognosis. Likewise, DC play an important role in shaping immune responses to tumors in addition to being a part of the complex network of mononuclear phagocyte infiltrate of tumor microenvironment and exhibiting non-immunologic effects on tumor cells. A thorough understanding of DC in tumor beds and tumor–DC interaction is therefore extremely important and may provide a rationale for specifically targeting this interaction as a novel approach to the therapy of human cancer.

Very little is known about the type, density and location of DC in the tumor bed and how tumors support the differentiation of DC in situ is yet to be established. A spectrum of new monoclonal antibodies will allow for the identification of distinct subsets of DC along with their functional properties in the inflammatory microenvironment. Presence of DC signature can be a sign of an active immune response or it may reflect a subversive effect of tumors. More studies in patients and experimental tumor models are needed to address these issues. These findings provide a foundation for the future work to thoroughly understand and characterize DC in the tumor bed and pave the way for the development of novel treatment modalities for cancer.

References

Albert, M. L., Pearce, S. F., Francisco, L. M., Sauter, B., Roy, P., Silverstein, R. L. and Bhardwaj, N. 1998. Immature dendritic cells phagocytose apoptotic cells via alphavbeta5 and CD36, and cross-present antigens to cytotoxic T lymphocytes. J Exp Med 188:1359–1368.

Arron, J. R. and Choi, Y. 2000. Bone versus immune system. Nature 408:535–536.

Aspord, C., Pedroza-Gonzalez, A., Gallegos, M., Tindle, S., Burton, E. C., Su, D., Marches, F., Banchereau, J. and Palucka, A. K. 2007. Breast cancer instructs dendritic cells to prime interleukin 13-secreting CD4+ T cells that facilitate tumor development. J Exp Med 204:1037–1047.

Bahlis, N. J., King, A. M., Kolonias, D., Carlson, L. M., Liu, H. Y., Hussein, M. A., Terebelo, H. R., Byrne, G. E., Jr., Levine, B. L., Boise, L. H. and Lee, K. P. 2007. CD28-mediated regulation of multiple myeloma cell proliferation and survival. Blood 109:5002–5010.

Banchereau, J., Briere, F., Caux, C., Davoust, J., Lebecque, S., Liu, Y. J., Pulendran, B. and Palucka, K. 2000. Immunobiology of dendritic cells. Annu Rev Immunol 18:767–811.

Bell, D., Chomarat, P., Broyles, D., Netto, G., Harb, G. M., Lebecque, S., Valladeau, J., Davoust, J., Palucka, K. A. and Banchereau, J. 1999. In breast carcinoma tissue, immature dendritic cells reside within the tumor, whereas mature dendritic cells are located in peritumoral areas. J Exp Med 190:1417–1426.

Boyle, W. J., Simonet, W. S. and Lacey, D. L. 2003. Osteoclast differentiation and activation. Nature 423:337–342.

Buell, J. F., Papaconstantinou, H. T., Skalow, B., Hanaway, M. J., Alloway, R. R. and Woodle, E. S. 2005. De novo colorectal cancer: five-year survival is markedly lower in transplant recipients compared with the general population. Transplant Proc 37:960–961.

Carlos, C. A., Dong, H. F., Howard, O. M., Oppenheim, J. J., Hanisch, F. G. and Finn, O. J. 2005. Human tumor antigen MUC1 is chemotactic for immature dendritic cells and elicits maturation but does not promote Th1 type immunity. J Immunol 175:1628–1635.

Caux, C., Ait-Yahia, S., Chemin, K., de Bouteiller, O., Dieu-Nosjean, M. C., Homey, B., Massacrier, C., Vanbervliet, B., Zlotnik, A. and Vicari, A. 2000. Dendritic cell biology and regulation of dendritic cell trafficking by chemokines. Springer Semin Immunopathol 22:345–369.

Chabot, V., Reverdiau, P., Iochmann, S., Rico, A., Senecal, D., Goupille, C., Sizaret, P. Y. and Sensebe, L. 2006. CCL5-enhanced human immature dendritic cell migration through the basement membrane in vitro depends on matrix metalloproteinase-9. J Leukoc Biol 79:767–778.

Chomarat, P., Banchereau, J., Davoust, J. and Palucka, A. K. 2000. IL-6 switches the differentiation of monocytes from dendritic cells to macrophages. Nat Immunol 1:510–514.

Condeelis, J. and Pollard, J. W. 2006. Macrophages: obligate partners for tumor cell migration, invasion, and metastasis. Cell 124:263–266.

Condeelis, J. and Segall, J. E. 2003. Intravital imaging of cell movement in tumours. Nat Rev Cancer 3:921–930.

Conejo-Garcia, J. R., Benencia, F., Courreges, M. C., Kang, E., Mohamed-Hadley, A., Buckanovich, R. J., Holtz, D. O., Jenkins, A., Na, H., Zhang, L., Wagner, D. S., Katsaros, D., Caroll, R. and Coukos, G. 2004. Tumor-infiltrating dendritic cell precursors recruited by a beta-defensin contribute to vasculogenesis under the influence of Vegf-A. Nat Med 10:950–958.

Coury, F., Annels, N., Rivollier, A., Olsson, S., Santoro, A., Speziani, C., Azocar, O., Flacher, M., Djebali, S., Tebib, J., Brytting, M., Egeler, R. M., Rabourdin-Combe, C., Henter, J. I., Arico, M. and Delprat, C. 2008. Langerhans cell histiocytosis reveals a new IL-17A-dependent pathway of dendritic cell fusion. Nat Med 14:81–87.

Coussens, L. M. and Werb, Z. 2002. Inflammation and cancer. Nature 420:860–867.

Curiel, T. J., Cheng, P., Mottram, P., Alvarez, X., Moons, L., Evdemon-Hogan, M., Wei, S., Zou, L., Kryczek, I., Hoyle, G., Lackner, A., Carmeliet, P. and Zou, W. 2004. Dendritic cell subsets differentially regulate angiogenesis in human ovarian cancer. Cancer Res 64:5535–5538.

D'Amico, G., Bianchi, G., Bernasconi, S., Bersani, L., Piemonti, L., Sozzani, S., Mantovani, A. and Allavena, P. 1998. Adhesion, transendothelial migration, and reverse transmigration of in vitro cultured dendritic cells. Blood 92:207–214.

da Costa, C. E., Annels, N. E., Faaij, C. M., Forsyth, R. G., Hogendoorn, P. C. and Egeler, R. M. 2005. Presence of osteoclast-like multinucleated giant cells in the bone and nonostotic lesions of Langerhans cell histiocytosis. J Exp Med 201:687–693.

de Visser, K. E., Eichten, A. and Coussens, L. M. 2006. Paradoxical roles of the immune system during cancer development. Nat Rev Cancer 6:24–37.

Delamarre, L., Pack, M., Chang, H., Mellman, I. and Trombetta, E. S. 2005. Differential lysosomal proteolysis in antigen-presenting cells determines antigen fate. Science 307:1630–1634.

Della Bella, S., Gennaro, M., Vaccari, M., Ferraris, C., Nicola, S., Riva, A., Clerici, M., Greco, M. and Villa, M. L. 2003. Altered maturation of peripheral blood dendritic cells in patients with breast cancer. Br J Cancer 89:1463–1472.

Dembic, Z., Schenck, K. and Bogen, B. 2000. Dendritic cells purified from myeloma are primed with tumor-specific antigen (idiotype) and activate CD4+ T cells. Proc Natl Acad Sci U S A 97:2697–2702.

Dunn, G. P., Bruce, A. T., Ikeda, H., Old, L. J. and Schreiber, R. D. 2002. Cancer immunoediting: from immunosurveillance to tumor escape. Nat Immunol 3:991–998.

Enk, A. H., Jonuleit, H., Saloga, J. and Knop, J. 1997. Dendritic cells as mediators of tumor-induced tolerance in metastatic melanoma. Int J Cancer 73:309–316.

Fernandez Pujol, B., Lucibello, F. C., Zuzarte, M., Lutjens, P., Muller, R. and Havemann, K. 2001. Dendritic cells derived from peripheral monocytes express endothelial markers and in the presence of angiogenic growth factors differentiate into endothelial-like cells. Eur J Cell Biol 80:99–110.

Ferrara, N. and Kerbel, R. S. 2005. Angiogenesis as a therapeutic target. Nature 438:967–974.

Finn, O. J., Jerome, K. R., Henderson, R. A., Pecher, G., Domenech, N., Magarian-Blander, J. and Barratt-Boyes, S. M. 1995. MUC-1 epithelial tumor mucin-based immunity and cancer vaccines. Immunol Rev 145:61–89.

Fujii, S., Shimizu, K., Kronenberg, M. and Steinman, R. M. 2002. Prolonged IFN-gamma-producing NKT response induced with alpha-galactosylceramide-loaded DCs. Nat Immunol 3:867–874.

Gabrilovich, D. I., Chen, H. L., Girgis, K. R., Cunningham, H. T., Meny, G. M., Nadaf, S., Kavanaugh, D. and Carbone, D. P. 1996. Production of vascular endothelial growth factor by human tumors inhibits the functional maturation of dendritic cells. Nat Med 2:1096–1103.

Ghiringhelli, F., Puig, P. E., Roux, S., Parcellier, A., Schmitt, E., Solary, E., Kroemer, G., Martin, F., Chauffert, B. and Zitvogel, L. 2005. Tumor cells convert immature myeloid dendritic cells into TGF-beta-secreting cells inducing CD4+CD25+ regulatory T cell proliferation. J Exp Med 202:919–929.

Giuliani, N., Bataille, R., Mancini, C., Lazzaretti, M. and Barille, S. 2001. Myeloma cells induce imbalance in the osteoprotegerin/osteoprotegerin ligand system in the human bone marrow environment. Blood 98:3527–3533.

Gottfried, E., Kreutz, M., Haffner, S., Holler, E., Iacobelli, M., Andreesen, R. and Eissner, G. 2007. Differentiation of human tumour-associated dendritic cells into endothelial-like cells: an alternative pathway of tumour angiogenesis. Scand J Immunol 65:329–335.

Grois, N., Prayer, D., Prosch, H. and Lassmann, H. 2005. Neuropathology of CNS disease in Langerhans cell histiocytosis. Brain 128:829–838.

Han, J. H., Choi, S. J., Kurihara, N., Koide, M., Oba, Y. and Roodman, G. D. 2001. Macrophage inflammatory protein-1alpha is an osteoclastogenic factor in myeloma that is independent of receptor activator of nuclear factor kappaB ligand. Blood 97:3349–3353.

Han, X., Sterling, H., Chen, Y., Saginario, C., Brown, E. J., Frazier, W. A., Lindberg, F. P. and Vignery, A. 2000. CD47, a ligand for the macrophage fusion receptor, participates in macrophage multinucleation. J Biol Chem 275:37984–37992.

Hanahan, D. and Weinberg, R. A. 2000. The hallmarks of cancer. Cell 100:57–70.

Hayashi, T., Hideshima, T., Akiyama, M., Raje, N., Richardson, P., Chauhan, D. and Anderson, K. C. 2003. Ex vivo induction of multiple myeloma-specific cytotoxic T lymphocytes. Blood 102:1435–1442.

Heath, W. R., Belz, G. T., Behrens, G. M., Smith, C. M., Forehan, S. P., Parish, I. A., Davey, G. M., Wilson, N. S., Carbone, F. R. and Villadangos, J. A. 2004. Cross-presentation, dendritic cell subsets, and the generation of immunity to cellular antigens. Immunol Rev 199:9–26.

Hiltbold, E. M., Vlad, A. M., Ciborowski, P., Watkins, S. C. and Finn, O. J. 2000. The mechanism of unresponsiveness to circulating tumor antigen MUC1 is a block in intracellular sorting and processing by dendritic cells. J Immunol 165:3730–3741.

Hollender, P., Ittelett, D., Villard, F., Eymard, J. C., Jeannesson, P. and Bernard, J. 2002. Active matrix metalloprotease-9 in and migration pattern of dendritic cells matured in clinical grade culture conditions. Immunobiology 206:441–458.

Iyengar, P., Combs, T. P., Shah, S. J., Gouon-Evans, V., Pollard, J. W., Albanese, C., Flanagan, L., Tenniswood, M. P., Guha, C., Lisanti, M. P., Pestell, R. G. and Scherer, P. E. 2003. Adipocyte-secreted factors synergistically promote mammary tumorigenesis through induction of anti-apoptotic transcriptional programs and proto-oncogene stabilization. Oncogene 22:6408–6423.

Klimp, A. H., Hollema, H., Kempinga, C., van der Zee, A. G., de Vries, E. G. and Daemen, T. 2001. Expression of cyclooxygenase-2 and inducible nitric oxide synthase in human ovarian tumors and tumor-associated macrophages. Cancer Res 61:7305–7309.

Koga, T., Inui, M., Inoue, K., Kim, S., Suematsu, A., Kobayashi, E., Iwata, T., Ohnishi, H., Matozaki, T., Kodama, T., Taniguchi, T., Takayanagi, H. and Takai, T. 2004.

Costimulatory signals mediated by the ITAM motif cooperate with RANKL for bone homeostasis. Nature 428:758–763.

Koide, N., Nishio, A., Sato, T., Sugiyama, A. and Miyagawa, S. 2004. Significance of macrophage chemoattractant protein-1 expression and macrophage infiltration in squamous cell carcinoma of the esophagus. Am J Gastroenterol 99:1667–1674.

Kouwenhoven, M., Ozenci, V., Tjernlund, A., Pashenkov, M., Homman, M., Press, R. and Link, H. 2002. Monocyte-derived dendritic cells express and secrete matrix-degrading metalloproteinases and their inhibitors and are imbalanced in multiple sclerosis. J Neuroimmunol 126:161–171.

Krtolica, A., Parrinello, S., Lockett, S., Desprez, P. Y. and Campisi, J. 2001. Senescent fibroblasts promote epithelial cell growth and tumorigenesis: a link between cancer and aging. Proc Natl Acad Sci U S A 98:12072–12077.

Kukita, T., Wada, N., Kukita, A., Kakimoto, T., Sandra, F., Toh, K., Nagata, K., Iijima, T., Horiuchi, M., Matsusaki, H., Hieshima, K., Yoshie, O. and Nomiyama, H. 2004. RANKL-induced DC-STAMP is essential for osteoclastogenesis. J Exp Med 200:941–946.

Kukreja, A., Hutchinson, A., Dhodapkar, K., Mazumder, A., Vesole, D., Angitapalli, R., Jagannath, S. and Dhodapkar, M. V. 2006. Enhancement of clonogenicity of human multiple myeloma by dendritic cells. J Exp Med 203:1859–1865.

Kukreja, A., Hutchinson, A., Mazumder, A., Vesole, D., Angitapalli, R., Jagannath, S., O'Connor O. A. and Dhodapkar, M. V. 2007. Bortezomib disrupts tumour-dendritic cell interactions in myeloma and lymphoma: therapeutic implications. Br J Haematol 136:106–110.

Laxmanan, S., Robertson, S. W., Wang, E., Lau, J. S., Briscoe, D. M. and Mukhopadhyay, D. 2005. Vascular endothelial growth factor impairs the functional ability of dendritic cells through Id pathways. Biochem Biophys Res Commun 334:193–198.

Lewis, C. E., Leek, R., Harris, A. and McGee, J. O. 1995. Cytokine regulation of angiogenesis in breast cancer: the role of tumor-associated macrophages. J Leukoc Biol 57:747–751.

Lin, E. Y. and Pollard, J. W. 2004. Role of infiltrated leucocytes in tumour growth and spread. Br J Cancer 90:2053–2058.

Lissbrant, I. F., Stattin, P., Wikstrom, P., Damber, J. E., Egevad, L. and Bergh, A. 2000. Tumor associated macrophages in human prostate cancer: relation to clinicopathological variables and survival. Int J Oncol 17:445–451.

Lissoni, P., Malugani, F., Bonfanti, A., Bucovec, R., Secondino, S., Brivio, F., Ferrari-Bravo, A., Ferrante, R., Vigore, L., Rovelli, F., Mandala, M., Viviani, S., Fumagalli, L. and Gardani, G. S. 2001. Abnormally enhanced blood concentrations of vascular endothelial growth factor (VEGF) in metastatic cancer patients and their relation to circulating dendritic cells, IL-12 and endothelin-1. J Biol Regul Homeost Agents 15:140–144.

Lucas, M., Schachterle, W., Oberle, K., Aichele, P. and Diefenbach, A. 2007. Dendritic cells prime natural killer cells by trans-presenting interleukin 15. Immunity 26:503–517.

MacLennan, I. and Vinuesa, C. 2002. Dendritic cells, BAFF, and APRIL: innate players in adaptive antibody responses. Immunity 17:235–238.

Menetrier-Caux, C., Montmain, G., Dieu, M. C., Bain, C., Favrot, M. C., Caux, C. and Blay, J. Y. 1998. Inhibition of the differentiation of dendritic cells from CD34(+) progenitors by tumor cells: role of interleukin-6 and macrophage colony-stimulating factor. Blood 92:4778–4791.

Moreaux, J., Cremer, F. W., Reme, T., Raab, M., Mahtouk, K., Kaukel, P., Pantesco, V., De Vos, J., Jourdan, E., Jauch, A., Legouffe, E., Moos, M., Fiol, G., Goldschmidt, H., Rossi, J. F., Hose, D. and Klein, B. 2005. The level of TACI gene expression in myeloma cells is associated with a signature of microenvironment dependence versus a plasmablastic signature. Blood 106:1021–1030.

Moreaux, J., Legouffe, E., Jourdan, E., Quittet, P., Reme, T., Lugagne, C., Moine, P., Rossi, J. F., Klein, B. and Tarte, K. 2004. BAFF and APRIL protect myeloma cells from apoptosis induced by interleukin 6 deprivation and dexamethasone. Blood 103:3148–3157.

Nestle, F. O., Burg, G., Fah, J., Wrone-Smith, T. and Nickoloff, B. J. 1997. Human sunlight-induced basal-cell-carcinoma-associated dendritic cells are deficient in T cell co-stimulatory molecules and are impaired as antigen-presenting cells. Am J Pathol 150:641–651.

Novak, A. J., Darce, J. R., Arendt, B. K., Harder, B., Henderson, K., Kindsvogel, W., Gross, J. A., Greipp, P. R. and Jelinek, D. F. 2004. Expression of BCMA, TACI, and BAFF-R in multiple myeloma: a mechanism for growth and survival. Blood 103:689–694.

Ohm, J. E. and Carbone, D. P. 2001. VEGF as a mediator of tumor-associated immunodeficiency. Immunol Res 23:263–272.

Osman, M., Tortorella, M., Londei, M. and Quaratino, S. 2002. Expression of matrix metalloproteinases and tissue inhibitors of metalloproteinases define the migratory characteristics of human monocyte-derived dendritic cells. Immunology 105:73–82.

Park, S. J., Nakagawa, T., Kitamura, H., Atsumi, T., Kamon, H., Sawa, S., Kamimura, D., Ueda, N., Iwakura, Y., Ishihara, K., Murakami, M. and Hirano, T. 2004. IL-6 regulates in vivo dendritic cell differentiation through STAT3 activation. J Immunol 173:3844–3854.

Pearse, R. N., Sordillo, E. M., Yaccoby, S., Wong, B. R., Liau, D. F., Colman, N., Michaeli, J., Epstein, J. and Choi, Y. 2001. Multiple myeloma disrupts the TRANCE/ osteoprotegerin cytokine axis to trigger bone destruction and promote tumor progression. Proc Natl Acad Sci U S A 98:11581–11586.

Pendl, G. G., Robert, C., Steinert, M., Thanos, R., Eytner, R., Borges, E., Wild, M. K., Lowe, J. B., Fuhlbrigge, R. C., Kupper, T. S., Vestweber, D. and Grabbe, S. 2002. Immature mouse dendritic cells enter inflamed tissue, a process that requires E- and P-selectin, but not P-selectin glycoprotein ligand 1. Blood 99:946–956.

Ratta, M., Fagnoni, F., Curti, A., Vescovini, R., Sansoni, P., Oliviero, B., Fogli, M., Ferri, E., Della Cuna, G. R., Tura, S., Baccarani, M. and Lemoli, R. M. 2002. Dendritic cells are functionally defective in multiple myeloma: the role of interleukin-6. Blood 100:230–237.

Rettig, M. B., Ma, H. J., Vescio, R. A., Pold, M., Schiller, G., Belson, D., Savage, A., Nishikubo, C., Wu, C., Fraser, J., Said, J. W. and Berenson, J. R. 1997. Kaposi's sarcoma-associated herpesvirus infection of bone marrow dendritic cells from multiple myeloma patients. Science 276:1851–1854.

Richardson, P. G. and Anderson, K. C. 2003. Bortezomib: a novel therapy approved for multiple myeloma. Clin Adv Hematol Oncol 1:596–600.

Richardson, P. G., Schlossman, R., Hideshima, T. and Anderson, K. C. 2005. New treatments for multiple myeloma. Oncology (Williston Park) 19:1781–1792; discussion 1792, 1795–1797.

Rivollier, A., Mazzorana, M., Tebib, J., Piperno, M., Aitsiselmi, T., Rabourdin-Combe, C., Jurdic, P. and Servet-Delprat, C. 2004. Immature dendritic cell transdifferentiation into osteoclasts: a novel pathway sustained by the rheumatoid arthritis microenvironment. Blood 104:4029–4037.

Said, J. W., Rettig, M. R., Heppner, K., Vescio, R. A., Schiller, G., Ma, H. J., Belson, D., Savage, A., Shintaku, I. P., Koeffler, H. P., Asou, H., Pinkus, G., Pinkus, J., Schrage, M., Green, E. and Berenson, J. R. 1997. Localization of Kaposi's sarcoma-associated herpesvirus in bone marrow biopsy samples from patients with multiple myeloma. Blood 90:4278–4282.

Saito, H., Tsujitani, S., Ikeguchi, M., Maeta, M. and Kaibara, N. 1998. Relationship between the expression of vascular endothelial growth factor and the density of dendritic cells in gastric adenocarcinoma tissue. Br J Cancer 78:1573–1577.

Sandel, M. H., Dadabayev, A. R., Menon, A. G., Morreau, H., Melief, C. J., Offringa, R., van der Burg, S. H., Janssen-van Rhijn, C. M., Ensink, N. G., Tollenaar, R. A., van de Velde, C. J. and Kuppen, P. J. 2005. Prognostic value of tumor-infiltrating dendritic cells in colorectal cancer: role of maturation status and intratumoral localization. Clin Cancer Res 11:2576–2582.

Sica, A., Allavena, P. and Mantovani, A. 2008. Cancer related inflammation: The macrophage connection. Cancer Lett 267:204–215.

Speziani, C., Rivollier, A., Gallois, A., Coury, F., Mazzorana, M., Azocar, O., Flacher, M., Bella, C., Tebib, J., Jurdic, P., Rabourdin-Combe, C. and Delprat, C. 2007. Murine dendritic cell transdifferentiation into osteoclasts is differentially regulated by innate and adaptive cytokines. Eur J Immunol 37:747–757.

Steinbrink, K., Graulich, E., Kubsch, S., Knop, J. and Enk, A. H. 2002. CD4(+) and CD8(+) anergic T cells induced by interleukin-10-treated human dendritic cells display antigen-specific suppressor activity. Blood 99:2468–2476.

Steinbrink, K., Jonuleit, H., Muller, G., Schuler, G., Knop, J. and Enk, A. H. 1999. Interleukin-10-treated human dendritic cells induce a melanoma-antigen-specific anergy in CD8(+) T cells resulting in a failure to lyse tumor cells. Blood 93:1634–1642.

Steinbrink, K., Wolfl, M., Jonuleit, H., Knop, J. and Enk, A. H. 1997. Induction of tolerance by IL-10-treated dendritic cells. J Immunol 159:4772–4780.

Steinman, R. M., Hawiger, D. and Nussenzweig, M. C. 2003. Tolerogenic dendritic cells. Annu Rev Immunol 21:685–711.

Sunderkotter, C., Goebeler, M., Schulze-Osthoff, K., Bhardwaj, R. and Sorg, C. 1991. Macrophage-derived angiogenesis factors. Pharmacol Ther 51:195–216.

Takahashi, A., Kono, K., Ichihara, F., Sugai, H., Fujii, H. and Matsumoto, Y. 2004. Vascular endothelial growth factor inhibits maturation of dendritic cells induced by lipopolysaccharide, but not by proinflammatory cytokines. Cancer Immunol Immunother 53:543–550.

Teitelbaum, S. L. 2000. Bone resorption by osteoclasts. Science 289:1504–1508.

Teitelbaum, S. L. and Ross, F. P. 2003. Genetic regulation of osteoclast development and function. Nat Rev Genet 4:638–649.

Trombetta, E. S. and Mellman, I. 2005. Cell biology of antigen processing in vitro and in vivo. Annu Rev Immunol 23:975–1028.

Tsutsui, S., Yasuda, K., Suzuki, K., Tahara, K., Higashi, H. and Era, S. 2005. Macrophage infiltration and its prognostic implications in breast cancer: the relationship with VEGF expression and microvessel density. Oncol Rep 14:425–431.

Uy, G. L., Trivedi, R., Peles, S., Fisher, N. M., Zhang, Q. J., Tomasson, M. H., DiPersio, J. F. and Vij, R. 2007. Bortezomib inhibits osteoclast activity in patients with multiple myeloma. Clin Lymphoma Myeloma 7:587–589.

Vermaelen, K. Y., Cataldo, D., Tournoy, K., Maes, T., Dhulst, A., Louis, R., Foidart, J. M., Noel, A. and Pauwels, R. 2003. Matrix metalloproteinase-9-mediated dendritic cell recruitment into the airways is a critical step in a mouse model of asthma. J Immunol 171:1016–1022.

Vicari, A. P., Treilleux, I. and Lebecque, S. 2004. Regulation of the trafficking of tumour-infiltrating dendritic cells by chemokines. Semin Cancer Biol 14:161–169.

Yagi, M., Miyamoto, T., Sawatani, Y., Iwamoto, K., Hosogane, N., Fujita, N., Morita, K., Ninomiya, K., Suzuki, T., Miyamoto, K., Oike, Y., Takeya, M., Toyama, Y. and Suda, T. 2005. DC-STAMP is essential for cell-cell fusion in osteoclasts and foreign body giant cells. J Exp Med 202:345–351.

Yang, A. S. and Lattime, E. C. 2003. Tumor-induced interleukin 10 suppresses the ability of splenic dendritic cells to stimulate CD4 and CD8 T-cell responses. Cancer Res 63:2150–2157.

Chapter 18
Dendritic Cell Maturation Versus Polarization in Tumor Escape

Michael W. Lipscomb, Walter J. Storkus, and Amy K. Wesa

Abstract Dendritic cells serve as key immunosurveillance agents throughout the body and orchestrate the coordinate innate and adaptive immune responses to antigenically complex cells and organisms that challenge the host. The ability of dendritic cells to promote beneficial versus irrelevant or even, counterproductive, immunity in the cancer setting depends to a large degree on the operational parameters displayed by the heterogeneous population of dendritic cells found in the tumor microenvironment. This chapter will discuss how tumors manipulate the state of maturation and type of functional polarization displayed by dendritic cells in order to affect immune escape.

18.1 Introduction

As described elsewhere in this volume, the tumor microenvironment (TME) has been characterized as an immunosuppressive one that impairs both innate and adaptive arms of the immune system (Kim et al. 2006b; Rabinovich et al. 2007; Yu et al. 2007). Given the ability of humoral- and T-cell-mediated immunity to promote tumor regression under optimal conditions, it is perhaps not surprising that the progressive TME has developed a myriad of immune-escape strategies (Yang and Carbone 2004; Kim et al. 2006b). Many of these are directed at host dendritic cells (DC) that operate at the nexus between innate and adaptive immunity, hence subversion of DC numbers/function by tumors have "ripple" effects on a diverse array of "downstream" host-protective mechanisms (Kim et al. 2006a, b; Shurin et al. 2006). As a result of regulating normal DC operations, tumors may indirectly subvert potent antitumor effector circuits including NK cells, T cells and B cells (antibodies) (Yang and Carbone 2004; Pinzon-Charry et al. 2005; Fricke and Gabrilovich 2006; Kim et al. 2006a; Shurin et al. 2006; Rabinovich et al. 2007). In this chapter, we will discuss

A.K. Wesa (✉)
Department of Dermatology, University of Pittsburgh Medical Center, W1045
Biomedical Sciences Tower, 200 Lothrop Street, Pittsburgh, PA 15213, USA
e-mail: wesaak@upmc.edu

M.R. Shurin, R.D. Salter (eds.), *Dendritic Cells in Cancer*,
DOI 10.1007/978-0-387-88611-4_18, © Springer Science+Business Media, LLC 2009

how the TME affects DC maturation and functional polarization and, consequently, how such alterations serve to shape evolving immune responses under chronic disease conditions, leading to tumor escape.

18.2 Tumor Manipulation of Dendritic Cell Biology

DC are a heterogeneous population of antigen-presenting cells (APC) that derive from hematopoietic progenitor cells in the bone marrow (Shortman and Naik 2007). DC seed the various barrier tissues of the body, where they serve as sentinels against pathogenic insult by constant surveillance (or sampling) of their microenvironment using a range of cytokine, chemokine and pattern-recognition receptors. DC exhibit functional plasticity, allowing them to rapidly respond to environmental cues (i.e., "danger signals", etc.) and transmit conditional signals dictating whether antigen-specific immunity or tolerance is promoted in tissue/tumor-draining lymph nodes (TDLN). An expanding range of functional DC subsets, and within these classifications, diverse states of DC differentiation exist in vivo (Shortman and Naik 2007). Notably, these distinct DC subsets display variable operational characteristics that are predictive of the functional nature of the adaptive immune response fostered by these cells. Hence, in a complex tissue microenvironment, the state of DC maturation, as well as the composition of DC subsets encountering antigen, has a cumulative and evolving bearing on the resultant magnitude and quality of corollary immune responses (Kapsenberg 2003; de Jong et al. 2005).

In this regard, tumors can regulate the state of DC maturation, as well as the type and differentiation of progenitor DC recruited into their microenvironment via elaborated soluble mediators (Fricke and Gabrilovich 2006; Shurin et al. 2006). It is clearly to the advantage of the tumor to recruit immature DC (iDC) or "tolerogenic" (or regulatory) DC (as well as myeloid suppressor cells, etc.) that yield immune tolerance/hyporesponsiveness (Yang and Carbone 2004; Condeelis and Pollard 2006; Kim et al. 2006a; Marigo et al. 2008) or to subvert normal DC development (Shurin and Gabrilovich 2001; Zou et al. 2001; Gabrilovich 2004; Nefedova et al. 2004; Wojas et al. 2004). Immune tolerance may be further refined upon consideration of DC-mediated mechanisms leading to the apoptotic deletion or functional restraint of tumor-specific T cells versus the functional suppression of T effector cells instigated by T regulatory cells (Treg) (Curiel 2007).

18.3 "Entry" Level Deviation of Dendritic Cells by Tumors

The TME is awash with a diverse array of factors that drive tumor proliferation and the establishment of a supportive stromal compartment capable of nourishing the progressive lesion. These factors may be produced directly by tumor cells themselves or by stromal cells under the influence of tumor-associated stimuli or the hypoxic conditions (i.e., hypoxia-induced factor (HIF)-associated) that are often observed within tumors (Table 18.1).

Growing evidence suggests a shift in the composition of tumor-associated DC subsets over time in the progressor TME, resulting in an overall decrease in DC numbers and accumulation of iDC and/or tolerogenic DC over functionally competent-mature DC (Yang and Carbone 2004; Kim et al. 2006a, b). Not surprisingly, the accumulation of iDC in the cancer setting has been associated with weak antitumor T-cell-mediated immunity, enhanced Treg frequencies/function and poor patient prognosis (Nagorsen et al. 2007; Stoitzner et al. 2008).

Clear compartmentalization of DC occurs within the TME of a broad range of cancer histologies, with accumulation of iDC, tolerogenic DC and tumor-associated macrophages (TAM) in the tumor core (most influenced by the hypoxia), whereas more mature DC are found at the normoxic, peripheral areas of the lesion (Fig. 18.1) (Troy et al. 1998; Bell et al. 1999; Zou et al. 2001; Vermi et al. 2003; Mantovani et al. 2006; Gulubova et al. 2008; Stoitzner et al. 2008). Interestingly, when they are found, perilesional DC appear to be composed of both myeloid (myDC) and plasmacytoid (pDC) subsets of DC and these cells exhibit a "semi-mature" phenotype characterized by the ability to cluster with infiltrating T cells and produce IL-12 and IFN-α, respectively,

Table 18.1 Tumor-associated "factors" impacting DC development, recruitment, maturation, and functional polarization

Factors	Sources	Impact on DC	References
ATP*	TME	Promotes tolerogenic DC	Wilkin et al. (2002)
Exosomes	Tumor cells	Prevent DC development Induction of MDSC	Iero et al. (2008)
Gangliosides	Tumor cells	Inhibit DC maturation	Shurin et al. (2001)
IL-6*	Tumor cells	Inhibits DC maturation/survival	Hegde et al. (2004)
IL-8*	TME	Inhibits DC migration in TME	Feijoo et al. (2005)
IL-10*	TME	Promotes tolerogenic DC	Steinbrink et al. (1999)
M-CSF*	Tumor cells	Inhibits DC maturation	Li et al. (2007)
MCP-1*	Tumor cells	Recruitment of MDSC, TAM	Huang et al. (2007)
MIP-3α	TME	Recruitment of immature DC	Caux et al. (2000)
MUC1	Tumor cells	Promotes tolerogenic DC	Carlos et al. (2005)
SDF-1*	Tumor cells	Recruitment of pDC, iDC	Zou et al. (2001)
TGF-β*	TME	Promotes tolerogenic DC	Rutella et al. (2006)
TSP-1*	Tumor cells	Inhibits DC maturation/survival	Doyen et al. (2003)
VEGF*	Tumor cells	Prevents DC development Promotes pro-angiogenic DC	Gabrilovich (2004) Murdoch et al. (2008)

ATP, adenosine triphosphate; iDC, immature DC; IL, interleukin; M-CSF, macrophage colony stimulating factor; MDSC, monocyte-derived suppressor cell; MIP, macrophage inhibitory protein; MUC, mucin; pDC, plasmacytoid DC; SDF-1, stromal cell differentiation factor-1; TAM, tumor-associated macrophage; TGF, tumor growth factor; TME, tumor microenvironment; TSP-1, thrombospondin-1; VEGF, vascular endothelial growth factor.
*Upregulated under hypoxic conditions.

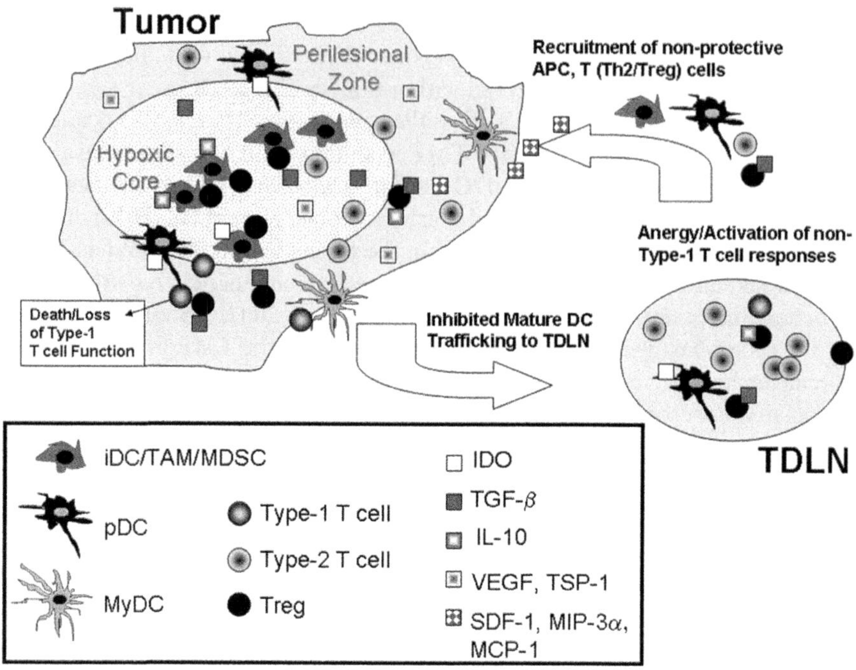

Fig. 18.1 Dendritic cells in the tumor microenvironment (TME). Factors produced within the TME promote the recruitment and regional localization of iDC, pDC and myDC within the tumor lesion. iDC are distributed throughout the tumor stroma and are the predominant form of DC in the hypoxic tumor core. Semi-mature pDC and myDC are typically seen at the periphery of the lesion, where they may interact with infiltrating T cells and promote effector silencing or Treg activity. While many "mature" DC may be retained in the tumor based on factors that limit DC expression of CCR7 or which serve as DC migration inhibitors, those DC that make their way to the tumor-draining lymph node (TDLN) typically promote non-Type-1 T-cell responses that are linked with poor prognosis. If full DC maturation does occur in the TME, migrating DC may effectively promote Type-1 T-cell responses within the TDLN that may mediate protective/regressor activity upon the migration of these effector T cells into the TME. However, if the suppressive environment of the TME is not therapeutically antagonized, these Type-1 cells may become functionally tolerized or deleted

albeit at modest and variable levels (Suzuki et al. 2002; Vermi et al. 2003; Colonna et al. 2004; Bergeron et al. 2006; Perrot et al. 2007). Nevertheless, the degree of perilesional infiltration by more mature DC-LAMP+ DC has been positively correlated with specific antitumor T-cell responses in the sentinel lymph node and with the lack of tumor metastases in human melanoma patients (Movassagh et al. 2004).

Tumor-derived cytokines and chemokines modulate the preferential recruitment of iDC and/or tolerogenic/regulatory DC into the TME (Bell et al. 1999). Vascular endothelial growth factor (VEGF) is over-expressed in a large proportion of tumor cell types and serves not only to support tumor angiogenesis but also to alter development of iDC from hematopoietic progenitor cells within the TME (Dikov et al. 2005). Levels of VEGF (and TGF-β) produced within tumor lesions have been linked to infiltration by iDC, but found to be inversely correlated with total numbers of CD83+ mature tumor-infiltrating DC (Takahashi et al. 2002; Iwamoto et al. 2003). Interestingly, beyond serving as hypostimulatory or tolerogenic APC to T cells, VEGF-conditioned iDC may function as stromal pericyte-like cells in driving tumor neovasculogenesis/neoangiogenesis (Fainaru et al. 2008; Murdoch et al. 2008). Additionally, tumor-shed microbodies (also known as exosomes) have been reported to blunt the development of DC from precursor cells (Valenti et al. 2007; Iero et al. 2008).

Tumor cells also produce the chemokine stromal-derived factor-1 (SDF-1; CXCL12) associated with the recruitment of CXCR4+ DC subtypes (pDC, but not myDC precursors) involved in immune tolerance and the blockade of infiltrating DC maturation/function (Zou et al. 2001). Additional TME-associated chemokines include MCP-1 (CCL2) and MIP-3α (CCL20), known to recruit myeloid-derived suppressor cells (MDSC) and TAM versus iDC, respectively (Huang et al. 2007).

18.4 Instruction Lost

Unfortunately, even after recruitment, iDC are influenced by additional TME-associated factors that prevent adaptive antitumor immune responses by blocking maturation and migration into TDLN. Typically, these agents inhibit nuclear translocation of NF-κB, a key step in DC maturation, which might normally occur via locoregional stimulation with pro-inflammatory cytokines or triggering of toll-like receptor (TLR) ligands (Hegde et al. 2004; Dikov et al. 2005). Classic examples of DC maturational inhibitors include gangliosides, IL-6, macrophage colony stimulating factor (M-CSF) and thromospodin-1 (TSP-1) (Table 18.1). In addition to their ability to prevent DC maturation, these factors may also sensitize iDC to apoptosis (Table 18.1) (Esche et al. 1999; Pirtskhalaishvili et al. 2000). Under such conditions, tumor-infiltrating iDC are non-durable APC for tumor-infiltrating T cells only, representing an ineffective "adjuvant" for the priming of adaptive immune responses. Interestingly, even if a cohort of infiltrating DC was able to take up tumor antigens and to become activated and transport-competent (i.e., via upregulation of lymph node homing receptors such as CCR7), they may fail to ever leave the TME. As recently reported for intratumoral administration of mature (gene-modified) DC in a phase I/II clinical trial, the vast majority of injected DC remain locked within the tumor lesion as a result of the influence of tumor-associated IL-8 that blocks their emigration (Feijoo et al. 2005).

18.5 Tumor Microenvironment-Associated Deviation of Dendritic Cell/T-Cell Functional Polarization

The fate of naïve T cells is determined by three critical signals provided by DC: cognate antigen in MHC complexes (signal 1), costimulatory/coinhibitory molecule interactions (signal 2) and polarizing cytokines/chemokines (signal 3). DC cytokines that promote effective Type-1 immunity (i.e., IFN-α, TNF-α, IL-12 family members and IL-15) (de Jong et al. 2005) are often limited by TME influences that restrict DC maturation, or, in some cases, enforce STAT3 or SOCS1 expression in DC (Gabrilovich 2004; Hanada et al. 2005; Evel-Kabler et al. 2006; Li et al. 2006).

Of the minor population of mature DC typically imaged at the tumor periphery or the infrequently matured DC originating from the tumor core, other TME-associated cytokines may affect the DC polarization state, and hence the type of adaptive immune response that may be prompted when priming T cells in the TDLN. Factors that are elevated in the TME and promote tolerogenic DC include ATP (via interaction with adenosine receptors expressed by DC), IL-10, MUC1, TGF-β, among others. In contrast to the other soluble polarizing agents, MUC1, a transmembrane mucin family member overexpressed in 90% of breast carcinomas, appears to mediate immunosuppression via direct tumor/DC contact, yielding IL-10 secreting DC (Carlos et al. 2005; Fricke and Gabrilovich 2006).

Indeed, regulatory/tolerogenic DC characteristically express low levels of T-cell "costimulatory" molecules, such as CD40, CD80 (B7.1), CD86 (B7.2), and high levels of "coinhibitory" molecules, such as B7-H1, B7-DC and B7-H4 (Chaux et al. 1997; Tirapu et al. 2006). These cells are also typically poor producers of Type-1-biasing cytokines (i.e., IL-12), strong producers of IL-10, and may express the tryptophan-catabolizing enzyme Indoleamine 2,3-dioxygenase (IDO) that might suppress T-cell responses (Mellor et al. 2003; Munn et al. 2004; Marteau et al. 2005; Popov and Schultze 2008). pDC within the tumor periphery and TDLN constitutively express IDO and represent tolerogenic DC (Munn et al. 2004). Overall, such conditioned DC are less able to promote protective Type-1 T-cell-mediated immunity and are far more likely to elicit Treg-type responses that benefit tumor escape and progression (Munn et al. 2004; Fricke and Gabrilovich 2006; Li et al. 2007; Sharma et al. 2007). The presence of Treg cells infiltrating the tumor tissue and surrounding peritumoral and TDLN has been shown to be directly involved in suppression of antitumor T effector cell responses and indirectly involved in modulating DC toward a tolerogenic fate; i.e., establishing a self-reinforcing DC → Treg → DC feedback loop (Curiel 2007; Larmonier et al. 2007; Sharma et al. 2007; Terme et al. 2008).

Tumor cells can regulate their own cytokine production and that of DC via modulating levels of suppressor of cytokine signaling (SOCS) proteins, STAT transcription factors, and/or regulating protein inhibitors of activated STAT (Jackson et al. 2004; Evel-Kabler et al. 2006; Yu et al. 2007). In particular, expression of SOCS1 in DC limits IL-12p70 production and hinders antitumor immune responses by CD8+ T cells, effectively ablating Type-1 immunity in cancer settings (Hanada et al. 2005; Evel-Kabler et al. 2006). Conversely, in models deficient in SOCS1 expression, inflammatory diseases were observed in association with unabated Type-I interferon (IFN-α/β) and IFN-γ production. The TME may also induce constitutive STAT3 activation in DC (Nefedova et al. 2004), thereby blocking DC maturation (Jackson et al. 2004) and upregulating DC expression of functional IDO, yielding tolerogenic DC (Kortylewski et al. 2005).

Cumulatively, the net balance of iDC and semi-mature pDC and myDC subsets in the TME and TDLN are consistent with the observed Type-2/Treg-biased antitumor T-cell repertoire in progressor cancer patients (Tatsumi et al. 2002, 2003). The tumor itself may directly suppress T cells (Rabinovich et al. 2007); however, most tumors, which lack costimulatory molecules, are not able to evoke primary tumor-specific responses. Since tumor-specific (albeit ineffective) T-cell responses are commonly observed in cancer patients, one must assume that DC have at some point influenced the priming of these effector cells. Indeed, T-cell priming by suppressive DC correlates with the induction of Treg or non-Type-1 responses (Nagorsen et al. 2007). In other patients, Type-1 T cells exhibiting limited functionality are observed, which may suggest that defective priming has occurred or that T cells succumb to the corollary suppressive influences of the TME. This is suggested by the finding that Type-1 antitumor T cells, when they can be found in such patients, typically exhibit a pro-apoptotic phenotype and limited functionality (Wesa et al. unpublished results). In these cases, iDC may only provide partial signals for T-cell responses to develop, resulting in premature T-cell senescence. Such cases are reminiscent of the so-called "helpless" CD8+ T-cell responses, where in the absence of CD40 ligand-mediated maturation of DC (i.e., dependent upon effectively primed Type-1 CD4+ responses), the CD8+ T cells initially proliferate and secrete cytokines, but fail to thrive and are unable to survive after several early cell divisions (Janssen et al. 2003). Such defects have been observed in patients, where tumor-specific effector and memory T cells appear phenotypically/functionally abnormal (Inokuma et al. 2007). Thus, TME-induced DC polarization may directly dictate the fate (i.e., effectively expanded or ultimately aborted) and inclination (i.e., polarization) of antitumor T-cell responses, thereby facilitating tumor escape.

18.6 Conclusions

A dynamic and synergistic network of tumor-derived immunosuppressive factors results in the suppressed maturation and altered functional polarization of DC within the TME and TDLN. Progressive tumor lesions exhibit an accumulation of non-Type-1 polarizing iDC and tolerogenic DC, as well as "semi-mature" pDC and myDC, that actively contribute to the poor induction of protective antitumor T effector cells. These tumor-conditioned DC may also support the generation of Treg cells that mediate peripheral tolerance of the antitumor immune response and establish a self-reinforcing feedback loop by conditioning additional tolerogenic DC. As DC dysfunction in the TME is multi-factorial, at the level of suppressive agents and the mechanisms by which they impact the various DC subsets, therapeutic solutions will invariably be combinational in nature. In this light, the results of recent combinational therapies using surgical debulking, DC-activating drugs and/or Treg antagonists (Vicari et al. 2002; Danna et al. 2004; Piconese et al. 2008) are intriguing and may provide clues for the development of translational approaches to mitigate TME-induced DC defects for optimizing the induction, as well as functionality and survival of therapeutic Type-1 antitumor T cells.

References

Bell, D., Chomarat, P., Broyles, D., Netto, G., Harb, G. M., Lebecque, S., Valladeau, J., Davoust, J., Palucka, K. A. and Banchereau, J. 1999. In breast carcinoma tissue, immature dendritic cells reside within the tumor, whereas mature dendritic cells are located in peritumoral areas. J Exp Med 190:1417–1426.

Bergeron, A., El-Hage, F., Kambouchner, M., Lecossier, D. and Tazi, A. 2006. Characterisation of dendritic cell subsets in lung cancer micro-environments. Eur Respir J 28:1170–1177.

Carlos, C. A., Dong, H. F., Howard, O. M., Oppenheim, J. J., Hanisch, F. G. and Finn, O. J. 2005. Human tumor antigen MUC1 is chemotactic for immature dendritic cells and elicits maturation but does not promote Th1 type immunity. J Immunol 175:1628–1635.

Caux, C., Ait-Yahia, S., Chemin, K., de Bouteiller, O., Dieu-Nosjean, M. C., Homey, B., Massacrier, C., Vanbervliet, B., Zlotnik, A. and Vicari, A. 2000. Dendritic cell biology and regulation of dendritic cell trafficking by chemokines. Springer Semin Immunopathol 22:345–369.

Chaux, P., Favre, N., Martin, M. and Martin, F. 1997. Tumor-infiltrating dendritic cells are defective in their antigen-presenting function and inducible B7 expression in rats. Int J Cancer 72:619–624.

Colonna, M., Trinchieri, G. and Liu, Y. J. 2004. Plasmacytoid dendritic cells in immunity. Nat Immunol 5:1219–1226.

Condeelis, J. and Pollard, J. W. 2006. Macrophages: obligate partners for tumor cell migration, invasion, and metastasis. Cell 124:263–266.

Curiel, T. J. 2007. Tregs and rethinking cancer immunotherapy. J Clin Invest 117:1167–1174.

Danna, E. A., Sinha, P., Gilbert, M., Clements, V. K., Pulaski, B. A. and Ostrand-Rosenberg, S. 2004. Surgical removal of primary tumor reverses tumor-induced immunosuppression despite the presence of metastatic disease. Cancer Res 64:2205–2211.

de Jong, E. C., Smits, H. H. and Kapsenberg, M. L. 2005. Dendritic cell-mediated T cell polarization. Springer Semin Immunopathol 26:289–307.

Dikov, M. M., Ohm, J. E., Ray, N., Tchekneva, E. E., Burlison, J., Moghanaki, D., Nadaf, S. and Carbone, D. P. 2005. Differential roles of vascular endothelial growth factor receptors 1 and 2 in dendritic cell differentiation. J Immunol 174:215–222.

Doyen, V., Rubio, M., Braun, D., Nakajima, T., Abe, J., Saito, H., Delespesse, G. and Sarfati, M. 2003. Thrombospondin 1 is an autocrine negative regulator of human dendritic cell activation. J Exp Med 198:1277–1283.

Esche, C., Lokshin, A., Shurin, G. V., Gastman, B. R., Rabinowich, H., Watkins, S. C., Lotze, M. T. and Shurin, M. R. 1999. Tumor's other immune targets: dendritic cells. J Leukoc Biol 66:336–344.

Evel-Kabler, K., Song, X. T., Aldrich, M., Huang, X. F. and Chen, S. Y. 2006. SOCS1 restricts dendritic cells' ability to break self tolerance and induce antitumor immunity by regulating IL-12 production and signaling. J Clin Invest 116:90–100.

Fainaru, O., Adini, A., Benny, O., Adini, I., Short, S., Bazinet, L., Nakai, K., Pravda, E., Hornstein, M. D., D'Amato, R. J. and Folkman, J. 2008. Dendritic cells support angiogenesis and promote lesion growth in a murine model of endometriosis. Faseb J 22:522–529.

Feijoo, E., Alfaro, C., Mazzolini, G., Serra, P., Penuelas, I., Arina, A., Huarte, E., Tirapu, I., Palencia, B., Murillo, O., Ruiz, J., Sangro, B., Richter, J. A., Prieto, J. and Melero, I. 2005. Dendritic cells delivered inside human carcinomas are sequestered by interleukin-8. Int J Cancer 116:275–281.

Fricke, I. and Gabrilovich, D. I. 2006. Dendritic cells and tumor microenvironment: a dangerous liaison. Immunol Invest 35:459–483.

Gabrilovich, D. 2004. Mechanisms and functional significance of tumour-induced dendritic-cell defects. Nat Rev Immunol 4:941–952.

Gulubova, M., Manolova, I., Cirowski, G., et al. 2008. Recruitment of dendritic cells in human liver with metastases. Clin Exp Metastasis 25:777–785, [Epub ahead of print].

Hanada, T., Tanaka, K., Matsumura, Y., Yamauchi, M., Nishinakamura, H., Aburatani, H., Mashima, R., Kubo, M., Kobayashi, T. and Yoshimura, A. 2005. Induction of hyper Th1 cell-type immune responses by dendritic cells lacking the suppressor of cytokine signaling-1 gene. J Immunol 174:4325–4332.

Hegde, S., Pahne, J. and Smola-Hess, S. 2004. Novel immunosuppressive properties of interleukin-6 in dendritic cells: inhibition of NF-kappaB binding activity and CCR7 expression. Faseb J 18:1439–1441.

Huang, B., Lei, Z., Zhao, J., Gong, W., Liu, J., Chen, Z., Liu, Y., Li, D., Yuan, Y., Zhang, G. M. and Feng, Z. H. 2007. CCL2/CCR2 pathway mediates recruitment of myeloid suppressor cells to cancers. Cancer Lett 252:86–92.

Iero, M., Valenti, R., Huber, V., Filipazzi, P., Parmiani, G., Fais, S. and Rivoltini, L. 2008. Tumour-released exosomes and their implications in cancer immunity. Cell Death Differ 15:80–88.

Inokuma, M., dela Rosa, C., Schmitt, C., Haaland, P., Siebert, J., Petry, D., Tang, M., Suni, M. A., Ghanekar, S. A., Gladding, D., Dunne, J. F., Maino, V. C., Disis, M. L. and Maecker, H. T. 2007. Functional T cell responses to tumor antigens in breast cancer patients have a distinct phenotype and cytokine signature. J Immunol 179:2627–2633.

Iwamoto, M., Shinohara, H., Miyamoto, A., Okuzawa, M., Mabuchi, H., Nohara, T., Gon, G., Toyoda, M. and Tanigawa, N. 2003. Prognostic value of tumor-infiltrating dendritic cells expressing CD83 in human breast carcinomas. Int J Cancer 104:92–97.

Jackson, S. H., Yu, C. R., Mahdi, R. M., Ebong, S. and Egwuagu, C. E. 2004. Dendritic cell maturation requires STAT1 and is under feedback regulation by suppressors of cytokine signaling. J Immunol 172:2307–2315.

Janssen, E. M., Lemmens, E. E., Wolfe, T., Christen, U., von Herrath, M. G. and Schoenberger, S. P. 2003. CD4+ T cells are required for secondary expansion and memory in CD8+ T lymphocytes. Nature 421:852–856.

Kapsenberg, M. L. 2003. Dendritic-cell control of pathogen-driven T-cell polarization. Nat Rev Immunol 3:984–993.

Kim, R., Emi, M. and Tanabe, K. 2006a. Functional roles of immature dendritic cells in impaired immunity of solid tumour and their targeted strategies for provoking tumour immunity. Clin Exp Immunol 146:189–196.

Kim, R., Emi, M., Tanabe, K. and Arihiro, K. 2006b. Tumor-driven evolution of immunosuppressive networks during malignant progression. Cancer Res 66:5527–5536.

Kortylewski, M., Jove, R. and Yu, H. 2005. Targeting STAT3 affects melanoma on multiple fronts. Cancer Metastasis Rev 24:315–327.

Larmonier, N., Marron, M., Zeng, Y., Cantrell, J., Romanoski, A., Sepassi, M., Thompson, S., Chen, X., Andreansky, S. and Katsanis, E. 2007. Tumor-derived CD4(+)CD25(+) regulatory T cell suppression of dendritic cell function involves TGF-beta and IL-10. Cancer Immunol Immunother 56:48–59.

Li, L., Li, S. P., Min, J. and Zheng, L. 2007. Hepatoma cells inhibit the differentiation and maturation of dendritic cells and increase the production of regulatory T cells. Immunol Lett 114:38–45.

Li, Y., Chu, N., Rostami, A. and Zhang, G. X. 2006. Dendritic cells transduced with SOCS-3 exhibit a tolerogenic/DC2 phenotype that directs type 2 Th cell differentiation in vitro and in vivo. J Immunol 177:1679–1688.

Mantovani, A., Schioppa, T., Porta, C., Allavena, P. and Sica, A. 2006. Role of tumor-associated macrophages in tumor progression and invasion. Cancer Metastasis Rev 25:315–322.

Marigo, I., Dolcetti, L., Serafini, P., Zanovello, P. and Bronte, V. 2008. Tumor-induced tolerance and immune suppression by myeloid derived suppressor cells. Immunol Rev 222:162–179.

Marteau, F., Gonzalez, N. S., Communi, D., Goldman, M., Boeynaems, J. M. and Communi, D. 2005. Thrombospondin-1 and indoleamine 2,3-dioxygenase are major targets of extracellular ATP in human dendritic cells. Blood 106:3860–3866.

Mellor, A. L., Baban, B., Chandler, P., Marshall, B., Jhaver, K., Hansen, A., Koni, P. A., Iwashima, M. and Munn, D. H. 2003. Cutting edge: induced indoleamine 2,3 dioxygenase expression in dendritic cell subsets suppresses T cell clonal expansion. J Immunol 171:1652–1655.

Movassagh, M., Spatz, A., Davoust, J., Lebecque, S., Romero, P., Pittet, M., Rimoldi, D., Lienard, D., Gugerli, O., Ferradini, L., Robert, C., Avril, M. F., Zitvogel, L. and Angevin, E. 2004. Selective accumulation of mature DC-Lamp+ dendritic cells in tumor sites is associated with efficient T-cell-mediated antitumor response and control of metastatic dissemination in melanoma. Cancer Res 64:2192–2198.

Munn, D. H., Sharma, M. D., Hou, D., Baban, B., Lee, J. R., Antonia, S. J., Messina, J. L., Chandler, P., Koni, P. A. and Mellor, A. L. 2004. Expression of indoleamine 2,3-dioxygenase by plasmacytoid dendritic cells in tumor-draining lymph nodes. J Clin Invest 114:280–290.

Murdoch, C., Muthana, M., Coffelt, S. B. and Lewis, C. E. 2008. The role of myeloid cells in the promotion of tumour angiogenesis. Nat Rev Cancer 8:618–631.

Nagorsen, D., Voigt, S., Berg, E., Stein, H., Thiel, E. and Loddenkemper, C. 2007. Tumor-infiltrating macrophages and dendritic cells in human colorectal cancer: relation to local regulatory T cells, systemic T-cell response against tumor-associated antigens and survival. J Transl Med 5:62.

Nefedova, Y., Huang, M., Kusmartsev, S., Bhattacharya, R., Cheng, P., Salup, R., Jove, R. and Gabrilovich, D. 2004. Hyperactivation of STAT3 is involved in abnormal differentiation of dendritic cells in cancer. J Immunol 172:464–474.

Perrot, I., Blanchard, D., Freymond, N., Isaac, S., Guibert, B., Pacheco, Y. and Lebecque, S. 2007. Dendritic cells infiltrating human non-small cell lung cancer are blocked at immature stage. J Immunol 178:2763–2769.

Piconese, S., Valzasina, B. and Colombo, M. P. 2008. OX40 triggering blocks suppression by regulatory T cells and facilitates tumor rejection. J Exp Med 205:825–839.

Pinzon-Charry, A., Maxwell, T. and Lopez, J. A. 2005. Dendritic cell dysfunction in cancer: a mechanism for immunosuppression. Immunol Cell Biol 83:451–461.

Pirtskhalaishvili, G., Shurin, G. V., Esche, C., Cai, Q., Salup, R. R., Bykovskaia, S. N., Lotze, M. T. and Shurin, M. R. 2000. Cytokine-mediated protection of human dendritic cells from prostate cancer-induced apoptosis is regulated by the Bcl-2 family of proteins. Br J Cancer 83:506–513.

Popov, A. and Schultze, J. L. 2008. IDO-expressing regulatory dendritic cells in cancer and chronic infection. J Mol Med 86:145–160.
Rabinovich, G. A., Gabrilovich, D. and Sotomayor, E. M. 2007. Immunosuppressive strategies that are mediated by tumor cells. Annu Rev Immunol 25:267–296.
Rutella, S., Danese, S. and Leone, G. 2006. Tolerogenic dendritic cells: cytokine modulation comes of age. Blood 108:1435–1440.
Sharma, M. D., Baban, B., Chandler, P., Hou, D. Y., Singh, N., Yagita, H., Azuma, M., Blazar, B. R., Mellor, A. L. and Munn, D. H. 2007. Plasmacytoid dendritic cells from mouse tumor-draining lymph nodes directly activate mature Tregs via indoleamine 2,3-dioxygenase. J Clin Invest 117:2570–2582.
Shortman, K. and Naik, S. H. 2007. Steady-state and inflammatory dendritic-cell development. Nat Rev Immunol 7:19–30.
Shurin, M. R. and Gabrilovich, D. I. 2001. Regulation of dendritic cell system by tumor. Cancer Res Ther Control 11:65–78.
Shurin, G. V., Shurin, M. R., Bykovskaia, S., Shogan, J., Lotze, M. T. and Barksdale, E. M., Jr. 2001. Neuroblastoma-derived gangliosides inhibit dendritic cell generation and function. Cancer Res 61:363–369.
Shurin, M. R., Shurin, G. V., Lokshin, A., Yurkovetsky, Z. R., Gutkin, D. W., Chatta, G., Zhong, H., Han, B. and Ferris, R. L. 2006. Intratumoral cytokines/chemokines/growth factors and tumor infiltrating dendritic cells: friends or enemies? Cancer Metastasis Rev 25:333–356.
Steinbrink, K., Jonuleit, H., Muller, G., Schuler, G., Knop, J. and Enk, A. H. 1999. Interleukin-10-treated human dendritic cells induce a melanoma-antigen-specific anergy in CD8(+) T cells resulting in a failure to lyse tumor cells. Blood 93:1634–1642.
Stoitzner, P., Green, L. K., Jung, J. Y., Price, K. M., Atarea, H., Kivell, B. and Ronchese, F. 2008. Inefficient presentation of tumor-derived antigen by tumor-infiltrating dendritic cells. Cancer Immunol Immunother 57:1665–1673.
Suzuki, A., Masuda, A., Nagata, H., Kameoka, S., Kikawada, Y., Yamakawa, M. and Kasajima, T. 2002. Mature dendritic cells make clusters with T cells in the invasive margin of colorectal carcinoma. J Pathol 196:37–43.
Takahashi, A., Kono, K., Itakura, J., Amemiya, H., Feng Tang, R., Iizuka, H., Fujii, H. and Matsumoto, Y. 2002. Correlation of vascular endothelial growth factor-C expression with tumor-infiltrating dendritic cells in gastric cancer. Oncology 62:121–127.
Tatsumi, T., Herrem, C. J., Olson, W. C., Finke, J. H., Bukowski, R. M., Kinch, M. S., Ranieri, E. and Storkus, W. J. 2003. Disease stage variation in CD4+ and CD8+ T-cell reactivity to the receptor tyrosine kinase EphA2 in patients with renal cell carcinoma. Cancer Res 63:4481–4489.
Tatsumi, T., Kierstead, L. S., Ranieri, E., Gesualdo, L., Schena, F. P., Finke, J. H., Bukowski, R. M., Mueller-Berghaus, J., Kirkwood, J. M., Kwok, W. W. and Storkus, W. J. 2002. Disease-associated bias in T helper type 1 (Th1)/Th2 CD4(+) T cell responses against MAGE-6 in HLA-DRB10401(+) patients with renal cell carcinoma or melanoma. J Exp Med 196:619–628.
Terme, M., Chaput, N., Combadiere, B., Ma, A., Ohteki, T. and Zitvogel, L. 2008. Regulatory T cells control dendritic cell/NK cell cross-talk in lymph nodes at the steady state by inhibiting CD4+ self-reactive T cells. J Immunol 180:4679–4686.
Tirapu, I., Huarte, E., Guiducci, C., Arina, A., Zaratiegui, M., Murillo, O., Gonzalez, A., Berasain, C., Berraondo, P., Fortes, P., Prieto, J., Colombo, M. P., Chen, L. and Melero, I. 2006. Low surface expression of B7-1 (CD80) is an immunoescape mechanism of colon carcinoma. Cancer Res 66:2442–2450.
Troy, A. J., Summers, K. L., Davidson, P. J., Atkinson, C. H. and Hart, D. N. 1998. Minimal recruitment and activation of dendritic cells within renal cell carcinoma. Clin Cancer Res 4:585–593.

Valenti, R., Huber, V., Iero, M., Filipazzi, P., Parmiani, G. and Rivoltini, L. 2007. Tumor-released microvesicles as vehicles of immunosuppression. Cancer Res 67:2912–2915.
Vermi, W., Bonecchi, R., Facchetti, F., Bianchi, D., Sozzani, S., Festa, S., Berenzi, A., Cella, M. and Colonna, M. 2003. Recruitment of immature plasmacytoid dendritic cells (plasmacytoid monocytes) and myeloid dendritic cells in primary cutaneous melanomas. J Pathol 200:255–268.
Vicari, A. P., Chiodoni, C., Vaure, C., Ait-Yahia, S., Dercamp, C., Matsos, F., Reynard, O., Taverne, C., Merle, P., Colombo, M. P., O'Garra, A., Trinchieri, G. and Caux, C. 2002. Reversal of tumor-induced dendritic cell paralysis by CpG immunostimulatory oligonucleotide and anti-interleukin 10 receptor antibody. J Exp Med 196:541–549.
Wilkin, F., Stordeur, P., Goldman, M., Boeynaems, J. M. and Robaye, B. 2002. Extracellular adenine nucleotides modulate cytokine production by human monocyte-derived dendritic cells: dual effect on IL-12 and stimulation of IL-10. Eur J Immunol 32:2409–2417.
Wojas, K., Tabarkiewicz, J., Jankiewicz, M. and Rolinski, J. 2004. Dendritic cells in peripheral blood of patients with breast and lung cancer – a pilot study. Folia Histochem Cytobiol 42:45–48.
Yang, L. and Carbone, D. P. 2004. Tumor-host immune interactions and dendritic cell dysfunction. Adv Cancer Res 92:13–27.
Yu, H., Kortylewski, M. and Pardoll, D. 2007. Crosstalk between cancer and immune cells: role of STAT3 in the tumour microenvironment. Nat Rev Immunol 7:41–51.
Zou, W., Machelon, V., Coulomb-L'Hermin, A., Borvak, J., Nome, F., Isaeva, T., Wei, S., Krzysiek, R., Durand-Gasselin, I., Gordon, A., Pustilnik, T., Curiel, D. T., Galanaud, P., Capron, F., Emilie, D. and Curiel, T. J. 2001. Stromal-derived factor-1 in human tumors recruits and alters the function of plasmacytoid precursor dendritic cells. Nat Med 7:1339–1346.

Part IV
Dendritic Cells in Cancer: Clinical Aspects

Understanding the crucial role of dendritic cells in tumor development and maintenance of antitumor immunity has resulted in immediate attempts to apply this knowledge to clinical practice in using dendritic cells for therapeutic purposes. Unexpected success in preclinical animal models and in in vitro human experiments led to initiation of multiple clinical trials in a variety of institutions focusing on safety and clinical applicability of dendritic cell vaccines for cancer therapy. Although the induction of tumor-specific immune responses is often evident, clinical responses in cancer patients were limited and inconsistent. A new generation of enhanced dendritic cell vaccines is under intense testing in preclinical settings and was designed based on growing understanding of dendritic cell biology and behavior in the tumor environment. The final section of the book addresses factors of importance to clinical aspects of cancer, with emphasis on issues relevant to immunotherapy. Chapter 19 describes and analyzes dendritic cell trafficking and homing in the tumor microenvironment, issues which play a key role in designing and determining the efficacy of dendritic cell vaccines after an adaptive transfer of cell in patients with cancer. A comprehensive functional and phenotypic characterization of dendritic cells in involved and non-involved draining lymph nodes is reviewed in Chapter 20. Chapter 21 discusses obstacles and problems for use of dendritic cell vaccines in clinical trials that continue to limit their effectiveness and feasibility for patients with cancer. Chapter 22 explains current optimal approaches for dendritic cell-based immunotherapy and presents multiple examples of their utilization for different tumor types in ongoing clinical trials. Chapter 23 focuses on the use of genetically modified dendritic cells in immunotherapy, as an illustration of the means available to improve the efficacy and value of therapeutic dendritic cell vaccines for cancer. Chapter 24 describes dendritic cell-derived tumors, their current classification, and available therapeutic modalities and prognosis of the diseases. This unusual perspective on dendritic cell biology is not widely discussed, but might be more appreciated in the future with the development of novel genetically engineered dendritic cell vaccines. Finally, Chapter 25 presents thoughts on future directions for research that may enhance the quality of dendritic cell-based therapy for cancer.

Chapter 19
Trafficking of Dendritic Cells in the Tumor Environment

Zoya R. Yurkovetsky and Ghena N. Yurkovetsky

Abstract To initiate immune responses, dendritic cells follow a migratory route from their recruitment as sentinels into tissues, including solid tumors, then to secondary lymphoid organs where they contour the immune response. Migratory capacities and chemokine responsiveness are therefore key elements in dendritic cell immunobiology. Tumor-derived factors alter dendritic cell maturation and function and thus markedly affect dendritic cell trafficking and distribution. Tumor-mediated down-regulation or redirection of dendritic cell migration and homing may markedly alter a normal pattern of dendritic cell localization and thus induction of antitumor immunity in tumor-bearing hosts, and might also strongly inhibit the efficacy of dendritic cell vaccines. Improved understanding of the regulation of DC trafficking might offer new opportunities for therapeutic interventions to control immune responses. Given the current interest in applying dendritic cell-based immunotherapy to cancer treatment, an understanding of the molecular defects responsible for dysfunction of dendritic cell trafficking and biodistribution in cancer is essential for designing and testing the next generation of dendritic cell vaccines.

19.1 Regulation of Dendritic Cell Trafficking: Introduction

A fundamental property of cells of the immune system, including dendritic cells (DC), is mobility, allowing them to navigate the body and contest microbial and tumor pathogens. DC are professional antigen-presenting cells which play a crucial role in initiation and regulation of both innate and adaptive immunity. In vivo, the capacity of DC to activate T cells depends on their ability to migrate to the T-cell areas of lymph nodes. This is modulated by a complex array of chemokines and their receptors, which provide the molecular signals to direct cells to where they are required. Chemokines have long been known to

Z.R. Yurkovetsky (✉)
University of Pittsburgh Medical Center, S730 Scaife Hall, 3550 Terrace Street, Pittsburgh, PA 15261, USA
e-mail: zoya100@hotmail.com

M.R. Shurin, R.D. Salter (eds.), *Dendritic Cells in Cancer*,
DOI 10.1007/978-0-387-88611-4_19, © Springer Science+Business Media, LLC 2009

orchestrate DC migration in the body (Caux et al. 2000). However, recent evidence has shown that chemokines can not only direct the trafficking of DC but also regulate their maturation status. This dual function of chemokines ensures that T cells and DC meet in T-cell regions of lymphoid organs and that antigen is presented in an immunologically optimal context for T-cell priming (Bachmann et al. 2006) (Fig. 19.1).

DC function generally depends on the level of cell maturation/activation and the local tissue milieu, especially cytokine and chemokine networks. DC differentiate from the hematopoietic precursors in the bone marrow or macrophages in the periphery and scattered throughout lymphoid and non-lymphoid tissues, where they act as immune sentinels by responding to invading pathogens. Immature DC reside at the sites of antigen capture and are active in internalization and processing of different antigens. After antigen capture, DC migrate to the lymphoid organs where, after a maturation step, they present processed antigenic peptides to antigen-specific T cells through MHC class I and II molecules, thereby inducing activation, differentiation, and expansion of naïve T cells into effector cells (Cavanagh and Weninger 2008). After accumulating in the lymph nodes through distinct trafficking pathways, DC interact with lymphocytes temporally and spatially to establish effective immune responses (Yoneyama et al. 2005). Interestingly, it has been recently reported that DC that migrate into the lymph nodes are not uniquely responsible for the

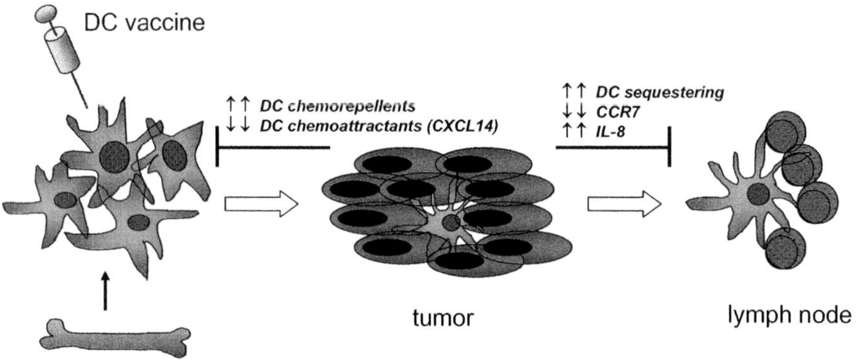

Fig. 19.1 Regulation of dendritic cell trafficking into and from the tumor mass as a mechanism of tumor escape from immune control. Tumor cells can prevent immigration of circulating dendritic cells (DC) into the tumor mass by producing chemorepellents that cause repulsion of DC by the mechanism termed chemorepulsion or fugetaxis. Tumor cells may also down-regulate expression of DC-attracting chemokins, such as CXCL14 (BRAK) and others. Furthermore, some tumors may effectively chemoattract DC, modulate their phenotype, and eventually damage DC mobility by inhibiting expression of CCR7 or producing IL-8. This mechanism prevents emigration of DC from the tumor site resulting in low number of DC reaching lymph nodes. These pathways operate for both endogenous or host DC and exogenous or injected DC, i.e., DC vaccines

activation of specific T cells. These DC can transfer antigens to resident DC and thereby spread antigens to a larger pool of DC to increase priming efficiency (Carbone et al. 2004; Allan et al. 2006).

DC precursors transmigrate through the blood vessel endothelia to enter the tissues under the direction of locally produced chemokines (McWilliam et al. 1996; Robert et al. 1999). DC trafficking from non-lymphoid tissues into the regional lymph nodes is promoted by locally produced homeostatic and inflammatory cytokines, chemokines, and other molecules that are generated or down-regulated in response to infection, inflammation, and the immune response development (Rescigno et al. 1999; Bertho et al. 2005; Mueller et al. 2007). Usually, circulating immature DC first enter the non-lymphoid tissues in response to constitutively expressed, like CXCL14, or inflammatory, like MIP-3α (CCL20), chemoattracting cytokines. After having ingested and processed incoming pathogens, they switch their chemokine receptor set, for instance from CCR6 to CCR7, and migrate to regional lymph nodes in response to homeostatic lymphoid chemokines, like CCL21, which also direct their position within lymphoid tissues so that DC can efficiently present processed antigens to lymphocytes, priming them for specific immune response (Yoshie 2000; McColl 2002). The trafficking of mature DC is primarily regulated by chemokine receptor CCR7 and a secondary lymphoid organ chemokine system operating at least in part through the afferent lymphatics (Dieu et al. 1998; Mantovani et al. 1998; Gunn et al. 1999). DC maturation signals drive DC trafficking and migration first to the lymphatic vessels and then to the draining lymph nodes via CCL21/6Ckine and other chemokines, including CCL19/MIP-3β and CCL20/MIP-3α (Lanzavecchia and Sallusto 2001; Bertho et al. 2005). Many cytokines including IL-10, IL-1β, CD40L, and TNF-α have been shown to regulate DC migratory responses via modulation of chemokine receptor expression (Takayama et al. 2001; Berthier-Vergnes et al. 2005). Infection agents also affect expression of homeostatic chemokines (Choi et al. 2003) and DC localization into secondary lymphoid tissues by altering expression of chemokines and chemokine receptors (e.g., CCR7 and CCR6) (Sallusto et al. 1998). Moreover, functional leukotriene B4 at low concentrations might promote the migration of immature and mature DC toward CCL19 and CCL21, which is associated with a rapid increase of CCR7 expression (Del Prete et al. 2007).

Regulation of migration and recruitment of different DC subpopulations are not alike, and different sets of chemokines and their receptors, as well as different trafficking pathways, are involved in a complex orchestration of the immune response by the DC system. For instance, conventional DC precursors are recruited to inflamed tissues in response to inflammatory chemokines and then remobilized to regional lymph nodes in response to CCL21. In contrast, plasmacytoid DC precursors directly transmigrate to regional lymph nodes via high endothelial venules in a CXCL9- and E-selectin-dependent manner (Randolph et al. 2005; Yoneyama et al. 2005). Although many molecules, pathways and mechanisms of DC recruitment, migration, and homing in a number of tissues have been identified, little is known about DC motility and trafficking in tumor-bearing hosts. The data on DC migratory routes have mostly been derived from observations in physiological conditions or infectious models. On the other hand, the DC migration network in the context of solid tumors, as well as the interactions between the tumor-associated DC and the tumor in terms of motility and activation or inhibition signals are still poorly understood. Yet, the understanding of these pathways could give us clues to improve or initiate effective antitumor immunity in preclinical and clinical settings (Vicari et al. 2004).

19.2 Trafficking of Dendritic Cells at the Tumor Site

Whether the immune system can actually target tumors has been debated for nearly a century (Burnet 1957; Keast 1970; Stagg et al. 2007). Compelling evidence now suggests that DC play a crucial role in limiting tumor progression and metastases, although the mechanisms regulating their motility and distribution in the tumor environment are mostly unknown. DC continuously migrate from the periphery into lymphatic tissues to orchestrate the immune system both in the presence and in the absence of "danger" signals (Allavena et al. 2000). Localization in tumor tissues and migration to lymphoid organs are essential steps in DC immunobiology that are linked to their maturation and T-cell stimulatory function (Shurin et al. 2006). Once penetrating the tumor, DC should move in response to local gradient of chemotactic and/or chemorepulsive factors to which they are sensitive (Vicari et al. 2004). Although the exact nature of those factors remains unclear, the presence and positioning of DC in the tumor bed reflect those influences. Studies in preclinical models and in clinical trials have demonstrated that DC infiltration of tumors is associated with improved survival of patients with a variety of cancers (Becker 1992; Shurin and Gabrilovich 2001). For instance, evaluating the prognostic significance of DC in 132 specimens from patients with primary head and neck squamous cell carcinoma (HNSCC), Reichert et al. reported that a low number of tumor-infiltrating DC were more predictive of poor survival than lymph node involvement (Reichert et al. 2001). Analyzing the distribution of DC in the primary tumor, adjacent tissue, and regional lymph nodes, Kikuchi et al. revealed that the number of DC in the tumor predicts overall survival, disease-free survival, and time to disease recurrence in patients with HNSCC (Kikuchi et al. 2002). Similarly, accumulation of DC in the tumor tissue has been associated with a better prognosis in patients with breast (Allan et al. 2004), gallbladder (Nakakubo et al. 2003), gastric (Huang et al. 2003), ovarian (Eisenthal et al. 2001), colorectal (Sandel et al. 2005) cancers, melanoma (Ladanyi et al. 2007), and other malignancies. However, the means by which tumors regulate and mechanisms of how tumors influence DC trafficking during tumor progression remain unclear.

Cancers exert local and systemic impact on immune cell function through various mechanisms. Among these, altered DC function is of utmost importance because DC are the most potent antigen-presenting cells of the immune system. Recent studies on the generation, maturation, longevity, and function of DC in cancer suggest that tumor-induced (i) apoptosis of DC; (ii) inhibition of DC capacity to uptake, process, and present tumor antigen(s); and (iii) redirection of DC migration are the three principle mechanisms employed by different tumor types to suppress the DC system and thus increase the likelihood of evading immune recognition (Shurin and Gabrilovich 2001; Yang and Carbone 2004; Bennaceur et al. 2008 #90). At present, it is clear that the defects of the DC system in cancer are systemic and are not always localized to the tumor tissues. Several reports proved that the high numbers of immature DC found in certain tumor masses were due to increased cell migration. Indeed, chemokines produced by transformed squamous cells might attract Langerhans cell precursors and thus regulate cell recruitment to the epithelial tumors (Halliday et al. 1992). Increased levels of immature DC in breast carcinoma were explained by high production of MIP-3α by tumor cells (Bell et al. 1999). Zou et al. reported that malignant human ovarian epithelial tumor cells express high levels of stromal-derived factor-1 (SDF-1 or CXCL12), which induces chemotaxis of plasmacytoid DC precursors and protects them from apoptosis (Zou et al. 2001). This contributed to increased IL-10 levels and poor T-cell activation. Thus, the quantity and quality of activation and inhibiting signals in the tumor milieu may dictate whether (i) tumor-associated DC remain immature and within the tumors, therefore preventing adaptive immune response or (ii) immature DC still reach the lymph node, resulting in tolerance induction (Vicari et al. 2004). Another important pathway is that tumor could secrete factors sequestering DC inside the tumor.

It has been reported that melanoma cell lines can effectively chemoattract DC, modulate their phenotype, and eventually damage DC mobility. Melanoma-treated DC exhibited an increased adhesion capacity to melanoma cells in vitro and did not migrate in response to DC chemokines (Remmel et al. 2001). In addition, ex vivo-generated DC failed to home into draining lymph nodes following intratumoral injection in patients with metastatic melanomas (Triozzi et al. 2000). One potential mechanism is the production of IL-8, which was reported to be expressed by hepatocellular, colorectal, and pancreatic cancers, and may contribute to the retention of DC inside malignant lesions and the destruction of DC migration toward CCR7-binding chemokines (Feijoo et al. 2005). In the context of tumors, CCR7 is likely involved in the migration of DC to draining lymph nodes, although may be impaired in certain cancers. Tumor-derived TGF-β, for instance, may increase expression of chemokine receptors CCR1, CCR2, CCR3, and CCR6 on immature DC and prevent expression of CCR7 and thus prevents DC migration toward lymph nodes keeping them immature at the tumor site (Bennaceur et al. 2008). Other data suggest that CCR7 expression on DC is enhanced with direct contact with apoptotic tumor cells and may have a critical role for DC migrating to regional lymph nodes (Hirao et al. 2000). Neuroblastoma, which is known to abrogate DC maturation via release of gangliosides (Shurin et al. 2001), has been recently shown to down-regulate

CCR7 expression on DC with tumor progression (Walker et al. 2005). Melanoma-derived gangliosides can also impair DC migratory function through the down-regulation of CCR7 expression (Bennaceur et al. 2006). Tumor-derived gangliosides alter spontaneous and chemokine-induced motility of DC by suppressing activity of small Rho GTPases in DC (Shurin et al. 2005b; Tourkova et al. 2007).

Other factors and conditions specific for the tumor microenvironment are also known to impede DC motility and alter DC distribution. For instance, it was reported that hypoxia, which is a characteristic of solid tumors, may inhibit DC migratory activity by regulating the balance between the expression of migration-related genes, such as matrix metalloproteinases (MMP) and their endogenous tissue inhibitors of matrix metalloproteinases (TIMP) (Qu et al. 2005). Because prostaglandin E2 (PgE2), which is overproduced in a wide variety of human malignancies, has been implicated in regulation of the balance between MMP and TIMP, it is possible that tumor-derived PgE2 might affect DC migratory capacity through the extracellular matrix (ECM) by altering MMP and TIMP equilibrium. In fact, it was reported that the treatment of DC with exogenous PgE2 was correlated with reduced DC migration through ECM (Baratelli et al. 2004). Similarly, DC cultured in supernatants from cyclooxygenase-2 overexpressing lung cancer cells that secrete high levels of PgE2, exhibited decreased migration through ECM (Baratelli et al. 2004). However, other results indicate that PgE2 significantly up-regulates MMP-9 expression and that in turn, DC-derived MMP-9 is essential for DC chemotaxis in response to the CCR7 ligand CCL19 (Yen et al. 2008). Consequently it is possible that DC matured within inflammatory sites require both CCR7 and PgE2-induced MMP-9 for their directional migration to draining lymph nodes. Moreover, HNSCC cell lines have been recently shown to autonomously express C-reactive protein, which could trigger the down-regulation of chemokine receptor CCR5 and led to a decreased migration of human DC (Frenzel et al. 2007). Thus sequestration of DC within tumor tissues and the subsequent inhibition of their migration are one of the several mechanisms by which tumors induce immunosuppression and escape from immune recognition and eradication.

Another related mechanism of tumor escape is the loss of expression of DC-attracting chemokines by malignant cells. For instance, DC chemokine CXCL14 was shown to be expressed in all normal tissues and cells, but its production in tumor tissues and by tumor cell lines was significantly down-regulated or blocked (Hromas et al. 1999; Frederick et al. 2000). Lost expression of CXCL14 protein in human HNSCC was associated with decreased attraction of DC to the tumor site both in vitro and in vivo and thus might result in suboptimal induction of antitumor immune responses (Shurin et al. 2005a). Using two murine HNSCC models, it was also demonstrated that migration and homing of DC into CXCL14-transfected tumors was associated with inhibition of tumor growth and induction of antitumor immunity. In agreement, overproduction of CCL20 (MIP-3α) by tumor cells causes the local accumulation of DC and activation of tumor-specific CTL in four murine tumor models (Fushimi et al. 2000). Expression of CCL20 was also found in the papillary carcinoma of the thyroid, a tumor heavily colonized by DC (Scarpino et al. 2000). In addition, the down-regulation of the expression of CXCL14 might be an important step in successful oncogenesis to prevent NK immune surveillance of the malignancy (Starnes et al. 2006).

Gaining a better understanding of the crosstalk between DC and tumor cells during the migration of DC to the tumor site and then to lymph nodes is essential for future advances in manipulating DC trafficking as a means to fine-tune immune responses in clinical settings, e.g., clinical trials utilizing DC vaccines for cancer treatment.

19.3 Dendritic Cell Vaccines and Dendritic Cell Trafficking in Cancer

The goal of immunotherapy in cancer patients is to activate and boost the immune system in order to destroy the tumor and to induce specific and long-lasting immune memory to protect against recurrent disease. After the discovery of DC's unique capacity to activate naive tumor antigen-specific T cells, numerous experimental treatments are now based on the vaccination of cancer patients with ex vivo prepared autologous DC loaded with tumor antigens (Palucka et al. 2005). Although immunological responses are observed in most clinical trials, clinical responses are limited to a minority of treated patients. Opportunities for improvement of clinically significant outcome lie in a number of variables, such as DC culture conditions, maturation stage of prepared DC, choice of DC subpopulations, means of antigen loading, route of DC administration, dose and frequency of vaccination, and others (Zhong et al. 2007).

One important aspect of DC-based immune therapy is appropriate delivery of the vaccine (Verdijk et al. 2008). For the delivery of DC three different options for administration are commonly distinguished: in peripheral tissue, such as skin [intradermal (i.d.) or subcutaneous (s.c.)] and mucosa, into the lymphatic system (intralymphatic (i.l.) or intranodal [i.n.]), and systemic intravenous (i.v.) or rare intraperitoneal (i.p.) administration. Vaccine delivery and distribution after injection has been studied in experimental animal models and clinical trials. In clinical protocols, DC prepared from blood-derived monocytes or hematopoietic progenitor cells were radiolabeled with indium or technetium and their localization after administration was analyzed by scintigraphic imaging of the patients. These studies revealed that after i.v. inoculation DC migrate to the liver, lung, spleen and bone marrow (Mackensen et al. 1999; Morse et al. 1999). Analyzing DC trafficking in patients with multiple myeloma, Prince et al. concluded that s.c. and i.d. routes of DC administration produced similar levels of DC migration to regional lymph nodes (Prince et al. 2008). The migration and homing patterns of tumor-specific peptide-pulsed DC administered via different routes to patients with HNSCC were studied employing single photon emission computed tomography (Horiguchi et al. 2007). DC administered directly into the nasal submucosa quickly migrated to the regional neck lymph nodes in the neck. However, after inoculation of the cells into the palatine tonsils, DC remained close to the site of administration and did not migrate to regional lymph nodes or to other mucosal regions.

In animals, analysis of DC labeled or transduced with fluorochromes proved that the route of DC administration was critical in determining the site of DC accumulation and time of DC persistence in vivo. The analysis demonstrates that i.d. injection increases the amount of labeled DC in the lymph node compared to s.c. injection (Eggert et al. 2003). DC-injected s.c. accumulated in the draining lymph node, and DC-injected i.v. in the spleen. DC appeared in the lymph node by 24 h after s.c. injection, their numbers peaked at 48 h and declined at 96 h. DC were found in the spleen at 3 and 24 h after i.v. injection, but their numbers were low and declined by 48 h (Huck et al. 2008). Depending on the tumor cell line used, DC-injected s.c. were as effective or more effective than DC-injected i.v. at inducing antitumor responses (Huck et al. 2008). However, s.c. but not i.v. injection of DC in a mouse melanoma model induced protective immunity against s.c. growing tumors, while, in contrast, metastatic lung lesions were controlled by i.v. immunization and only partially by s.c. immunization of DC vaccines (Eggert et al. 1999; Mullins et al. 2003; Okada et al. 2005). Interestingly, tumor antigen-loaded DC migrated more efficiently to the lymph nodes and were more effective activators of local (but not systemic) cellular immune response than were unloaded DC in a murine colon carcinoma model (Rossowska et al. 2007).

To target peripheral lymph nodes, DC can be administered into the skin or directly into the skin-draining lymphatics or lymph node. In mice, s.c. injections of DC are effective in both targeting DC to draining lymph nodes and inducing immune responses. In human, no or little migration of DC to skin-draining lymph nodes was found after s.c. injection of DC. Intradermal injection was more effective although the number of DC that reach the regional lymph nodes did not exceed 4% (Verdijk et al. 2008). Investigating the influence of indium and technetium labeling on the motility of antigen-loaded DC in humans, the authors did not demonstrate migration of DC from i.d. or s.c. sites of injection to regional lymph nodes and concluded that a large proportion of antigen-loaded DC, as used in current human trials for cancer therapy, may not reach regional lymph nodes (Blocklet et al. 2003). However, another study demonstrated that the rate of migration of antigen-bearing DC in situ from the skin to the lymph node is 100-fold higher than previously estimated (Adema et al. 2005). In contrast, evaluation of in vivo migratory capacity of i.d. administered DC during a phase I/II clinical trial for metastatic melanoma revealed that technetium-labeled DC traveled to the draining inguinal lymph nodes within 10 min, and the draining lymph nodes were clearly outlined up to 4 h after injection (Thomas et al. 1999). Furthermore, in a recent clinical trial in patients with multiple myeloma utilizing anti-idiotype (Id) vaccination with DC-administered s.c. and i.v., the results revealed that s.c. injections of Id-pulsed DC were safe and, in contrast with i.v. administrations, induced antitumor T-cell responses (Curti et al. 2007).

Alternatively, ex vivo-generated DC can be injected directly into the lymph nodes or lymphatic vessels draining the skin. However, initial analyses showed that i.n. injection is not always successful, even when it is performed under ultrasound guidance by a highly experienced radiologist (de Vries et al. 2005). Intralymphatic (i.l.) injection is an even more challenging technique that requires a highly skilled surgeon and hospital admission (Quillien et al. 2005). Importantly, despite the fact that most of the injected cells reach one or in most cases multiple lymph nodes, DC still need to migrate into the T-cell-rich areas (Verdijk et al. 2008).

Thus, one can conclude that i.d. and i.v. administration of tumor-antigen-loaded DC is preferred in clinical protocols, although one important drawback is the low efficiency in targeting DC to peripheral lymph nodes. In addition, there are several potential problems with the above strategies, namely, (i) the homing efficiency of cytotoxic T lymphocytes (CTL) induced by antigen-loaded DC to the tumor sites may be relatively low (Grover et al. 2006); (ii) the tumor antigens are not being recognized in many malignancies; and (iii) it is difficult to induce antigen-specific CTL suitable for the full repertoire of tumor antigens expressed by tumors. Furthermore, despite eliciting vigorous immune responses, to date, most DC vaccines have had limited clinical impact justifying the rationale for developing alternative DC-based immunotherapeutic strategies. Although DC vaccination occasionally leads to tumor regression, clinical efficacy and immunogenicity of DC in clinical trials have not yet been clarified. One explanation may be the small rate of DC reaching the afferent lymph nodes (less than 10%) (Schuler et al. 2003). Interestingly, it has recently been suggested that DC migration from the site of injection to the lymph nodes and the subsequent immune response may be enhanced by the previous topical administration of inflammatory factors. This should lead to up-regulation of the CCR7 ligand CCL21 in the local lymphoid vessels, thus favoring DC trafficking through the vessels to the lymph nodes (MartIn-Fontecha et al. 2003). Accordantly, CCR7-tranduced DC have been shown to migrate to the regional lymph nodes more efficiently then control cells (Okada et al. 2005).

In view of these findings, it seems that the development of a strategy to enhance and direct DC migration in the tumor environment, particularly when using cancer vaccines, is highly desirable. An interesting alternative approach that is currently being explored is to target tumor antigens directly to DC in situ and boost DC mobilization from tumor tissues.

19.4 Intratumoral Dendritic Cell Vaccines

Immune cells, such as cytotoxic T lymphocytes, NK cells, B cells, and DC, have a central role in cancer immunotherapy. Conventional studies of antitumor therapy have focused mainly on the search for an efficient means to prime/boost tumor antigen-specific immunity. A systematic understanding of the molecular basis of the trafficking and biodistribution of immune cells, however, is essential for the development of more efficacious cancer immunotherapies. It is well recognized that the basis and premise of immunotherapy is the accumulation of effective immune cells in tumor tissues. In animal tumor models, chemokine-induced recruitment of DC to the site of the tumor has typically been correlated with amplified antitumor immune responses. Examples include the administration of tumor cells genetically engineered to express chemokines CCL7, CCL16, CCL21, CXCL14, and others (Fioretti et al. 1998; Giovarelli et al. 2000; Vicari et al. 2000), viral delivery of the chemokine genes, e.g., CCL20 (Fushimi et al. 2000), in growing tumors or the intratumoral/intralesional injection of recombinant chemokines, e.g., CCL21 (Kirk et al. 2001). Furthermore, the clinical benefits from intratumoral administration of DC, including tumor rejection, prolonged survival, and induction of immune memory, have been reported in mice and rats (Melcher et al. 1999; Pirtskhalaishvili et al. 2000b; Candido et al. 2001; Tong et al. 2001; Ehtesham et al. 2003; Ahmed et al. 2004). A pilot study in patients with melanoma and breast carcinoma demonstrated marked antitumor potential of intratumoral delivery of DC without addition of tumor antigens (Triozzi et al. 2000). Other clinical trials, evaluating intralesional administration of IL-12-transfected DC in patients with hepatocellular carcinomas and metastatic pancreatic and colorectal malignancies, concluded that intratumoral injection of DC is feasible and well tolerated (Feijoo et al. 2005; Mazzolini et al. 2005). Similarly, injection of autologous immature DC into tumor in conjunction with radiotherapy in patients with advanced hepatoma revealed its safety and also induction of tumor-specific and innate immunities (Chi et al. 2005). Evidence has been obtained showing that intratumoral DC can capture and process tumor antigens to be presented to T lymphocytes (Melero et al. 2000). Interestingly, DC were recently reported to be cytotoxic for several tumor cell lines suggesting that this may have important consequences for their ability to pick up antigens from dying tumor cells and thus stimulate tumor-specific CTL (Vanderheyde et al. 2001; Yang et al. 2001).

Thus, intratumoral approaches are appealing because they may direct antigen-specific responses at the tumor site, exploit the presence of multiple undefined tumor antigens present in the endogenous tumor, and would be expected to reduce systemic toxicity (Crittenden et al. 2005). However, clinical efficacy of intratumoral delivery of DC might be limited by suppression of DC longevity and function in the tumor microenvironment. Tumor-induced DC apoptosis and suppression of DC function are the key mechanisms mediating suppression of the DC system in the local tumor microenvironment (Gabrilovich et al. 1997; Esche et al. 1999; Shurin and Gabrilovich 2001; Bennaceur et al. 2008). Of special interest is the fact that tumor-induced apoptosis of DC or inhibition of DC function may be tumor-specific: melanoma, lung carcinoid, and several prostate adenocarcinoma cell lines were strong inhibitors of DC survival (Esche et al. 1999; Katsenelson et al. 2001; Pirtskhalaishvili et al. 2001), while neuroblastoma and HNSCC did not induce significant DC apoptosis but suppressed antigen presentation by DC (Shurin et al. 2001; Tourkova et al. 2005). Therefore, in the intratumoral DC delivery strategy, for the purpose of enhancing antigen-specific antitumor immunity in the absence of define tumor antigens, it is important to increase antigen processing in DC and prolong survival of DC at the local tumor sites.

Several approaches can be used to increase the antitumor efficacy of intratumorally delivered DC vaccines. For instance, transfection of DC with the anti-apoptotic gene *Bcl-xL* significantly increased their resistance to prostate cancer-induced apoptosis and was associated with significant antitumor potential upon intratumoral delivery in vivo in animal models (Pirtskhalaishvili et al. 2000b). Pretreatment of DC with certain cytokines or transfection of DC with the cytokine genes, including TNF-α, CD40L, IL-12, and IL-15, also protects DC from tumor-induced apoptosis (Pirtskhalaishvili et al. 2000a; Esche et al. 2001; Pirtskhalaishvili et al. 2001; Tourkova et al. 2002). This effect is mediated by up-regulation of expression of the anti-apoptotic proteins Bcl-2 and Bcl-xL and down-regulation of the pro-apoptotic protein Bax. For instance, it was recently reported that intratumoral injection of adenovectors encoding 4-1BBL and CD40L resulted in a significantly stronger inhibition of murine breast cancer progression than the treatment with RANKL (Yurkovetsky et al. 2006). The authors also demonstrated that intratumoral administration of DC transduced to express the TNF-related ligands resulted in a significant inhibition of colon adenocarcinoma growth as compared with control groups treated with control DC. Interestingly, DC expressing CD40L, 4-1BBL and RANKL survived significantly longer at the tumor site than control DC (Yurkovetsky et al. 2006). Many agents, including TNF-α, IFN-γ, CD40L, TRAIL, LIGHT, CpG, TLR ligands, IL-12, IL-15, and IL-1, have been shown to stimulate DC maturation and function and might protect DC longevity and function in the tumor microenvironment. For instance, the magnitude of T-cell response, which was proportional to the number of antigen-carrying DC that reached the lymph node, could be boosted up to 40-fold by preinjection of TNF-α that conditioned the tissue for increased DC migration (MartIn-Fontecha et al. 2003). Furthermore, all-trans retinoic acid has been shown to enhance DC migration from the tumor bed to draining lymph nodes due to an increase in MMP production with a simultaneous decrease in the production of tissue inhibitors of metalloproteinases, allowing for enhanced DC mobilization (Darmanin et al. 2007).Thus, in summary, together with clinical evidence demonstrating that infiltration of tumor mass by DC is associated with better patient survival, administration of DC vaccines into the tumor site might provoke proficient antitumor immune responses.

In conclusion, optimal DC migration for cellular therapy against tumors may be achieved by generating a protective and stimulatory environment together with making DC with a high migratory capacity and immune activation potential. Controlling DC distribution, homing and trafficking to lymph nodes to a level that is sufficient for inducing proficient immune response against the tumor and long-lasting protection against recurrence of disease is essential for the development of DC-based immunotherapy into a valid treatment.

19.5 Conclusions

The critical role of DC in immune responses suggests that the ability to manipulate DC migration might have therapeutic potential for cancer treatment. Before this potential can be realized, however, additional studies are required for understanding molecular pathways that direct DC trafficking and homing in different physiologic and pathophysiologic settings (Randolph et al. 2008). For instance, there are virtually no data available regarding trafficking requirements for DC in patients with different types of cancer. The ability to alter recruitment of DC precursors from the blood to the tumor site might have therapeutic benefit in humans, but was not yet evaluated. In an effort to maximize the migration of DC for cellular therapy, it is indispensable to know how DC migration is initiated and what factors are essential for the cell to move toward the draining lymph node and enter the T-cell-rich areas in the tumor environment. Clearly, as our understanding of DC trafficking becomes more developed, there will be an increased likelihood of successfully exploiting DC migration for therapeutic intervention.

References

Adema, G. J., de Vries, I. J., Punt, C. J. and Figdor, C. G. 2005. Migration of dendritic cell based cancer vaccines: in vivo veritas? Curr Opin Immunol 17:170–174.

Ahmed, S. U., Okamoto, M., Oshikawa, T., Tano, T., Sasai, A., Kan, S., Hiroshima, T., Ohue, H., Moriya, Y., Ryoma, Y., Saito, M. and Sato, M. 2004. Anti-tumor effect of an intratumoral administration of dendritic cells in combination with TS-1, an oral fluoropyrimidine anti-cancer drug, and OK-432, a streptococcal immunopotentiator: involvement of toll-like receptor 4. J Immunother 27:432–441.

Allan, C. P., Turtle, C. J., Mainwaring, P. N., Pyke, C. and Hart, D. N. 2004. The immune response to breast cancer, and the case for DC immunotherapy. Cytotherapy 6:154–163.

Allan, R. S., Waithman, J., Bedoui, S., Jones, C. M., Villadangos, J. A., Zhan, Y., Lew, A. M., Shortman, K., Heath, W. R. and Carbone, F. R. 2006. Migratory dendritic cells transfer antigen to a lymph node-resident dendritic cell population for efficient CTL priming. Immunity 25:153–162.

Allavena, P., Sica, A., Vecchi, A., Locati, M., Sozzani, S. and Mantovani, A. 2000. The chemokine receptor switch paradigm and dendritic cell migration: its significance in tumor tissues. Immunol Rev 177:141–149.

Bachmann, M. F., Kopf, M. and Marsland, B. J. 2006. Chemokines: more than just road signs. Nat Rev Immunol 6:159–164.

Baratelli, F. E., Heuze-Vourc'h, N., Krysan, K., Dohadwala, M., Riedl, K., Sharma, S. and Dubinett, S. M. 2004. Prostaglandin E2-dependent enhancement of tissue inhibitors of metalloproteinases-1 production limits dendritic cell migration through extracellular matrix. J Immunol 173:5458–5466.

Becker, Y. 1992. Anticancer role of dendritic cells (DC) in human and experimental cancers – a review. Anticancer Res 12:511–520.

Bell, D., Chomarat, P., Broyles, D., Netto, G., Harb, G. M., Lebecque, S., Valladeau, J., Davoust, J., Palucka, K. A. and Banchereau, J. 1999. In breast carcinoma tissue, immature dendritic cells reside within the tumor, whereas mature dendritic cells are located in peritumoral areas. J Exp Med 190:1417–1426.

Bennaceur, K., Chapman, J., Brikci-Nigassa, L., Sanhadji, K., Touraine, J. L. and Portoukalian, J. 2008. Dendritic cells dysfunction in tumour environment. Cancer Lett 272(2):186–196.

Bennaceur, K., Popa, I., Portoukalian, J., Berthier-Vergnes, O. and Peguet-Navarro, J. 2006. Melanoma-derived gangliosides impair migratory and antigen-presenting function of human epidermal Langerhans cells and induce their apoptosis. Int Immunol 18:879–886.

Berthier-Vergnes, O., Bermond, F., Flacher, V., Massacrier, C., Schmitt, D. and Peguet-Navarro, J. 2005. TNF-alpha enhances phenotypic and functional maturation of human epidermal Langerhans cells and induces IL-12 p40 and IP-10/CXCL-10 production. FEBS Lett 579:3660–3668.

Bertho, N., Adamski, H., Toujas, L., Debove, M., Davoust, J. and Quillien, V. 2005. Efficient migration of dendritic cells toward lymph node chemokines and induction of T(H)1 responses require maturation stimulus and apoptotic cell interaction. Blood 106:1734–1741.

Blocklet, D., Toungouz, M., Kiss, R., Lambermont, M., Velu, T., Duriau, D., Goldman, M. and Goldman, S. 2003. 111In-oxine and 99mTc-HMPAO labelling of antigen-loaded dendritic cells: in vivo imaging and influence on motility and actin content. Eur J Nucl Med Mol Imaging 30:440–447.

Burnet, M. 1957. Cancer; a biological approach. I. The processes of control. Br Med J 1:779–786.

Candido, K. A., Shimizu, K., McLaughlin, J. C., Kunkel, R., Fuller, J. A., Redman, B. G., Thomas, E. K., Nickoloff, B. J. and Mule, J. J. 2001. Local administration of dendritic cells inhibits established breast tumor growth: implications for apoptosis-inducing agents. Cancer Res 61:228–236.

Carbone, F. R., Belz, G. T. and Heath, W. R. 2004. Transfer of antigen between migrating and lymph node-resident DCs in peripheral T-cell tolerance and immunity. Trends Immunol 25:655–658.

Caux, C., Ait-Yahia, S., Chemin, K., de Bouteiller, O., Dieu-Nosjean, M. C., Homey, B., Massacrier, C., Vanbervliet, B., Zlotnik, A. and Vicari, A. 2000. Dendritic cell biology and regulation of dendritic cell trafficking by chemokines. Springer Semin Immunopathol 22:345–369.

Cavanagh, L. L. and Weninger, W. 2008. Dendritic cell behaviour in vivo: lessons learned from intravital two-photon microscopy. Immunol Cell Biol 86:428–438.

Chi, K. H., Liu, S. J., Li, C. P., Kuo, H. P., Wang, Y. S., Chao, Y. and Hsieh, S. L. 2005. Combination of conformal radiotherapy and intratumoral injection of adoptive dendritic cell immunotherapy in refractory hepatoma. J Immunother 28:129–135.

Choi, Y. K., Fallert, B. A., Murphey-Corb, M. A. and Reinhart, T. A. 2003. Simian immunodeficiency virus dramatically alters expression of homeostatic chemokines and dendritic cell markers during infection in vivo. Blood 101:1684–1691.

Crittenden, M. R., Thanarajasingam, U., Vile, R. G. and Gough, M. J. 2005. Intratumoral immunotherapy: using the tumour against itself. Immunology 114:11–22.

Curti, A., Tosi, P., Comoli, P., Terragna, C., Ferri, E., Cellini, C., Massaia, M., D'Addio, A., Giudice, V., Di Bello, C., Cavo, M., Conte, R., Gugliotta, G., Baccarani, M. and Lemoli, R. M. 2007. Phase I/II clinical trial of sequential subcutaneous and intravenous delivery of dendritic cell vaccination for refractory multiple myeloma using patient-specific tumour idiotype protein or idiotype (VDJ)-derived class I-restricted peptides. Br J Haematol 139:415–424.

Darmanin, S., Chen, J., Zhao, S., Cui, H., Shirkoohi, R., Kubo, N., Kuge, Y., Tamaki, N., Nakagawa, K., Hamada, J., Moriuchi, T. and Kobayashi, M. 2007. All-trans retinoic acid enhances murine dendritic cell migration to draining lymph nodes via the balance of matrix metalloproteinases and their inhibitors. J Immunol 179:4616–4625.

de Vries, I. J., Lesterhuis, W. J., Barentsz, J. O., Verdijk, P., van Krieken, J. H., Boerman, O. C., Oyen, W. J., Bonenkamp, J. J., Boezeman, J. B., Adema, G. J., Bulte, J. W., Scheenen, T. W., Punt, C. J., Heerschap, A. and Figdor, C. G. 2005. Magnetic resonance tracking of dendritic cells in melanoma patients for monitoring of cellular therapy. Nat Biotechnol 23:1407–1413.

Del Prete, A., Shao, W. H., Mitola, S., Santoro, G., Sozzani, S. and Haribabu, B. 2007. Regulation of dendritic cell migration and adaptive immune response by leukotriene B4 receptors: a role for LTB4 in up-regulation of CCR7 expression and function. Blood 109:626–631.

Dieu, M. C., Vanbervliet, B., Vicari, A., Bridon, J. M., Oldham, E., Ait-Yahia, S., Briere, F., Zlotnik, A., Lebecque, S. and Caux, C. 1998. Selective recruitment of immature and mature dendritic cells by distinct chemokines expressed in different anatomic sites. J Exp Med 188:373–386.

Eggert, A. A., Schreurs, M. W., Boerman, O. C., Oyen, W. J., de Boer, A. J., Punt, C. J., Figdor, C. G. and Adema, G. J. 1999. Biodistribution and vaccine efficiency of murine dendritic cells are dependent on the route of administration. Cancer Res 59:3340–3345.

Eggert, A. A., van der Voort, R., Torensma, R., Moulin, V., Boerman, O. C., Oyen, W. J., Punt, C. J., Diepstra, H., de Boer, A. J., Figdor, C. G. and Adema, G. J. 2003. Analysis of dendritic cell trafficking using EGFP-transgenic mice. Immunol Lett 89:17–24.

Ehtesham, M., Kabos, P., Gutierrez, M. A., Samoto, K., Black, K. L. and Yu, J. S. 2003. Intratumoral dendritic cell vaccination elicits potent tumoricidal immunity against malignant glioma in rats. J Immunother 26:107–116.

Eisenthal, A., Polyvkin, N., Bramante-Schreiber, L., Misonznik, F., Hassner, A. and Lifschitz-Mercer, B. 2001. Expression of dendritic cells in ovarian tumors correlates with clinical outcome in patients with ovarian cancer. Hum Pathol 32:803–807.

Esche, C., Lokshin, A., Shurin, G. V., Gastman, B. R., Rabinowich, H., Watkins, S. C., Lotze, M. T. and Shurin, M. R. 1999. Tumor's other immune targets: dendritic cells. J Leukoc Biol 66:336–344.

Esche, C., Shurin, G. V., Kirkwood, J. M., Wang, G. Q., Rabinowich, H., Pirtskhalaishvili, G. and Shurin, M. R. 2001. Tumor necrosis factor-alpha-promoted expression of Bcl-2 and inhibition of mitochondrial cytochrome c release mediate resistance of mature dendritic cells to melanoma-induced apoptosis. Clin Cancer Res 7:974s–979.

Feijoo, E., Alfaro, C., Mazzolini, G., Serra, P., Penuelas, I., Arina, A., Huarte, E., Tirapu, I., Palencia, B., Murillo, O., Ruiz, J., Sangro, B., Richter, J. A., Prieto, J. and Melero, I. 2005. Dendritic cells delivered inside human carcinomas are sequestered by interleukin-8. Int J Cancer 116:275–281.

Fioretti, F., Fradelizi, D., Stoppacciaro, A., Ramponi, S., Ruco, L., Minty, A., Sozzani, S., Garlanda, C., Vecchi, A. and Mantovani, A. 1998. Reduced tumorigenicity and augmented leukocyte infiltration after monocyte chemotactic protein-3 (MCP-3) gene transfer: perivascular accumulation of dendritic cells in peritumoral tissue and neutrophil recruitment within the tumor. J Immunol 161:342–346.

Frederick, M. J., Henderson, Y., Xu, X., Deavers, M. T., Sahin, A. A., Wu, H., Lewis, D. E., El-Naggar, A. K. and Clayman, G. L. 2000. In vivo expression of the novel CXC chemokine BRAK in normal and cancerous human tissue. Am J Pathol 156:1937–1950.

Frenzel, H., Pries, R., Brocks, C. P., Jabs, W. J., Wittkopf, N. and Wollenberg, B. 2007. Decreased migration of myeloid dendritic cells through increased levels of C-reactive protein. Anticancer Res 27:4111–4115.

Fushimi, T., Kojima, A., Moore, M. A. and Crystal, R. G. 2000. Macrophage inflammatory protein 3alpha transgene attracts dendritic cells to established murine tumors and suppresses tumor growth. J Clin Invest 105:1383–1393.

Gabrilovich, D. I., Corak, J., Ciernik, I. F., Kavanaugh, D. and Carbone, D. P. 1997. Decreased antigen presentation by dendritic cells in patients with breast cancer. Clin Cancer Res 3:483–490.

Giovarelli, M., Cappello, P., Forni, G., Salcedo, T., Moore, P. A., LeFleur, D. W., Nardelli, B., Carlo, E. D., Lollini, P. L., Ruben, S., Ullrich, S., Garotta, G. and Musiani, P. 2000. Tumor rejection and immune memory elicited by locally released LEC chemokine are associated with an impressive recruitment of APCs, lymphocytes, and granulocytes. J Immunol 164:3200–3206.

Grover, A., Kim, G. J., Lizee, G., Tschoi, M., Wang, G., Wunderlich, J. R., Rosenberg, S. A., Hwang, S. T. and Hwu, P. 2006. Intralymphatic dendritic cell vaccination induces tumor antigen-specific, skin-homing T lymphocytes. Clin Cancer Res 12:5801–5808.

Gunn, M. D., Kyuwa, S., Tam, C., Kakiuchi, T., Matsuzawa, A., Williams, L. T. and Nakano, H. 1999. Mice lacking expression of secondary lymphoid organ chemokine have defects in lymphocyte homing and dendritic cell localization. J Exp Med 189:451–460.

Halliday, G. M., Lucas, A. D. and Barnetson, R. S. 1992. Control of Langerhans' cell density by a skin tumour-derived cytokine. Immunology 77:13–18.

Hirao, M., Onai, N., Hiroishi, K., Watkins, S. C., Matsushima, K., Robbins, P. D., Lotze, M. T. and Tahara, H. 2000. CC chemokine receptor-7 on dendritic cells is induced after interaction with apoptotic tumor cells: critical role in migration from the tumor site to draining lymph nodes. Cancer Res 60:2209–2217.

Horiguchi, S., Matsuoka, T., Okamoto, Y., Sakurai, D., Kobayashi, K., Chazono, H., Hanazawa, T. and Tanaka, Y. 2007. Migration of tumor antigen-pulsed dendritic cells after mucosal administration in the human upper respiratory tract. J Clin Immunol 27:598–604.

Hromas, R., Broxmeyer, H. E., Kim, C., Nakshatri, H., Christopherson, K., II, Azam, M. and Hou, Y. H. 1999. Cloning of BRAK, a novel divergent CXC chemokine preferentially expressed in normal versus malignant cells. Biochem Biophys Res Commun 255:703–706.

Huang, H. L., Wu, B. Y., You, W. D., Shen, M. S. and Wang, W. J. 2003. Influence of dendritic cell infiltration on prognosis and biologic characteristics of progressing gastric cancer. Zhonghua Zhong Liu Za Zhi 25:468–471.

Huck, S. P., Tang, S. C., Andrew, K. A., Yang, J., Harper, J. L. and Ronchese, F. 2008. Activation and route of administration both determine the ability of bone marrow-derived dendritic cells to accumulate in secondary lymphoid organs and prime CD8+ T cells against tumors. Cancer Immunol Immunother 57:63–71.

Katsenelson, N. S., Shurin, G. V., Bykovskaia, S. N., Shogan, J. and Shurin, M. R. 2001. Human small cell lung carcinoma and carcinoid tumor regulate dendritic cell maturation and function. Mod Pathol 14:40–45.

Keast, D. 1970. Immunosurveillance and cancer. Lancet 2:710–712.

Kikuchi, K., Kusama, K., Taguchi, K., Ishikawa, F., Okamoto, M., Shimada, J., Sakashita, H. and Yamamo, Y. 2002. Dendritic cells in human squamous cell carcinoma of the oral cavity. Anticancer Res 22:545–557.

Kirk, C. J., Hartigan-O'Connor, D., Nickoloff, B. J., Chamberlain, J. S., Giedlin, M., Aukerman, L. and Mule, J. J. 2001. T cell-dependent antitumor immunity mediated by secondary lymphoid tissue chemokine: augmentation of dendritic cell-based immunotherapy. Cancer Res 61:2062–2070.

Ladanyi, A., Kiss, J., Somlai, B., Gilde, K., Fejos, Z., Mohos, A., Gaudi, I. and Timar, J. 2007. Density of DC-LAMP(+) mature dendritic cells in combination with activated T lymphocytes infiltrating primary cutaneous melanoma is a strong independent prognostic factor. Cancer Immunol Immunother 56:1459–1469.

Lanzavecchia, A. and Sallusto, F. 2001. The instructive role of dendritic cells on T cell responses: lineages, plasticity and kinetics. Curr Opin Immunol 13:291–298.

Mackensen, A., Krause, T., Blum, U., Uhrmeister, P., Mertelsmann, R. and Lindemann, A. 1999. Homing of intravenously and intralymphatically injected human dendritic cells generated in vitro from CD34+ hematopoietic progenitor cells. Cancer Immunol Immunother 48:118–122.

Mantovani, A., Allavena, P., Vecchi, A. and Sozzani, S. 1998. Chemokines and chemokine receptors during activation and deactivation of monocytes and dendritic cells and in amplification of Th1 versus Th2 responses. Int J Clin Lab Res 28:77–82.

MartIn-Fontecha, A., Sebastiani, S., Hopken, U. E., Uguccioni, M., Lipp, M., Lanzavecchia, A. and Sallusto, F. 2003. Regulation of dendritic cell migration to the draining lymph node: impact on T lymphocyte traffic and priming. J Exp Med 198:615–621.

Mazzolini, G., Alfaro, C., Sangro, B., Feijoo, E., Ruiz, J., Benito, A., Tirapu, I., Arina, A., Sola, J., Herraiz, M., Lucena, F., Olague, C., Subtil, J., Quiroga, J., Herrero, I., Sadaba, B., Bendandi, M., Qian, C., Prieto, J. and Melero, I. 2005. Intratumoral injection of dendritic cells engineered to secrete interleukin-12 by recombinant adenovirus in patients with metastatic gastrointestinal carcinomas. J Clin Oncol 23:999–1010.

McColl, S. R. 2002. Chemokines and dendritic cells: a crucial alliance. Immunol Cell Biol 80:489–496.

McWilliam, A. S., Napoli, S., Marsh, A. M., Pemper, F. L., Nelson, D. J., Pimm, C. L., Stumbles, P. A., Wells, T. N. and Holt, P. G. 1996. Dendritic cells are recruited into the airway epithelium during the inflammatory response to a broad spectrum of stimuli. J Exp Med 184:2429–2432.

Melcher, A., Todryk, S., Bateman, A., Chong, H., Lemoine, N. R. and Vile, R. G. 1999. Adoptive transfer of immature dendritic cells with autologous or allogeneic tumor cells generates systemic antitumor immunity. Cancer Res 59:2802–2805.

Melero, I., Vile, R. G. and Colombo, M. P. 2000. Feeding dendritic cells with tumor antigens: self-service buffet or a la carte? Gene Ther 7:1167–1170.

Morse, M. A., Coleman, R. E., Akabani, G., Niehaus, N., Coleman, D. and Lyerly, H. K. 1999. Migration of human dendritic cells after injection in patients with metastatic malignancies. Cancer Res 59:56–58.

Mueller, S. N., Hosiawa-Meagher, K. A., Konieczny, B. T., Sullivan, B. M., Bachmann, M. F., Locksley, R. M., Ahmed, R. and Matloubian, M. 2007. Regulation of homeostatic chemokine expression and cell trafficking during immune responses. Science 317:670–674.

Mullins, D. W., Sheasley, S. L., Ream, R. M., Bullock, T. N., Fu, Y. X. and Engelhard, V. H. 2003. Route of immunization with peptide-pulsed dendritic cells controls the distribution of memory and effector T cells in lymphoid tissues and determines the pattern of regional tumor control. J Exp Med 198:1023–1034.

Nakakubo, Y., Miyamoto, M., Cho, Y., Hida, Y., Oshikiri, T., Suzuoki, M., Hiraoka, K., Itoh, T., Kondo, S. and Katoh, H. 2003. Clinical significance of immune cell infiltration within gallbladder cancer. Br J Cancer 89:1736–1742.

Okada, N., Mori, N., Koretomo, R., Okada, Y., Nakayama, T., Yoshie, O., Mizuguchi, H., Hayakawa, T., Nakagawa, S., Mayumi, T., Fujita, T. and Yamamoto, A. 2005. Augmentation of the migratory ability of DC-based vaccine into regional lymph nodes by efficient CCR7 gene transduction. Gene Ther 12:129–139.

Palucka, A. K., Laupeze, B., Aspord, C., Saito, H., Jego, G., Fay, J., Paczesny, S., Pascual, V. and Banchereau, J. 2005. Immunotherapy via dendritic cells. Adv Exp Med Biol 560:105–114.

Pirtskhalaishvili, G., Shurin, G. V., Esche, C., Cai, Q., Salup, R. R., Bykovskaia, S. N., Lotze, M. T. and Shurin, M. R. 2000a. Cytokine-mediated protection of human dendritic cells from prostate cancer-induced apoptosis is regulated by the Bcl-2 family of proteins. Br J Cancer 83:506–513.

Pirtskhalaishvili, G., Shurin, G. V., Esche, C., Trump, D. L. and Shurin, M. R. 2001. TNF-alpha protects dendritic cells from prostate cancer-induced apoptosis. Prostate Cancer Prostatic Dis 4:221–227.

Pirtskhalaishvili, G., Shurin, G. V., Gambotto, A., Esche, C., Wahl, M., Yurkovetsky, Z. R., Robbins, P. D. and Shurin, M. R. 2000b. Transduction of dendritic cells with Bcl-xL increases their resistance to prostate cancer-induced apoptosis and antitumor effect in mice. J Immunol 165:1956–1964.

Prince, H. M., Wall, D. M., Ritchie, D., Honemann, D., Harrrison, S., Quach, H., Thompson, M., Hicks, R., Lau, E., Davison, J., Loudovaris, M., Moloney, J., Loveland, B., Bartholeyns, J., Katsifis, A. and Mileshkin, L. 2008. In vivo tracking of dendritic cells in patients with multiple myeloma. J Immunother 31:166–179.

Qu, X., Yang, M. X., Kong, B. H., Qi, L., Lam, Q. L., Yan, S., Li, P., Zhang, M. and Lu, L. 2005. Hypoxia inhibits the migratory capacity of human monocyte-derived dendritic cells. Immunol Cell Biol 83:668–673.

Quillien, V., Moisan, A., Carsin, A., Lesimple, T., Lefeuvre, C., Adamski, H., Bertho, N., Devillers, A., Leberre, C. and Toujas, L. 2005. Biodistribution of radiolabelled human dendritic cells injected by various routes. Eur J Nucl Med Mol Imaging 32:731–741.

Randolph, G. J., Angeli, V. and Swartz, M. A. 2005. Dendritic-cell trafficking to lymph nodes through lymphatic vessels. Nat Rev Immunol 5:617–628.

Randolph, G. J., Ochando, J. and Partida-Sanchez, S. 2008. Migration of dendritic cell subsets and their precursors. Annu Rev Immunol 26:293–316.

Reichert, T. E., Scheuer, C., Day, R., Wagner, W. and Whiteside, T. L. 2001. The number of intratumoral dendritic cells and zeta-chain expression in T cells as prognostic and survival biomarkers in patients with oral carcinoma. Cancer 91:2136–2147.

Remmel, E., Terracciano, L., Noppen, C., Zajac, P., Heberer, M., Spagnoli, G. C. and Padovan, E. 2001. Modulation of dendritic cell phenotype and mobility by tumor cells in vitro. Hum Immunol 62:39–49.

Rescigno, M., Granucci, F. and Ricciardi-Castagnoli, P. 1999. Dendritic cells at the end of the millennium. Immunol Cell Biol 77:404–410.

Robert, C., Fuhlbrigge, R. C., Kieffer, J. D., Ayehunie, S., Hynes, R. O., Cheng, G., Grabbe, S., von Andrian, U. H. and Kupper, T. S. 1999. Interaction of dendritic cells with skin endothelium: A new perspective on immunosurveillance. J Exp Med 189:627–636.

Rossowska, J., Pajtasz-Piasecka, E., Szyda, A., Zietara, N. and Dus, D. 2007. Tissue localization of tumor antigen-loaded mouse dendritic cells applied as an anti-tumor vaccine and their influence on immune response. Folia Histochem Cytobiol 45:349–355.

Sallusto, F., Schaerli, P., Loetscher, P., Schaniel, C., Lenig, D., Mackay, C. R., Qin, S. and Lanzavecchia, A. 1998. Rapid and coordinated switch in chemokine receptor expression during dendritic cell maturation. Eur J Immunol 28:2760–2769.

Sandel, M. H., Dadabayev, A. R., Menon, A. G., Morreau, H., Melief, C. J., Offringa, R., van der Burg, S. H., Janssen-van Rhijn, C. M., Ensink, N. G., Tollenaar, R. A., van de Velde, C. J. and Kuppen, P. J. 2005. Prognostic value of tumor-infiltrating dendritic cells in colorectal cancer: role of maturation status and intratumoral localization. Clin Cancer Res 11:2576–2582.

Scarpino, S., Stoppacciaro, A., Ballerini, F., Marchesi, M., Prat, M., Stella, M. C., Sozzani, S., Allavena, P., Mantovani, A. and Ruco, L. P. 2000. Papillary carcinoma of the thyroid: hepatocyte growth factor (HGF) stimulates tumor cells to release chemokines active in recruiting dendritic cells. Am J Pathol 156:831–837.

Schuler, G., Schuler-Thurner, B. and Steinman, R. M. 2003. The use of dendritic cells in cancer immunotherapy. Curr Opin Immunol 15:138–147.

Shurin, G. V., Ferris, R., Tourkova, I. L., Perez, L., Lokshin, A., Balkir, L., Collins, B., Chatta, G. S. and Shurin, M. R. 2005a. Loss of new chemokine CXCL14 in tumor tissue is associated with low infiltration by dendritic cells (DC), while restoration of human

CXCL14 expression in tumor cells causes attraction of DC both in vitro and in vivo. J Immunol 174:5490–5498.
Shurin, G. V., Shurin, M. R., Bykovskaia, S., Shogan, J., Lotze, M. T. and Barksdale, E. M., Jr. 2001. Neuroblastoma-derived gangliosides inhibit dendritic cell generation and function. Cancer Res 61:363–369.
Shurin, G. V., Tourkova, I. L., Chatta, G. S., Schmidt, G., Wei, S., Djeu, J. Y. and Shurin, M. R. 2005b. Small rho GTPases regulate antigen presentation in dendritic cells. J Immunol 174:3394–3400.
Shurin, M. R. and Gabrilovich, D. I. 2001. Regulation of dendritic cell system by tumor. Cancer Res Ther Control 11:65–78.
Shurin, M. R., Shurin, G. V., Lokshin, A., Yurkovetsky, Z. R., Gutkin, D. W., Chatta, G., Zhong, H., Han, B. and Ferris, R. L. 2006. Intratumoral cytokines/chemokines/growth factors and tumor infiltrating dendritic cells: friends or enemies? Cancer Metastasis Rev 25:333–356.
Stagg, J., Johnstone, R. W. and Smyth, M. J. 2007. From cancer immunosurveillance to cancer immunotherapy. Immunol Rev 220:82–101.
Starnes, T., Rasila, K. K., Robertson, M. J., Brahmi, Z., Dahl, R., Christopherson, K. and Hromas, R. 2006. The chemokine CXCL14 (BRAK) stimulates activated NK cell migration: implications for the downregulation of CXCL14 in malignancy. Exp Hematol 34:1101–1105.
Takayama, T., Morelli, A. E., Onai, N., Hirao, M., Matsushima, K., Tahara, H. and Thomson, A. W. 2001. Mammalian and viral IL-10 enhance C–C chemokine receptor 5 but down-regulate C–C chemokine receptor 7 expression by myeloid dendritic cells: impact on chemotactic responses and in vivo homing ability. J Immunol 166:7136–7143.
Thomas, R., Chambers, M., Boytar, R., Barker, K., Cavanagh, L. L., MacFadyen, S., Smithers, M., Jenkins, M. and Andersen, J. 1999. Immature human monocyte-derived dendritic cells migrate rapidly to draining lymph nodes after intradermal injection for melanoma immunotherapy. Melanoma Res 9:474–481.
Tong, Y., Song, W. and Crystal, R. G. 2001. Combined intratumoral injection of bone marrow-derived dendritic cells and systemic chemotherapy to treat pre-existing murine tumors. Cancer Res 61:7530–7535.
Tourkova, I. L., Shurin, G. V., Chatta, G. S., Perez, L., Finke, J., Whiteside, T. L., Ferrone, S. and Shurin, M. R. 2005. Restoration by IL-15 of MHC class I antigen-processing machinery in human dendritic cells inhibited by tumor-derived gangliosides. J Immunol 175:3045–3052.
Tourkova, I. L., Shurin, G. V., Wei, S. and Shurin, M. R. 2007. Small rho GTPases mediate tumor-induced inhibition of endocytic activity of dendritic cells. J Immunol 178:7787–7793.
Tourkova, I. L., Yurkovetsky, Z. R., Gambotto, A., Makarenkova, V. P., Perez, L., Balkir, L., Robbins, P. D., Shurin, M. R. and Shurin, G. V. 2002. Increased function and survival of IL-15-transduced human dendritic cells are mediated by up-regulation of IL-15Ralpha and Bcl-2. J Leukoc Biol 72:1037–1045.
Triozzi, P. L., Khurram, R., Aldrich, W. A., Walker, M. J., Kim, J. A. and Jaynes, S. 2000. Intratumoral injection of dendritic cells derived in vitro in patients with metastatic cancer. Cancer 89:2646–2654.
Vanderheyde, N., Aksoy, E., Amraoui, Z., Vandenabeele, P., Goldman, M. and Willems, F. 2001. Tumoricidal activity of monocyte-derived dendritic cells: evidence for a caspase-8-dependent, Fas-associated death domain-independent mechanism. J Immunol 167:3565–3569.
Verdijk, P., Aarntzen, E. H., Punt, C. J., de Vries, I. J. and Figdor, C. G. 2008. Maximizing dendritic cell migration in cancer immunotherapy. Expert Opin Biol Ther 8:865–874.

Vicari, A. P., Ait-Yahia, S., Chemin, K., Mueller, A., Zlotnik, A. and Caux, C. 2000. Antitumor effects of the mouse chemokine 6Ckine/SLC through angiostatic and immunological mechanisms. J Immunol 165:1992–2000.

Vicari, A. P., Treilleux, I. and Lebecque, S. 2004. Regulation of the trafficking of tumour-infiltrating dendritic cells by chemokines. Semin Cancer Biol 14:161–169.

Walker, S. R., Aboka, A., Ogagan, P. D. and Barksdale, E. M., Jr. 2005. Murine neuroblastoma attenuates dendritic cell cysteine cysteine receptor 7 (CCR7) expression. J Pediatr Surg 40:983–987.

Yang, L. and Carbone, D. P. 2004. Tumor-host immune interactions and dendritic cell dysfunction. Adv Cancer Res 92:13–27.

Yang, R., Xu, D., Zhang, A. and Gruber, A. 2001. Immature dendritic cells kill ovarian carcinoma cells by a FAS/FASL pathway, enabling them to sensitize tumor-specific CTLs. Int J Cancer 94:407–413.

Yen, J. H., Khayrullina, T. and Ganea, D. 2008. PGE2-induced metalloproteinase-9 is essential for dendritic cell migration. Blood 111:260–270.

Yoneyama, H., Matsuno, K. and Matsushimaa, K. 2005. Migration of dendritic cells. Int J Hematol 81:204–207.

Yoshie, O. 2000. Role of chemokines in trafficking of lymphocytes and dendritic cells. Int J Hematol 72:399–407.

Yurkovetsky, Z. R., Shurin, G. V., Barry, D. A., Schuh, A. C., Shurin, M. R. and Robbins, P. D. 2006. Comparative analysis of antitumor activity of CD40L, RANKL, and 4-1BBL in vivo following intratumoral administration of viral vectors or transduced dendritic cells. J Gene Med 8:129–137.

Zhong, H., Shurin, M. R. and Han, B. 2007. Optimizing dendritic cell-based immunotherapy for cancer. Expert Rev Vaccines 6:333–345.

Zou, W., Machelon, V., Coulomb-L'Hermin, A., Borvak, J., Nome, F., Isaeva, T., Wei, S., Krzysiek, R., Durand-Gasselin, I., Gordon, A., Pustilnik, T., Curiel, D. T., Galanaud, P., Capron, F., Emilie, D. and Curiel, T. J. 2001. Stromal-derived factor-1 in human tumors recruits and alters the function of plasmacytoid precursor dendritic cells. Nat Med 7:1339–1346.

Chapter 20
Dendritic Cells in Tumor-Draining Lymph Nodes

Alistair J. Cochran, Rong-Rong Huang, Eijun Itakura, Jonathan H. Lee, and Barbara G. Molenkamp

Abstract Dendritic cells are neither the most obvious nor the most abundant cells in lymph nodes, but are certainly among the most immunologically influential, acting primarily as highly potent professional antigen-presenting cells that control complex sequences of innate and adaptive immune responses. Immune responses against tumors, immune resistance, and tolerance are initiated in tumor-draining lymph nodes, including sentinel lymph nodes where mature dendritic cells present processed tumor antigens to naïve T cells generating tumor-specific cytotoxic and helper T cells. Dendritic cells in sentinel lymph nodes of cancer patients have maturation defects that influence their function and phenotype. Lymph nodes contain follicular dendritic cells and paracortical dendritic cells (interdigitating dendritic cells). This chapter deals exclusively with the latter. The topic is a research interest for the authors who have published previously on the topic and this review draws extensively on these prior publications.

20.1 Normal Physiologic Flow of Dendritic Cells to Lymph Nodes

Dendritic cells (DC) are neither the most obvious nor the most abundant cells in lymph nodes, but are certainly among the most immunologically influential, acting primarily as highly potent professional antigen-presenting cells that control complex sequences of innate and adaptive immune responses. They are characterized by elegant attenuated cell processes (dendrites) that dramatically increase the surface area for their contacts with other cells. DC were first identified in the epidermis by Paul Langerhans (Langerhans 1868). Langerhans considered incorrectly that epidermal DC were nerve cells: "branched skin cells resembling neurons". DC were given their current name by Steinman and Cohn (Steinman and Cohn 1973). Lymph nodes contain follicular DC and

A.J. Cochran (✉)
Department of Pathology and Laboratory Medicine, David Geffen School of Medicine at UCLA, 10833 LeConte Avenue, Los Angeles, CA 90095, USA
e-mail: acochran@mednet.ucla.edu

M.R. Shurin, R.D. Salter (eds.), *Dendritic Cells in Cancer*,
DOI 10.1007/978-0-387-88611-4_20,

paracortical DC (interdigitating DC). This chapter deals exclusively with the latter. The topic is a research interest for the authors who have published previously on the topic and this review draws extensively on these prior publications (Cochran et al. 1987; Huang et al. 2000; Cochran et al. 2001; Vuylsteke et al. 2002; Lee et al. 2005; Cochran et al. 2006; Molenkamp et al. 2007).

DC develop from bone marrow stem cells and evolve via hematopoietic precursors of either common myeloid/monocytic lineage (myeloid DC/mDC) or common lymphocytic lineage (plasmacytoid DC/pDC) under the influence of cytokines. Identification of mDC and pDC in lymph nodes is based in part on the extent of their HLA-DR expression (Cox et al. 2005). mDC populate peripheral tissues as immature DC and migrate to lymph nodes via afferent lymphatics where they constitute the mature DC population and stimulate antigen-specific T cells (Banchereau et al. 2000; Randolph et al. 2005). pDC migrate directly from blood to lymph nodes across the high endothelial venules and are phenotypically immature (Cella et al. 1999; Summers et al. 2001).

In peripheral tissues immature DC, highly competent endocytic cells, but poorly competent antigen presenters (for T cells), sample their environment for microbial, tumor, and other antigens, using pattern recognition molecules such as Toll-like receptors (TLR). When appropriate antigens are encountered, immature DC mature to antigen-presenting DC by upregulating their co-stimulatory profile, developing the capacity to present processed antigens with MHC molecules to T cells. Mature DC can also interact, directly and/or indirectly via CD1c or CD1d, with NK cells and some B cells (Munz et al. 2005). After acquisition of appropriate antigens and under the influence of chemokine receptor CCR7, mature DC migrate via afferent lymphatics (as veiled cells) to regional lymph nodes. In lymph nodes DC settle in the paracortical (T-dependent areas) and after upregulating co-receptors for T-cell activation (CD80, CD86 and CD40) interact with naïve and memory T lymphocytes. In the steady state, most DC with a mature phenotype in lymphoid organs have migrated from epithelial surfaces. Mature DC in peripheral lymph nodes originate as either dermal DC or epidermal Langerhans cells, the latter expressing CD1a strongly (Wilson et al. 2003).

Antigen capture, processing, and presentation by DC are key steps in the antigen-specific control of adaptive immunogenic and tolerogenic immune responses (Steinman and Banchereau 2007). Immunogenic mature DC have upregulated expression of the co-stimulatory molecules CD80, CD86, and CD40 and markers of maturity such as CD83 and CCR7 and provide resistance to infection and tumor progression. They induce activation, proliferation and/or expansion of T-cell responses via MHC class molecules. Tolerogenic DC lead to silencing of immune responses by deleting T cells (Hawiger et al. 2001; Probst et al. 2005), suppressing T-cell responses to produce anergy (Luo et al. 2007), or by generating T regulatory cells that restrict T-cell activity (Shortman and Naik 2007). Silencing is desirable in transplantation, autoimmunity, and allergy, but undesirable in tumor immunity where it facilitates escape from host immune surveillance. Tolerogenic DC are resistant to maturation in response to "danger signals", such as HMGB1, TLR ligands, and CD40L, and express low levels of

MHC and co-stimulatory molecules (Morelli and Thomson 2007). Tolerogenic DC can generate, select, and expand alloantigen-specific, naturally occurring or adaptive regulatory T cells ($CD4^{+}CD25^{+}Foxp3^{+}$Tregs). Mature/activated DC stimulate T-cell proliferation, whereas immature DC can induce $CD4^{+}CD25^{+}Foxp3^{+}$Tregs. Tregs can in turn inhibit the evolution of DC, preventing their maturation and inducing downregulation of the cytokine IL-10, and immunosuppressive molecules of the B7-H family (Mahnke et al. 2007). $CD4^{+}$ $Foxp3^{+}$ T cells are more frequent in the peripheral blood of melanoma patients (relative to healthy donors), tumor-infiltrated lymph nodes (relative to normal nodes), and in tumor sites (Jandus et al. 2008). In animal models, tumor-draining lymph nodes (TDLN) contain increased $CD4^{+}CD25^{+}Foxp3^{+}$ Tregs (Liyanage et al. 2006) and/or indoleamine 2,3-dioxigenase (IDO)-expressing pDC (Munn et al. 2004), key cells in regulation of T-cell-mediated immunity. Maturation, activation, and functioning of these regulatory cells are governed in part by the composition of and variations in the local cytokine microenvironment. For example, $Foxp3^{+}$ Tregs, which suppress responses to self and non-self antigens, are inducible by a TGF-β-mediated mechanism (Liyanage et al. 2006; Liu et al. 2008). TGF-β produced by the cells of many solid tumors suppresses $CD4^{+}$ and $CD8^{+}$ T-cell responses, inhibits activation of naïve $CD4^{+}$ T cells, suppresses Th1 but not Th2 memory cell activation and cytokine production, and enhances IL-10 production by macrophages (Maeda et al. 1995; Maeda and Shiraishi 1996; Ludviksson et al. 2000).

DC evolution requires a continuing supply of immature DC to peripheral tissues, their continued capacity to seek and acquire relevant antigens, migrate to lymph nodes and there present processed antigens to recipient T cells. Interference with any stage of this process impacts the differentiation, proliferation, migration, and maturation of DC and significantly down regulates immune responses against microbial and tumor antigens with potentially disastrous clinical results. In studies of DC around primary melanomas, we (Stene et al. 1988) and others (Facchetti et al. 1984; Gatter et al. 1984) demonstrated a decreased frequency of Langerhans cells in the epidermis overlying and adjacent to invasive primary melanomas. This raised the possibility that melanomas might generate and secrete soluble molecules that regulated local DC and, via the lymphatics, nodal DC, thus influencing primary melanoma progression and metastases. While Vermi et al. found increased numbers of intratumoral and peritumoral pDC in primary melanomas (Vermi et al. 2003), such cells are incompletely activated/matured as evidenced by their expression of CD86, but neither CD80 nor CD83 suggesting a possible tolerogenic role for these pDC in peripheral lymphoid tissue. These "defective" pDC can still produce IFN-α, but expression of the IFN-α inducible protein MxA is extremely inconstant and very limited in melanomas. The paucity of intratumoral DC and the predominantly immature phenotype of peritumoral dermal DC indicate defective maturation of DC associated with primary cutaneous melanoma, a defect that is likely to reduce T-cell priming (Vermi et al. 2003).

20.2 Heterogeneity of Immune Reactivity from Node to Node

We evaluated individual tumor-draining lymph nodes and found differences in reaction intensity and in indices of immune function from node to node (Cochran et al. 1987). Nodes were evaluated histologically for paracortical, follicular, and sinusoidal reactions. Paracortical DC frequency, density and dendrite formation were evaluated using antibodies to S-100 protein, CD1a, HLA-DR, and fascin. To assess density and distribution of T lymphocytes, we used antibodies to CD3, CD43RA, CD45RO, and CD25. DC and T-cell frequency and density were analyzed by computer-linked morphometry. Lymph nodes containing metastases, nodes closest to primary tumors and nodes adjacent to metastatic nodes, were least reactive: with reduced area and density of paracortical DC compared to more remotely located nodes. Follicular and sinusoidal activity was similar regardless of nodal position relative to tumor (Cochran et al. 1987).

Functional studies of nodal lymphocytes (Hoon et al. 1987) evaluated spontaneous uptake of tritiated thymidine and uptake in response to phytomitogens, cytokines, and alloantigens. Internodal variations in spontaneous and mitogen- or alloantigen-induced proliferation were detected in a majority of patients: nodes nearest tumor being less reactive than more remote nodes. Differences in isotope uptake between unstimulated cell suspensions from individual nodes suggested that the alterations detected were present in vivo. Similar internodal differences were observed with assays of inhibition of leukocyte motility in supernatants of short-term cultures of cells from metastasis-free nodes at differing distances from primary tumors (Wen et al. 1989). Nodes near tumor generated supernatants that were least inhibitory of leukocyte migration and lymphoid cells that proliferated weakly in response to standard stimuli.

Two-color flow cytometry of nodal lymphocytes from melanoma patients revealed significant variations in lymphocyte subpopulations (Farzad et al. 1990) that correlated with clinical stage, proximity to the primary tumor, and tumor status of the node, confirming prior data from breast cancer patients (Morton et al. 1986). There was a striking reduction in $CD4^+$ helper/inducer T cells, with associated alterations in the CD4/CD8 ratio and an increase in $CD56^+$ NK cells with increasing clinical stage (Farzad et al. 1990). Lymphocytes from tumor-oriented nodes varied in their effect (inhibitory or enhancing) on melanoma cell growth in vitro (Farzad et al. 1997). Lymph nodes close to primary tumors or nodally metastatic melanoma harbor increased ConA inducible suppressor cells relative to more remote nodes (Hoon et al. 1990).

These data demonstrate that lymph nodes close to cancer are less reactive than more remote nodes, but the studies were limited by technical considerations. Exact lymphatic connections were unknowable, the extent of internodal lymphatic connections could not be assessed and the specific lymphatics that originated in the skin around a primary tumor could not be identified. The need for an accurate, dynamic approach to identify the node most influenced by tumor was one stimulus to our development of dye-directed lymphatic mapping and sentinel node biopsy.

20.3 Sentinel Node Concept

Minimally invasive intraoperative lymphatic mapping and sentinel node biopsy (LM/SNB) are widely used to stage the regional lymph nodes of patients with early-stage melanoma (Morton et al. 1992; Cochran et al. 2000; Morton et al. 2006). LM/SNB accurately detects clinically occult metastases. If the sentinel lymph node (SLN) is tumor-free (ca. 80% of patients), it is likely that the remaining nodes in the basin will also be tumor free. If the SLN contains tumor (ca. 20% of patients), additional regional nodes may also be tumor positive. On this basis complete lymph node dissection (with its associated morbidity and cost) can be reserved for patients with a tumor-positive SLN. In addition to the many clinical benefits associated with LM/SNB, the technique provides for the first time the opportunity to study physical interactions between a primary tumor and the specific regional lymph nodes that receives lymph directly from the tumor and its immediate environs. The balance of this chapter describes studies of nodal pathophysiology based on LM/SNB and exploration of the thesis that tumor cells create a nodal microenvironment that facilitates the survival and expansion of metastasis-competent tumor cells that arrive in the lymph node. We report studies that show that tumor-induced nodal immune down-regulation may be reversed in vivo and consider whether such reversal may inhibit or reverse the establishment of nodal metastases.

20.4 Alterations in Dendritic Cell Number, Phenotype and Activation Status

Immune responses against tumors, immune resistance, and tolerance are initiated in TDLN, including SLN where mature DC present processed tumor antigens to naïve T cells generating tumor-specific cytotoxic and/or helper T cells. DC in SLN of cancer patients have maturation defects that influence their function and phenotype. Compared to nodes from non-cancer patients, TDLN contain significantly fewer CD1a^{+}DC, S100^{+}DC, and (possibly) CD86^{+}DC (Laguens et al. 2002). SLN had a lower frequency and density of S100^{+}, HLA-DR^{+}, and fascin^{+} DC compared to NSLN (Huang et al. 2000; Lana et al. 2001; Huang et al. 2004a). In breast cancer patients, cellular immune responses, DC maturation, and Th1 responses are less active in SLN (relative to NSLN) with significantly lower expression of CD83 and IFN-γ mRNA, and decreased populations of HLA-DR^{+},

$CD80^+$, $CD83^+$, and $CD40^+$DC, alterations detectable before development of nodal metastases (Matsuura et al. 2006). We demonstrated a predominance of oligodendritic DC in SLN contrasting with the abundant polydendritic DC seen in NSLN (Huang et al. 2000, 2004a, b). From a microscopic perspective, oligodendritic morphology is a correlate of DC immaturity. Oligodendritic DC express S100, HLA-DR, CD1a, and fascin, as well as markers of immaturity, such as Langerin. Meshwork formation by $fascin^+$DC was lower in SLN than in NSLN, an alteration that correlated with the presence of fewer $CD45RO^+$ activated T cells, raising the possibility that reduced DC surface area secondarily downregulates dendrite-dependent interactions between DC and T cells (Huang et al. 2004a).

Immature DC can transmigrate across resting endothelia in vitro, whereas mature DC cannot (Wethmar et al. 2006). Mature DC migrate to secondary lymphoid organs through afferent lymphatics, dependent on CCR7 (Ohl et al. 2004). Oligodendritic $Langerin^+$DC locate preferentially around high endothelial venules (HEV) in melanoma-draining SLN (Huang et al. 2008). It is possible that these oligodendritic DC may be either apoptotic DC (apopDC) or pDC. Circulating DC from patients with early-stage breast cancer certainly show increased rates of spontaneous apoptosis (Pinzon-Charry et al. 2006). Melanoma-derived gangliosides impair Langerhans cell maturation by downregulating co-stimulatory markers CD40, CD54, CD80, and CD86 and maturation markers CD83 and CCR7, which correlate with impairment of the capacity of these cells to induce allogeneic T-cell proliferation and with reduced Langerhans cell migration to lymph nodes. Melanoma-derived gangliosides also enhance spontaneous apoptosis of LC (Bennaceur et al. 2006). Using terminal deoxynucleotidyl transferase biotin-dUTP nick end labeling (TUNEL) and double and/or triple cocktail staining with CD68, fascin, and CD25 antibodies, we found that most TUNEL-positive "ApopDC" are $CD68^+$/$fascin^-$ macrophages (some co-expressing CD25) with a minority of $fascin^+$/$CD68^-$ DC containing engulfed TUNEL-positive apoptotic bodies. Apoptotic $fascin^+$DC were not seen in SLN and NSLN from melanoma patients (Huang et al. 2005). pDC express CD123, CD68, CLA/HECA452, and BDCA-2 (Facchetti et al. 1988; Dzionek et al. 2002), but are infrequent in "healthy" lymph nodes in the steady state. Using immunohistochemical double staining, we found most oligodendritic $Langerin^+$ DC to be $CD68^-$, as are some $HLA\text{-}DR^+$ and fascin oligodendritic DC in both SLN and NSLN, suggesting that not all oligodendritic DC in TDLN are pDC. These findings support our hypothesis that DC with no dendrites or short dendrites are in some cases defective mature DC. While accumulation of immature DC in lymph nodes may indicate inhibition of DC maturation (Vermi et al. 2003), accumulation of mature $DC\text{-}LAMP^+$ DC in SLN is reportedly associated with continued generation of effective cytotoxic T cells and a favorable prognosis (Movassagh et al. 2004; Elliott et al. 2007; Ladanyi et al. 2007).

It is possible that all areas of a TDLN are not uniformly exposed to tumor-derived downregulatory molecules. To evaluate this possibility we examined SLN where the point of entry of tumor-derived lymph was indicated by the presence of carbon particles derived from intradermal injection (with marker dye) in the skin around the primary tumor (Lucci et al. 1999; Haigh et al. 2001; Morton et al. 2003). Tumor-free SLN were divided into four quadrants and we compared the frequency and density of DC and activated T cells in the different quadrants oriented to carbon deposits that indicate the point of entry of lymph from the area of the primary melanoma. There were significantly fewer DC in the quadrant adjacent to the point of entry of the afferent lymphatics, i.e., the area of the node most exposed to tumor lymph and its molecular contents (Huang et al. 2008).

20.5 Molecular and Signaling Alterations in the Sentinel Lymph Nodes

The SLN also provides a unique opportunity to study molecular aspects of tumor-associated immunological alterations. Downregulation of the SLN is believed to be caused by soluble immunosuppressive factors released by tumor cells and associated inflammatory cells, including regulatory immune cells within the microenvironment of the primary tumor and draining lymph nodes. IL-10 is considered to have a prime role in this process. Melanoma cells, especially metastatic melanoma cells frequently express IL-10 (Chen et al. 1994; Krüger-Krasagakes et al. 1994), which is also elevated in the serum of patient with advanced melanoma (Dummer et al. 1995; Boyano et al. 2000). Melanoma-derived IL-10 impedes the differentiation and function of DC downregulating CD1 molecules on DC (Gerlini et al. 2004). IL-10 expression levels are higher in tumor-positive SLN than in non-SLN (Lee et al. 2005). Increased expression of IL-10 in SLN from melanoma patients has been confirmed (Torisu-Itakura et al. 2007b). While the cellular source(s) of IL-10 within the SLN remain to be determined, IL-10 is expressed by $CD14^{+}$ monocytes in the peripheral blood of melanoma patients (Torisu-Itakura et al. 2007a). Tumor-infiltrating lymphocytes (Biggs and Eiselein 2001; Seo et al. 2001; Polak et al. 2007a) and macrophages (Pollard 2004) are also sources of IL-10 in the tumor microenvironment.

TGF-β is an immune modulatory cytokine that has dual roles as an immune suppressor (and in this role may actually be oncogenic) and as a tumor suppressor that inhibits proliferation of tumor cells. In melanoma, TGF-β produced by tumor cells induces immune suppression at the primary site acting through tolerogenic DC and regulatory T cells that express TGF-β receptor 1 (Polak et al. 2007a, b). In contrast, melanoma cells are reported to be generally resistant to the tumor-suppressive effects of TGF-β (Hussein 2005). Other tumor-derived factors may assist tumor cells to escape cytolysis by cytotoxic T cells (Kim et al. 2006). Soluble Fas (CD95) and FasL are detected in advanced melanoma, and these molecules are associated with poor prognosis (Mouawad et al. 2000; Ugurel et al. 2001; Redondo et al. 2002) possibly as a result of impairment of functions of the Fas receptor-ligand system. The tryptophan-degrading enzyme indoleamine 2,3-dioxygenase (IDO) is expressed by antigen-presenting cells and is also considered to influence acquired tolerance to tumors by inhibiting T-cell function. In animal models, IDO-expressing pDC activate mature resting Tregs in an MHC restricted fashion via the GCN2 pathway (Sharma et al. 2007). IDO-positive cells are present in the perisinusoidal regions of lymph nodes from melanoma patients (Lee et al. 2003). Expression of IDO is high in tumor-positive SLN increases with melanoma progression (Polak et al. 2007a) and is associated with significantly decreased survival (Munn et al. 2004). In a comparison of tumor-positive and tumor-negative SLN, IL-13, leptin, lymphotoxin β receptor (LTβR), and CCL4 (MIP-1β, macrophage inflammatory protein-1β) were all upregulated, while IL-11Rα was down-regulated (Torisu-Itakura et al. 2007b).

CCL2 (MCP-1: monocyte chemotactic protein-1) plays a key role in recruiting monocytes and enriching tumor-associated macrophages (TAM) in the tumor microenvironment (Varney et al. 2005a, b). In the tumor microenvironment, TAM are educated by tumor cells have immunosuppressive characteristics and produce immune suppressive factors (Kim et al. 2006). Another mechanism is postulated by which TAM may influence tumor angiogenesis by generating angiogenic factors, VEGF and IL-8 (CXCL8) (Torisu et al. 2000), both of which are identifiable in tumor cells in advanced melanoma (Lacal et al. 2000; Varney et al. 2006). Targeting TAM by inhibition of CCL2 reduces angiogenesis and the growth of melanoma xenografts (Gazzaniga et al. 2007).

In addition, individual chemokine receptors expressed by melanoma cells are reportedly associated with metastasis to specific organs, possibly acting through chemotactic mechanisms (Payne and Cornelius 2002; Murakami et al. 2004). For example, CCR7 expression by melanoma cells is associated with lymphatic metastasis in working with CCL21, produced by lymphatic endothelial cells (Wiley et al. 2001; Takeuchi et al. 2004; Shields et al. 2007). Also CXCR4 is associated with metastases to lung and liver in response to CXCL12, but CXCR4 is also present in some nodal metastasis (Scala et al. 2006). CCR9 expression by melanoma cells is associated with metastasis to the small intestine in response to CCL25 (Amersi et al. 2008).

Thus, secretion of immune modulatory molecules by tumor cells and tumor-associated leukocytes drives the evolution of immunosuppressive networks within which tumor can escape attack by the host immune system. Such immune suppressive networks contribute to tumor progression and the development and progression of metastases.

20.6 Reversal of Immunomodulation of the Sentinel Lymph Nodes: A Novel Adjuvant Therapeutic Option?

The role of GM-CSF, a prototypical pro-inflammatory cytokine, in determining and shaping the SLN immune milieu is of considerable interest. In an in vitro setting, IL-4 and GM-CSF can induce the differentiation of monocytes into $CD1a^+$ DC (Palucka et al. 1998). Exogenous dermal injection of GM-CSF has resulted in reversal of the histological abnormalities noted in SLN from melanoma patients (Lee et al. 2005) restoring SLN T-cell and DC areas and density back to "normal" levels. Animal studies, on the other hand, suggest that increased expression of GM-CSF within TDLN results in increased recruitment of IDO-expressing DC (Munn et al. 2004). A similar functional dichotomy is seen with IFN-γ, another prototypical pro-inflammatory cytokine. In the in vitro setting, IFN-γ itself can mature the DC that induce cytotoxic T cells; however, with co-expression of IL-10, IFN-γ can induce maturation of IDO-expressing DC (Munn et al. 2002).

To study the effect of local immunotherapy on immune effector cells in the SLN, a special technique was developed to investigate DC and T-cell functions without interfering with routine diagnostic pathology procedures (Vuylsteke et al. 2002). A small phase II study was performed in which Stage I melanoma patients were randomized to receive intracutaneous injections of either GM-CSF or plain saline around the scar of the excision of the primary melanoma, prior to local re-excision and SLN dissection (Vuylsteke et al. 2004). GM-CSF was chosen because it has powerful stimulatory effects in vivo on recruitment, activation and survival of mDC (Dranoff 2002).

Flow cytometric analysis showed a significant increase in the number and maturation status of $CD1a^+$ mDC in the SLN of patients exposed to preoperative GM-CSF and a concomitant increase in DC/T-cell clustering (Vuylsteke et al. 2004). In view of the critical role of mDC in the initiation of T-cell-mediated immunity, it was hypothesized that potentiated mDC functions in the GM-CSF-administered group might be associated with a higher number of tumor-specific $CD8^+$ T cells in the SLN. We tested this and the results supported our hypothesis, showing that local priming of melanoma-specific $CD8^+$ T cells was associated with a high frequency of mDC in the SLN, best observed in patients receiving locally administered preoperative GM-CSF (Vuylsteke et al. 2006). Together, these two investigations demonstrate that locally primed antitumor T-cell responses in the SLN are detectable as early as Stage I of melanoma development and may be enhanced by GM-CSF-induced increases in SLN-DC frequencies.

Unmethylated cytosine–phosphate–guanine oligodeoxynucleotides (CpG ODN) directly stimulate pDC through intracellular TLR9 triggering. Activated pDC preferentially release large amounts of IFN-α (Krug et al. 2001; Duramad et al. 2003; Ito et al. 2006), which may facilitate direct activation of $CD8^+$ T cells and NK cells, as well as promote the differentiation and maturation of neighboring mDC or their precursors, and thus provide indirect as well as direct stimulation of T-cell activation (Patterson 2000; Dzionek et al. 2001; Kawarada et al. 2001; Gursel et al. 2002; Salio et al. 2003). To investigate the effects of local administration of the CpG B-type ODN, PF-3512676 (formerly known as CpG 7909) applied prior to SNB, on DC and T-cell subsets in the SLN, clinical Stage I–II melanoma patients were randomized to receive either PF-3512676 or saline (Molenkamp et al. 2007). We demonstrated that intradermal PF-3512676 administered around the primary tumor excision site prior to SNB resulted in increased activation status of pDC and mDC. In the group who received PF-3512676, we found induction of a newly identified TRAIL$^+$ mDC subset with a mature T-cell stimulatory phenotype, an increased pro-inflammatory Type-1 T-cell cytokine profile and a reduction in the frequency of immunosuppressive CD4$^+$CD25$^+$FoxP3$^+$ Tregs (Molenkamp et al. 2007).

We hypothesize that these PF-3512676-induced immunostimulatory effects affecting both DC and T-cell subsets in the SLN would translate into higher frequencies of melanoma-associated antigens (MAA)-specific CD8$^+$ T cells. This hypothesis was supported by our finding of increased CD8$^+$ T cell responsiveness to melanoma-associated epitopes in the PF-3512676 recipient group (Molenkamp et al. 2008). CD8$^+$ T cells from SLN and peripheral blood were tested for reactivity in an IFN-α ELISPOT-assay against a number of HLA-A1/-A2/-A3-restricted epitopes derived from a range of MAA. Melanoma-specific CD8$^+$ T-cell responses against more than one MAA epitope in either the SLN or the post-injection blood samples were seen in 0 of 11 patients in the saline control group and 6 of 10 patients who received PF-3512676. Cells derived from both post-injection blood and the SLN from 4 of the 6 responding patients had a significant response to more than one MAA epitope. A clear relationship was found between increased frequencies of melanoma-specific CD8$^+$ T cells and NK cells in the SLN and PF-3512676-induced pDC maturation. These data demonstrate a simultaneous increase in melanoma-specific CD8$^+$ T cells and innate NK effector cells following PF-3512676 treatment and support PF-3512676 as a possible candidate agent for adjuvant treatment of early-stage melanoma patients.

We have demonstrated that both GM-CSF and PF-3512676 are potent boosters of the immune system. Even without the separate simultaneous administration of MAA, we have found increased frequencies of melanoma-specific CD8$^+$ T cells in both the SLN and the peripheral blood after tumor-site administration of GM-CSF or PF-3512676. This clearly indicates that primed T cells are present in the SLN at relatively early stages of melanoma development and that these cells that can be (re)activated by an immunostimulatory "push" such as that provided by pro-inflammatory cytokines and/or TLR ligands leading to increased local and systemic frequencies of functionally active CD8$^+$ T cells.

To definitively demonstrate the benefit of non-specific and generally applicable immune potentiation, large-scale, randomized phase III trials are needed. Until then, more feasible small-scale phase II trials are appropriate to test the relative efficacy of different cytokines, TLR ligands or novel modulators of tumor-related immune suppression (e.g., small-molecule inhibitors of IDO or STAT3 phosphorylation), in terms of DC and melanoma-specific T-cell activation. These studies may ultimately lead to the identification of the most promising immunopotentiating cocktails to be further developed for routine clinical application in the adjuvant treatment of early stage and possibly more advanced melanoma patients. In the absence of effective chemotherapy, time invested in this relatively novel approach to management of early-stage melanoma may be highly rewarding.

References

Amersi, F. F., Terando, A. M., Goto, Y., Scolyer, R. A., Thompson, J. F., Tran, A. N., Faries, M. B., Morton, D. L. and Hoon, D. S. B. 2008. Activation of CCR9/CCL25 in cutaneous melanoma mediates preferential metastasis to the small intestine. Clin Cancer Res 14:638–645.

Banchereau, J., Briere, F., Caux, C., Davoust, J., Lebecque, S., Liu, Y. J., Pulendran, B. and Palucka, K. 2000. Immunobiology of dendritic cells. Annu Rev Immunol 18:767–811.

Bennaceur, K., Popa, I., Portoukalian, J., Berthier-Vergnes, O. and Peguet-Navarro, J. 2006. Melanoma-derived gangliosides impair migratory and antigen-presenting function of human epidermal Langerhans cells and induce their apoptosis. Int Immunol 18:879–886.

Biggs, M. W. and Eiselein, J. E. 2001. Suppression of immune surveillance in melanoma. Med Hypotheses 56:648–652.

Boyano, M. D., Garcia-Vazquez, M. D., Lopez-Michelena, T., Gardeazabal, J., Bilbao, J., Canavate, M. L., Galdeano, A. G., Izu, R., Diaz-Ramon, L., Raton, J. A. and Diaz-Perez, J. L. 2000. Soluble interleukin-2 receptor, intercellular adhesion molecule-1 and interleukin-10 serum levels in patients with melanoma. Br J Cancer 83:847–852.

Cella, M., Jarrossay, D., Facchetti, F., Alebardi, O., Nakajima, H., Lanzavecchia, A. and Colonna, M. 1999. Plasmacytoid monocytes migrate to inflamed lymph nodes and produce large amounts of type I interferon. Nat Med 5:919–923.

Chen, Q., Daniel, V., Maher, D. W. and Hersey, P. 1994. Production of IL-10 by melanoma cells: examination of its role in immunosuppression mediated by melanoma. Int J Cancer 56:755–760.

Cochran, A. J., Balda, B. R., Starz, H., Bachter, D., Krag, D. N., Cruse, C. W., Pijpers, R. and Morton, D. L. 2000. The Augsburg Consensus. Techniques of lymphatic mapping, sentinel lymphadenectomy, and completion lymphadenectomy in cutaneous malignancies. Cancer 89:236–241.

Cochran, A. J., Huang, R. R., Lee, J., Itakura, E., Leong, S. P. L. and Essner, R. 2006. Tumour-induced immune modulation of sentinel lymph nodes. Nat Rev Immunol 6:659–670.

Cochran, A. J., Morton, D. L., Stern, S., Lana, A. M. A., Essner, R. and Wen, D. R. 2001. Sentinel lymph nodes show profound downregulation of antigen-presenting cells of the paracortex: implications for tumor biology and treatment. Mod Pathol 14:604–608.

Cochran, A. J., Pihl, E., Wen, D. R., Hoon, D. S. B. and Korn, E. L. 1987. Zoned immune suppression of lymph nodes draining malignant melanoma: histologic and immunohistologic studies. J Natl Cancer Inst 78:399–405.

Cox, K., North, M., Burke, M., Singhal, H., Renton, S., Aqel, N., Islam, S. and Knight, S. C. 2005. Plasmacytoid dendritic cells (PDC) are the major DC subset innately producing cytokines in human lymph nodes. J Leukoc Biol 78:1142–1152.

Dranoff, G. 2002. GM-CSF-based cancer vaccines. Immunol Rev 188:147–154.

Dummer, W., Becker, J. C., Schwaaf, A., Leverkus, M., Moll, T. and Brocker, E. B. 1995. Elevated serum levels of interleukin-10 in patients with metastatic malignant melanoma. Melanoma Res 5:67–68.

Duramad, O., Fearon, K. L., Chan, J. H., Kanzler, H., Marshall, J. D., Coffman, R. L. and Barrat, F. J. 2003. IL-10 regulates plasmacytoid dendritic cell response to CpG-containing immunostimulatory sequences. Blood 102:4487–4492.

Dzionek, A., Inagaki, Y., Okawa, K., Nagafune, J., Rock, J., Sohma, Y., Winkels, G., Zysk, M., Yamaguchi, Y. and Schmitz, J. 2002. Plasmacytoid dendritic cells: from specific surface markers to specific cellular functions. Hum Immunol 63:1133–1148.

Dzionek, A., Sohma, Y., Nagafune, J., Cella, M., Colonna, M., Facchetti, F., Gunther, G., Johnston, I., Lanzavecchia, A., Nagasaka, T., Okada, T., Vermi, W., Winkels, G., Yamamoto, T., Zysk, M., Yamaguchi, Y. and Schmitz, J. 2001. BDCA-2, a novel plasmacytoid dendritic cell-specific type II C-type lectin, mediates antigen capture and is a potent inhibitor of interferon alpha/beta induction. J Exp Med 194:1823–1834.

Elliott, B., Scolyer, R. A., Suciu, S., Lebecque, S., Rimoldi, D., Gugerli, O., Musat, E., Sharma, R. N., Lienard, D., Keilholz, U., Testori, A., Eggermont, A., MacKie, R., Robert, C., Cook, M., Thompson, J. F., Angevin, E. and Spatz, A. 2007. Long-term protective effect of mature DC-LAMP$^+$ dendritic cell accumulation in sentinel lymph nodes containing micrometastatic melanoma. Clin Cancer Res 13:3825–3830.

Facchetti, F., de Wolf-Peeters, C., de Greef, H. and Desmet, V. J. 1984. Langerhans cells in various benign and malignant pigment-cell lesions of the skin. Arch Dermatol Res 276:283–287.

Facchetti, F., de Wolf-Peeters, C., Mason, D. Y., Pulford, K., van den Oord, J. J. and Desmet, V. J. 1988. Plasmacytoid T cells. Immunohistochemical evidence for their monocyte/macrophage origin. Am J Pathol 133:15–21.

Farzad, Z., Cochran, A. J., McBride, W. H., Gray, J. D., Wong, V. and Morton, D. L. 1990. Lymphocyte subset alterations in nodes regional to human melanoma. Cancer Res 50:3585–3588.

Farzad, Z., McBride, W. H., Ogbechi, H., Asnong-Holthoff, C., Morton, D. L. and Cochran, A. J. 1997. Lymphocytes from lymph nodes at different distances from human melanoma vary in their capacity to inhibit/enhance tumor cell growth in vitro. Melanoma Res 7 Suppl 2:S59–65.

Gatter, K. C., Pulford, K. A., Vanstapel, M. J., Roach, B., Mortimer, P., Woolston, R. E., Taylor-Papadimitriou, J., Lane, E. B. and Mason, D. Y. 1984. An immunohistological study of benign and malignant skin tumours: epithelial aspects. Histopathology 8:209–227.

Gazzaniga, S., Bravo, A. I., Guglielmotti, A., van Rooijen, N., Maschi, F., Vecchi, A., Mantovani, A., Mordoh, J. and Wainstok, R. 2007. Targeting tumor-associated macrophages and inhibition of MCP-1 reduce angiogenesis and tumor growth in a human melanoma xenograft. J Invest Dermatol 127:2031–2041.

Gerlini, G., Tun-Kyi, A., Dudli, C., Burg, G., Pimpinelli, N. and Nestle, F. O. 2004. Metastatic melanoma secreted IL-10 down-regulates CD1 molecules on dendritic cells in metastatic tumor lesions. Am J Pathol 165:1853–1863.

Gursel, M., Verthelyi, D. and Klinman, D. M. 2002. CpG oligodeoxynucleotides induce human monocytes to mature into functional dendritic cells. Eur J Immunol 32:2617–2622.

Haigh, P. I., Lucci, A., Turner, R. R., Bostick, P. J., Krasne, D. L., Stern, S. L. and Morton, D. L. 2001. Carbon dye histologically confirms the identity of sentinel lymph nodes in cutaneous melanoma. Cancer 92:535–541.

Hawiger, D., Inaba, K., Dorsett, Y., Guo, M., Mahnke, K., Rivera, M., Ravetch, J. V., Steinman, R. M. and Nussenzweig, M. C. 2001. Dendritic cells induce peripheral T cell unresponsiveness under steady state conditions in vivo. J Exp Med 194:769–779.

Hoon, D. S. B., Foshag, L. J., Nizze, A. S., Bohman, R. and Morton, D. L. 1990. Suppressor cell activity in a randomized trial of patients receiving active specific immunotherapy with melanoma cell vaccine and low dosages of cyclophosphamide. Cancer Res 50:5358–5364.

Hoon, D. S. B., Korn, E. L. and Cochran, A. J. 1987. Variations in functional immunocompetence of individual tumor-draining lymph nodes in humans. Cancer Res 47:1740–1744.
Huang, R. R., Paul, E., Lin, L. L., Wen, D. R., Itakura, E. and Cochran, A. J. 2005. Sentinel lymph nodes are immunosuppressed whether or not they contain metastatic melanoma. Mod Pathol 18:84A–85A.
Huang, R. R., Turner, R. and Cochran, A. J. 2008. Tumor-free sentinel nodes show immune modulation. Ann Surg Oncol 15:S1.
Huang, R. R., Wen, D. R., Guo, J., Giuliano, A. E., Nguyen, M., Offodile, R., Stern, S., Turner, R. and Cochran, A. J. 2000. Selective modulation of paracortical dendritic cells and T-lymphocytes in breast cancer sentinel lymph nodes. Breast J 6:225–232.
Huang, R. R., Wen, D. R., Itakura, E., Essner, R. and Cochran, A. J. 2004a. Loss of dendritic cell complexity in sentinel nodes correlates with reduced activated T cells. Ann Surg Oncol 11:S3.
Huang, R. R., Wen, D. R., Itakura, E., Turner, R. and Cochran, A. J. 2004b. Does the pathologist have a role in conforming that a node submitted as sentinel is true sentinel? Mod Pathol 17:93A.
Hussein, M. R. 2005. Transforming growth factor-beta and malignant melanoma: molecular mechanisms. J Cutan Pathol 32:389–395.
Ito, T., Kanzler, H., Duramad, O., Cao, W. and Liu, Y. J. 2006. Specialization, kinetics, and repertoire of type 1 interferon responses by human plasmacytoid predendritic cells. Blood 107:2423–2431.
Jandus, C., Bioley, G., Speiser, D. E. and Romero, P. 2008. Selective accumulation of differentiated FOXP3^{+} CD4^{+} T cells in metastatic tumor lesions from melanoma patients compared to peripheral blood. Cancer Immunol Immunother 57:1795–1805.
Kawarada, Y., Ganss, R., Garbi, N., Sacher, T., Arnold, B. and Hammerling, G. J. 2001. NK- and CD8^{+} T cell-mediated eradication of established tumors by peritumoral injection of CpG-containing oligodeoxynucleotides. J Immunol 167:5247–5253.
Kim, R., Emi, M., Tanabe, K. and Arihiro, K. 2006. Tumor-driven evolution of immunosuppressive networks during malignant progression. Cancer Res 66:5527–5536.
Krug, A., Towarowski, A., Britsch, S., Rothenfusser, S., Hornung, V., Bals, R., Giese, T., Engelmann, H., Endres, S., Krieg, A. M. and Hartmann, G. 2001. Toll-like receptor expression reveals CpG DNA as a unique microbial stimulus for plasmacytoid dendritic cells which synergizes with CD40 ligand to induce high amounts of IL-12. Eur J Immunol 31:3026–3037.
Krüger-Krasagakes, S., Krasagakis, K., Garbe, C., Schmitt, E., Hüls, C., Blankenstein, T. and Diamantstein, T. 1994. Expression of interleukin 10 in human melanoma. Br J Cancer 70:1182–1185.
Lacal, P. M., Failla, C. M., Pagani, E., Odorisio, T., Schietroma, C., Falcinelli, S., Zambruno, G. and D'Atri, S. 2000. Human melanoma cells secrete and respond to placenta growth factor and vascular endothelial growth factor. J Invest Dermatol 115:1000–1007.
Ladanyi, A., Kiss, J., Somlai, B., Gilde, K., Fejos, Z., Mohos, A., Gaudi, I. and Timar, J. 2007. Density of DC-LAMP^{+} mature dendritic cells in combination with activated T lymphocytes infiltrating primary cutaneous melanoma is a strong independent prognostic factor. Cancer Immunol Immunother 56:1459–1469.
Laguens, G., Coronato, S., Laguens, R., Portiansky, E. and Di Girolamo, V. 2002. Human regional lymph nodes draining cancer exhibit a profound dendritic cell depletion as comparing to those from patients without malignancies. Immunol Lett 84:159–162.
Lana, A. M. A., Wen, D. R. and Cochran, A. J. 2001. The morphology, immunophenotype and distribution of paracortical dendritic leucocytes in lymph nodes regional to cutaneous melanoma. Melanoma Res 11:401–410.
Langerhans, P. 1868. Ueber die Nerven der menschlichen Haut. Virchows Archives [Pathol Anat] 44:325–337.
Lee, J. H., Torisu-Itakara, H., Cochran, A. J., Kadison, A., Huynh, Y., Morton, D. L. and Essner, R. 2005. Quantitative analysis of melanoma-induced cytokine-mediated immunosuppression in melanoma sentinel nodes. Clin Cancer Res 11:107–112.

Lee, J. R., Dalton, R. R., Messina, J. L., Sharma, M. D., Smith, D. M., Burgess, R. E., Mazzella, F., Antonia, S. J., Mellor, A. L. and Munn, D. H. 2003. Pattern of recruitment of immunoregulatory antigen-presenting cells in malignant melanoma. Lab Invest 83:1457–1466.

Liu, Y., Zhang, P., Li, J., Kulkarni, A. B., Perruche, S. and Chen, W. 2008. A critical function for TGF-β signaling in the development of natural $CD4^+CD25^+Foxp3^+$ regulatory T cells. Nat Immunol 9:632–640.

Liyanage, U. K., Goedegebuure, P. S., Moore, T. T., Viehl, C. T., Moo-Young, T. A., Larson, J. W., Frey, D. M., Ehlers, J. P., Eberlein, T. J. and Linehan, D. C. 2006. Increased prevalence of regulatory T cells (T_{reg}) is induced by pancreas adenocarcinoma. J Immunother 29:416–424.

Lucci, A., Turner, R. R. and Morton, D. L. 1999. Carbon dye as an adjunct to isosulfan blue dye for sentinel lymph node dissection. Surgery 126:48–53.

Ludviksson, B. R., Seegers, D., Resnick, A. S. and Strober, W. 2000. The effect of TGF-β on immune responses of naive versus memory $CD4^+$ Th1/Th2 T cells. Eur J Immunol 30:2101–2111.

Luo, X., Tarbell, K. V., Yang, H., Pothoven, K., Bailey, S. L., Ding, R., Steinman, R. M. and Suthanthiran, M. 2007. Dendritic cells with TGF-β differentiate naïve $CD4^+CD25^-$ T cells into islet-protective $Foxp3^+$ regulatory T cells. Proc Natl Acad Sci U S A 104:2821–2826.

Maeda, H., Kuwahara, H., Ichimura, Y., Ohtsuki, M., Kurakata, S. and Shiraishi, A. 1995. TGF-β enhances macrophage ability to produce IL-10 in normal and tumor-bearing mice. J Immunol 155:4926–4932.

Maeda, H. and Shiraishi, A. 1996. TGF-beta contributes to the shift toward Th2-type responses through direct and IL-10-mediated pathways in tumor-bearing mice. J Immunol 156:73–78.

Mahnke, K., Bedke, T. and Enk, A. H. 2007. Regulatory conversation between antigen presenting cells and regulatory T cells enhance immune suppression. Cell Immunol 250:1–13.

Matsuura, K., Yamaguchi, Y., Ueno, H., Osaki, A., Arihiro, K. and Toge, T. 2006. Maturation of dendritic cells and T-cell responses in sentinel lymph nodes from patients with breast carcinoma. Cancer 106:1227–1236.

Molenkamp, B. G., Sluijter, B. J., van Leeuwen, P. A., Santegoets, S. J. A. M., Meijer, S., Wijnands, P. G. J. T. B., Haanen, J. B. A. G., van den Eertwegh, A. J. M., Scheper, R. J. and de Gruijl, T. D. 2008. Local administration of PF-3512676 CpG-B instigates tumor-specific $CD8^+$ T-cell reactivity in melanoma patients. Clin Cancer Res 14:4532–4542.

Molenkamp, B. G., van Leeuwen, P. A. M., Meijer, S., Sluijter, B. J. R., Wijnands, P. G. J. T. B., Baars, A., van den Eertwegh, A. J. M., Scheper, R. J. and de Gruijl, T. D. 2007. Intradermal CpG-B activates both plasmacytoid and myeloid dendritic cells in the sentinel lymph node of melanoma patients. Clin Cancer Res 13:2961–2969.

Morelli, A. E. and Thomson, A. W. 2007. Tolerogenic dendritic cells and the quest for transplant tolerance. Nat Rev Immunol 7:610–621.

Morton, D. L., Hoon, D. S. B., Cochran, A. J., Turner, R. R., Essner, R., Takeuchi, H., Wanek, L. A., Glass, E., Foshag, L. J., Hsueh, E. C., Bilchik, A. J., Elashoff, D. and Elashoff, R. 2003. Lymphatic mapping and sentinel lymphadenectomy for early-stage melanoma: therapeutic utility and implications of nodal microanatomy and molecular staging for improving the accuracy of detection of nodal micrometastases. Ann Surg 238:538–549.

Morton, B. A., Ramey, W. G., Paderon, H. and Miller, R. E. 1986. Monoclonal antibody-defined phenotypes of regional lymph node and peripheral blood lymphocyte subpopulations in early breast cancer. Cancer Res 46:2121–2126.

Morton, D. L., Thompson, J. F., Cochran, A. J., Mozzillo, N., Elashoff, R., Essner, R., Nieweg, O. E., Roses, D. F., Hoekstra, H. J., Karakousis, C. P., Reintgen, D. S., Coventry, B. J., Glass, E. C. and Wang, H. J. 2006. Sentinel-node biopsy or nodal observation in melanoma. N Engl J Med 355:1307–1317.

Morton, D. L., Wen, D. R., Wong, J. H., Economou, J. S., Cagle, L. A., Storm, F. K., Foshag, L. J. and Cochran, A. J. 1992. Technical details of intraoperative lymphatic mapping for early stage melanoma. Arch Surg 127:392–399.

Mouawad, R., Khayat, D. and Soubrane, C. 2000. Plasma Fas ligand, an inducer of apoptosis, and plasma soluble Fas, an inhibitor of apoptosis, in advanced melanoma. Melanoma Res 10:461–467.

Movassagh, M., Spatz, A., Davoust, J., Lebecque, S., Romero, P., Pittet, M., Rimoldi, D., Lienard, D., Gugerli, O., Ferradini, L., Robert, C., Avril, M. F., Zitvogel, L. and Angevin, E. 2004. Selective accumulation of mature DC-Lamp$^+$ dendritic cells in tumor sites is associated with efficient T-cell-mediated antitumor response and control of metastatic dissemination in melanoma. Cancer Res 64:2192–2198.

Munn, D. H., Sharma, M. D., Hou, D., Baban, B., Lee, J. R., Antonia, S. J., Messina, J. L., Chandler, P., Koni, P. A. and Mellor, A. L. 2004. Expression of indoleamine 2,3-dioxygenase by plasmacytoid dendritic cells in tumor-draining lymph nodes. J Clin Invest 114:280–290.

Munn, D. H., Sharma, M. D., Lee, J. R., Jhaver, K. G., Johnson, T. S., Keskin, D. B., Marshall, B., Chandler, P., Antonia, S. J., Burgess, R., Slingluff, C. L., Jr. and Mellor, A. L. 2002. Potential regulatory function of human dendritic cells expressing indoleamine 2,3-dioxygenase. Science 297:1867–1870.

Munz, C., Steinman, R. M. and Fujii, S. 2005. Dendritic cell maturation by innate lymphocytes: coordinated stimulation of innate and adaptive immunity. J Exp Med 202:203–207.

Murakami, T., Cardones, A. R. and Hwang, S. T. 2004. Chemokine receptors and melanoma metastasis. J Dermatol Sci 36:71–78.

Ohl, L., Mohaupt, M., Czeloth, N., Hintzen, G., Kiafard, Z., Zwirner, J., Blankenstein, T., Henning, G. and Forster, R. 2004. CCR7 governs skin dendritic cell migration under inflammatory and steady-state conditions. Immunity 21:279–288.

Palucka, K. A., Taquet, N., Sanchez-Chapuis, F. and Gluckman, J. C. 1998. Dendritic cells as the terminal stage of monocyte differentiation. J Immunol 160:4587–4595.

Patterson, S. 2000. Flexibility and cooperation among dendritic cells. Nat Immunol 1:273–274.

Payne, A. S. and Cornelius, L. A. 2002. The role of chemokines in melanoma tumor growth and metastasis. J Invest Dermatol 118:915–922.

Pinzon-Charry, A., Maxwell, T., McGuckin, M. A., Schmidt, C., Furnival, C. and Lopez, J. A. 2006. Spontaneous apoptosis of blood dendritic cells in patients with breast cancer. Breast Cancer Res 8:R5.

Polak, M. E., Borthwick, N. J., Gabriel, F. G., Johnson, P., Higgins, B., Hurren, J., McCormick, D., Jager, M. J. and Cree, I. A. 2007a. Mechanisms of local immunosuppression in cutaneous melanoma. Br J Cancer 96:1879–1887.

Polak, M. E., Borthwick, N. J., Johnson, P., Hungerford, J. L., Higgins, B., Di Palma, S., Jager, M. J. and Cree, I. A. 2007b. Presence and phenotype of dendritic cells in uveal melanoma. Br J Ophthalmol 91:971–976.

Pollard, J. W. 2004. Tumour-educated macrophages promote tumour progression and metastasis. Nat Rev Cancer 4:71–78.

Probst, H. C., McCoy, K., Okazaki, T., Honjo, T. and van den Broek, M. 2005. Resting dendritic cells induce peripheral CD8$^+$ T cell tolerance through PD-1 and CTLA-4. Nat Immunol 6:280–286.

Randolph, G. J., Angeli, V. and Swartz, M. A. 2005. Dendritic-cell trafficking to lymph nodes through lymphatic vessels. Nat Rev Immunol 5:617–628.

Redondo, P., Solano, T., B, V. A., Bauza, A. and Idoate, M. 2002. Fas and Fas ligand: expression and soluble circulating levels in cutaneous malignant melanoma. Br J Dermatol 147:80–86.

Salio, M., Cella, M., Vermi, W., Facchetti, F., Palmowski, M. J., Smith, C. L., Shepherd, D., Colonna, M. and Cerundolo, V. 2003. Plasmacytoid dendritic cells prime IFN-γ-secreting melanoma-specific CD8 lymphocytes and are found in primary melanoma lesions. Eur J Immunol 33:1052–1062.

Scala, S., Giuliano, P., Ascierto, P. A., Ierano, C., Franco, R., Napolitano, M., Ottaiano, A., Lombardi, M. L., Luongo, M., Simeone, E., Castiglia, D., Mauro, F., De Michele, I., Calemma, R., Botti, G., Caraco, C., Nicoletti, G., Satriano, R. A. and Castello, G. 2006. Human melanoma metastases express functional CXCR4. Clin Cancer Res 12:2427–2433.

Seo, N., Hayakawa, S., Takigawa, M. and Tokura, Y. 2001. Interleukin-10 expressed at early tumour sites induces subsequent generation of $CD4^+$ T-regulatory cells and systemic collapse of antitumour immunity. Immunology 103:449–457.

Sharma, M. D., Baban, B., Chandler, P., Hou, D. Y., Singh, N., Yagita, H., Azuma, M., Blazar, B. R., Mellor, A. L. and Munn, D. H. 2007. Plasmacytoid dendritic cells from mouse tumor-draining lymph nodes directly activate mature Tregs via indoleamine 2,3-dioxygenase. J Clin Invest 117:2570–2582.

Shields, J. D., Emmett, M. S., Dunn, D. B., Joory, K. D., Sage, L. M., Rigby, H., Mortimer, P. S., Orlando, A., Levick, J. R. and Bates, D. O. 2007. Chemokine-mediated migration of melanoma cells towards lymphatics – a mechanism contributing to metastasis. Oncogene 26:2997–3005.

Shortman, K. and Naik, S. H. 2007. Steady-state and inflammatory dendritic-cell development. Nat Rev Immunol 7:19–30.

Steinman, R. M. and Banchereau, J. 2007. Taking dendritic cells into medicine. Nature 449:419–426.

Steinman, R. M. and Cohn, Z. A. 1973. Identification of a novel cell type in peripheral lymphoid organs of mice. I. Morphology, quantitation, tissue distribution. J Exp Med 137:1142–1162.

Stene, M. A., Babajanians, M., Bhuta, S. and Cochran, A. J. 1988. Quantitative alterations in cutaneous Langerhans cells during the evolution of malignant melanoma of the skin. J Invest Dermatol 91:125–128.

Summers, K. L., Hock, B. D., McKenzie, J. L. and Hart, D. N. 2001. Phenotypic characterization of five dendritic cell subsets in human tonsils. Am J Pathol 159:285–295.

Takeuchi, H., Fujimoto, A., Tanaka, M., Yamano, T., Hsueh, E. and Hoon, D. S. B. 2004. CCL21 chemokine regulates chemokine receptor CCR7 bearing malignant melanoma cells. Clin Cancer Res 10:2351–2358.

Torisu, H., Ono, M., Kiryu, H., Furue, M., Ohmoto, Y., Nakayama, J., Nishioka, Y., Sone, S. and Kuwano, M. 2000. Macrophage infiltration correlates with tumor stage and angiogenesis in human malignant melanoma: possible involvement of TNFα and IL-1α. Int J Cancer 85:182–188.

Torisu-Itakura, H., Lee, J. H., Huynh, Y., Ye, X., Essner, R. and Morton, D. L. 2007a. Monocyte-derived IL-10 expression predicts prognosis of stage IV melanoma patients. J Immunother 30:831–838.

Torisu-Itakura, H., Lee, J. H., Scheri, R. P., Huynh, Y., Ye, X., Essner, R. and Morton, D. L. 2007b. Molecular characterization of inflammatory genes in sentinel and nonsentinel nodes in melanoma. Clin Cancer Res 13:3125–3132.

Ugurel, S., Rappl, G., Tilgen, W. and Reinhold, U. 2001. Increased soluble CD95 (sFas/CD95) serum level correlates with poor prognosis in melanoma patients. Clin Cancer Res 7:1282–1286.

Varney, M. L., Johansson, S. L. and Singh, R. K. 2005a. Tumour-associated macrophage infiltration, neovascularization and aggressiveness in malignant melanoma: role of monocyte chemotactic protein-1 and vascular endothelial growth factor-A. Melanoma Res 15:417–425.

Varney, M. L., Johansson, S. L. and Singh, R. K. 2006. Distinct expression of CXCL8 and its receptors CXCR1 and CXCR2 and their association with vessel density and aggressiveness in malignant melanoma. Am J Clin Pathol 125:209–216.

Varney, M. L., Olsen, K. J., Mosley, R. L. and Singh, R. K. 2005b. Paracrine regulation of vascular endothelial growth factor – a expression during macrophage-melanoma cell interaction: role of monocyte chemotactic protein-1 and macrophage colony-stimulating factor. J Interferon Cytokine Res 25:674–683.

Vermi, W., Bonecchi, R., Facchetti, F., Bianchi, D., Sozzani, S., Festa, S., Berenzi, A., Cella, M. and Colonna, M. 2003. Recruitment of immature plasmacytoid dendritic cells (plasmacytoid monocytes) and myeloid dendritic cells in primary cutaneous melanomas. J Pathol 200:255–268.

Vuylsteke, R. J. C. L. M., Molenkamp, B. G., Gietema, H. A., van Leeuwen, P. A. M., Wijnands, P. G. J. T. B., Vos, W., van Diest, P. J., Scheper, R. J., Meijer, S. and de Gruijl, T. D. 2004. Local administration of granulocyte/macrophage colony-stimulating factor increases the number and activation state of dendritic cells in the sentinel lymph node of early-stage melanoma. Cancer Res 64:8456–8460.

Vuylsteke, R. J. C. L. M., Molenkamp, B. G., van Leeuwen, P. A. M., Meijer, S., Wijnands, P. G. J. T. B., Haanen, J. B. A. G., Scheper, R. J. and de Gruijl, T. D. 2006. Tumor-specific $CD8^+$ T cell reactivity in the sentinel lymph node of GM-CSF-treated stage I melanoma patients is associated with high myeloid dendritic cell content. Clin Cancer Res 12:2826–2833.

Vuylsteke, R. J. C. L. M., van Leeuwen, P. A. M., Meijer, S., Wijnands, P. G. J. T. B., Statius Muller, M. G., Busch, D. H., Scheper, R. J. and de Gruijl, T. D. 2002. Sampling tumor-draining lymph nodes for phenotypic and functional analysis of dendritic cells and T cells. Am J Pathol 161:19–26.

Wen, D. R., Hoon, D. S. B., Chang, C. and Cochran, A. J. 1989. Variations in lymphokine generation by individual lymph nodes draining human malignant tumors. Cancer Immunol Immunother 30:277–282.

Wethmar, K., Helmus, Y., Luhn, K., Jones, C., Laskowska, A., Varga, G., Grabbe, S., Lyck, R., Engelhardt, B., Bixel, M. G., Butz, S., Loser, K., Beissert, S., Ipe, U., Vestweber, D. and Wild, M. K. 2006. Migration of immature mouse DC across resting endothelium is mediated by ICAM-2 but independent of β_2-integrins and murine DC-SIGN homologues. Eur J Immunol 36:2781–2794.

Wiley, H. E., Gonzalez, E. B., Maki, W., Wu, M. T. and Hwang, S. T. 2001. Expression of CC chemokine receptor-7 and regional lymph node metastasis of B16 murine melanoma. J Natl Cancer Inst 93:1638–1643.

Wilson, N. S., El-Sukkari, D., Belz, G. T., Smith, C. M., Steptoe, R. J., Heath, W. R., Shortman, K. and Villadangos, J. A. 2003. Most lymphoid organ dendritic cell types are phenotypically and functionally immature. Blood 102:2187–2194.

Chapter 21
Dendritic Cell Vaccines in Cancer: Obstacles to Overcome

Ezra D. Mirvish, Rebecca G. Pomerantz, Louis D. Falo, and Larisa J. Geskin

Abstract Significant progress has been made in the development of cancer immunotherapies. However, as advances continue to unravel the complexity of the immune system and processes involved in the induction of antitumor immunity and tolerance, new questions increasingly arise. Various approaches have been undertaken in an effort to induce efficient antitumor immunity. Dendritic cells are the most potent antigen-presenting cells known, and this capacity of dendritic cells has been exploited against various malignancies under different conditions. This chapter reviews various controversies in dendritic cell-based vaccine development, including sources of dendritic cells, their differentiation, maturation and antigen loading, vaccine delivery, as well as effective monitoring of the immune and clinical responses in vivo.

21.1 Introduction

Dendritic cells (DC) are the most potent known antigen-presenting cells (APC) capable of effectively taking up, processing, and presenting antigens to induce and modulate immune responses, including antitumor immunity. This capacity of DC has been widely explored for use in cancer vaccination approaches. However, despite significant progress in the development of DC-based vaccines and growing knowledge about the immunology and biology of DC, a number of unresolved questions remain. This chapter will address some of the issues surrounding the design of DC vaccines for cancer immunotherapy, including, but not limited to

1. Which subset of DC should be used in order to elicit potent antitumor responses in various malignancies?

L.J. Geskin (✉)
Cutaneous Oncology and Photopheresis Center, University of Pittsburgh School of Medicine, 190 Lothrop Street, Lothrop Hall, Suite 101, Pittsburgh, PA 15213, USA
e-mail: geskinlj@upmc.edu

M.R. Shurin, R.D. Salter (eds.), *Dendritic Cells in Cancer*,
DOI 10.1007/978-0-387-88611-4_21,

2. What is the most efficient way to generate effective DC for use in anticancer vaccines?
3. What types and sources of antigens should be utilized in order to ensure optimum uptake by DC and stimulation of immune effectors?
4. What is the most effective technique for loading DC with antigens?
5. How do the conditions of DC maturation affect the nature of the immune responses induced by DC vaccination?
6. After DC are generated and loaded with antigen, what is the optimal route of administration? How does the method of vaccine administration affect the type and intensity of the elicited immune response?
7. What agents can be used as adjuvants to enhance the anticancer immune responses stimulated by DC vaccines?
8. How should the effects of DC vaccination be evaluated?

Growing body of evidence further expands our understanding of mechanisms of immune responses; daily many more issues arise and indicate ever increasing complexity and intricacy of the immune system and processes involved in the control of immunity.

21.2 Use of Dendritic Cell Subsets for Immunotherapy

DC arise from stem cell precursors in the bone marrow; they share certain defining features, such as high levels of HLA-DR major histocompatibility complex (MHC) antigens and a lack of other leukocyte lineage markers (CD3, CD14, CD19, CD20, CD16, and CD56). However, DC constitute a heterogeneous cell population, with variability in immunophenotype, function and, potentially, hematopoietic lineage (Yang et al. 2005). Two major subsets of peripheral blood DC have been described: myeloid DC and lymphoid DC (Table 21.1).

Myeloid DC ($CD14^{-}/CD11c^{+}/CD123^{-}$) are potent activators of T cells and inducers of cytotoxic T lymphocyte (CTL) responses. Lymphoid DC ($CD14^{-}/CD11c^{-}/CD123^{+}$), also called "plasmacytoid" DC due to their morphologic similarities to plasma cells, are primarily known for their ability to produce high levels of IFN-α. Early studies showed that in comparison to myeloid DC, lymphoid DC have weaker capacities for phagocytosis and T-cell stimulation (O'Doherty et al. 1994) and that upon stimulation, lymphoid DC favor a Th2 rather than Th1 cytokine secretion profile (Rissoan et al. 1999). Due to the well-characterized role of CTL in the destruction of tumor cells and the more potent CTL responses induced by myeloid DC than by lymphoid DC, the former subset has been employed in current DC vaccination approaches. However, IFN-secreting lymphoid DC can also be induced to drive Th1 polarization (Cella et al. 2000), and recent evidence demonstrates synergy between lymphoid DC and myeloid DC in stimulating T-cell activation and antitumor immune responses (Lou et al. 2007). Moreover, appropriately stimulated plasmacytoid

Table 21.1 Myeloid and plasmacytoid dendritic cell subsets

DC type	Cell-surface markers	Cytokine secretion	Exogenous activation signal	Morphology/phenotype	DC subtypes (location)	Toll-like receptors
Myeloid*	CD14(–) CD11b(+) CD11c(+) CD13(+) CD123(–)	TNF-α IL-12 IL-6 (IFN-α/β) IL-23 IL-27	Bacterial (and viral) infection	– Large – Extensive processes – Low nuclear-to-cytoplasmic ratio – Able to produce higher levels of IL-12 in response to antigen challenge	Langerhans cells (epidermis) Interstitial DC (all other tissue)	(TLR 1) TLR 2 TLR 3 (TLR 4 (moDC)) (TLR 5) (TLR 8 (moDC))
Plasmacytoid†	CD14(–) CD11c(–) CD123(+) BDCA-2 BDCA-4	IFN-α IFN-β (IL-6)	Viral infection($\rightarrow$ $\uparrow$IFN-α secretion, as well as IFN-β, ω)	– Small, round – No processes – Higher nuclear-to-cytoplasmic ratio – Poorer endocytic capability, but capable of priming effector T cells – Key effectors of innate immunity (due to IFN) – Precursor pDC‡ have capacity to induce T-helper anergy, CD4+ Tregs, and regulatory CTL	–	TLR 7 TLR 9

*The exact relationship between myeloid DC and monocyte-derived DC (moDC) is unclear, but both are often grouped within the term "myeloid DC".
†Plasmacytoid DC derive from both lymphoid and myeloid progenitor cells.
‡pDC in vitro will die without cytokines, e.g., IL-3 , and are incapable of inducing allogeneic T-cell response sans cytokines, unlike fresh myeloid DC (mDC); thus, pDC appear to be "precursors", while cognate mDC are "immature".

DC may play a more significant role in anticancer immune responses than once believed (Basner-Tschakarjan et al. 2006). Taken together, emerging evidence supports the idea that the interplay between DC subsets is important in eliciting appropriate immune responses and that any subset, given the appropriate conditions, may produce potent antitumor immune responses. Investigations of combined vaccination approaches incorporating multiple DC subsets are underway.

21.3 Generation of Dendritic Cells for Therapeutic Vaccines

Methods for generating DC have changed significantly within the past two decades. DC are rare in peripheral blood, and sufficient numbers of DC need to be collected for effective immunotherapy to take place. The simplest and most commonly utilized approach to obtaining DC in large amounts is by utilizing monocyte progenitors. Induction of DC differentiation in vitro from isolated monocyte precursors has enabled the generation of large numbers of mouse (Inaba et al. 1992) and human(Romani et al. 1996) DC from bone marrow as well as peripheral blood in as little as 48 h. Autologous or allogeneic monocyte progenitors can be used for DC vaccine generation. Because cancer patients have decreased numbers of DC precursors and functionally deficient circulating DC (Romani et al. 1996; Gabrilovich et al. 1997; Gottfried et al. 2008), allogeneic approaches may be considered preferable, though they pose a risk of HLA-mismatch and graft-vs-host disease (GVHD). All the same, DC generated from the monocyte precursors of cancer patients have been shown to be capable of inducing potent immune responses (Babatz et al. 2003; Pedersen et al. 2005; Schaft et al. 2005); accordingly, autologous DC have been widely utilized in the generation of anticancer vaccines (Fig. 21.1).

Currently, the most widely employed technique for preparing DC is differentiation of peripheral blood monocytes by short-term in vitro culture with interleukin-4 (IL-4) and granulocyte–macrophage colony-stimulating factor (GM-CSF). Cytokines other than IL-4 can also be used for ex vivo DC differentiation; IL-15 cultured monocytic DC, for example, have been shown to induce a more robust CTL response in vitro than their IL-4-cultured counterparts (Mohamadzadeh et al. 2001). Immature myeloid DC obtained via in vitro differentiation (CD14+/CD11c+/CD123–) are efficient in antigen uptake. However, immature DC are thought to promote immune tolerance rather than active immunity, leading to the proliferation of regulatory T cells (Treg) and induction of other immune-suppressive processes. This tolerogenic property of immature DC may be related to their low expression of costimulatory molecules (e.g., CD80, CD86) or their high expression of various inhibitory cell-surface proteins, such as CTLA-4 (Steinman et al. 2003). Although not a desired attribute of anticancer vaccines, the tolerogenic quality of immature DC has

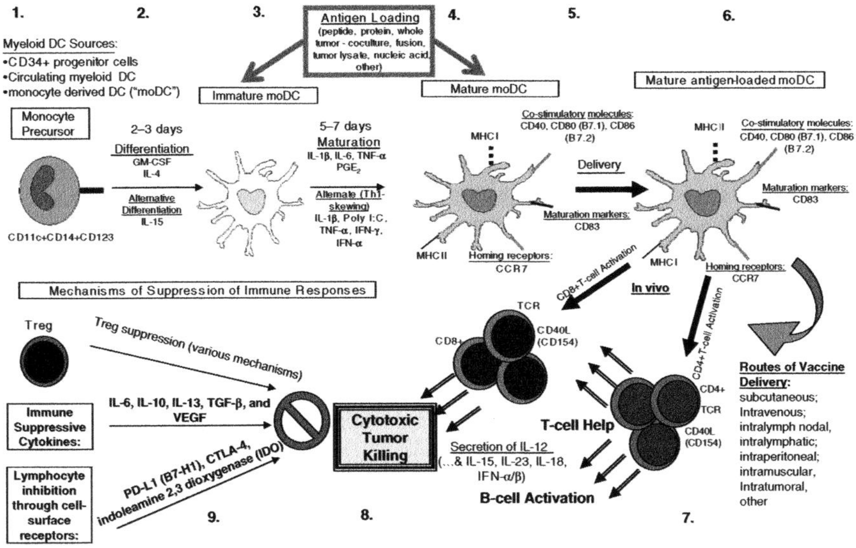

Fig. 21.1 Generation of dendritic cell-based vaccines and induction of antitumor immune response. Several steps are required for DC vaccine preparation. **1.** DC for therapeutic use can be derived from various sources, Monocyte-derived DC (moDC) are the most commonly used subtype of DC. **2.** Differentiation of DC from DC precursors is most commonly accomplished by using IL-4 and GM-CSF. Alternatively, IL-15 may be used. **3**. Maturation of DC induces loss of endocytotic/phagocytic receptors, which is associated with a decrease in antigen capture; concomitant changes that increase DC motility and LN homing (e.g., increased CCR7 expression); pMHC translocation to cell surface, and increased expression of costimulatory molecules (e.g., CD80, CD86, Tumor Necrosis Factor Receptor (TNFR) ligands like OX40L and 4-1BBL). Use of TLR ligands like poly I:C (p-I:C) in DC maturation cocktails may induce DC polarization to the DC1 phenotype, with an associated increase in secretion of Th1-favoring cytokines like IL-12 (Hokey et al. 2005). PgE2 induces increased CCR7 expression, which in turn increases lymph node homing, but also decreased IL-12 expression (Kalinski et al. 2001), leading to decreased CTL response. **4.** Antigen loading is most commonly carried out in immature DC due to their increased ability to endocytose antigen, though mature DC can also be loaded with antigen successfully depending on the type of antigen utilized. **5.** Mature and antigen-loaded DC-based vaccine expressing high levels of costimulatory molecules can then be delivered in vivo. **6.** Numerous routes of immunization can be utilized (*curved arrow*), depending on the desired immune effect (see text). **7.** In vivo, DC vaccine can stimulate appropriate lymphocyte populations. Depending on the antigen used for loading, this stimulation may be restricted by MHC class I or class II presentation. In some cases, such as when the whole tumor is used for DC loading, antigen presentation can occur through both MHC class I and class II molecules, which results in synergistic stimulation of both CD8 + and CD4 + lymphocytes. **8.** DC-stimulated cytotoxic T cells can kill tumor cells with or without the assistance of CD4 + T-helper cells. **9.** Immune responses to DC vaccines can be hampered by various immune mechanisms, including the induction of regulatory T cells, secretion of immunosuppressive cytokines, or action of other suppressive molecules

been explored for the treatment of autoimmune diseases and in transplantation medicine (Popov et al. 2006).

Characteristic qualities of immature DC include low levels of costimulatory molecules and maturation markers (CD40; CD80 (B7.1) and CD86 (B7.2); CD83; and various members of the TNF-receptor superfamily, including OX40L and 4-1BBL) (O'Neill et al. 2004) coupled with a notable ability to internalize and process antigens (Morelli et al. 2004). The high endocytic activity of immature DC can be harnessed for a number of qualitatively different antigens in cases where antigen must be taken up before endocytic receptors are lost. However, in cases utilizing genetic material, such as RNA, antigen loading may be successfully accomplished before or after the maturation process is completed (Morse et al. 1998; Schaft et al. 2005).

Following their in vitro differentiation and loading with antigen, DC can be induced to mature without exposure to additional external stimuli (Stober et al. 2002; Gamvrellis et al. 2004) or in a 48–72 h process involving exposure to supplemental cytokines. These maturation processes yield stable, homogeneous, and mature antigen-loaded DC populations capable of inducing tumoricidal immune responses. Mature DC lose endocytic receptors and gain motility (which correlates with an increase in expression of CCR7) as well as surface expression of MHC receptors plus antigen peptide (pMHC). The standard cytokine cocktail used for maturation includes IL-1β, IL-6, tumor necrosis factor (TNF)-α, and PgE_2(Jonuleit et al. 1997). While PgE_2 increases DC expression of the CCR7 chemokine receptor, which increases lymph node homing in response to the MIP-3β chemokine (Scandella et al. 2002), it may also inhibit secretion of IL-12, a cytokine fundamental to the induction of a Th1 immune response (Kalinski et al. 2001). These competing concerns have led to the development of a number of different approaches to in vitro DC maturation, including transduction of DC with an IL-12-encoding vector in order to promote expression of the cytokine (Nishimura et al. 2001). Mailliard and colleagues developed a PgE_2-free cocktail containing interferon (IFN)-α, IFN-γ, polyinosinic:polycytidylic acid (p-I:C), IL-1β, and TNF-α (Mailliard et al. 2004). The synthetic ligand p-I:C, which binds to toll-like receptor-3 (TLR-3), is utilized due to the important role of particular TLR ligations in the maturation of DC (Sporri and Reis e Sousa 2005) and the polarization of DC to DC1, which induce the Th1 T-cell phenotype that effects cytotoxicity (Hokey et al. 2005).

DC can also be generated from CD34+ hematopoietic stem cells from bone marrow or peripheral blood. These progenitors are cultured with TNF-α, GM-CSF, and Flt3 ligand (Flt3-L) in the presence or absence of IL-4 to yield differentiated DC. In contrast to monocyte-derived DC, $CD34^+$-progenitor-derived DC comprise multiple DC subtypes including dermal DC and Langerhans cells (LC). The inclusion of LC in $CD34^+$-derived vaccines has been hypothesized to increase their immunogenicity, and $CD34^+$-derived DC vaccines have been employed in a number of clinical trials in melanoma (Di Nicola et al. 2004; Banchereau et al. 2005; Fay et al. 2006). While a comparative analysis of $CD34^+$-derived DC and monocyte-derived DC revealed similar antigen uptake and presentation abilities (Herbst et al. 1997), other studies have shown that the former DC type stimulates more potent in vitro CTL responses after induction with certain experimental stimuli (Mortarini et al. 1997; Ferlazzo et al. 1999; Ratzinger et al. 2004). Functional differences between monocyte-derived and $CD34^+$ progenitor-derived DC remain incompletely characterized; however, and further investigation is needed to assess their relative utility for vaccination approaches.

An alternative approach in obtaining DC for vaccination is to isolate circulating immature DC populations (CD14–/CD11c+/CD123–) from human blood. As discussed previously, the paucity of DC in peripheral blood is a major limitation; however, in vivo administration of granulocyte colony-stimulating factor (G-CSF) and Flt3-L to donors expands populations of circulating DC by up to 20-fold (Pulendran et al. 2000; Fong et al. 2001). DC generated in this manner have the theoretical advantage of greater similarity to physiologic DC, and increased responsiveness to maturation stimuli has been observed in DC generated in vivo with Flt3-L compared to DC generated in vitro from monocytes (Luft et al. 2002). However, the limited DC yield remains a major obstacle to the implementation of this approach, and functionally significant differences between circulating DC and CD14+ DC differentiated in vitro are not apparent.

21.4 Antigen Selection for Therapeutic Vaccination

One of the most significant obstacles in the development of DC-based immunotherapy is the identification and selection of appropriate antigens for DC loading. In most human cancers either specific dominant antigens are not known, or the roles of known antigens are not well defined. Identification of tumor-specific antigens is of utmost importance to avoid damage to nonmalignant cells when anticancer immune responses are elicited, and to prevent induction of tolerance by presenting "self-antigen".

To date, most trials of DC vaccines for cancer have utilized DC charged with a single tumor antigen (TA). Most commonly, these vaccines are produced by pulsing DC with synthetic peptides derived from a defined TA. Alternatively, specific TA components can be delivered to DC by genetic integration approaches, including viral and nonviral (passive uptake of naked nucleic acids, lipofection, particle-mediated transfection, and electroporation) techniques (Breckpot et al. 2004). Loading DC with one or a few TA components, by either peptide pulsing or genetic modification, enables specific targeting of validated markers known to be expressed by a particular tumor.

Vaccination approaches in which DC are loaded with one or a few defined TA confer certain advantages. For some cancer types, TAs that are expressed on the majority of tumors are well characterized and contain short, immunogenic peptide sequences that can be synthesized with relative ease. Defined-TA DC vaccination approaches may avert the need for autologous tumor material, which can be limited or difficult to extract. Furthermore, since DC are primed to present only one or a few specific epitopes, the risk of inducing autoimmune disease in patients is low. Another advantage of utilizing DC vaccines loaded with a defined TA is the opportunity to serially monitor levels of that antigen as an assessment of treatment response.

However, the use of defined-TA vaccines has several important limitations. First, there are many cancer types for which specific TAs have not been identified or sufficiently well characterized. Additionally, DC with a single antigenic target do not induce immunity against tumor cells that fail to express that particular epitope; administration of such a vaccine could lead to clonal expansion of antigen-negative cells and the consequent development of a vaccine-resistant tumor. Similarly, these TA may not represent the dominant antigen in a particular tumor type and immunization using such TA may result in observation of immune responses alone, without clinically significant correlates. Some smaller single peptide antigens can be targeted to enter MHC class I pathways (Falo et al. 1995), which results in CD8+ T-cell activation without concurrent MHCII activation of CD4+ T cells, and accordingly, the elicited antitumor response may not be optimally robust (Celluzzi et al. 1996). Other longer peptides go through MHC class II processing pathways and elicit CD4+ T-helper responses without activation of CD8+ cytotoxic cells (Kudela et al. 2007; Adams et al. 2008).

To overcome these obstacles in antigen selection, autologous whole tumor cell products, containing a broad repertoire of undefined antigens, have been used to prime DC. This approach averts the need for identification of an appropriate TA, while enabling targeting of uniquely expressed (polymorphic or mutated) tumor antigens that may be important components of immune evasion mechanisms. Unlike DC primed by a single antigen, DC loaded with undefined tumor cell components present multiple epitopes in MHCI as well as MHCII, triggering activation of both CD4+ and CD8+ T cells (Hokey et al. 2002; Kingston et al. 2003). Theoretically, DC charged with a broad repertoire of tumor epitopes are more likely to effect a polyclonal immune response, reducing the opportunity for tumor immune escape by antigenic variation.

21.5 Optimizing Conditions for Antigen Loading

Whether DC are to be charged with a single TA or an entire tumor cell repertoire, loading can be achieved via peptide/protein-based approaches or by genetic transfection of nucleic acids. That is, DC can be pulsed with a single peptide or with whole tumor cell lysate, and, similarly, DC can be transfected with a short genetic sequence encoding a single known antigen or with the entire genetic content of a tumor cell. Utilizing tumor mRNA or cDNA may hold particular appeal, as array technology or subtractive hybridization could be used to compare tumor DNA to normal genomic material from the same patient; such high-throughput technologies could provide a powerful means of charging DC with a wide selection of highly specific TAs that are unlikely to induce autoimmunity. While the protein products of transfected mRNA can potentially induce strong CTL immunity through MHC class I presentation, they tend not to undergo endosome trafficking and consequent MHCII presentation, thus limiting the potential for induction of a strong CD4+ T-cell response. Given this constraint, TA mRNA can be modified to encode a C-terminal domain that targets the peptide to the lysosome, with the goal of simultaneously inducing antigen presentation in both MHCII and MHCI molecules (Wu et al. 1995). Genetic modification approaches may entail significant technical complexity, but they may also reduce the quantity of required tumor tissue, since nucleic acids can be highly amplified prior to delivery. Concerns regarding genetic approaches include potential effects of genetic modification on DC function as well as the biosafety of DC with incorporated viral vectors.

Tumor lysates represent only one of many available protein sources for DC loading: live tumor cells, necrotic debris, apoptotic bodies, heat shock proteins (HSP), and exosomes derived from tumors have been utilized in clinical vaccination approaches (Mosca et al. 2007). Another related method involves fusing DC with tumor cells to form DC-tumor hybridomas, which have been generated with polyethylene glycol or, more recently, through the use of electrofusion techniques, in which cells are mechanically aligned and application of electric pulses leads to fusion of the cell membranes of juxtaposed cells. Though capable of inducing a robust immune response, autologous tumor lysate as a source of DC antigen requires a large amount of tumor material. Allogeneic tumor lysate may be more widely available in the form of banked specimens or cultured tumor cell lines, but this approach suffers from a lack of specificity for the repertoire of epitopes expressed by the tumor cells of a given patient.

Considerable controversy surrounds the issue of the relative usefulness of necrotic vs apoptotic tumor cells for loading DC. Contradictory findings of the effects of these two approaches on antigen uptake, DC maturation, and release of HSP have been reported (Sauter et al. 2000; Kotera et al. 2001). The complexity of the problem has been further demonstrated by studies which show that standard procedures used to induce apoptosis actually produce a mixed population of viable, apoptotic, and necrotic tumor cells (Kotera et al. 2001). Multiple lines of evidence suggest that DC pulsed with tumor-derived apoptotic bodies or lysates, or fused or co-cultured with live tumor cells, can be effective immunogens in vivo; direct mechanistic analyses of tumor antigen internalization, processing, and presentation are currently under study (Hokey et al. 2001). While little data are currently available regarding the comparative efficiency of antigen delivery and presentation using these different delivery strategies, a randomized clinical trial is currently in progress at the University of Pittsburgh Cancer Institute to assess the biological and clinical merits of various loading techniques (L. Geskin, personal communication). The results of this trial will be available within the next year.

While the vaccination approaches described above rely on the in vitro generation of mature, antigen-loaded DC, a number of alternative methods have been employed. Vaccines have been prepared from antigen-loaded immature DC, which are injected in combination with an inflammatory adjuvant that enhances DC trafficking to lymph nodes and immune response induction (Nair et al. 2003). Approaches utilizing immature DC bypass the complex, expensive process of inducing DC maturation in vitro, but their efficacy relies on effective stimulation of endogenous DC maturation mechanisms, with the risk of inducing an immune-suppressing response if DC fail to mature. Alternatively, tumor epitopes can be delivered to DC in vivo by injection of antigens conjugated to DC-targeting elements, such as nanobeads, antibodies, or specific proteins (Tacken et al. 2006). Similarly, this method depends critically upon a functional endogenous system of DC activation, maturation, and antigen presentation, processes which may not be fully intact in individuals with cancer.

21.6 Enhancing Immune Responses to Dendritic Cell Vaccination

Most clinical trials to date have demonstrated that DC-based vaccines elicit antigen-specific immune responses in the majority of immunized patients. However, these immune responses have not been associated with significant clinical responses in Phase III clinical trials. Various hypotheses have been proposed in order to explain this phenomenon; studies have explored the roles of regulatory T cells and immunosuppressive cytokines, inhibition of lymphocyte function by T-cell receptors, and promotion of DC survival and function by a broad array of molecules and receptors. It is evident that the regulation of immune responses is a multifactorial and highly complex process, and optimization of desired responses may involve a variety of interventions involving DC, lymphocytes, and general modifications of the tumor microenvironment.

Reduction of Treg populations in cancer patients may enable the recovery or induction of tumor-specific immune responses capable of limiting disease progression (Yang 2008). Selective and non-selective approaches to Treg depletion have been attempted. Cyclophosphamide was found to deplete Treg cells (Ghiringhelli et al. 2004) and decrease the activity of remaining Treg (Lutsiak et al. 2005) in rodents. Alternatively, CD25 (the alpha-chain of the IL-2 receptor) may be targeted with antibodies or the recombinant IL-2 diptheria toxin conjugate denileukin diftitox, which has been shown to reduce circulating Treg and enhance T-cell antitumor response when administered prior to DC injection (Dannull et al. 2005).

The timing of DC-based immunotherapy coupled with Treg depletion is also of major importance. Since cyclophosphamide-induced Treg depletion is transient, DC vaccines should be delivered within a specific window following cyclophosphamide administration, during which Treg are maximally depleted and have not yet begun to significantly regenerate. In a murine model of melanoma, pre-treatment with cyclophosphamide enhanced the response of tumor-primed effector T cells to a DC-based exosome vaccine (Taieb et al. 2006). In the case of CD25-targeting approaches for Treg depletion, activated T cells transiently express high levels of CD25; therefore, Treg-depleting, CD25-targeting agents should be administered 2–10 days before DC vaccination (Litzinger et al. 2007).

Immune-suppressing cytokines – including IL-6, IL-10, IL-13, TGF-β, and VEGF (Rabinovich et al. 2007) – are also potential targets for antibody-mediated reduction in conjunction with DC vaccination. It has been shown, for instance, that IFN-γ production and CTL response to injected tumor cells are enhanced in the presence of anti-IL-10 and anti-TGF-β antibodies in mouse models (Jarnicki et al. 2006). Similarly, transgenic mice whose T cells are insensitive to TGF-β signaling are more resistant to exogenous tumor cell infiltration than control mice (Gorelik and Flavell 2001). Further, in vitro studies of human cancer cells have shown that TGF-β suppresses CTL activation and function (Ahmadzadeh and Rosenberg 2005), while anti-TGF-β mechanisms, including antibodies, soluble receptors, and antisense oligonucleotides, can induce antitumor immune responses. In a trial of a DC vaccine for glioblastoma, tumor expression of TGF-β was associated with poorer clinical outcome and was inversely correlated with vaccine-induced T-cell infiltration of tumor (Liau et al. 2005).

Similarly, cell-surface receptors that mediate lymphocyte inhibition (e.g., PD-L1 (B7-H1), CTLA-4, indoleamine 2,3-dioxygenase (IDO)) can be targeted with antibodies(O'Neill et al. 2004). Such cell-surface markers, much like the inhibitory cytokines, create immune-tolerant milieus conducive to tumor growth. PD-L1, for example, binds PD-1 on T cells and promotes immune tolerance; blocking this interaction (e.g., through the use of an anti-PD-L1 antibody) leads to increased activation of T cells, increased secretion of cytotoxicity-promoting cytokines such as IFN-γ, and decreased secretion of tolerance-inducing cytokines such as IL-10 in mice (Curiel et al. 2003; Strome et al. 2003; Hirano et al. 2005). NSAID (Ulrich et al. 2006), certain chemotherapeutic agents, or other small molecules could also be utilized to alter the immunotolerant tumor microenvironment that decreases DC vaccine effectiveness. The COX-2 inhibitor celecoxib, for instance, increases the effectiveness of DC-based (Basu et al. 2006) and other (Zeytin et al. 2004) antitumor vaccines in mice.

Factors that contribute to DC suppression can also be reduced through the use of RNA interference (RNAi) technology. For instance, blocking DC expression of pro-apoptotic proteins (e.g., BAX, BAK) and immunosuppressive factors (e.g., PD-L1, IDO, IL-10, SOCS1) via RNAi mechanisms may help effect more robust DC and CTL activity (Shen et al. 2004; Mao et al. 2007). Analogously, DC can be transfected or transduced with anti-apoptotic genes, such as the Bcl family members (Pirtskhalaishvili et al. 2000; Kim et al. 2003).

In addition to abrogating immune-signaling pathways that oppose DC effector responses, immune targeting approaches that promote DC-activating pathways may also be employed as adjuvant therapies. Cytokines that promote Th1 differentiation and CTL activity, including IL-12, IFN-γ, and TLR ligands, can be co-transfected or co-transduced with TA or administered as adjuvants with DC vaccination. Similarly, antibodies that agonize members of the TNF-receptor superfamily, including CD40, CD27, CD137 (4-1BB), and CD134 (OX40) (Barr et al. 2006; Yurkovetsky et al. 2006), may help effect robust DC responses. CD40 in particular has been implicated as playing an important role in DC activation, stimulating the secretion of Th1-promoting cytokines (e.g., TNF-α and IL-12) as well as the expression of various costimulatory molecules (Grewal and Flavell 1998). In clinical trials, CD40L (CD154)-transduced leukemic T cells have been shown to induce immune response to tumor cells and to decrease tumor burden (Wierda et al. 2000). Alternatively, soluble CD40L can be utilized ex vivo, along with various cytokines, in order to activate DC before administering them to patients (Davis et al. 2006; Palucka et al. 2006). Thus, the CD40-CD40L-signaling pathway is an appealing target for manipulation.

Many other cytokines and/or cell-surface receptors have been proposed as potential adjuvants or agonist targets to enhance DC-based immunotherapy, but the involvement of many of these molecules in complex and apparently paradoxical immune-signaling pathways may complicate their suitability for use as adjuvant therapy. For example, IL-15 is involved in the generation of long-term, robust cellular immunity (Oh et al. 2003) yet has also been implicated as being important in the maintenance of Treg suppressive function (Yates et al. 2007). This highlights a need for careful study in selecting agents to administer along with DC vaccines, especially when these agents may have multiple or incompletely elucidated roles in immune cascades.

21.7 Routes, Dose, and Timing of Vaccine Administration

Route of DC administration is one of the critical factors influencing vaccination outcome. Intravenous (i.v.), intradermal (i.d.), subcutaneous (s.c.), intralymphatic (i.l.)/intranodal (i.n.) (Nestle et al. 1998), and intratumoral (i.t.) routes have been utilized in clinical trials. A study comparing i.v., i.n., and i.d. administration showed that the i.n. route was associated with superior delayed-type hypersensitivity and T-cell cytokine secretion responses to novel peptide exposure (Bedrosian et al. 2003). Additional studies have demonstrated that intranodal DC vaccination can lead to enduring T-helper cell activity and Th1-skewed cytokine production (Gilliet et al. 2003), as well as objective clinical response (Maier et al. 2003). However, questions exist regarding the maintenance of lymphatic cytoarchitecture after intranodal injection, a factor that may increase the appeal of intralymphatic injection. Yet another option is intratumoral DC injection; this has been tried with encouraging results in animal models (Nishioka et al. 1999; Satoh et al. 2002; Song and Levy 2005; Zhong et al. 2007a) and in clinical trials (Chi et al. 2005; Yamanaka et al. 2005; Guo et al. 2007), but is limited by tumor accessibility. Identifying the most appropriate route of administration is thus in itself a complex issue, and it is further complicated by the finding that based on site of injection, DC can induce T cells with different homing patterns that may in effect confer regional immunity (Geskin et al. 2001; Mullins et al. 2003). This observation suggests that depending on the localization of tumor and metastases, it may even be useful to combine various administration routes in individual patients.

In addition to route of administration, vaccine dose and frequency are important considerations. Tracking radiolabeled DC has shown that neither i.v.- nor s.c.-administered DC localize to lymph nodes, while perhaps less than 1% of i.d.-administered DC do in fact drain to local lymph nodes (Morse et al. 1999); another study that differentiated between lymph node homing of immature and mature DC estimated this number at less than 3% for mature DC and less than 0.3% for immature DC (De Vries et al. 2003). Clinical trials have tended to use DC doses averaging approximately 10^6–10^8 cells per inoculation (Ridgway 2003), with outcomes suggesting that higher-range doses do not affect greater immune responses (Banchereau et al. 2001; Bedrosian et al. 2003).

There is insufficient empirical evidence at present to define optimum time intervals between initial DC vaccinations and injections. However, the practice of giving boosters at 2-week intervals has been bolstered by the discovery that DC can generate memory-like T cells in 4–6 days (Badovinac et al. 2005), as well as the consensus that repeated injections are required to retain the ability to carry out antitumor activities. Pre-injection with TNF-α or other DC at the site of vaccination induces greater than ten-fold increases in lymph node homing in mice (MartIn-Fontecha et al. 2003); therefore, prior administration of these agents may increase the proportion of injected DC that stimulate T cells and decrease the required dose of vaccine. TLR-ligand pre-treatment also increases DC lymph node migration, and even stimulates DC maturation, thus averting the need for ex vivo DC maturation (Nair et al. 2003).

21.8 Evaluating Efficacy of Dendritic Cell Vaccines

While the goal of DC vaccination is achievement of objective clinical responses and ultimately, improved survival, clinical trials to date have shown limited efficacy with respect to these endpoints. In order to allow for meaningful interpretation of study results and comparisons between different vaccination approaches, defining immunologic outcomes for assessment of response to DC vaccination is an important goal. Various methods of immunomonitoring have been developed or adapted to gauge the reactivity of peripheral CD8+ T cells to tumor antigens after DC vaccination. These include a chromium release assay, in which a patient's CD8+ T cells induce apoptosis in antigen- and Cr-pulsed cells; IFN-γ Enzyme Immunospot (ELISPOT) and flow cytometry detection, which monitor IFN-γ production in response to antigen presentation (Berard et al. 2000); and detection of MHC-peptide tetramer bound to reactive T cells (Altman et al. 1996; Paczesny et al. 2004). An assay for measurement of CTL induction by whole tumor-loaded cells, when the specific antigen and MHC are not known, has also been developed (Palucka et al. 2006). In a study comparing the ELISPOT, flow cytometry, and tetramer techniques, the three methods yielded disparate results (Whiteside et al. 2003), highlighting the limitations of commonly employed immunomonitoring techniques. One plausible explanation for this discordant data is the fact that CTL capable of responding to tumor antigen may be concentrated in lymph nodes (Romero et al. 1998) or bone marrow (Feuerer et al. 2001a, b), limiting the presence of these populations in peripheral blood.

An alternate approach to assessing the induction of immune responses by DC vaccines is with delayed-type hypersensitivity (DTH) skin tests (Bedrosian et al. 2003). One investigation, however, found that DTH indurations could be elicited by DC alone; in the same study, when antigen-loaded DC were injected and biopsies of the resultant indurations were performed, positive staining for antigen-specific T cells was associated with clinical response to vaccination. These results suggest that immunohistologic characterization of DTH-induced indurations, but not the simple presence or absence of DTH response, is a pertinent immunologic outcome (de Vries et al. 2005). Other immune measures that may serve as correlates of clinical response to DC vaccination are concentrations of activated immune effectors other than CD8+ T cells, including NK cells, CD4+ T cells, and B cells.

21.9 Conclusion

There are a multitude of variables surrounding DC vaccination that affect antitumor outcomes; ongoing studies are addressing various limitations of DC-based vaccination approaches in order to achieve optimization of these complex parameters (Zhong et al. 2007b). While we describe some of the many issues surrounding the development, administration, and evaluation of DC vaccines in this chapter, space constraints limit our ability to discuss all of the pertinent hypotheses and approaches that have been proposed by investigators.

DC vaccines have been evaluated in over 150 phase I-III clinical trials in at least 30 histologically distinct malignancies. With respect to achievement of objective clinical responses, the results of DC vaccines for cancer have been disappointing thus far. Modest rates of objective clinical responses have been reported in clinical trials, with some small phase II studies showing improved survival in select patients with corresponding CTL responses (Wheeler et al. 2008). However, recent phase III trials have failed to replicate some of these positive findings, and no phase IV trials have yet been performed. Despite low overall rates of response in clinical trials to date, complete responses have been observed in some patients. These individual successes, combined with the achievement of intermediate immunologic endpoints in many patients as well as numerous preclinical reports of therapeutic efficacy in murine models, demonstrate the antitumor potential of these agents. Given that successful, precise manipulation of a wide array of variables is necessary in order to produce an effective vaccine, it is to be expected that unsuccessful initial attempts will precede simultaneous optimization of all relevant parameters.

DC vaccines have been remarkably well tolerated in clinical trials. The most commonly reported adverse events in clinical trials are fever and local injection site reactions, and there have been no reports of serious side effects. Given their excellent safety profile, it might be reasonable to investigate certain DC vaccines in earlier stages of disease, or after standard interventions have been employed to eradicate the bulk of the tumor burden. DC vaccination may be more successful in destroying residual disease or in preventing recurrence than in leading to tumor regression in patients with advanced, disseminated disease, and compromised immunity. Furthermore, refinement of patient selection criteria for individual vaccines, as well as standardization of vaccination parameters in order to allow for more meaningful comparisons between trials, will help to direct the development of the next generation of anticancer DC vaccines.

References

Adams, S., O'Neill, D. W., Nonaka, D., Hardin, E., Chiriboga, L., Siu, K., Cruz, C. M., Angiulli, A., Angiulli, F., Ritter, E., Holman, R. M., Shapiro, R. L., Berman, R. S., Berner, N., Shao, Y., Manches, O., Pan, L., Venhaus, R. R., Hoffman, E. W., Jungbluth, A., Gnjatic, S., Old, L., Pavlick, A. C. and Bhardwaj, N. 2008. Immunization of malignant melanoma patients with full-length NY-ESO-1 protein using TLR7 agonist imiquimod as vaccine adjuvant. J Immunol 181:776–784.

Ahmadzadeh, M. and Rosenberg, S. A. 2005. TGF-beta 1 attenuates the acquisition and expression of effector function by tumor antigen-specific human memory CD8 T cells. J Immunol 174:5215–5223.

Altman, J. D., Moss, P. A., Goulder, P. J., Barouch, D. H., McHeyzer-Williams, M. G., Bell, J. I., McMichael, A. J. and Davis, M. M. 1996. Phenotypic analysis of antigen-specific T lymphocytes. Science 274:94–96.

Babatz, J., Rollig, C., Oelschlagel, U., Zhao, S., Ehninger, G., Schmitz, M. and Bornhauser, M. 2003. Large-scale immunomagnetic selection of CD14+ monocytes to generate dendritic cells for cancer immunotherapy: a phase I study. J Hematother Stem Cell Res 12:515–523.

Badovinac, V. P., Messingham, K. A., Jabbari, A., Haring, J. S. and Harty, J. T. 2005. Accelerated CD8+ T-cell memory and prime-boost response after dendritic-cell vaccination. Nat Med 11:748–756.

Banchereau, J., Palucka, A. K., Dhodapkar, M., Burkeholder, S., Taquet, N., Rolland, A., Taquet, S., Coquery, S., Wittkowski, K. M., Bhardwaj, N., Pineiro, L., Steinman, R. and Fay, J. 2001. Immune and clinical responses in patients with metastatic melanoma to CD34(+) progenitor-derived dendritic cell vaccine. Cancer Res 61:6451–6458.

Banchereau, J., Ueno, H., Dhodapkar, M., Connolly, J., Finholt, J. P., Klechevsky, E., Blanck, J. P., Johnston, D. A., Palucka, A. K. and Fay, J. 2005. Immune and clinical outcomes in patients with stage IV melanoma vaccinated with peptide-pulsed dendritic cells derived from CD34 + progenitors and activated with type I interferon. J Immunother 28:505–516.

Barr, T. A., Carlring, J. and Heath, A. W. 2006. Co-stimulatory agonists as immunological adjuvants. Vaccine 24:3399–3407.

Basner-Tschakarjan, E., Gaffal, E., O'Keeffe, M., Tormo, D., Limmer, A., Wagner, H., Hochrein, H. and Tuting, T. 2006. Adenovirus efficiently transduces plasmacytoid dendritic cells resulting in TLR9-dependent maturation and IFN-alpha production. J Gene Med 8:1300–1306.

Basu, G. D., Tinder, T. L., Bradley, J. M., Tu, T., Hattrup, C. L., Pockaj, B. A. and Mukherjee, P. 2006. Cyclooxygenase-2 inhibitor enhances the efficacy of a breast cancer vaccine: role of IDO. J Immunol 177:2391–2402.

Bedrosian, I., Mick, R., Xu, S., Nisenbaum, H., Faries, M., Zhang, P., Cohen, P. A., Koski, G. and Czerniecki, B. J. 2003. Intranodal administration of peptide-pulsed mature dendritic cell vaccines results in superior CD8 + T-cell function in melanoma patients. J Clin Oncol 21:3826–3835.

Berard, F., Blanco, P., Davoust, J., Neidhart-Berard, E. M., Nouri-Shirazi, M., Taquet, N., Rimoldi, D., Cerottini, J. C., Banchereau, J. and Palucka, A. K. 2000. Cross-priming of naive CD8 T cells against melanoma antigens using dendritic cells loaded with killed allogeneic melanoma cells. J Exp Med 192:1535–1544.

Breckpot, K., Heirman, C., Neyns, B. and Thielemans, K. 2004. Exploiting dendritic cells for cancer immunotherapy: genetic modification of dendritic cells. J Gene Med 6:1175–1188.

Cella, M., Facchetti, F., Lanzavecchia, A. and Colonna, M. 2000. Plasmacytoid dendritic cells activated by influenza virus and CD40L drive a potent TH1 polarization. Nat Immunol 1:305–310.

Celluzzi, C. M., Mayordomo, J. I., Storkus, W. J., Lotze, M. T. and Falo, L. D., Jr. 1996. Peptide-pulsed dendritic cells induce antigen-specific CTL-mediated protective tumor immunity. J Exp Med 183:283–287.

Chi, K. H., Liu, S. J., Li, C. P., Kuo, H. P., Wang, Y. S., Chao, Y. and Hsieh, S. L. 2005. Combination of conformal radiotherapy and intratumoral injection of adoptive dendritic cell immunotherapy in refractory hepatoma. J Immunother 28:129–135.

Curiel, T. J., Wei, S., Dong, H., Alvarez, X., Cheng, P., Mottram, P., Krzysiek, R., Knutson, K. L., Daniel, B., Zimmermann, M. C., David, O., Burow, M., Gordon, A., Dhurandhar, N., Myers, L., Berggren, R., Hemminki, A., Alvarez, R. D., Emilie, D., Curiel, D. T., Chen, L. and Zou, W. 2003. Blockade of B7-H1 improves myeloid dendritic cell-mediated antitumor immunity. Nat Med 9:562–567.

Dannull, J., Su, Z., Rizzieri, D., Yang, B. K., Coleman, D., Yancey, D., Zhang, A., Dahm, P., Chao, N., Gilboa, E. and Vieweg, J. 2005. Enhancement of vaccine-mediated antitumor immunity in cancer patients after depletion of regulatory T cells. J Clin Invest 115:3623–3633.

Davis, I. D., Chen, Q., Morris, L., Quirk, J., Stanley, M., Tavarnesi, M. L., Parente, P., Cavicchiolo, T., Hopkins, W., Jackson, H., Dimopoulos, N., Tai, T. Y., MacGregor, D., Browning, J., Svobodova, S., Caron, D., Maraskovsky, E., Old, L. J., Chen, W. and Cebon, J. 2006. Blood dendritic cells generated with Flt3 ligand and CD40 ligand prime CD8 + T cells efficiently in cancer patients. J Immunother 29:499–511.

de Vries, I. J., Bernsen, M. R., Lesterhuis, W. J., Scharenborg, N. M., Strijk, S. P., Gerritsen, M. J., Ruiter, D. J., Figdor, C. G., Punt, C. J. and Adema, G. J. 2005. Immunomonitoring tumor-specific T cells in delayed-type hypersensitivity skin biopsies after dendritic cell vaccination correlates with clinical outcome. J Clin Oncol 23:5779–5787.

De Vries, I. J., Krooshoop, D. J., Scharenborg, N. M., Lesterhuis, W. J., Diepstra, J. H., Van Muijen, G. N., Strijk, S. P., Ruers, T. J., Boerman, O. C., Oyen, W. J., Adema, G. J., Punt,

C. J. and Figdor, C. G. 2003. Effective migration of antigen-pulsed dendritic cells to lymph nodes in melanoma patients is determined by their maturation state. Cancer Res 63:12–17.

Di Nicola, M., Carlo-Stella, C., Mortarini, R., Baldassari, P., Guidetti, A., Gallino, G. F., Del Vecchio, M., Ravagnani, F., Magni, M., Chaplin, P., Cascinelli, N., Parmiani, G., Gianni, A. M. and Anichini, A. 2004. Boosting T cell-mediated immunity to tyrosinase by vaccinia virus-transduced, CD34(+)-derived dendritic cell vaccination: a phase I trial in metastatic melanoma. Clin Cancer Res 10:5381–5390.

Falo, L. D., Jr., Kovacsovics-Bankowski, M., Thompson, K. and Rock, K. L. 1995. Targeting antigen into the phagocytic pathway in vivo induces protective tumour immunity. Nat Med 1:649–653.

Fay, J. W., Palucka, A. K., Paczesny, S., Dhodapkar, M., Johnston, D. A., Burkeholder, S., Ueno, H. and Banchereau, J. 2006. Long-term outcomes in patients with metastatic melanoma vaccinated with melanoma peptide-pulsed CD34(+) progenitor-derived dendritic cells. Cancer Immunol Immunother 55:1209–1218.

Ferlazzo, G., Wesa, A., Wei, W. Z. and Galy, A. 1999. Dendritic cells generated either from CD34+ progenitor cells or from monocytes differ in their ability to activate antigen-specific CD8+ T cells. J Immunol 163:3597–3604.

Feuerer, M., Beckhove, P., Bai, L., Solomayer, E. F., Bastert, G., Diel, I. J., Pedain, C., Oberniedermayr, M., Schirrmacher, V. and Umansky, V. 2001a. Therapy of human tumors in NOD/SCID mice with patient-derived reactivated memory T cells from bone marrow. Nat Med 7:452–458.

Feuerer, M., Rocha, M., Bai, L., Umansky, V., Solomayer, E. F., Bastert, G., Diel, I. J. and Schirrmacher, V. 2001b. Enrichment of memory T cells and other profound immunological changes in the bone marrow from untreated breast cancer patients. Int J Cancer 92:96–105.

Fong, L., Hou, Y., Rivas, A., Benike, C., Yuen, A., Fisher, G. A., Davis, M. M. and Engleman, E. G. 2001. Altered peptide ligand vaccination with Flt3 ligand expanded dendritic cells for tumor immunotherapy. Proc Natl Acad Sci U S A 98:8809–8814.

Gabrilovich, D. I., Corak, J., Ciernik, I. F., Kavanaugh, D. and Carbone, D. P. 1997. Decreased antigen presentation by dendritic cells in patients with breast cancer. Clin Cancer Res 3:483–490.

Gamvrellis, A., Leong, D., Hanley, J. C., Xiang, S. D., Mottram, P. and Plebanski, M. 2004. Vaccines that facilitate antigen entry into dendritic cells. Immunol Cell Biol 82:506–516.

Geskin, L., Watkins, S., Spencer, L., Plowey, J., Celluzzi, C. and Falo, L. D. J. 2001. Dendritic-cell based immunotherapies: route of administration effects dendritic cell trafficking and regional immune responses. J Invest Dermatol 2:449.

Ghiringhelli, F., Larmonier, N., Schmitt, E., Parcellier, A., Cathelin, D., Garrido, C., Chauffert, B., Solary, E., Bonnotte, B. and Martin, F. 2004. CD4+CD25+ regulatory T cells suppress tumor immunity but are sensitive to cyclophosphamide which allows immunotherapy of established tumors to be curative. Eur J Immunol 34:336–344.

Gilliet, M., Kleinhans, M., Lantelme, E., Schadendorf, D., Burg, G. and Nestle, F. O. 2003. Intranodal injection of semimature monocyte-derived dendritic cells induces T helper type 1 responses to protein neoantigen. Blood 102:36–42.

Gorelik, L. and Flavell, R. A. 2001. Immune-mediated eradication of tumors through the blockade of transforming growth factor-beta signaling in T cells. Nat Med 7:1118–1122.

Gottfried, E., Kreutz, M. and Mackensen, A. 2008. Tumor-induced modulation of dendritic cell function. Cytokine Growth Factor Rev 19:65–77.

Grewal, I. S. and Flavell, R. A. 1998. CD40 and CD154 in cell-mediated immunity. Annu Rev Immunol 16:111–135.

Guo, J., Zhu, J., Sheng, X., Wang, X., Qu, L., Han, Y., Liu, Y., Zhang, H., Huo, L., Zhang, S., Lin, B. and Yang, Z. 2007. Intratumoral injection of dendritic cells in combination with local hyperthermia induces systemic antitumor effect in patients with advanced melanoma. Int J Cancer 120:2418–2425.

Herbst, B., Kohler, G., Mackensen, A., Veelken, H., Mertelsmann, R. and Lindemann, A. 1997. CD34+ peripheral blood progenitor cell and monocyte derived dendritic cells: a comparative analysis. Br J Haematol 99:490–499.

Hirano, F., Kaneko, K., Tamura, H., Dong, H., Wang, S., Ichikawa, M., Rietz, C., Flies, D. B., Lau, J. S., Zhu, G., Tamada, K. and Chen, L. 2005. Blockade of B7-H1 and PD-1 by monoclonal antibodies potentiates cancer therapeutic immunity. Cancer Res 65:1089–1096.

Hokey, D., Geskin, L. and Falo, L. D. J. 2001. Comparative analysis of strategies designed to deliver tumor-derived proteins to dendritic cells. J Invest Dermatol 2:451.

Hokey, D., Geskin, L. and Falo, L. D. J. 2002. Differential MHC class I presentation of cytosolic and membrane-bound tumor antigens by dendritic cells. J Invest Dermatol 119:306.

Hokey, D. A., Larregina, A. T., Erdos, G., Watkins, S. C. and Falo, L. D., Jr. 2005. Tumor cell loaded type-1 polarized dendritic cells induce Th1-mediated tumor immunity. Cancer Res 65:10059–10067.

Inaba, K., Inaba, M., Romani, N., Aya, H., Deguchi, M., Ikehara, S., Muramatsu, S. and Steinman, R. M. 1992. Generation of large numbers of dendritic cells from mouse bone marrow cultures supplemented with granulocyte/macrophage colony-stimulating factor. J Exp Med 176:1693–1702.

Jarnicki, A. G., Lysaght, J., Todryk, S. and Mills, K. H. 2006. Suppression of antitumor immunity by IL-10 and TGF-beta-producing T cells infiltrating the growing tumor: influence of tumor environment on the induction of CD4+ and CD8+ regulatory T cells. J Immunol 177:896–904.

Jonuleit, H., Kuhn, U., Muller, G., Steinbrink, K., Paragnik, L., Schmitt, E., Knop, J. and Enk, A. H. 1997. Pro-inflammatory cytokines and prostaglandins induce maturation of potent immunostimulatory dendritic cells under fetal calf serum-free conditions. Eur J Immunol 27:3135–3142.

Kalinski, P., Vieira, P. L., Schuitemaker, J. H., de Jong, E. C. and Kapsenberg, M. L. 2001. Prostaglandin E(2) is a selective inducer of interleukin-12 p40 (IL-12p40) production and an inhibitor of bioactive IL-12p70 heterodimer. Blood 97:3466–3469.

Kim, T. W., Hung, C. F., Ling, M., Juang, J., He, L., Hardwick, J. M., Kumar, S. and Wu, T. C. 2003. Enhancing DNA vaccine potency by coadministration of DNA encoding anti-apoptotic proteins. J Clin Invest 112:109–117.

Kingston, A., Geskin, L., Plowey, J., Robb, A., McCann, S. and Falo, L. D., Jr. 2003. Development of dendritic cell therapies for cutaneous T cell lymphoma. J Invest Dermatol 121:A1001.

Kotera, Y., Shimizu, K. and Mule, J. J. 2001. Comparative analysis of necrotic and apoptotic tumor cells as a source of antigen(s) in dendritic cell-based immunization. Cancer Res 61:8105–8109.

Kudela, P., Janjic, B., Fourcade, J., Castelli, F., Andrade, P., Kirkwood, J. M., El-Hefnawy, T., Amicosante, M., Maillere, B. and Zarour, H. M. 2007. Cross-reactive CD4+ T cells against one immunodominant tumor-derived epitope in melanoma patients. J Immunol 179:7932–7940.

Liau, L. M., Prins, R. M., Kiertscher, S. M., Odesa, S. K., Kremen, T. J., Giovannone, A. J., Lin, J. W., Chute, D. J., Mischel, P. S., Cloughesy, T. F. and Roth, M. D. 2005. Dendritic cell vaccination in glioblastoma patients induces systemic and intracranial T-cell responses modulated by the local central nervous system tumor microenvironment. Clin Cancer Res 11:5515–5525.

Litzinger, M. T., Fernando, R., Curiel, T. J., Grosenbach, D. W., Schlom, J. and Palena, C. 2007. IL-2 immunotoxin denileukin diftitox reduces regulatory T cells and enhances vaccine-mediated T-cell immunity. Blood 110:3192–3201.

Lou, Y., Liu, C., Kim, G. J., Liu, Y. J., Hwu, P. and Wang, G. 2007. Plasmacytoid dendritic cells synergize with myeloid dendritic cells in the induction of antigen-specific antitumor immune responses. J Immunol 178:1534–1541.

Luft, T., Jefford, M., Luetjens, P., Toy, T., Hochrein, H., Masterman, K. A., Maliszewski, C., Shortman, K., Cebon, J. and Maraskovsky, E. 2002. Functionally distinct dendritic cell (DC) populations induced by physiologic stimuli: prostaglandin E(2) regulates the migratory capacity of specific DC subsets. Blood 100:1362–1372.

Lutsiak, M. E., Semnani, R. T., De Pascalis, R., Kashmiri, S. V., Schlom, J. and Sabzevari, H. 2005. Inhibition of CD4(+)25+ T regulatory cell function implicated in enhanced immune response by low-dose cyclophosphamide. Blood 105:2862–2868.

Maier, T., Tun-Kyi, A., Tassis, A., Jungius, K. P., Burg, G., Dummer, R. and Nestle, F. O. 2003. Vaccination of patients with cutaneous T-cell lymphoma using intranodal injection of autologous tumor-lysate-pulsed dendritic cells. Blood 102:2338–2344.

Mailliard, R. B., Wankowicz-Kalinska, A., Cai, Q., Wesa, A., Hilkens, C. M., Kapsenberg, M. L., Kirkwood, J. M., Storkus, W. J. and Kalinski, P. 2004. alpha-type-1 polarized dendritic cells: a novel immunization tool with optimized CTL-inducing activity. Cancer Res 64:5934–5937.

Mao, C. P., Hung, C. F. and Wu, T. C. 2007. Immunotherapeutic strategies employing RNA interference technology for the control of cancers. J Biomed Sci 14:15–29.

MartIn-Fontecha, A., Sebastiani, S., Hopken, U. E., Uguccioni, M., Lipp, M., Lanzavecchia, A. and Sallusto, F. 2003. Regulation of dendritic cell migration to the draining lymph node: impact on T lymphocyte traffic and priming. J Exp Med 198:615–621.

Mohamadzadeh, M., Berard, F., Essert, G., Chalouni, C., Pulendran, B., Davoust, J., Bridges, G., Palucka, A. K. and Banchereau, J. 2001. Interleukin 15 skews monocyte differentiation into dendritic cells with features of Langerhans cells. J Exp Med 194:1013–1020.

Morelli, A. E., Larregina, A. T., Shufesky, W. J., Sullivan, M. L., Stolz, D. B., Papworth, G. D., Zahorchak, A. F., Logar, A. J., Wang, Z., Watkins, S. C., Falo, L. D., Jr. and Thomson, A. W. 2004. Endocytosis, intracellular sorting, and processing of exosomes by dendritic cells. Blood 104:3257–3266.

Morse, M. A., Coleman, R. E., Akabani, G., Niehaus, N., Coleman, D. and Lyerly, H. K. 1999. Migration of human dendritic cells after injection in patients with metastatic malignancies. Cancer Res 59:56–58.

Morse, M. A., Lyerly, H. K., Gilboa, E., Thomas, E. and Nair, S. K. 1998. Optimization of the sequence of antigen loading and CD40-ligand-induced maturation of dendritic cells. Cancer Res 58:2965–2968.

Mortarini, R., Anichini, A., Di Nicola, M., Siena, S., Bregni, M., Belli, F., Molla, A., Gianni, A. M. and Parmiani, G. 1997. Autologous dendritic cells derived from CD34+ progenitors and from monocytes are not functionally equivalent antigen-presenting cells in the induction of melan-A/Mart-1(27–35)-specific CTLs from peripheral blood lymphocytes of melanoma patients with low frequency of CTL precursors. Cancer Res 57:5534–5541.

Mosca, P. J., Lyerly, H. K., Clay, T. M., Morse, M. A. and Lyerly, H. K. 2007. Dendritic cell vaccines. Front Biosci 12:4050–4060.

Mullins, D. W., Sheasley, S. L., Ream, R. M., Bullock, T. N., Fu, Y. X. and Engelhard, V. H. 2003. Route of immunization with peptide-pulsed dendritic cells controls the distribution of memory and effector T cells in lymphoid tissues and determines the pattern of regional tumor control. J Exp Med 198:1023–1034.

Nair, S., McLaughlin, C., Weizer, A., Su, Z., Boczkowski, D., Dannull, J., Vieweg, J. and Gilboa, E. 2003. Injection of immature dendritic cells into adjuvant-treated skin obviates the need for ex vivo maturation. J Immunol 171:6275–6282.

Nestle, F. O., Alijagic, S., Gilliet, M., Sun, Y., Grabbe, S., Dummer, R., Burg, G. and Schadendorf, D. 1998. Vaccination of melanoma patients with peptide- or tumor lysate-pulsed dendritic cells. Nat Med 4:328–332.

Nishimura, N., Nishioka, Y., Shinohara, T. and Sone, S. 2001. Enhanced efficiency by centrifugal manipulation of adenovirus-mediated interleukin 12 gene transduction into human monocyte-derived dendritic cells. Hum Gene Ther 12:333–346.

Nishioka, Y., Hirao, M., Robbins, P. D., Lotze, M. T. and Tahara, H. 1999. Induction of systemic and therapeutic antitumor immunity using intratumoral injection of dendritic cells genetically modified to express interleukin 12. Cancer Res 59:4035–4041.

O'Doherty, U., Peng, M., Gezelter, S., Swiggard, W. J., Betjes, M., Bhardwaj, N. and Steinman, R. M. 1994. Human blood contains two subsets of dendritic cells, one immunologically mature and the other immature. Immunology 82:487–493.

O'Neill, D. W., Adams, S. and Bhardwaj, N. 2004. Manipulating dendritic cell biology for the active immunotherapy of cancer. Blood 104:2235–2246.

Oh, S., Berzofsky, J. A., Burke, D. S., Waldmann, T. A. and Perera, L. P. 2003. Coadministration of HIV vaccine vectors with vaccinia viruses expressing IL-15 but not IL-2 induces long-lasting cellular immunity. Proc Natl Acad Sci U S A 100:3392–3397.

Paczesny, S., Banchereau, J., Wittkowski, K. M., Saracino, G., Fay, J. and Palucka, A. K. 2004. Expansion of melanoma-specific cytolytic CD8+ T cell precursors in patients with metastatic melanoma vaccinated with CD34+ progenitor-derived dendritic cells. J Exp Med 199:1503–1511.

Palucka, A. K., Ueno, H., Connolly, J., Kerneis-Norvell, F., Blanck, J. P., Johnston, D. A., Fay, J. and Banchereau, J. 2006. Dendritic cells loaded with killed allogeneic melanoma cells can induce objective clinical responses and MART-1 specific CD8+ T-cell immunity. J Immunother 29:545–557.

Pedersen, A. E., Thorn, M., Gad, M., Walter, M. R., Johnsen, H. E., Gaarsdal, E., Nikolajsen, K., Buus, S., Claesson, M. H. and Svane, I. M. 2005. Phenotypic and functional characterization of clinical grade dendritic cells generated from patients with advanced breast cancer for therapeutic vaccination. Scand J Immunol 61:147–156.

Popov, I., Li, M., Zheng, X., San, H., Zhang, X., Ichim, T. E., Suzuki, M., Feng, B., Vladau, C., Zhong, R., Garcia, B., Strejan, G., Inman, R. D. and Min, W. P. 2006. Preventing autoimmune arthritis using antigen-specific immature dendritic cells: a novel tolerogenic vaccine. Arthritis Res Ther 8:R141.

Pulendran, B., Banchereau, J., Burkeholder, S., Kraus, E., Guinet, E., Chalouni, C., Caron, D., Maliszewski, C., Davoust, J., Fay, J. and Palucka, K. 2000. Flt3-ligand and granulocyte colony-stimulating factor mobilize distinct human dendritic cell subsets in vivo. J Immunol 165:566–572.

Rabinovich, G. A., Gabrilovich, D. and Sotomayor, E. M. 2007. Immunosuppressive strategies that are mediated by tumor cells. Annu Rev Immunol 25:267–296.

Ratzinger, G., Baggers, J., de Cos, M. A., Yuan, J., Dao, T., Reagan, J. L., Munz, C., Heller, G. and Young, J. W. 2004. Mature human Langerhans cells derived from CD34+ hematopoietic progenitors stimulate greater cytolytic T lymphocyte activity in the absence of bioactive IL-12p70, by either single peptide presentation or cross-priming, than do dermal-interstitial or monocyte-derived dendritic cells. J Immunol 173:2780–2791.

Ridgway, D. 2003. The first 1000 dendritic cell vaccinees. Cancer Invest 21:873–886.

Rissoan, M. C., Soumelis, V., Kadowaki, N., Grouard, G., Briere, F., de Waal Malefyt, R. and Liu, Y. J. 1999. Reciprocal control of T helper cell and dendritic cell differentiation. Science 283:1183–1186.

Romani, N., Reider, D., Heuer, M., Ebner, S., Kampgen, E., Eibl, B., Niederwieser, D. and Schuler, G. 1996. Generation of mature dendritic cells from human blood. An improved method with special regard to clinical applicability. J Immunol Methods 196:137–151.

Romero, P., Dunbar, P. R., Valmori, D., Pittet, M., Ogg, G. S., Rimoldi, D., Chen, J. L., Lienard, D., Cerottini, J. C. and Cerundolo, V. 1998. Ex vivo staining of metastatic lymph nodes by class I major histocompatibility complex tetramers reveals high numbers of antigen-experienced tumor-specific cytolytic T lymphocytes. J Exp Med 188:1641–1650.

Satoh, Y., Esche, C., Gambotto, A., Shurin, G. V., Yurkovetsky, Z. R., Robbins, P. D., Watkins, S. C., Todo, S., Herberman, R. B., Lotze, M. T. and Shurin, M. R. 2002. Local administration of IL-12-transfected dendritic cells induces antitumor immune responses to colon adenocarcinoma in the liver in mice. J Exp Ther Oncol 2:337–349.

Sauter, B., Albert, M. L., Francisco, L., Larsson, M., Somersan, S. and Bhardwaj, N. 2000. Consequences of cell death: exposure to necrotic tumor cells, but not primary tissue cells or apoptotic cells, induces the maturation of immunostimulatory dendritic cells. J Exp Med 191:423–434.

Scandella, E., Men, Y., Gillessen, S., Forster, R. and Groettrup, M. 2002. Prostaglandin E2 is a key factor for CCR7 surface expression and migration of monocyte-derived dendritic cells. Blood 100:1354–1361.

Schaft, N., Dorrie, J., Thumann, P., Beck, V. E., Muller, I., Schultz, E. S., Kampgen, E., Dieckmann, D. and Schuler, G. 2005. Generation of an optimized polyvalent monocyte-derived dendritic cell vaccine by transfecting defined RNAs after rather than before maturation. J Immunol 174:3087–3097.

Shen, L., Evel-Kabler, K., Strube, R. and Chen, S. Y. 2004. Silencing of SOCS1 enhances antigen presentation by dendritic cells and antigen-specific anti-tumor immunity. Nat Biotechnol 22:1546–1553.

Song, W. and Levy, R. 2005. Therapeutic vaccination against murine lymphoma by intratumoral injection of naive dendritic cells. Cancer Res 65:5958–5964.

Sporri, R. and Reis e Sousa, C. 2005. Inflammatory mediators are insufficient for full dendritic cell activation and promote expansion of CD4+ T cell populations lacking helper function. Nat Immunol 6:163–170.

Steinman, R. M., Hawiger, D. and Nussenzweig, M. C. 2003. Tolerogenic dendritic cells. Annu Rev Immunol 21:685–711.

Stober, D., Trobonjaca, Z., Reimann, J. and Schirmbeck, R. 2002. Dendritic cells pulsed with exogenous hepatitis B surface antigen particles efficiently present epitopes to MHC class I-restricted cytotoxic T cells. Eur J Immunol 32:1099–1108.

Strome, S. E., Dong, H., Tamura, H., Voss, S. G., Flies, D. B., Tamada, K., Salomao, D., Cheville, J., Hirano, F., Lin, W., Kasperbauer, J. L., Ballman, K. V. and Chen, L. 2003. B7-H1 blockade augments adoptive T-cell immunotherapy for squamous cell carcinoma. Cancer Res 63:6501–6505.

Tacken, P. J., Torensma, R. and Figdor, C. G. 2006. Targeting antigens to dendritic cells in vivo. Immunobiology 211:599–608.

Taieb, J., Chaput, N., Schartz, N., Roux, S., Novault, S., Menard, C., Ghiringhelli, F., Terme, M., Carpentier, A. F., Darrasse-Jeze, G., Lemonnier, F. and Zitvogel, L. 2006. Chemoimmunotherapy of tumors: cyclophosphamide synergizes with exosome based vaccines. J Immunol 176:2722–2729.

Ulrich, C. M., Bigler, J. and Potter, J. D. 2006. Non-steroidal anti-inflammatory drugs for cancer prevention: promise, perils and pharmacogenetics. Nat Rev Cancer 6:130–140.

Wheeler, C. J., Black, K. L., Liu, G., Mazer, M., Zhang, X. X., Pepkowitz, S., Goldfinger, D., Ng, H., Irvin, D. and Yu, J. S. 2008. Vaccination elicits correlated immune and clinical responses in glioblastoma multiforme patients. Cancer Res 68:5955–5964.

Whiteside, T. L., Zhao, Y., Tsukishiro, T., Elder, E. M., Gooding, W. and Baar, J. 2003. Enzyme-linked immunospot, cytokine flow cytometry, and tetramers in the detection of T-cell responses to a dendritic cell-based multipeptide vaccine in patients with melanoma. Clin Cancer Res 9:641–649.

Wierda, W. G., Cantwell, M. J., Woods, S. J., Rassenti, L. Z., Prussak, C. E. and Kipps, T. J. 2000. CD40-ligand (CD154) gene therapy for chronic lymphocytic leukemia. Blood 96:2917–2924.

Wu, T. C., Guarnieri, F. G., Staveley-O' Carroll, K. F., Viscidi, R. P., Levitsky, H. I., Hedrick, L., Cho, K. R., August, J. T. and Pardoll, D. M. 1995. Engineering an intracellular pathway for major histocompatibility complex class II presentation of antigens. Proc Natl Acad Sci U S A 92:11671–11675.

Yamanaka, R., Homma, J., Yajima, N., Tsuchiya, N., Sano, M., Kobayashi, T., Yoshida, S., Abe, T., Narita, M., Takahashi, M. and Tanaka, R. 2005. Clinical evaluation of dendritic

cell vaccination for patients with recurrent glioma: results of a clinical phase I/II trial. Clin Cancer Res 11:4160–4167.
Yang, X. F. 2008. Factors regulating apoptosis and homeostasis of CD4+ CD25(high) FOXP3+ regulatory T cells are new therapeutic targets. Front Biosci 13:1472–1499.
Yang, G. X., Lian, Z. X., Kikuchi, K., Moritoki, Y., Ansari, A. A., Liu, Y. J., Ikehara, S. and Gershwin, M. E. 2005. Plasmacytoid dendritic cells of different origins have distinct characteristics and function: studies of lymphoid progenitors versus myeloid progenitors. J Immunol 175:7281–7287.
Yates, J., Rovis, F., Mitchell, P., Afzali, B., Tsang, J. Y., Garin, M., Lechler, R. I., Lombardi, G. and Garden, O. A. 2007. The maintenance of human CD4+ CD25+ regulatory T cell function: IL-2, IL-4, IL-7 and IL-15 preserve optimal suppressive potency in vitro. Int Immunol 19:785–799.
Yurkovetsky, Z. R., Shurin, G. V., Barry, D. A., Schuh, A. C., Shurin, M. R. and Robbins, P. D. 2006. Comparative analysis of antitumor activity of CD40L, RANKL, and 4-1BBL in vivo following intratumoral administration of viral vectors or transduced dendritic cells. J Gene Med 8:129–137.
Zeytin, H. E., Patel, A. C., Rogers, C. J., Canter, D., Hursting, S. D., Schlom, J. and Greiner, J. W. 2004. Combination of a poxvirus-based vaccine with a cyclooxygenase-2 inhibitor (celecoxib) elicits antitumor immunity and long-term survival in CEA.Tg/MIN mice. Cancer Res 64:3668–3678.
Zhong, H., Han, B., Tourkova, I. L., Lokshin, A., Rosenbloom, A., Shurin, M. R. and Shurin, G. V. 2007a. Low-dose paclitaxel prior to intratumoral dendritic cell vaccine modulates intratumoral cytokine network and lung cancer growth. Clin Cancer Res 13:5455–5462.
Zhong, H., Shurin, M. R. and Han, B. 2007b. Optimizing dendritic cell-based immunotherapy for cancer. Expert Rev Vaccines 6:333–345.

Chapter 22
How Best to Generate Dendritic Cells from Patients with Cancer and How Best to Use them for Immunotherapeutic Purposes

Mark M. Aloysius, Richard A. Robins, and Oleg Eremin

Abstract The purpose of vaccinating patients with cancer is to achieve objective tumor regression or stasis, in contrast to vaccination strategy in infectious diseases, where it is primarily preventive in nature. Anti-cancer vaccines are designed to maximally induce and sustain host immune responses targeted against tumor antigens, leading to subsequent cancer cell death. Most cancer immunotherapy trials have been performed in patients with advanced cancers, in whom established therapies are usually ineffective or the patient has run out of therapeutic options. Such a therapeutic immunity is clearly more challenging to generate and sustain, compared with preventive immunity, as active immune surveillance and constant targeting of the tumor is required. As dendritic cells are the most potent antigen-presenting cells that sensitize naive T cells to antigen, designing cancer vaccines using the patient's own autologous dendritic cells is a logical and core concept in cancer immunotherapy, which continues to be an area of major interest and therapeutic challenge for both clinicians and scientists.

22.1 Naturally Occurring and Ex Vivo-Generated Dendritic Cells

Naturally occurring DC are found in low numbers in the circulation (0.1–2% in healthy individuals) and may be even lower in cancer patients who have received chemotherapy and immunotherapy. Thus, in order to study these cells significant volumes of blood have to be obtained by venipuncture and substantial volumes (usually via leukapheresis) when harvesting for immunotherapeutic purposes (Savary et al. 1998). Decreased numbers of DC in patients

O. Eremin (✉)
Research and Development Department, United Lincolnshire Hospitals NHS Trust, Lincoln County Hospital, Greetwell Road, Lincoln LN2 5QY, UK
e-mail: val.elliott@ulh.nhs.uk

M.R. Shurin, R.D. Salter (eds.), *Dendritic Cells in Cancer*,
DOI 10.1007/978-0-387-88611-4_22, © Springer Science+Business Media, LLC 2009

with cancer may be due to tumor-induced apoptosis, to inhibition of their differentiation in vivo by soluble factors released by the tumor or both biological mechanisms (Freudenthal and Steinman 1990). Furthermore, tumor-induced release of cytokines, such as IL-10, may retard DC development (O'Doherty et al. 1993; Thomas et al. 1993; Savary et al. 1998).

Circulating DC in patients with cancer are phenotypically and functionally hyporeactive, rendering them ineffective inducers of T-cell responses (Hoffmann et al. 2002; Gabrilovich 2004). We have shown that DC, even in women with early operable breast cancer, are dysfunctional and anergic (Satthaporn et al. 2004). We have also demonstrated that this dysfunction can be overcome by ex vivo generation of autologous monocyte-derived DC in the presence of specific cytokine cocktails (Satthaporn et al. 2008).

Sources of DC for clinical trial use include CD34+ blood or bone marrow-derived precursors, or CD14+ monocytes. CD14+ monocyte-derived DC are the most widely used in clinical trials as it is difficult to isolate circulating DC due to their very low numbers in peripheral blood and the lack of a DC-specific marker (Van Voorhis et al. 1983). Both approaches using either CD34+ or CD14+ cells and ex vivo DC generation have recently been found to be equivalent, with comparable functional and phenotypic profiles. There may, however, be a slightly reduced yield from CD34+ stem cells, compared with CD14+ monocyte precursors (Syme et al. 2005).

DC can be cultured from CD34+ cells mobilized by granulocyte colony-stimulating factor (G-CSF) or granulocyte macrophage colony-stimulating factor (GM-CSF) from the bone marrow. In the presence of GM-CSF, interleukin (IL)-4, IL-13, tumor necrosis factor (TNF)-α, and stem cell factor and 10 days of culture in vitro, the generated DC are $CD14^-$ and human leukocyte antigen (HLA) DR^- (Szabolcs et al. 1995a, b; Young et al. 1995). If transforming growth factor (TGF)-β is included in this culture system, the DC express CD1a, which indicates a predominant Langerhans cell (LC) phenotype, thereby, possessing good T-cell allostimulatory capacity (Strobl et al. 1996). Culture of CD34+ human cord blood cells with GM-CSF and TNF-α generates DC with LC phenotype, expressing CD1a, HLA-DR, and co-stimulatory molecules, but the yield is only 5–15% (Caux et al. 1992). Mobilized peripheral blood CD34+ cells cultured under similar conditions provide a better yield of CD1a expressing cells (30–55%) (Strunk et al. 1996). Another approach used to double the yield of CD1a expressing DC is to add Flt-3 ligand to the culture system (Strobl et al. 1997).

CD14+ monocytes have been shown to develop into DC when cultured in the presence of IL-4 and GM-CSF (Sallusto and Lanzavecchia 1994). However, these DC are immature, induce tolerance and are, therefore, unsuitable for cancer immunotherapy. However, an additional maturation stimulus may be provided through a variety of cytokines and adjuvants, which can make these DC mature, active, and abundant for immunotherapy. Some of these maturation factors are discussed below.

22.2 Dendritic Cell Generation for Clinical Trials

The 'gold standard' approach, commonly used in many pilot studies, to derive active and mature DC from CD14+ precursor cells, is to culture these cells in the presence of IL-4, GM-CSF, IL-1β, TNF-α, and prostaglandin E-2 (PgE2) (Figdor et al. 2004). However, this approach has been criticized for its low yield and absence of production of IL-12 (Kalinski et al. 2001). Moreover, there is speculation that lack of response may be due to terminally mature DC (expressing CD83), incapable of inducing efficient cytotoxic T lymphocytes (CTL) against tumor antigens (TA) (Langenkamp et al. 2000). Alternative DC activation and maturation approaches have been described using IL-13 instead of IL-4, CD40 ligand, poly-I:C, and Flt3 ligand (Figdor et al. 2004). The relatively low clinical responses with all these differentially derived DC, underpins the current need for novel ex vivo DC generation and reactivation techniques, yielding optimally functional and mature DC.

Different combinations of cytokines have been described in generating DC from CD14+ monocytes obtained from healthy volunteers. Various combinations with IFN-α have been shown to produce DC of activated and mature phenotype and enhanced function (Radvanyi et al. 1999; Dauer et al. 2003; Tamir et al. 2005; Dauer et al. 2006). We have recently demonstrated that IFN-α is critical for optimal phenotypic maturity of monocyte-derived DC from patients with cancer (Satthaporn et al. 2008). Cutaneous recurrences of primary breast cancers when treated with injections of natural IFN-α and -γ, delivered in combination or separately, resulted in complete regression of some of these lesions and was shown to correlate with enhanced expression of HLA-DR in the DC and concurrent increased activation and infiltration of T cells locally. Such an enhanced DC maturation and activation response supports our findings that IFN-α is a pre-requisite for optimal DC maturation and activation in patients with breast cancer (Ozzello et al. 1992). Traditionally, all methods of DC ex vivo generation from monocyte precursors required 7 days of culture in vitro. However, by culturing monocytes with GM-CSF and IL-4 for 48 h and IFN-α for the subsequent 24 h, it is possible to generate "fast DC" within 72 h from monocytes (Dauer et al. 2006). In advanced breast cancer, TNF-α and polyI:C have also been used previously as maturation stimuli and to induce secretion of IL-12p70, as this is the chief functional determinant of a T helper-1 (Th1) T-cell response (Pedersen et al. 2005).

Triggering toll-like receptors (TLR) on DC induces high IL-12p70 production, which is critical for a Th1 response in the context of cancer immunotherapy (Napolitani et al. 2005). Maturation of DC by pro-inflammatory cytokines alone is insufficient to produce DC that support T-cell clonal expansion; they fail to efficiently direct effector T-cell differentiation (Sporri and Reis e Sousa 2005). Interestingly, DC matured in the presence of TLR ligands are able to enhance T-cell effector function (Sato and Iwasaki 2004). Moreover, cross-presentation of exogenous antigen in major histocompatibility complex (MHC) class I proteins is facilitated in the presence of TLR ligands (Datta and Raz 2005). DC matured with poly-I:C (TLR3 ligand) and/or R848 (TLR7/8 ligand) are able to produce large amounts of IL-12p70, but exhibit a reduced migratory capacity, which can be overcome by the addition of PgE_2 (Boullart et al. 2008).

22.3 Quality Control Required for Dendritic Cell Produced for Immunotherapy

Validated standards have yet to be established for clinical trials using ex vivo generated DC for immunotherapy.

22.3.1 Purity

Purity (percentage of cells that are DC) determinations use flow cytometry to evaluate the presence of lymphocytes (CD3 +, CD19+ or CD56/16+) and/or monocytes (CD14+) in the final DC cellular product. The absence of the CD14+ surface marker may be helpful for defining DC differentiation. However, standards for the level of purity of DC depend on the source of progenitor cells (i.e., CD34+ or CD14+ cells), in vitro maturation conditions used and duration of culture, and have yet to be fully established and validated. Moreover, DC lack a single unique identification phenotypic marker and are usually characterized by multiple surface markers, which makes the characterization and standardization of the cells produced by different groups and processes difficult.

22.3.2 Sterility

Sterility testing, including a 7–14-day incubation of DC products, ensures the absence of microorganisms including mycoplasma. This is determined by polymerase chain reaction (PCR) or a 28-day in vitro culture, and bacterial products (e.g., endotoxins) are assayed by using a validated chromogenic technique; this is done in most Good Manufacturing Practice (GMP)-accredited laboratories. Clearly, a 100% sterile product is optimal for all clinical trials.

22.3.3 Viability

Viability of DC can be measured microscopically by the trypan blue dye test or by flow cytometry using propidium iodide (PI). A viability of 95% or more is acceptable to most regulatory agencies.

22.3.4 Maturation

DC upon activation, typically express high levels of HLA molecules, and co-stimulatory molecules, such as CD80, CD86 and maturation marker CD83, in addition to enhanced expression of the chemokine receptor CCR7. This is routinely assessed by flow cytometry. The degree of maturation and activation will depend on the culture conditions used. Before exposure to TA DC should be sufficiently mature and activated to be non-tolerogenic but not terminally differentiated to lose the capability to take up antigen and present it to naïve T cells in the paracortical compartments of the lymph node.

22.3.5 Stability

Stability assays evaluate the ability of DC to sustain viability and activity over varying chronological and thermal conditions to ensure stability over time, and freeze-thaw cycles.

22.3.6 Potency

DC are known to secrete multiple cytokines, however, the key cytokine that determines an effective Th1 response is IL-12p70. This cytokine is spontaneously secreted by DC or in response to CD40L stimulation, with or without the addition of innate immune signals (e.g., lipopolysaccharide [LPS]). This property of DC has been recently used to standardize a DC potency assay defined as a quantitative measure of DC biologic function(s) in vitro and in vivo (Butterfield et al. 2008).

Most DC generation protocols require patients to undergo apheresis to obtain precursor white blood cells exposing the patients to the attendant risks of this procedure. Subsequent immunomagnetic isolation of CD14+ or CD34+ cells and their culture under controlled conditions requires GMP grade facilities, which makes the whole procedure laborious and expensive. There have been considerable advances in the development of closed bag systems which are alternatives to sterile GMP environments (Pullarkat et al. 2002; Rouard et al. 2003; Elias et al. 2005). The entire DC generation process is carried out in a closed system of sterile bags, from the apheresis sample to the final DC product (Sorg et al. 2003). Closed-bag systems reduce the costs considerably for large-scale production of DC for use in cancer immunotherapy trials.

Once DC are prepared, the next step is to load them with suitable TA enabling the TA-loaded DC to prime naïve CD8 + T cells to generate TA-specific CTL in the paracortical zone of the regional lymph node draining the inoculation site. Procedures used to load DC with TA are diverse and each technique has its own advantages and disadvantages, which are outlined in Table 22.1.

Table 22.1 Different tumor antigen-loading procedures: advantages and disadvantages

Tumor antigens	Advantages	Disadvantages
Peptides	Antigen-specific responses can be evaluated	HLA-restriction, rendering non-HLA-restricted types to be unsuitable
Tumor proteins	No HLA restriction	Antigen-specific responses cannot be assessed
Tumor tissue lysate	No HLA restriction	Many antigens unknown, risk of autoimmunity, compared with defined proteins or peptides
mRNA/DNA encoding antigens	Generation of CD4+cognate help along with CD8+T-cell priming from the same DC	Risk of viral contamination and pathogenicity of viruses used for transfecting DC
Tumor-derived total RNA	Generation CD4+ cognate help to enhance CD8+ CTL responses	Lack of specificity to monitor responses
Dendritoma (DC–tumor cell hybrid)	Tailored to the patient's cancer	Lack of specificity to monitor responses

22.4 Dendritic Cell Vaccines in Specific Cancers: In vitro and In vivo Assessment

22.4.1 Esophageal Carcinoma

There have been very few studies, investigating the role of immunotherapy in esophageal cancer. This has been due to the inability to identify specific TA in this malignancy. Nagao et al.showed that whole tumor (esophageal cancer)-pulsed DC could activate CTL to target esophageal cancer cells in vitro (Nagao et al. 1999). When these DC were injected locally into skin lesions in a patient with metastatic esophageal squamous cell cancer, the skin lesions disappeared (Nagao et al. 1999). Another study by Kanaoka et al. showed that DC pulsed with HLA-A2-restricted peptide MAGE-3 were able to elicit specific CTL responses against an esophageal cancer cell line bearing MAGE-3 (Kanaoka et al. 1999). This has not yet been confirmed in vivo. There have been some promising results from preliminary studies. However, adequately powered clinical trials are required before such interventions can be used as possible therapy in advanced esophageal cancer.

22.4.2 Gastric Carcinoma

Various TA have been used in immunotherapy against gastric cancer. Sun et al. demonstrated that wild-type p53 in an adenovirus vector transfected into DC and stimulated with gastric cancer lysates induced antigen-specific CTL (Sun et al. 2005). These DC, when injected into mice, showed a reduction in the volume of implanted gastric cancer xenografts (Liu et al. 2004; Sun et al. 2004). RNA from gastric cancer has been used to pulse DC to induce antigen-specific CTL and antitumor responses in mice bearing tumor xenografts (Liu et al. 2004). A novel technique of introducing gastric cancer cell RNA into DC by electroporation was described by Ohshita et al. Using this method, tumor RNA was amplified before being introduced into DC; the CTL generated from these DC were able to produce more tumor-specific IFN-γ than conventionally stimulated CTL (Ohshita et al. 2003). Survivin, an apoptotic inhibitor, and HLA-A.2.1-restricted CEA peptides (CEA(9)671 and CEA(9)652) are other TA from gastric cancer that have been used to pulse DC in order to induce antigen-specific CTL (Kim et al. 2000; Ohta et al. 2002; Sun et al. 2005). However, these interesting in vitro findings need to be confirmed in vivo.

Human epidermal growth factor receptor 2 (EGFR-2) was shown to induce antigen-specific CTL in vitro when introduced through the MHC class I protein pathway (Shiku et al. 2000). The only clinical study, to date, has been the use of DC pulsed with HLA-A2-specific EGFR-2 (p369)-derived peptides. These DC were able to generate peptide-specific CTL in six out of nine patients in vivo, two of whom had a partial clinical response (Kono et al. 2002). A few novel techniques exist, which can incorporate TA into DC without selection for HLA-A2. Hybrid cell formation by electrofusion of a DC with a gastric cancer cell is one such approach. This effectively produces an activated DC capable of eliciting tumor-specific CTL responses (Imura et al. 2004). Another approach is to use purified complexes of heat shock proteins linked to TA (HSPPC). This stimulated a strong antitumor response, longer than that generated by expression of the same TA by the tumor cell. Phase I/II trials, using HSP-based tumor vaccine (HSPPC-96, Oncophage), are being carried out in patients with gastric and pancreatic cancers, and results are awaited with interest (BioDrugs 2002).

Biological response modifiers (BRM) have been tested successfully in many in vitro gastric cancer studies. They are peptides that regulate the cytotoxic activity of various effector cells by the induction of response augmenting cytokines and inhibition of response inhibiting cytokines. Streptococcal antigen OK-432 and fungal antigen PSK have been tested in gastric cancers. OK-432-incubated DC produced more IL-12, IFN-γ, and expressed more co-stimulatory molecules than controls (Itoh et al. 2003). PSK, a protein bound polysaccharide from Basidiomycetes, was able to recover DC function and maturation by blocking their inhibitory stimuli in the cancer mileu (Okuzawa et al. 2002).

22.4.3 Cholangiocarcinoma

An in vitro study demonstrated that apoptotic cholangiocarcinoma cells (apoptosis induced by mitomycin) were able to prime DC to activate tumor-specific

CTL (Wu et al. 2003). Immunotherapy in cholangiocarcinoma is an unexplored area, possibly due to the low incidence of this tumor in Europe and the United States. However, this should not prevent exploration of newer tumor vaccination strategies as the prognosis is poor in these patients despite conventional therapy. Many present late and are inoperable and incurable.

22.4.4 Pancreatic Carcinoma

In pancreatic cancer, a wide range of TA has been tried in vitro or in vivo. A pilot study was carried out using DC loaded with CEA mRNA and administered to three patients following curative resection (Morse et al. 2002). All three patients were disease-free for more than 30 months following surgery (Morse et al. 2002). In another clinical study, autologous DC pulsed with tumor lysates were injected into a patient with Stage IV pancreatic adenocarcinoma leading to the generation of tumor-specific CTL and disease stabilization for 6 months (Stift et al. 2003).

A novel DC-tumor cell hybrid vaccine with staphylococcal enterotoxin B (T-cell stimulant) was able to stimulate tumor-specific CTL and enhance survival in transgenic mice with spontaneous pancreatic carcinoma (McConnell et al. 2002). In an in vitro study, DC transfected with total tumor pancreatic RNA were potent activators of tumor-specific CTL (Kalady et al. 2004). Tumor-specific CTL could also be generated in vitro by pulsing DC with CA19-9 and K-ras (Marten et al. 2000). DC transfected with HLA-A2-restricted peptide 652–662, from proto-oncogene HER2/neu, were able to produce similar results (Peiper et al. 1997). As there are no specific TA for pancreatic cancer, tumor lysates, tumor RNA, and DC-tumor hybridomas have been used in various vaccines. Irradiated pancreatic cells, that have undergone apoptosis and necrosis, have been found to be a good source of peptides for pulsing DC (Shimamura et al. 2005). Although the results are promising, the safety and efficacy of many of these vaccines await validation through appropriate clinical trials.

22.4.5 Hepatocellular Carcinoma (HCC)

Various TA have been used to pulse DC in immunotherapy in patients with HCC. Hyperthermia-induced tumor lysates, were found to enhance CD8+/CD4+ cytotoxicity against the HCC cell line HEPG2. This was related to expression of HSP-70 and HSP-90 following heat treatment (Schueller et al. 2003). In another study, tumor-derived HSP-70 was used to pulse DC, which induced the generation of tumor-specific CTL (Wang et al. 2005a, b). Transplantation of NKT pulsed with HCC lysates in mice with HCC showed tumor suppressive properties in mice (Margalit et al. 2005). In a related human study,

15 post-hepatectomy patients injected with autologous DC pulsed with HCC lysates demonstrated lower recurrence rates and better survival, compared with a similar number who received only chemotherapy (Gao et al. 2005). In a not dissimilar clinical study, 17 patients received DC pulsed with tumor lysates, followed by booster doses. Their survival was compared with a group of 14 patients who received only a single dose of DC pulsed with tumor lysates. Pulsed DC vaccination with tumor lysates followed by boosters was shown to confer a survival advantage, compared with pulsed DC vaccination alone (Lee et al. 2005). Similarly, pulsed autologous DC with autologous HCC antigens injected in two advanced HCC patients was found to be safe and effective (Ladhams et al. 2002). The most effective mode of delivering autologous-pulsed DC is to inoculate directly into the tumor, as it is apparent from published evidence that the tumor microenvironment directly inhibits DC infiltration.

Intratumoral injection of DC into the liver, following injection of 100% ethanol, has been found to be safe with no adverse effects (Kumagi et al. 2005). Various other TA-pulsed DC have been shown in vitro to induce tumor-specific CTL. These are α-fetoprotein, human cancer antigen (HCA-587), HSP-70, and glycoprotein-96 (gp-96) (Wang et al. 2005a, b). Tranfection of DC with the tumor cell genome is another approach being explored in activating DC. Transfection of DC with the *HBsAg* gene is capable of evoking an effective CTL response in vitro against HCC originating from livers which had chronic hepatitis B infection (Yang et al. 2004). A similar response was observed when DC were transfected with an adenovirus-encoding human cancer antigen HCA-661 (Chan et al. 2004). Recently, dendritomas from tumor cell–DC fusions have been shown to be successful in generating mature and functional DC, able to induce tumor-specific CTL in vitro (Guan et al. 2004, 2005).

22.4.6 *Colorectal Carcinoma (CRC)*

CRC lack well-characterized TA. CEA-pulsed DC or CEA mRNA-transfected DC appear to evoke tumor-specific CTL against CEA-expressing cells in vivo and in vitro (Nakamori et al. 2000; Eppler et al. 2002). However, not all CRC cells express CEA. HLA–A2-restricted HER-2/neu-derived peptides, E75 and GP2, have been used to pulse DC. These DC generate CTL which are capable of lysing CRC cells bearing these epitopes in vitro (Brossart et al. 1998). The use of whole tumor mRNA to transfect DC overcomes the requirement of HLA restriction and the identification of TA expressed by CRC. This concept was demonstrated in vitro by plasmid transfection of whole tumor mRNA into DC, which seem to activate CTL against multiple cancer cell lines (Nencioni et al. 2003). Moreover, levels of CRC cell lysis obtained with tumor-specific CTL (blood mononuclear cells stimulated with DC transfected with total tumor RNA) were higher than the lysis obtained with CEA-specific CTL (blood mononuclear cells stimulated with DC transfected with CEA RNA) (Nair et al. 2002). Such enhanced tumor lysis may be due to unidentified non-CEA antigens, priming DC. Arming DC with tumor lysates is another approach to overcome HLA restriction. A phase I/II clinical trial (30 patients) of autologous DC-based whole tumor vaccine (DC pulsed with whole tumor mRNA or DC pulsed with whole tumor lysates) has demonstrated the safety of this approach with no adverse effects (Rains et al. 2001). The majority of patients had stable disease but no significant tumor regression was observed.

Murine studies have shown that CRC treated with mitomycin C are rich sources of apoptotic cell products; when exposed to DC they are capable of inducing tumor regression (Inoue et al. 2003). Allogeneic dendritomas can be produced by fusion of DC with CRC cells using 50% polyethylene glycol (Xu et al. 2004). Such dendritomas have been shown to elicit tumor-specific CTL and helper T-cell responses against metastatic CRC cells in vitro (Xu et al. 2005). Tumor-derived TGF-β has been shown to reduce the effectiveness of dendritoma-based immunotherapy in vivo (Kao et al. 2003). This highlights the complex requirements for neutralizing tumor-derived or dendritoma-derived TGF-β before vaccine delivery. Similar to gastric cancer, PSK, a BRM from Basidiomycetes, has been shown to reactivate in vitro, dysfunctional DC from CRC patients (Okuzawa et al. 2002). This may represent another approach to optimize autologous DC function in CRC patients before immunotherapy.

22.5 Future Directions

Molecular targeting of tumors with a strategy to disrupt antigens in the cancer cell survival pathways is a clear advantage over targeting antigens which may escape immune surveillance, immune editing, and escape. Examples of such novel antigens are survivin and human telomerase reverse transcriptase (hTERT). Survivin is an anti-apoptotic molecule in cancer cells. hTERT is the catalytic unit of telomerase which continually elongates telomeres in cancer cells rendering them "immortal". Survivin has been used to pulse DC before vaccinating prostate and colorectal cancer patients and has demonstrated strong immunological responses (Schmitz et al. 2000; Casati et al. 2003; Reed and Wilson 2003; Fuessel et al. 2006). Similarly, hTERT has been used in a much wider spectrum of malignancies with promising immunological and some clinical responses in patients with advanced malignancies (Aloysius et al. 2006; Aloysius 2008).

Cancer immunotherapy is at a crucial stage of development, with evidence accruing on safety, immunogenicity, and limited clinical efficacy for well-designed novel immunization protocols with cancer generic antigens like survivin and hTERT. It is also crucial to understand the scientific basis of negative clinical responses in individual studies. Such an approach would enable investigators to revisit their studies and design protocols to selectively block tumor-immune escape pathways. For example, inhibition or deletion of CD4+ 25+Foxp3+

T regulatory cells (which induce tolerance to TA) by receptor-specific antibodies or selectively enhance cytotoxic pathways by viral transfection of co-stimulatory molecules on tumor cells. Continually emerging new knowledge and a better understanding of basic immunology will lead to further new and novel strategies, and hopefully clinically beneficial therapeutic outcomes.

Acknowledgments MMA is a recipient of a research grant from the Royal College of Surgeons of Edinburgh. We gratefully acknowledge financial support from the following: Candles (Lincolnshire), Friends of Lincoln hospital, Boston Leukaemia Research Fund, and Rosetrees Trust.

References

Aloysius, M. M., Robins, R. A., Eremin, J. M. and Eremin, O. 2006. Vaccination therapy in malignant disease. Surgeon 4:309–320.

Aloysius, M., Verma C. and Eremin, O. 2008. Targeting human telomerase reverse transcriptase: a promising anticancer immunotherapeutic approach. Cancer Chemother Rev 3:35.

BioDrugs 2002. Cancer vaccine – antigenics. BioDrugs 16:72–74.

Boullart, A. C., Aarntzen, E. H., Verdijk, P., Jacobs, J. F., Schuurhuis, D. H., Benitez-Ribas, D., Schreibelt, G., van de Rakt, M. W., Scharenborg, N. M., de Boer, A., Kramer, M., Figdor, C. G., Punt, C. J., Adema, G. J. and de Vries, I. J. 2008. Maturation of monocyte-derived dendritic cells with Toll-like receptor 3 and 7/8 ligands combined with prostaglandin E(2) results in high interleukin-12 production and cell migration. Cancer Immunol Immunother 57:1589–1597.

Brossart, P., Stuhler, G., Flad, T., Stevanovic, S., Rammensee, H. G., Kanz, L. and Brugger, W. 1998. Her-2/neu-derived peptides are tumor-associated antigens expressed by human renal cell and colon carcinoma lines and are recognized by in vitro induced specific cytotoxic T lymphocytes. Cancer Res 58:732–736.

Butterfield, L. H., Gooding, W. and Whiteside, T. L. 2008. Development of a potency assay for human dendritic cells: IL-12p70 production. J Immunother 31:89–100.

Casati, C., Dalerba, P., Rivoltini, L., Gallino, G., Deho, P., Rini, F., Belli, F., Mezzanzanica, D., Costa, A., Andreola, S., Leo, E., Parmiani, G. and Castelli, C. 2003. The apoptosis inhibitor protein survivin induces tumor-specific CD8+ and CD4+ T cells in colorectal cancer patients. Cancer Res 63:4507–4515.

Caux, C., Dezutter-Dambuyant, C., Schmitt, D. and Banchereau, J. 1992. GM-CSF and TNF-alpha cooperate in the generation of dendritic Langerhans cells. Nature 360: 258–261.

Chan, R. C., Pang, X. W., Wang, Y. D., Chen, W. F. and Xie, Y. 2004. Transduction of dendritic cells with recombinant adenovirus encoding HCA661 activates autologous cytotoxic T lymphocytes to target hepatoma cells. Br J Cancer 90:1636–1643.

Datta, S. K. and Raz, E. 2005. Induction of antigen cross-presentation by Toll-like receptors. Springer Semin Immunopathol 26:247–255.

Dauer, M., Obermaier, B., Herten, J., Haerle, C., Pohl, K., Rothenfusser, S., Schnurr, M., Endres, S. and Eigler, A. 2003. Mature dendritic cells derived from human monocytes within 48 hours: a novel strategy for dendritic cell differentiation from blood precursors. J Immunol 170:4069–4076.

Dauer, M., Schad, K., Junkmann, J., Bauer, C., Herten, J., Kiefl, R., Schnurr, M., Endres, S. and Eigler, A. 2006. IFN-alpha promotes definitive maturation of dendritic cells generated by short-term culture of monocytes with GM-CSF and IL-4. J Leukoc Biol 80:278–286.

Elias, M., van Zanten, J., Hospers, G. A., Setroikromo, A., de Jong, M. A., de Leij, L. F. and Mulder, N. H. 2005. Closed system generation of dendritic cells from a single blood volume for clinical application in immunotherapy. J Clin Apher 20:197–207.

Eppler, E., Horig, H., Kaufman, H. L., Groscurth, P. and Filgueira, L. 2002. Carcinoembryonic antigen (CEA) presentation and specific T cell-priming by human dendritic cells transfected with CEA-mRNA. Eur J Cancer 38:184–193.

Figdor, C. G., de Vries, I. J., Lesterhuis, W. J. and Melief, C. J. 2004. Dendritic cell immunotherapy: mapping the way. Nat Med 10:475–480.

Freudenthal, P. S. and Steinman, R. M. 1990. The distinct surface of human blood dendritic cells, as observed after an improved isolation method. Proc Natl Acad Sci USA 87:7698–7702.

Fuessel, S., Meye, A., Schmitz, M., Zastrow, S., Linne, C., Richter, K., Lobel, B., Hakenberg, O. W., Hoelig, K., Rieber, E. P. and Wirth, M. P. 2006. Vaccination of hormone-refractory prostate cancer patients with peptide cocktail-loaded dendritic cells: results of a phase I clinical trial. Prostate 66:811–821.

Gabrilovich, D. 2004. Mechanisms and functional significance of tumour-induced dendritic-cell defects. Nat Rev Immunol 4:941–952.

Gao, J., Chen, M. and Ren, H. 2005. [Clinical effects of dendritic cells pulsed with autologous hepatoma cell lysates on the postoperative recurrence and metastasis of hepatocellular carcinoma]. Zhonghua Gan Zang Bing Za Zhi 13:432–435.

Guan, X., Peng, J. R. and Leng, X. S. 2005. [Establishment of dendritomas by fusion of human dendritic cells with human hepatocellular carcinoma cell line HLE cells]. Zhonghua Zhong Liu Za Zhi 27:465–467.

Guan, X., Peng, J. R., Yuan, L., Wang, H., Wei, Y. H. and Leng, X. S. 2004. A novel, rapid strategy to form dendritomas from human dendritic cells and hepatocellular carcinoma cell line HCCLM3 cells using mature dendritic cells derived from human peripheral blood CD14+ monocytes within 48 hours of in vitro culture. World J Gastroenterol 10:3564–3568.

Hoffmann, T. K., Muller-Berghaus, J., Ferris, R. L., Johnson, J. T., Storkus, W. J. and Whiteside, T. L. 2002. Alterations in the frequency of dendritic cell subsets in the peripheral circulation of patients with squamous cell carcinomas of the head and neck. Clin Cancer Res 8:1787–1793.

Imura, K., Hayashi, T., Yano, Y., Naito, K., Kouhara, J., Ueda, Y., Nakane, K., Matsuura, Y., Takeda, A., Takeda, T., Kawai, K. and Yamagishi, H. 2004. Immunogenic reactivity of CTLs induced by electrofusion cells of human dendritic cells and gastric cancer cells. Gan To Kagaku Ryoho 31:1797–1799.

Inoue, N., Yamasaki, S., Kondo, K., Kan, T., Furumoto, K. and Imamura, M. 2003. Dendritic cells coinjected with tumor cells treated with an anticancer drug to induce tumor rejection. Surg Today 33:269–276.

Itoh, T., Ueda, Y., Okugawa, K., Fujiwara, H., Fuji, N., Yamashita, T., Fujiki, H., Harada, S., Yoshimura, T. and Yamagishi, H. 2003. Streptococcal preparation OK432 promotes functional maturation of human monocyte-derived dendritic cells. Cancer Immunol Immunother 52:207–214.

Kalady, M. F., Onaitis, M. W., Emani, S., Abdul-Wahab, Z., Pruitt, S. K. and Tyler, D. S. 2004. Dendritic cells pulsed with pancreatic cancer total tumor RNA generate specific antipancreatic cancer T cells. J Gastrointest Surg 8:175–181; discussion 181–182.

Kalinski, P., Vieira, P. L., Schuitemaker, J. H., de Jong, E. C. and Kapsenberg, M. L. 2001. Prostaglandin E(2) is a selective inducer of interleukin-12 p40 (IL-12p40) production and an inhibitor of bioactive IL-12p70 heterodimer. Blood 97:3466–3469.

Kanaoka, S., Yamasaki, S., Okino, T., Inoue, N., Shimada, Y., Kaneko, M., Otaka, A., Fujii, N. and Imamura, M. 1999. Induction of human leukocyte antigen (HLA)-A2-restricted and MAGE-3-gene-derived peptide-specific cytolytic T lymphocytes using cultured dendritic cells from an HLA-A2 esophageal cancer patient. J Surg Oncol 71:16–21.

Kao, J. Y., Gong, Y., Chen, C. M., Zheng, Q. D. and Chen, J. J. 2003. Tumor-derived TGF-beta reduces the efficacy of dendritic cell/tumor fusion vaccine. J Immunol 170:3806–3811.

Kim, C. H., Todoroki, T., Matsumura, M. and Ohno, T. 2000. Eligibility of antigenic-peptide-pre-loaded and fixed adhesive peripheral blood cells for induction of cytotoxic T lymphocytes from cancer patients with elevated serum levels of carcinoembryonic antigen. J Cancer Res Clin Oncol 126:383–390.

Kono, K., Takahashi, A., Sugai, H., Fujii, H., Choudhury, A. R., Kiessling, R. and Matsumoto, Y. 2002. Dendritic cells pulsed with HER-2/neu-derived peptides can induce specific T-cell responses in patients with gastric cancer. Clin Cancer Res 8:3394–3400.

Kumagi, T., Akbar, S. M., Horiike, N., Kurose, K., Hirooka, M., Hiraoka, A., Hiasa, Y., Michitaka, K. and Onji, M. 2005. Administration of dendritic cells in cancer nodules in hepatocellular carcinoma. Oncol Rep 14:969–973.

Ladhams, A., Schmidt, C., Sing, G., Butterworth, L., Fielding, G., Tesar, P., Strong, R., Leggett, B., Powell, L., Maddern, G., Ellem, K. and Cooksley, G. 2002. Treatment of non-resectable hepatocellular carcinoma with autologous tumor-pulsed dendritic cells. J Gastroenterol Hepatol 17:889–896.

Langenkamp, A., Messi, M., Lanzavecchia, A. and Sallusto, F. 2000. Kinetics of dendritic cell activation: impact on priming of TH1, TH2 and nonpolarized T cells. Nat Immunol 1:311–316.

Lee, W. C., Wang, H. C., Hung, C. F., Huang, P. F., Lia, C. R. and Chen, M. F. 2005. Vaccination of advanced hepatocellular carcinoma patients with tumor lysate-pulsed dendritic cells: a clinical trial. J Immunother 28:496–504.

Liu, B. Y., Chen, X. H., Gu, Q. L., Li, J. F., Yin, H. R., Zhu, Z. G. and Lin, Y. Z. 2004. Antitumor effects of vaccine consisting of dendritic cells pulsed with tumor RNA from gastric cancer. World J Gastroenterol 10:630–633.

Margalit, M., Shibolet, O., Klein, A., Elinav, E., Alper, R., Thalenfeld, B., Engelhardt, D., Rabbani, E. and Ilan, Y. 2005. Suppression of hepatocellular carcinoma by transplantation of ex-vivo immune-modulated NKT lymphocytes. Int J Cancer 115:443–449.

Marten, A., Schottker, B., Ziske, C., Weineck, S., Buttgereit, P., Huhn, D., Sauerbruch, T. and Schmidt-Wolf, I. G. 2000. Increase of the immunostimulatory effect of dendritic cells by pulsing with CA 19-9 protein. J Immunother 23:464–472.

McConnell, E. J., Pathangey, L. B., Madsen, C. S., Gendler, S. J. and Mukherjee, P. 2002. Dendritic cell-tumor cell fusion and staphylococcal enterotoxin B treatment in a pancreatic tumor model. J Surg Res 107:196–202.

Morse, M. A., Nair, S. K., Boczkowski, D., Tyler, D., Hurwitz, H. I., Proia, A., Clay, T. M., Schlom, J., Gilboa, E. and Lyerly, H. K. 2002. The feasibility and safety of immunotherapy with dendritic cells loaded with CEA mRNA following neoadjuvant chemoradiotherapy and resection of pancreatic cancer. Int J Gastrointest Cancer 32:1–6.

Nagao, N., Katoh, M., Kumazawa, I., Tomita, H., Sugiyama, Y., Kunieda, K., Miya, K. and Saji, S. 1999. A recurrent case of esophageal cancer in which metastatic skin tumor disappeared after local injection of activated lymphocytes with tumor-pulsed dendritic cells. Gan To Kagaku Ryoho 26:1937–1939.

Nair, S. K., Morse, M., Boczkowski, D., Cumming, R. I., Vasovic, L., Gilboa, E. and Lyerly, H. K. 2002. Induction of tumor-specific cytotoxic T lymphocytes in cancer patients by autologous tumor RNA-transfected dendritic cells. Ann Surg 235:540–549.

Nakamori, M., Iwahashi, M., Tani, M., Yamaue, H., Ueda, K., Matsuda, K. and Tanimura, H. 2000. New therapeutic strategy against colon cancer based on a tumor-specific approach. Gan To Kagaku Ryoho 27:2209–2215.

Napolitani, G., Rinaldi, A., Bertoni, F., Sallusto, F. and Lanzavecchia, A. 2005. Selected Toll-like receptor agonist combinations synergistically trigger a T helper type 1-polarizing program in dendritic cells. Nat Immunol 6:769–776.

Nencioni, A., Muller, M. R., Grunebach, F., Garuti, A., Mingari, M. C., Patrone, F., Ballestrero, A. and Brossart, P. 2003. Dendritic cells transfected with tumor RNA for the induction of antitumor CTL in colorectal cancer. Cancer Gene Ther 10:209–214.

O'Doherty, U., Steinman, R. M., Peng, M., Cameron, P. U., Gezelter, S., Kopeloff, I., Swiggard, W. J., Pope, M. and Bhardwaj, N. 1993. Dendritic cells freshly isolated from human blood express CD4 and mature into typical immunostimulatory dendritic cells after culture in monocyte-conditioned medium. J Exp Med 178:1067–1076.

Ohshita, A., Yamaguchi, Y., Minami, K., Miyahara, E., Kawabuchi, Y., Shimizu, K., Ohta, K., Hihara, J. and Toge, T. 2003. Generation of TRiDAK (tumor RNA-introduced dendritic cell-activated killer) cells. Gan To Kagaku Ryoho 30:1809–1812.

Ohta, K., Yamaguchi, Y., Shimizu, K., Miyahara, E. and Toge, T. 2002. Novel system for generating cytotoxic effector lymphocytes using carcinoembryonic antigen (CEA) peptide and cultured dendritic cells. Anticancer Res 22:2597–2606.

Okuzawa, M., Shinohara, H., Kobayashi, T., Iwamoto, M., Toyoda, M. and Tanigawa, N. 2002. PSK, a protein-bound polysaccharide, overcomes defective maturation of dendritic cells exposed to tumor-derived factors in vitro. Int J Oncol 20:1189–1195.

Ozzello, L., Habif, D. V., De Rosa, C. M. and Cantell, K. 1992. Cellular events accompanying regression of skin recurrences of breast carcinomas treated with intralesional injections of natural interferons alpha and gamma. Cancer Res 52:4571–4581.

Pedersen, A. E., Thorn, M., Gad, M., Walter, M. R., Johnsen, H. E., Gaarsdal, E., Nikolajsen, K., Buus, S., Claesson, M. H. and Svane, I. M. 2005. Phenotypic and functional characterization of clinical grade dendritic cells generated from patients with advanced breast cancer for therapeutic vaccination. Scand J Immunol 61:147–156.

Peiper, M., Goedegebuure, P. S. and Eberlein, T. J. 1997. Generation of peptide-specific cytotoxic T lymphocytes using allogeneic dendritic cells capable of lysing human pancreatic cancer cells. Surgery 122:235–241; discussion 241–242.

Pullarkat, V., Lau, R., Lee, S. M., Bender, J. G. and Weber, J. S. 2002. Large-scale monocyte enrichment coupled with a closed culture system for the generation of human dendritic cells. J Immunol Methods 267:173–183.

Radvanyi, L. G., Banerjee, A., Weir, M. and Messner, H. 1999. Low levels of interferon-alpha induce CD86 (B7.2) expression and accelerates dendritic cell maturation from human peripheral blood mononuclear cells. Scand J Immunol 50:499–509.

Rains, N., Cannan, R. J., Chen, W. and Stubbs, R. S. 2001. Development of a dendritic cell (DC)-based vaccine for patients with advanced colorectal cancer. Hepatogastroenterology 48:347–351.

Reed, J. C. and Wilson, D. B. 2003. Cancer immunotherapy targeting survivin: commentary re: V. Pisarev et al., full-length dominant-negative survivin for cancer immunotherapy. Clin Cancer Res 9:6523–6533. Clin Cancer Res 9:6310–6315.

Rouard, H., Leon, A., De Reys, S., Taylor, L., Logan, J., Marquet, J., Jouault, H., Loper, K., Maison, P., Delfau-Larue, M. H., Beaujean, F., Farcet, J. P. and Noga, S. J. 2003. A closed and single-use system for monocyte enrichment: potential for dendritic cell generation for clinical applications. Transfusion 43:481–487.

Sallusto, F. and Lanzavecchia, A. 1994. Efficient presentation of soluble antigen by cultured human dendritic cells is maintained by granulocyte/macrophage colony-stimulating factor plus interleukin 4 and downregulated by tumor necrosis factor alpha. J Exp Med 179:1109–1118.

Sato, A. and Iwasaki, A. 2004. Induction of antiviral immunity requires Toll-like receptor signaling in both stromal and dendritic cell compartments. Proc Natl Acad Sci U S A 101:16274–16279.

Satthaporn, S., Aloysius, M. M., Robins, R. A., Verma, C., Chuthapisith, S., McKechnie, A. J., El-Sheemy, M., Vassanasiri, W., Valerio, D., Clark, D., Jibril, J. A. and Eremin, O. 2008. Ex vivo recovery and activation of dysfunctional, anergic, monocyte-derived dendritic cells from patients with operable breast cancer: critical role of IFN-alpha. BMC Immunol 9:32.

Satthaporn, S., Robins, A., Vassanasiri, W., El-Sheemy, M., Jibril, J. A., Clark, D., Valerio, D. and Eremin, O. 2004. Dendritic cells are dysfunctional in patients with operable breast cancer. Cancer Immunol Immunother 53:510–518.

Savary, C. A., Grazziutti, M. L., Melichar, B., Przepiorka, D., Freedman, R. S., Cowart, R. E., Cohen, D. M., Anaissie, E. J., Woodside, D. G., McIntyre, B. W., Pierson, D. L., Pellis, N. R. and Rex, J. H. 1998. Multidimensional flow-cytometric analysis of dendritic cells in peripheral blood of normal donors and cancer patients. Cancer Immunol Immunother 45:234–240.

Schmitz, M., Diestelkoetter, P., Weigle, B., Schmachtenberg, F., Stevanovic, S., Ockert, D., Rammensee, H. G. and Rieber, E. P. 2000. Generation of survivin-specific CD8+ T effector cells by dendritic cells pulsed with protein or selected peptides. Cancer Res 60:4845–4849.

Schueller, G., Stift, A., Friedl, J., Dubsky, P., Bachleitner-Hofmann, T., Benkoe, T., Jakesz, R. and Gnant, M. 2003. Hyperthermia improves cellular immune response to human hepatocellular carcinoma subsequent to co-culture with tumor lysate pulsed dendritic cells. Int J Oncol 22:1397–1402.

Shiku, H., Wang, L., Ikuta, Y., Okugawa, T., Schmitt, M., Gu, X., Akiyoshi, K., Sunamoto, J. and Nakamura, H. 2000. Development of a cancer vaccine: peptides, proteins, and DNA. Cancer Chemother Pharmacol 46Suppl:S77–82.

Shimamura, H., Sunamura, M., Tsuchihara, K., Egawa, S., Takeda, K. and Matsuno, S. 2005. Irradiated pancreatic cancer cells undergo both apoptosis and necrosis, and could be phagocytized by dendritic cells. Eur Surg Res 37:228–234.

Sorg, R. V., Ozcan, Z., Brefort, T., Fischer, J., Ackermann, R., Muller, M. and Wernet, P. 2003. Clinical-scale generation of dendritic cells in a closed system. J Immunother 26:374–383.

Sporri, R. and Reis e Sousa, C. 2005. Inflammatory mediators are insufficient for full dendritic cell activation and promote expansion of CD4+ T cell populations lacking helper function. Nat Immunol 6:163–170.

Stift, A., Friedl, J., Dubsky, P., Bachleitner-Hofmann, T., Benkoe, T., Brostjan, C., Jakesz, R. and Gnant, M. 2003. In vivo induction of dendritic cell-mediated cytotoxicity against allogeneic pancreatic carcinoma cells. Int J Oncol 22:651–656.

Strobl, H., Riedl, E., Scheinecker, C., Bello-Fernandez, C., Pickl, W. F., Majdic, O. and Knapp, W. 1997. TGF-beta 1 dependent generation of LAG+ dendritic cells from CD34+ progenitors in serum-free medium. Adv Exp Med Biol 417:161–165.

Strobl, H., Riedl, E., Scheinecker, C., Bello-Fernandez, C., Pickl, W. F., Rappersberger, K., Majdic, O. and Knapp, W. 1996. TGF-beta 1 promotes in vitro development of dendritic cells from CD34+ hemopoietic progenitors. J Immunol 157:1499–1507.

Strunk, D., Rappersberger, K., Egger, C., Strobl, H., Kromer, E., Elbe, A., Maurer, D. and Stingl, G. 1996. Generation of human dendritic cells/Langerhans cells from circulating CD34+ hematopoietic progenitor cells. Blood 87:1292–1302.

Sun, H. W., Tang, C., Tang, Q. B., Zou, S. Q. and Qiu, F. Z. 2005. Study of immunological effect of dendritic cell transfected with survivin gene on the specific anti-alimentary tract tumor. Zhonghua Wai Ke Za Zhi 43:313–316.

Sun, H. W., Tang, Q. B., Cheng, Y. J. and Zou, S. Q. 2004. Effects of dendritic cells transfected with full-length wild-type p53 and stimulated by gastric cancer lysates on immune response. World J Gastroenterol 10:2595–2597.

Syme, R., Bajwa, R., Robertson, L., Stewart, D. and Gluck, S. 2005. Comparison of CD34 and monocyte-derived dendritic cells from mobilized peripheral blood from cancer patients. Stem Cells 23:74–81.

Szabolcs, P., Feller, E. D., Moore, M. A. and Young, J. W. 1995a. Progenitor recruitment and in vitro expansion of immunostimulatory dendritic cells from human CD34+ bone marrow cells by c-kit-ligand, GM-CSF, and TNF alpha. Adv Exp Med Biol 378:17–20.

Szabolcs, P., Moore, M. A. and Young, J. W. 1995b. Expansion of immunostimulatory dendritic cells among the myeloid progeny of human CD34+ bone marrow precursors cultured with c-kit ligand, granulocyte-macrophage colony-stimulating factor, and TNF-alpha. J Immunol 154:5851–5861.

Tamir, A., Jordan, W. J., Ritter, M., Habib, N., Lechler, R. I., Foster, G. R. and Lombardi, G. 2005. Interferon-alpha2a is sufficient for promoting dendritic cell immunogenicity. Clin Exp Immunol 142:471–480.

Thomas, R., Davis, L. S. and Lipsky, P. E. 1993. Isolation and characterization of human peripheral blood dendritic cells. J Immunol 150:821–834.

Van Voorhis, W. C., Valinsky, J., Hoffman, E., Luban, J., Hair, L. S. and Steinman, R. M. 1983. Relative efficacy of human monocytes and dendritic cells as accessory cells for T cell replication. J Exp Med 158:174–191.

Wang, X. H., Qin, Y., Hu, M. H. and Xie, Y. 2005a. Dendritic cells pulsed with gp96-peptide complexes derived from human hepatocellular carcinoma (HCC) induce specific cytotoxic T lymphocytes. Cancer Immunol Immunother 54:971–980.

Wang, X. H., Qin, Y., Hu, M. H. and Xie, Y. 2005b. Dendritic cells pulsed with hsp70-peptide complexes derived from human hepatocellular carcinoma induce specific anti-tumor immune responses. World J Gastroenterol 11:5614–5620.

Wu, G., Han, B. L. and Pei, X. T. 2003. Immune response of dendritic cells acquiring antigens from apoptotic cholangiocarcinoma cells induced by mitomycin. Zhonghua Gan Zang Bing Za Zhi 11:149–152.

Xu, F., Ye, Y. J., Cui, Z. R. and Wang, S. 2005. Allogeneic dendritomas induce anti-tumour immunity against metastatic colon cancer. Scand J Immunol 61:364–369.

Xu, F., Ye, Y. J. and Wang, S. 2004. In vitro antitumor immune response induced by fusion of dendritic cells and colon cancer cells. World J Gastroenterol 10:1162–1166.

Yang, J. Y., Ren, J., Bai, J., Liu, D. H., Fan, L., Si, X. M., Teng, Z. H. and Yang, W. T. 2004. Cytotoxicity induced by HBsAg gene modified-dendritic cells against hepatocellular carcinoma cell HepG2.2.15. Ai Zheng 23:914–917.

Young, J. W., Szabolcs, P. and Moore, M. A. 1995. Identification of dendritic cell colony-forming units among normal human CD34+ bone marrow progenitors that are expanded by c-kit-ligand and yield pure dendritic cell colonies in the presence of granulocyte/macrophage colony-stimulating factor and tumor necrosis factor alpha. J Exp Med 182:1111–1119.

Chapter 23
Genetically Modified Dendritic Cells in Cancer Immunotherapy

Brian J. Morrison, Jason C. Steel, Melissa Gregory, John C. Morris, and Anatoli M. Malyguine

Abstract Dendritic cells are powerful antigen-presenting cells that can generate primary cytolytic T lymphocyte responses against tumors. Consequently, there has been much interest in their application as antitumor vaccines. A number of dendritic cell-based vaccine trials targeting a variety of tumors have been conducted in different countries; however, the rate of clinical responses remains low. The majority of these studies have administered dendritic cells loaded with synthetic peptide epitopes or tumor lysates. Genetic modification of dendritic cells to express tumor antigens or immunostimulatory molecules through gene transfer or mRNA transfection offers a logical alternative with potential advantages over antigen loading in dendritic cells. In this chapter, we review the current and future prospects for genetically modified dendritic cell vaccines for cancer therapy.

23.1 Introduction

Studies of antitumor vaccination using tumor antigen(s) (TA) loaded or modified DC have gained increasing popularity. This approach is based on the high efficiency of DC, as powerful professional antigen-presenting cells in the regulation of tolerance and immunity, and induction-specific antitumor responses. DC pulsed with TA peptides, proteins, tumor cell lysates, apoptotic or necrotic tumor cells, or fused with tumor cells, as well as genetically modified DC, have been studied as potential anti-cancer vaccines. Numerous publications describe different approaches for improving efficacy of DC vaccines for the treatment of cancer. This includes such critical parameters as the source of DC, methods of preparation, maturation and activation, methods of loading tumor antigens, source of tumor antigens, route of administration, and immunomonitoring of subjects (Gilboa 2007; Zhong et al. 2007; Nencioni et al. 2008; Vulink et al. 2008). To further enhance antigen presentation and immunostimulation, DC can be genetically modified. In this chapter, we focus on this interesting and promising approach for cancer immunotherapy.

A.M. Malyguine (✉)
Laboratory of Cell Mediated Immunity, CSP, SAIC-Frederick, Inc., National Cancer Institute at Frederick, P.O. Box B, Frederick, MD 21702-1201, USA
e-mail: amalyguine@ncifcrf.gov

M.R. Shurin, R.D. Salter (eds.), *Dendritic Cells in Cancer*,
DOI 10.1007/978-0-387-88611-4_23, © Springer Science+Business Media, LLC 2009

23.2 Genetic Modification of Dendritic Cells

Genetic modification of DC for anti-cancer vaccination can take two general, non-mutually exclusive approaches: (i) the transfer of genes-encoding TA or antigen-expressing mRNA into DC and (ii) the transfer of genes-encoding immunostimulatory molecules, cytokines, chemokines, anti-apoptotic molecules, and other factors to enhance DC function. Genetically modified DC offer a number of potential advantages over tumor antigen loading or cell-fusion strategies. No knowledge of individual MHC molecule or peptide-binding affinities is required. The expression of complete antigen sequences within genetically modified DC supplies greater number of epitopes to which an immune response may be generated. Intracellular antigen processing by genetically modified DC may lead to more effective antigen presentation compared to exogenous antigen loading. Gene transfer may permit long-term antigen expression. Foreign antigens expressed by viral gene transfer vectors may stimulate the maturation and activation of DC, resulting in a more robust immune response, and finally, the possibility of an unwanted autoimmune response may be decreased since an array of normal antigens is not loaded, as in the tumor lysate or cell-fusion strategies.

Genetic modification of DC can be accomplished by viral or non-viral gene transfer vectors (Fig. 23.1). Non-viral vectors consist mainly of plasmid DNA constructs expressing the cDNA of TA or extracted mRNA. Biophysical techniques such as electroporation, lipofection, or chemical transfection are used to facilitate the uptake of the plasmid. Unfortunately, plasmid DNA transfection is a relatively inefficient process and transfection efficiency varies between DC types (Arthur et al. 1997). Questions have also been raised as to the specificity of the CTL and memory response achieved with DNA transfection. In the other approach, transfer of mRNA-encoding TA, including unidentified antigens using total tumor cell RNA is used. Alternatively, a specific antigen can be expressed by cloning and in vitro mRNA, transcription to generate mRNA encoding only the TA for transfection (Van Tendeloo et al. 2001; Gilboa and Vieweg 2004). Using this approach, mRNA can be obtained from small clinical samples such as micro-dissected tumor cells and subsequently amplified (Caretti et al. 2008).

Recombinant viral vectors have proven to be very efficient at transferring genetic material into a large variety of cells, including DC. Several types of recombinant viruses have been employed for this purpose. These viruses are genetically engineered to carry genes encoding TA and are at the same time made replication deficient by deleting critical genes required for their reproduction. The most widely used vectors are replication-deficient adenoviruses, which exhibit high-level gene expression and can infect DC at relatively low concentrations. Adenoviruses also infect non-dividing mature DC. In addition, infection with adenovirus has been shown to increase expression of MHC class I and II, B7 co-stimulatory molecules, CD40 and enhance IL-12 and IFN-α production by DC. Molony-based retroviruses, lentiviruses, vaccinia virus, fowlpox, and canarypox, Herpes simplex virus, adeno-associated virus, and alphaviruses

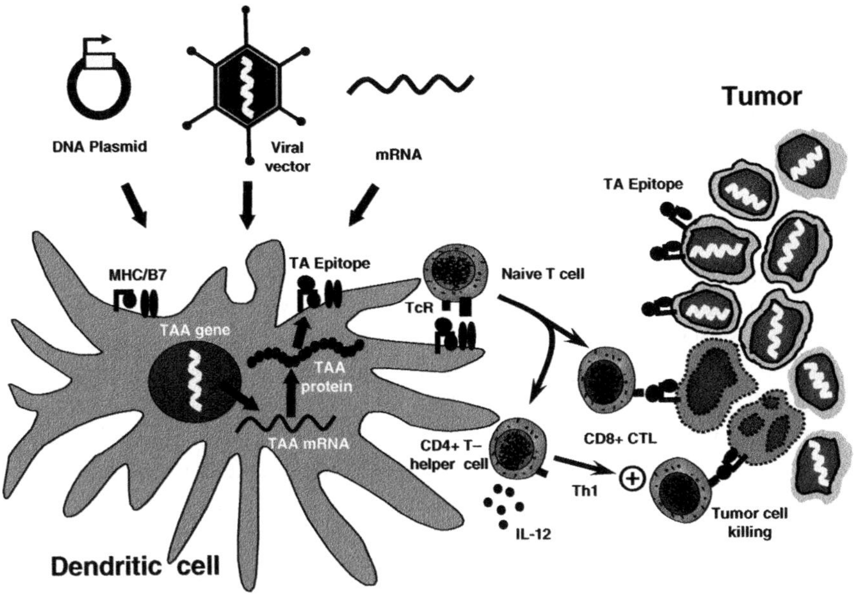

Fig. 23.1 Techniques for generating genetically modified DC for antitumor vaccination. DC are modified to express TA through nucleic acid transfection with DNA plasmids, TA mRNA or viral-mediated gene transfer. After gene transfer and expression, the TA is processed, loaded and presented by an MHC complex molecule as peptide epitopes to the receptor on T cells. This results in the activation and expansion of CD4+ T-helper cells and tumor-specific CD8+ CTLs that recognize and eliminate tumors expressing the antigen

have also been employed for gene transfer into DC. A comparison of gene transfer methods and the use of various non-viral and viral vectors in the transfection of human DC have been extensively reviewed (Arthur et al. 1997; Van Tendeloo et al. 2001; Ribas 2005; Kikuchi 2006).

Genetic modification may also be used to engineer DC that are capable of overcoming functional inhibition in the tumor environment, enhance DC survival and provide additional activation signals for T and NK cells (Ribas 2005; Kikuchi 2006; Vulink et al. 2008) (Table 23.1).

Table 23.1 Genetic modification of DC function

Groups of factors	Examples
Cytokines	IL-2, IL-7, IL-12, IL-15, IL-18, IL-21, IL-23, GM-CSF, IFN-α, β, γ, TNF-α
Chemokines	SLC/CCL21, MDC/CCL22, lymphotactin/XCL1, Mig/CXCL9, CXCL19, fractalkine/CX3CL1
Adhesion and co-stimulatory molecules	ICAM-1, LFA-3, B7.1, B7.2, 4-1BBL, CD40, CD40L
Pro-inflammatory and anti-apoptotic molecules	NF-κB, Bcl-xL, bcl-2, FLIP, XIAP/hILP, dominant-negative procaspase-9, HSP70, Akt-1

23.3 Pre-clinical Models of Genetically Modified Dendritic Cell Vaccines

Studies of antitumor vaccination using gene-modified DC in mouse models have demonstrated activation of both humoral and cell-mediated immunities, protection from tumor challenge, and to cause the regression of established tumors. Vaccination of *neu* transgenic mice with DC modified with an adenoviral vector expressing a truncated *neu* oncogene, induced protective antibodies, enhanced IFN-γ production and prevented the development of mammary cancers (Sakai et al. 2004). A lentiviral vector encoding the murine tyrosinase-related protein-2 (TRP-2) efficiently transduced mouse DC and induced the rejection of established B16 melanoma (Metharom et al. 2001). In another study using a murine B16 melanoma brain tumor model, mice vaccinated with DC modified with a recombinant Semliki forest virus vector expressing IL-12 showed marked survival improvement (Yamanaka et al. 2002). Genetically modified DC transfected with an adenoviral vector expressing IL-12 were shown to eradicate established tumors when administered intratumorally to subcutaneous murine CT26 colon carcinoma (Melero et al. 1999). Vaccination of carcinoembryonic antigen (CEA) transgenic mice with genetically modified DC expressing CEA, granulocyte-macrophage colony-stimulating factor (GM-CSF), and IL-12 elicited a more potent therapeutic response than vaccination with CEA-transfected DC alone (Ojima et al. 2007a).

The possibility of utilizing genetically modified human DC from healthy donors and cancer patients was explored in a number of studies. Using normal donors, Grünebach et al. (2005) demonstrated that electroporation of DC with mRNA coding for the HER-2/neu antigen elicited antigen-specific effector CTL and that co-expression of 4-1BBL (CD137L) enhanced DC ability to elicit primary CTL responses. In another study (Ojima et al. 2007b), CEA-specific CTL were generated from donor peripheral blood mononuclear cells (PBMC) and stimulated with genetically modified CEA-expressing DC co-transducted with GM-CSF cDNA. Co-transduction of the DC with the GM-CSF gene inhibited apoptosis by upregulating Bcl-X_L expression. Another method for extending DC survival was demonstrated by Balkir et al. (2004). DC transduced with anti-apoptotic molecules FLIP, XIAP/hILP, a dominant-negative procaspase-9, and HSP-70 showed resistance to melanoma cell-induced apoptosis. It has been shown that DC transfected with a fusion recombinant human telomerase reverse transcriptase (hTERT)-human IL-18 (hIL18) cDNA (hTERT-IL18) showed greater telomerase activity and elicited an hTERT-specific CTL response in vitro (Tong et al. 2008). Bonehill et al. (2008) showed that immature DC electroporated with CD40L and/or constitutively active toll-like receptor-4 (TLR4) mRNA acquired a mature phenotype along with enhanced secretion of several cytokines and chemokines. These DC were found to be potent in inducing naive CD4+ T cells to differentiate into IFN-γ-secreting Th1-type cells. When all three molecules were transfected, a >500-fold increase in Melan A-specific CD8+ T cells was observed when compared with immature DC, and a >200-fold increase was seen when compared with cytokine-matured DC. The authors concluded that immature DC genetically modified to express stimulating molecules can induce tumor antigen-specific T cells in vitro and could be a significant improvement over DC matured with the methods currently in use.

González-Carmona et al. (2006) attempted to break immunological tolerance toward the hepatocellular carcinoma (HCC)-associated α-fetoprotein (AFP) antigen. When DC from healthy donors or HCC patients were transduced with hAFP encoding adenovirus and co-cultured with cytokine-induced killer (CIK) cells, these CIK lysed approximately 70% of AFP-expressing HCC cells. Interestingly, with this approach, CIK cells from immunosuppressed patients with AFP-positive HCC could be stimulated to lyse AFP-expressing HCC cells as effectively as CIK cells derived from healthy individuals. Zhang et al. (2006) transfected AFP mRNA into DC and evaluated the ability of the DC to induce AFP-specific CD4+ and CD8+ T cells. They found that this strategy could generate fully functional DC that could induce T cells to recognize AFP-expressing HCC cells. In another study (Morandi et al. 2006), the generation of neuroblastoma-specific CTL from T cells primed by tumor mRNA-transfected DC was investigated both in healthy donors and in patients with neuroblastoma. The expanded CTL expressed an effector/memory phenotype and a Th1-like cytokine-secretion profile. CTL specificity was demonstrated by IFN-γ release on incubation with HLA-matched tumor cell lines and by CTL-mediated HLA-restricted lysis. This study supports a possible role for DC immunotherapy in the treatment of extracranial solid tumors.

Ni et al. (2008) studied Sézary syndrome, the leukemic variant of cutaneous T-cell lymphoma. Autologous DC, generated from peripheral blood monocytes from 10 cancer patients, were transfected with total RNA amplified by in vitro transcription of mRNA extracted from Sézary cells. The RNA-transfected DC induced antitumor IFN-γ-producing CTL in 7 of 10 subjects and granzyme B-producing CTL in 6 of 9 subjects, as measured by ELISPOT. Both CD3+CD8+ T cells and CD4+CD25+ T cells were expanded without induction of regulatory T cells. Vujanovic et al. (2006) demonstrated that IL-12p70 and IL-18 gene-modified DC loaded with MAGE-A6 recombinant protein effectively stimulate specific type 1 CD4+ T-cell responses in normal donors and melanoma patients in vitro. DC co-infected with Adeno/IL-12 and Adeno/IL-18 were more effective at stimulating MAGE-A6-specific Th1-type CD4+ T-cell responses than DC infected with either of the cytokine vectors alone. The superiority of co-infected DC-based stimulation in melanoma patients was independent of initial disease stage or current disease status.

23.4 Genetically Modified Dendritic Cell Vaccines in Clinical Trials

Pre-clinical data provided a strong rationale for the translation of genetically modified DC antitumor vaccines to clinical studies. Table 23.2 summarizes the results of a number of these trials. In a phase I trial performed in 13 patients with metastatic prostate cancer, no dose-limiting toxicity, adverse effects, or autoimmunity was seen when prostate-specific antigen (PSA) mRNA-transfected DC were administered. Induction of PSA-specific T-cell responses was detected in all patients. Vaccination was associated with a significant decrease in the log-slope of the serum PSA elevation in six of seven subjects examined. Three analyzed patients exhibited a transient molecular clearance of circulating tumor cells (Heiser et al. 2002). Morse et al. (2003) reported a phase I study of patients with advanced carcinoembryonic antigen (CEA)-expressing malignancies followed by a phase II study in patients with colon cancer undergoing resection of hepatic metastases. The immunizations with autologous DC transduced with CEA encoding mRNA were well tolerated. Of the 24 evaluated phase I patients, there was 1 complete response measured by tumor markers, 2 minor responses, and 3 patients had stable disease. In the phase II study, 9 of 13 patients recurred at a median of 122 days. However, since this was not a controlled study, the significance of this is unknown. Evidence of an immunologic response was demonstrated in biopsies of the vaccine injection sites and in the peripheral blood of selected patients.

A clinical trial was initiated by Su et al. (2005), in which telomerase reverse transcriptase (hTERT) mRNA-transfected DC were administered to 20 patients with metastatic prostate cancer. hTERT is an attractive target for cancer immunotherapy, since it is activated in most human tumors but not expressed in most normal cells. Nine of 20 patients receiving autologous DC transfected with mRNA encoding a chimeric lysosome-associated membrane protein-1 (LAMP)-hTERT fusion protein, developed hTERT-specific CD8+ and CD4+ T-cell responses. After repeated vaccination, hTERT-specific T cells were found at the intradermal injection site. In 19 of 20 patients, expansion of hTERT-specific CD8+ T cells was measured in the peripheral blood. Immunization with the chimeric LAMP-hTERT vaccine resulted in higher frequencies of hTERT-specific CD4+ T cells over immunization with only hTERT-transfected DC. CTL-mediated killing of hTERT+ targets was enhanced in the LAMP-hTERT-vaccinated group, suggesting that an improved CD4+ response augmented the CTL response. Vaccination was associated with a transient reduction of PSA velocity in 5 of 12 patients and molecular clearance of circulating tumor cells in 9 of 10 patients.

Mazzolini et al. (2005) studied the effect of the intratumoral injection of DC modified with a recombinant adenovirus engineered to secrete IL-12 in 17 patients with metastatic gastrointestinal carcinomas. Fifteen of the 17 patients were evaluable for toxicity and 11 for clinical and immune responses. Treatment

Table 23.2 Summary of clinical trials using genetically modified DC vaccines for cancer

Vector	Antigen[†]	Additional treatment	Cancer: no. of patient	Immune response*	Clinical response**	Reference
mRNA	Total tumor	Co-transfection with KLH	Colorectal: 15	11/13 patients had a response to KHL	Decrease in serum CEA levels in 7/13 patients	Rains et al. (2001)
mRNA	CEA	None	Pancreatic: 3	Infiltrate of CD4 + cells at injection site in 3/3 patients	All patients alive 2.5 years after diagnosis	Morse et al. (2002)
Adenovirus	Gp100/ MART-1	None	Melanoma: 12	3/12 patients developed leukoderma	Disease progression in all	Tsao et al. (2002)
cDNA	MUC1	None	Breast: 7 Pancreatic: 2 Papillary: 1	2/10 patients positive DTH; increase in IFN-γ producing MUC1-specific CD8 + T cells in 4/10 patients	1/10 patients with stable disease for 3 months	Pecher et al. (2002)
mRNA	PSA	None	Prostate: 13	PSA-specific T-cell response in all patients	Log-scale decrease of PSA in 6/7 patients. Transient clearance of circulating micrometastasis in 3/3 patients	Heiser et al. (2002)
mRNA	CEA	None	CEA-expressing cancers: Phase I: 29 Phase II: 13	Increase in CEA-specific T-cell activity in 3/3 patients	Phase I: 1/24 CEA decreased to 0; 1 CA-15-3 decrease, 1 clinical improvement, 3 prolonged survival; Phase II: 3/13 patients disease free >500 days	Morse et al. (2003)
mRNA	Total tumor	None	Renal: 10	Expansion of tumor-specific T cells in 6/7 patients	Decrease in mortality (7/10 patients alive on average 20 months)	Su et al. (2003)
Vaccinia virus	Human tyrosinase gene	None	Melanoma: 6	Increase in tumor-specific T-cell frequency in 4/5 patients	1 PR (RECIST)/6 patients	DiNicola et al. (2004)
mRNA	Total tumor	None	Pediatric brain tumors: 9	Tumor-specific response in 2/7 patients	1 partial response and 2 with stable disease/7 patients	Caruso et al. (2004)

Table 23.2 (continued)

Vector	Antigen[†]	Additional treatment	Cancer: no. of patient	Immune response*	Clinical response**	Reference
mRNA	Allogeneic total tumor	Depletion of CD25-expressing Tregs	Renal: 10	Increase in tumor-specific T-cell response including a CTL response with combination treatment compared to vaccine alone.	7/10 patients alive at the conclusion of the trial	Dannull et al. (2005)
mRNA	Allogeneic total tumor	None	Prostate: 20	Specific immune response in 12/20 patients	PSA log-slope decrease in 13/20 patients	Mu et al. (2005)
Fowlpox virus	CEA	Co-transfection with co-stimulatory molecules (TRICOM)	Colorectal: 11 Lung : 3	CD8 + and CD4 + T-cell responses in 11/14 patients. NK activity increase in 4/9 patients	Decreased CEA level in 1 patient and 5 with stable disease/13 patients. Correlation with NK activity among 9 tested patients	Morse et al. (2005) and Osada et al. (2006)
mRNA	Total tumor	None	Neuroblastoma: 11	Tumor-specific antibodies seen in 2/3 patients after treatment	No objective response	Caruso et al. (2005)
mRNA	hTERT	Co-transfection with LAMP-1	Prostate: 11 w/hTERT, 9 w/hTERT and LAMP-1	Increase of frequencies and activity of CD8 + and CD4 + T cells in 19/20 patients; in 6/6 patients CD4 + and CD8 + T-cell infiltration of injection site	Transient reduction of PSA velocity in 5/12 patients. Clearance of circulating micro-metastasis in 9/10 patients	Su et al. (2005)
Adenovirus	None	IL-12 gene	Pancreatic: 3 Colorectal: 5 Liver: 9	Increase in serum IFN-γ and IL-6 in all tested patients. Increase in NK activity in 5/5 patients. Increased infiltration with CD8 + cells in 3/11 tumors analyzed	1 PR (RECIST) and 2 stable disease/11 patients	Mazzolini et al. (2005)

Table 23.2 (continued)

Vector	Antigen†	Additional treatment	Cancer: no. of patient	Immune response*	Clinical response**	Reference
mRNA	Autologous & allogeneic total tumor	None	Melanoma: 22 Prostate: 19	Specific T-cell responses: 9/19 melanoma patients and 12/19 prostate patients	Log-scale PSA decrease in 13/19 patients	Kyte et al. (2006 a,b)
Adenovirus	P53 gene	Chemotherapy	Small-cell lung cancer: 29	P53-specific T-cell response in 16/28 patients	1 PR (RECIST), seven stable disease/29 patients. Increase in response to chemotherapy after vaccination: 61.9% objective clinical response (RECIST)/21 patients, associated immune response	Antonia et al. (2006)
mRNA	Folate-receptor type α (FR-α)	None	Ovarian: 1	Antigen-specific T-cell responses	PR (RECIST)	Hernando et al. (2007)
mRNA	Total tumor	None	Melanoma: 2	Sustained CD4+ and CD8+ T-cell response	23 and 31 months survival	Kyte et al. (2007)
Adenovirus	MART-1	None	Melanoma: 18	MART-1-specific CD8+ response in 6/11 patients and CD4+ in 2/4 patients	4/17 patients with stable disease	Butterfield et al. (2008)

†All vaccines used autologous dendritic cells; *Immune responses/Number of evaluable patients; **Clinical responses/number of evaluable patients. DTH, delayed-type hypersensitivity response; IL-12, interleukin-12; KLH, keyhole limpet hemocyanin; LAMP-1, lysosome-associated membrane protein-1; MUC1, mammary-type apomucin tumor antigen; PSA, prostate-specific antigen; RECIST, response evaluation criteria in solid tumors; Tregs, regulatory T cells; TRICOM, a recombinant vaccinia virus-based antitumor vaccine expressing a triad of co-stimulatory molecules including B7-1, ICAM-1, and LFA-3; hTERT, human telomerase reverse transcriptase

was well tolerated, and concentrations of IFN-γ and IL-6 were increased in the serum of 15 patients. Increase in NK cell activity in five patients' peripheral blood was documented. A potent antibody response against adenoviral capsid proteins was seen, along with a marked increase of infiltrating CD8+ T lymphocytes in 3 of 11 tumor biopsies. A partial response was observed in one patient with pancreatic carcinoma, and stable disease was observed in two patients.

A combination of a p53 gene-modified DC vaccine with chemotherapy in patients with advanced stage small-cell lung cancer was studied by Antonia et al. (2006). DC transduced with the full-length wild-type p53 gene delivered via an adenoviral vector were administered to 29 patients. The p53-specific T cells were observed in 57.1% of patients and were associated with a moderate increase in the titer of anti-adenovirus antibodies. One patient had a partial response to vaccination and seven patients had stable disease.

All together, 302 patients with various cancers were treated in the clinical trials presented in Table 23.2. The majority had advanced prostate cancer ($n = 72$) or malignant melanoma ($n = 60$). Of the 302 patients, 216 were studied for tumor-specific CD8+ T cells' responses to vaccination. In 153 (70.8%), there was an increase of specific CD8+ T-cell frequency and/or activity. Nine of 14 (64%) tested displayed an increase in NK activity. Despite these immunological responses, the clinical results have been disappointing. Of the 265 evaluated patients, 22 (8.3%) demonstrated some survival benefits based on comparison with historical numbers, and stable disease was seen in 21 (7.9%) patients. Since these clinical trials were not randomized, the significance of these findings is uncertain. Forty-six patients (17.3%) demonstrated an improvement in some biochemical marker of tumor burden or growth rate, including indirect measures such as decreased PSA log-slope or velocity, or a decrease in serum CEA levels. In fact, PSA changes were detected in 37 of 58 vaccinated prostate cancer patients. However, overall, only five patients (2%) showed response by tumor measurement. When analyzing these trials, it should be kept in mind that the majority of these studies enrolled heavily pre-treated, immunocompromised patients with advanced stage disease. In addition, comparison of the data between studies is made difficult due to inconsistent response criteria between investigators, insufficient information provided by authors and often inadequate evaluation of immunological and clinical responses.

To date, it is estimated that about 300 DC-based vaccine clinical trials involving more then 3000 patients have been conducted worldwide (Zhong et al. 2007). A large number of different DC vaccine formulations utilizing different generation methods, antigen sources, tumor cell fusions, and genetic modifications in combination with various adjuvants and carriers have been evaluated (Gilboa 2007; Palucka et al. 2007; Zhong et al. 2007; Nencioni et al. 2008; Vulink et al. 2008). While DC vaccines have been successful in inducing tumor-specific T-cell responses in cancer patients, the clinical outcomes have been disappointing (Rosenberg et al. 2004; Zhong et al. 2007).

The limited clinical benefit can be attributed to many reasons including, suboptimal DC preparations, lack of uniform DC production standards, flaws in protocol design, and response monitoring. Other explanations to consider are the attenuation of DC function in cancer patients and the difficulties in inducing an immune response in patients with advanced cancer, especially with treatment-induced immunosuppression. These issues have been addressed in several recent reviews (Gilboa 2007; Palucka et al. 2007; Zhong et al. 2007; Nencioni et al. 2008; Vulink et al. 2008) and highlighted the need for additional research in improving DC-based vaccines. Another unresolved issue is the observation that tumor-specific cellular immune response that follows vaccination, including DC-based vaccines, does not lead to clinical cancer regression in spite of the ability to stimulate T cells to produce immunostimulatory cytokines, to generate tumor-specific CTL and kill tumor cells ex vivo. Based on the analysis presented in Table 23.2, 56% of vaccinated patients exhibited a tumor-specific T-cell response, while few patients experienced tumor regression. This lack of correlation between the immune responses and the clinical efficacy may be explained by (1) lack of universal standards for immunological monitoring of clinical trials; (2) low observed clinical response rates not allowing meaningful correlations of immune response with clinical response; (3) typically only the peripheral blood components are accessible for serial analysis when in fact, there is no convincing evidence that the peripheral blood immune responses are representative of those occurring in the tumor; (4) CTL usually maintain effector function for a relatively short period that quickly tapers into a memory phase, still recognizing target cells, but lacking meaningful effector activity; (5) unlikelihood that the study of single parameters is enough to provide reliable information about extremely complex CTL–tumor interactions; and (6) suitable animal models for studying immune monitoring are lacking. Thus, data from ex vivo T-cell assays have not yet defined surrogate markers for clinical efficacy. Analyzing effector cell responses after vaccination currently may or may not help to detect factors that determine success or failure of the vaccine and its immunological potency.

23.5 Strategies for Increasing Efficacy of Genetically Modified Dendritic Cell Vaccines

In an effort to enhance the antitumor activity of genetically modified DC vaccines and to improve tumor responses, a number of additional strategies are under investigation. These include co-transfection of DC with immunomodulating factors to enhance antigen presentation and antitumor effector cell activation. Modifications to enhance DC maturation, decrease tolerogenic activity and prolong survival are also being researched. Along with TA, DC can be engineered to express various immunomodulating cytokines, chemokines, co-stimulatory and adhesion molecules, pro-inflammatory and anti-apoptotic proteins, or a combination of these (Tables 23.1 and 23.2).

Methods for the depletion of T-regulatory cells (Treg) that attenuate immune responses are under investigation. Studies have shown elevated numbers of Tregs in the peripheral blood and tumor microenvironment in cancer patients (Woo et al. 2002; Dannull et al. 2005; Ménard et al. 2008). Strategies for eliminating these cells involve lymphoablative conditioning or targeted elimination of CD4 + CD25 + Tregs using monoclonal antibodies or immunotoxins. Lymphoablation prior to adoptive T-cell transfer in patients with metastatic melanoma significantly improved clinical response rates (Rosenberg et al. 2008). Recently, Dannull et al. (2005) demonstrated that elimination of CD4 + /CD25 + cells using the recombinant IL-2-diphtheria toxin conjugate, denileukin difitox (Ontak^{TM}), enhanced the immunostimulatory effect of a tumor RNA-transfected DC vaccine. Denileukin difitox reduced the number of Tregs in the peripheral blood of metastatic renal cell carcinoma (RCC) patients and abrogated Treg-mediated immunosuppressive activity in vivo. Moreover, treatment with denileukin difitox followed by vaccination with RNA-transfected DC improved the tumor-specific T-cell responses in RCC patients compared to vaccine alone.

The blockade of tumor-associated immune suppressor factors is another potential approach. Tumors, their stromal cells and a variety of tumor-associated immune regulatory cells, can produce a number of soluble factors that influence NK, T cell and DC development and function. The efficacy of DC vaccines may be improved by blocking these tumor-produced immune suppressor factors, which include vascular endothelial growth factor (VEGF), transforming growth factor-β (TGF-β), IL-10, IL-13, indoleamine-2, 3-dioxygenase (IDO), inducible nitrous oxide synthase (iNOS), prostagladin E2 and arginase-1 (Lizee et al. 2006; Rabinovich et al. 2007; Mantovani et al. 2008; Ménard et al. 2008). Neutralization or blockage of immunosuppressive factors by specific monoclonal antibodies, soluble receptors or antagonists may be used to enhance vaccines and adoptive immune transfer. Consequently, bevacizumab (Avastin^{TM}), a humanized monoclonal antibody directed against VEGF and approved for the treatment of metastatic colon, lung, and breast cancers (Midgley and Kerr 2005; Tyagi 2005), is associated with a decrease in the accumulation of immature progenitor cells and induces a modest increase in the DC population in peripheral blood. Moreover, anti-VEGF antibody treatment enhances allostimulatory capacity of DC and T-cell proliferation against recall antigens (Osada et al. 2008). The targeting of two enzymes expressed by myeloid suppressor cells and some tumor cells, iNOS and arginase-1, may also thwart immune suppression (Bronte et al. 2005; De Santo et al. 2005).

Other potential targets include CTLA-4, an immunoattenuating transmembrane protein expressed by Treg and effector T cells, and the immunoinhibitory receptor programmed death 1 (PD-1), along with its ligand B7-H1 (Lizee et al. 2006; Ménard et al. 2008). Significant increases in antitumor immunity and objective tumor regression following treatment with an anti-CTLA-4 monoclonal antibody have been observed in vaccinated melanoma patients (Phan et al. 2003). Antibody blockade of PD-1 and its ligand, B7-H1 (PD-L), has also been shown to augment vaccine immunotherapy in mouse tumor models (Hirano et al. 2005).

A new strategy is the use of small interfering RNA (siRNA) to silence immunosuppressive gene expression. For example, silencing of the inhibitory molecules IDO and SOCS1 (suppressor of cytokine signaling 1) by siRNA, enhanced the antitumor immune response from antigen-pulsed DC vaccines (Mao et al. 2007). The siRNA knockdown of DC PD-L (programmed death ligand), DIgR2 (immunoglobulin receptor 2) and notch ligand resulted in higher levels of tumor-specific CD4+ and CD8+ T cells, as well as improved immunity to tumors (Mao et al. 2007). The silencing of endogenous IL-10 in DC preferentially activates Th1 type cells and subsequently produces a more robust CTL response against human tumor antigens (Chhabra et al. 2008).

Another approach is to improve DC interactions with NK cells, which play an important role in antitumor immunity (Woo et al. 2006). The combined activation of TA-specific CTL and NK cells may be crucial for eradication of both MHC-I-positive and negative tumors. In pre-clinical studies, immunization with MART-1 antigen-engineered DC in C57Bl/6 mice resulted in the generation of antigen-specific CTL. Protective responses against murine B16 melanoma were demonstrated in vivo and found to be dependent on the presence of functional NK cells, although NK cells alone did not generate a fully protective response (Wargo et al. 2005). In vitro vaccination with mature human DC transfected with MART-1 and pro-inflammatory cytokines IL-12 and IL-18, increased NK cell and TA-specific CTL activities (Bontkes et al. 2008). In a phase I clinical trial, NK cell responses were analyzed after vaccination with autologous DC transducted with a fowlpox vector encoding CEA. Among nine patients studied, NK cytolytic activity increased in four, including three who also had increased NK cell frequency. These NK responses correlated more closely with clinical outcome than did T-cell responses (Morse et al. 2005; Osada et al. 2006). Intratumoral injection of autologous DC transfected with an adenovirus encoding IL-12 activated NK cells in five patients with metastatic gastrointestinal carcinomas (Mazzolini et al. 2005).

Combining DC vaccinations with other therapeutic modalities, such as surgery, radiation or chemotherapy, offers an attractive strategy. Combination DC vaccine and chemotherapy may augment tumor cell killing or indirectly generate additional pro-inflammatory signals and may overcome inhibitory immune signals to enhance antitumor responses. Antonia et al. (2006) reported an unexpectedly high rate (61.9%) of objective clinical responses to chemotherapy immediately following vaccination with p53-transduced DC. The clinical response to chemotherapy was closely associated with an immunological response to the DC vaccination.

23.6 Conclusions

The enormous progress in our understanding of DC biology and of the role DC play in generating immune responses has translated into their study as therapeutic antitumor vaccines. Current DC vaccination strategies are limited by the low affinity of most tumor epitopes for MHC, the low availability of “free”

MHC molecules for occupancy on the surface of DC, the limited numbers of different epitopes that may be exogenously loaded, suboptimal DC maturation and activation protocols, and the inability to routinely generate clinically relevant specific CTL responses to tumor. These problems may be overcome in part by the use of gene-modified DC expressing optimized tumor antigens and immunostimulatory molecules or cytokines that enhance and appropriately direct the vaccine response. From analysis of available clinical trials in which genetically modified DC were utilized (Table 23.2), two important lessons may be learned. First, DC vaccination against cancer is safe, and second, a need for significant improvement in efficacy remains. Despite this, considering the complexity of the host–tumor relationship and multilevel regulatory control of immune response, it remains surprising that single intervention can ever lead to objective clinical responses in patients with advanced cancer. Thus, even low rates of response to DC vaccine may still be considered as an encouraging proof of principal for DC-based cancer immunotherapy. Genetically modified DC-based anti-cancer vaccination holds promise, perhaps being employed in the adjuvant setting with minimal residual disease after primary treatment, or in combination with other antitumor or immune-enhancing therapies.

Acknowledgments This research was supported by the Intramural Research Program of the NIH, National Cancer Institute, Center for Cancer Research.

This project has been funded in whole or in part with federal funds from the National Cancer Institute, National Institutes of Health, under contract no. N01-CO-12400. The content of this publication does not necessarily reflect the views or policies of the Department of Health and Human Services, nor does mention of trade names, commercial products or organizations imply endorsement by the US Government.

References

Antonia, S. J., Mirza, N., Fricke, I., Chiappori, A., Thompson, P., Williams, N., Bepler, G., Simon, G., Janssen, W., Lee, J. H., Menander, K., Chada, S. and Gabrilovich, D. I. 2006. Combination of p53 cancer vaccine with chemotherapy in patients with extensive stage small cell lung cancer. Clin Cancer Res 12:878–887.

Arthur, J. F., Butterfield, L. H., Roth, M. D., Bui, L. A., Kiertscher, S. M., Lau, R., Dubinett, S., Glaspy, J., McBride, W. H. and Economou, J. S. 1997. A comparison of gene transfer methods in human dendritic cells. Cancer Gene Ther 4:17–25.

Balkir, L., Tourkova, I. L., Makarenkova, V. P., Shurin, G. V., Robbins, P. D., Yin, X. M., Chatta, G. and Shurin, M. R. 2004. Comparative analysis of dendritic cells transduced with different anti-apoptotic molecules: sensitivity to tumor-induced apoptosis. J Gene Med 6:537–544.

Bonehill, A., Tuyaerts, S., Van Nuffel, A. M., Heirman, C., Bos, T. J., Fostier, K., Neyns, B. and Thielemans, K. 2008. Enhancing the T-cell stimulatory capacity of human dendritic cells by co-electroporation with CD40L, CD70 and constitutively active TLR4 encoding mRNA. Mol Ther 16:1170–1180.

Bontkes, H. J., Kramer, D., Ruizendaal, J. J., Meijer, C. J. and Hooijberg, E. 2008. Tumor associated antigen and interleukin-12 mRNA transfected dendritic cells enhance effector function of natural killer cells and antigen specific T-cells. Clin Immunol 127:375–384.

Bronte, V., Kasic, T., Gri, G., Gallana, K., Borsellino, G., Marigo, I., Battistini, L., Iafrate, M., Prayer-Galetti, T., Pagano, F. and Viola, A. 2005. Boosting antitumor responses of T lymphocytes infiltrating human prostate cancers. J Exp Med 201:1257–1268.
Caretti, E., Devarajan, K., Coudry, R., Ross, E., Clapper, M. L., Cooper, H. S. and Bellacosa, A. 2008. Comparison of RNA amplification methods and chip platforms for microarray analysis of samples processed by laser capture microdissection. J Cell Biochem 103:556–563.
Chhabra, A., Chakraborty, N. G. and Mukherji, B. 2008. Silencing of endogenous IL-10 in human dendritic cells leads to the generation of an improved CTL response against human melanoma associated antigenic epitope, MART-1 27-35. Clin Immunol 126:251–259.
Dannull, J., Su, Z., Rizzieri, D., Yang, B. K., Coleman, D., Yancey, D., Zhang, A., Dahm, P., Chao, N., Gilboa, E. and Vieweg, J. 2005. Enhancement of vaccine-mediated antitumor immunity in cancer patients after depletion of regulatory T cells. J Clin Invest 115:3623–3633.
De Santo, C., Serafini, P., Marigo, I., Dolcetti, L., Bolla, M., Del Soldato, P., Melani, C., Guiducci, C., Colombo, M. P., Iezzi, M., Musiani, P., Zanovello, P. and Bronte, V. 2005. Nitroaspirin corrects immune dysfunction in tumor-bearing hosts and promotes tumor eradication by cancer vaccination. Proc Natl Acad Sci USA 102:4185–4190.
Gilboa, E. 2007. DC-based cancer vaccines. J Clin Invest 117:1195–1203.
Gilboa, E. and Vieweg, J. 2004. Cancer immunotherapy with mRNA-transfected dendritic cells. Immunol Rev 199:251–263.
González-Carmona, M., Märten, A., Hoffmann, P., Schneider, C., Sievers, E., Schmidt-Wolf, I., Sauerbruch, T. and Caselmann, W. 2006. Patient-derived dendritic cells transduced with an a-fetoprotein-encoding adenovirus and co-cultured with autologous cytokine-induced lymphocytes induce a specific and strong immune response against hepatocellular carcinoma cells. Liver Int 26:369–379.
Grunebach, F., Kayser, K., Weck, M. M., Muller, M. R., Appel, S. and Brossart, P. 2005. Cotransfection of dendritic cells with RNA coding for HER-2/neu and 4-1BBL increases the induction of tumor antigen specific cytotoxic T lymphocytes. Cancer Gene Ther 12:749–756.
Heiser, A., Coleman, D., Dannull, J., Yancey, D., Maurice, M. A., Lallas, C. D., Dahm, P., Niedzwiecki, D., Gilboa, E. and Vieweg, J. 2002. Autologous dendritic cells transfected with prostate-specific antigen RNA stimulate CTL responses against metastatic prostate tumors. J Clin Invest 109:409–417.
Hirano, F., Kaneko, K., Tamura, H., Dong, H., Wang, S., Ichikawa, M., Rietz, C., Flies, D. B., Lau, J. S., Zhu, G., Tamada, K. and Chen, L. 2005. Blockade of B7-H1 and PD-1 by monoclonal antibodies potentiates cancer therapeutic immunity. Cancer Res 65:1089–1096.
Kikuchi, T. 2006. Genetically modified dendritic cells for therapeutic immunity. Tohoku J Exp Med 208:1–8.
Lizee, G., Radvanyi, L. G., Overwijk, W. W. and Hwu, P. 2006. Improving antitumor immune responses by circumventing immunoregulatory cells and mechanisms. Clin Cancer Res 12:4794–4803.
Mantovani, A., Romero, P, Palucka, AK, Marincola, FM 2008. Tumour immunity: effector response to tumour and role of the microenvironment. Lancet 371:771–783.
Mao, C. P., Hung, C. F. and Wu, T. C. 2007. Immunotherapeutic strategies employing RNA interference technology for the control of cancers. J Biomed Sci 14:15–29.
Mazzolini, G., Alfaro, C., Sangro, B., Feijoo, E., Ruiz, J., Benito, A., Tirapu, I., Arina, A., Sola, J., Herraiz, M., Lucena, F., Olague, C., Subtil, J., Quiroga, J., Herrero, I., Sadaba, B., Bendandi, M., Qian, C., Prieto, J. and Melero, I. 2005. Intratumoral injection of dendritic cells engineered to secrete interleukin-12 by recombinant adenovirus in patients with metastatic gastrointestinal carcinomas. J Clin Oncol 23:999–1010.
Melero, I., Duarte, M., Ruiz, J., Sangro, B., Galofre, J., Mazzolini, G., Bustos, M., Qian, C. and Prieto, J. 1999. Intratumoral injection of bone-marrow derived dendritic cells engineered to produce interleukin-12 induces complete regression of established murine transplantable colon adenocarcinomas. Gene Ther 6:1779–1784.

Ménard, C., Martin, F., Apetoh, L., Bouyer, F. and Ghiringhelli, F. 2008. Cancer chemotherapy: not only a direct cytotoxic effect, but also an adjuvant for antitumor immunity. Cancer Immunol Immunother. Epub ahead of print.

Metharom, P., Ellem, K. A., Schmidt, C. and Wei, M. Q. 2001. Lentiviral vector-mediated tyrosinase-related protein 2 gene transfer to dendritic cells for the therapy of melanoma. Hum Gene Ther 12:2203–2213.

Midgley, R. and Kerr, D. 2005. Bevacizumab–current status and future directions. Ann Oncol 16:999–1004.

Morandi, F., Chiesa, S., Bocca, P., Millo, E., Salis, A., Solari, M., Pistoia, V. and Prigione, I. 2006. Tumor mRNA-transfected dendritic cells stimulate the generation of CTL that recognize neuroblastoma-associated antigens and kill tumor cells: immunotherapeutic implications. Neoplasia 8:833–842.

Morse, M. A., Clay, T. M., Hobeika, A. C., Osada, T., Khan, S., Chui, S., Niedzwiecki, D., Panicali, D., Schlom, J. and Lyerly, H. K. 2005. Phase I study of immunization with dendritic cells modified with fowlpox encoding carcinoembryonic antigen and costimulatory molecules. Clin Cancer Res 11:3017–3024.

Morse, M. A., Nair, S. K., Mosca, P. J., Hobeika, A. C., Clay, T. M., Deng, Y., Boczkowski, D., Proia, A., Neidzwiecki, D., Clavien, P. A., Hurwitz, H. I., Schlom, J., Gilboa, E. and Lyerly, H. K. 2003. Immunotherapy with autologous, human dendritic cells transfected with carcinoembryonic antigen mRNA. Cancer Invest 21:341–349.

Nencioni, A., Grunebach, F., Schmidt, S., Muller, M., Boy, D., Patrone, F., Ballestrero, A. and Brossart, P. 2008. The use of dendritic cells in cancer immunotherapy. Crit Rev Oncol Hematol 65:191–199.

Ni, X., Richmond, H. M., Liao, X. M., Decker, W. K., Shiue, L. H., Shpall, E. J. and Duvic, M. 2008. Induction of T-cell responses against cutaneous T-cell lymphomas ex vivo by autologous dendritic cells transfected with amplified tumor mRNA. J Invest Dermatol 128:2631–2639.

Ojima, T., Iwahashi, M., Nakamura, M., Matsuda, K., Nakamori, M., Ueda, K., Naka, T., Ishida, K., Primus, F. J. and Yamaue, H. 2007a. Successful cancer vaccine therapy for carcinoembryonic antigen (CEA)-expressing colon cancer using genetically modified dendritic cells that express CEA and T helper-type 1 cytokines in CEA transgenic mice. Int J Cancer 120:585–593.

Ojima, T., Iwahashi, M., Nakamura, M., Matsuda, K., Nakamori, M., Ueda, K., Naka, T., Katsuda, M., Miyazawa, M. and Yamaue, H. 2007b. Benefits of gene transduction of granulocyte macrophage colony-stimulating factor in cancer vaccine using genetically modified dendritic cells. Int J Oncol 31:931–939.

Osada, T., Chong, G., Tansik, R., Hong, T., Spector, N., Kumar, R., Hurwitz, H. I., Dev, I., Nixon, A. B., Lyerly, H. K., Clay, T. and Morse, M. A. 2008. The effect of anti-VEGF therapy on immature myeloid cell and dendritic cells in cancer patients. Cancer Immunol Immunother 57:1115–1124.

Osada, T., Clay, T., Hobeika, A., Lyerly, H. K. and Morse, M. A. 2006. NK cell activation by dendritic cell vaccine: a mechanism of action for clinical activity. Cancer Immunol Immunother 55:1122–1131.

Palucka, A. K., Ueno, H., Fay, J. W. and Banchereau, J. 2007. Taming cancer by inducing immunity via dendritic cells. Immunol Rev 220:129–150.

Phan, G. Q., Yang, J. C., Sherry, R. M., Hwu, P., Topalian, S. L., Schwartzentruber, D. J., Restifo, N. P., Haworth, L. R., Seipp, C. A., Freezer, L. J., Morton, K. E., Mavroukakis, S. A., Duray, P. H., Steinberg, S. M., Allison, J. P., Davis, T. A. and Rosenberg, S. A. 2003. Cancer regression and autoimmunity induced by cytotoxic T lymphocyte-associated antigen 4 blockade in patients with metastatic melanoma. Proc Natl Acad Sci USA 100:8372–8377.

Rabinovich, G. A., Gabrilovich, D. and Sotomayor, E. M. 2007. Immunosuppressive strategies that are mediated by tumor cells. Annu Rev Immunol 25:267–296.

Ribas, A. 2005. Genetically modified dendritic cells for cancer immunotherapy. Cur Gene Ther 5:619–628.

Rosenberg, S. A., Restifo, N. P., Yang, J. C., Morgan, R. A. and Dudley, M. E. 2008. Adoptive cell transfer: a clinical path to effective cancer immunotherapy. Nat Rev Cancer 8:299–308.

Rosenberg, S. A., Yang, J. C. and Restifo, N. P. 2004. Cancer immunotherapy: moving beyond current vaccines. Nat Med 10:909–915.

Sakai, Y., Morrison, B. J., Burke, J. D., Park, J. M., Terabe, M., Janik, J. E., Forni, G., Berzofsky, J. A. and Morris, J. C. 2004. Vaccination by genetically modified dendritic cells expressing a truncated neu oncogene prevents development of breast cancer in transgenic mice. Cancer Res 64:8022–8028.

Su, Z., Dannull, J., Yang, B. K., Dahm, P., Coleman, D., Yancey, D., Sichi, S., Niedzwiecki, D., Boczkowski, D., Gilboa, E. and Vieweg, J. 2005. Telomerase mRNA-transfected dendritic cells stimulate antigen-specific CD8+ and CD4+ T-cell responses in patients with metastatic prostate cancer. J Immunol 174:3798–3807.

Tong, X. M., Zheng, S. E., Bader, A., Yao, H. P., Wu, N. P., Altmeyer, P., Brockmeyer, N. H. and Jin, J. 2008. Construction of expression vector of hTERT- hIL18 fusion gene and induction of cytotoxic T lymphocyte response against hTERT. Eur J Med Res 13:7–14.

Tyagi, P. 2005. First-line treatment with bevacizumab and paclitaxel prolongs progression-free survival in metastatic breast cancer. Clin Breast Cancer 6:105–107.

Van Tendeloo, V. F., Ponsaerts, P., Lardon, F., Nijs, G., Lenjou, M., Van Broeckhoven, C., Van Bockstaele, D. R. and Berneman, Z. N. 2001. Highly efficient gene delivery by mRNA electroporation in human hematopoietic cells: superiority to lipofection and passive pulsing of mRNA and to electroporation of plasmid cDNA for tumor antigen loading of dendritic cells. Blood 98:49–56.

Vujanovic, L., Ranieri, E., Gambotto, A., Olson, W. C., Kirkwood, J. M. and Storkus, W. J. 2006. IL-12p70 and IL-18 gene-modified dendritic cells loaded with tumor antigen-derived peptides or recombinant protein effectively stimulate specific Type-1 CD4+ T-cell responses from normal donors and melanoma patients in vitro. Cancer Gene Ther 13:798–805.

Vulink, A., Radford, K. J., Melief, C. and Hart, D. N. 2008. Dendritic cells in cancer immunotherapy. Adv Cancer Res 99:363–407.

Wargo, J., Schumacher, L., Comin-Anduix, B., Dissette, V., Glaspy, J., McBride, W., Butterfield, L., Economou, J. and Ribas, A. 2005. Natural killer cells play a critical role in the immune response following immunization with melanoma-antigen-engineered dendritic cells. Cancer Gene Ther 12:516–527.

Woo, C. Y., Clay, T. M., Lyerly, H. K., Morse, M. A. and Osada, T. 2006. Role of natural killer cell function in dendritic cell-based vaccines. Expert Rev Vaccines 5:55–65.

Woo, E. Y., Yeh, H., Chu, C. S., Schlienger, K., Carroll, R. G., Riley, J. L., Kaiser, L. R. and June, C. H. 2002. Cutting edge: Regulatory T cells from lung cancer patients directly inhibit autologous T cell proliferation. J Immunol 168:4272–4276.

Yamanaka, R., Zullo, S. A., Ramsey, J., Yajima, N., Tsuchiya, N., Tanaka, R., Blaese, M. and Xanthopoulos, K. G. 2002. Marked enhancement of antitumor immune responses in mouse brain tumor models by genetically modified dendritic cells producing Semliki Forest virus-mediated interleukin-12. J Neurosurg 97:611–618.

Zhang, H. M., Zhang, L. W., Ren, J., Fan, L., Si, X. M. and Liu, W. C. 2006. Induction of alpha-fetoprotein-specific CD4- and CD8-mediated T-cell response using RNA-transfected dendritic cells. Cell Immunol 239:144–150.

Zhong, H., Shurin, M. R. and Han, B. 2007. Optimizing dendritic cell-based immunotherapy for cancer. Expert Rev Vaccines 6:333–345.

Chapter 24
Dendritic Cell Tumors

Lourdes R. Ylagan

Abstract Neoplasms of histiocytes and dendritic cells are rare, and their phenotypic and biological definition is incomplete. This chapter deals with the morphology of dendritic cell tumors as seen by pathologists and cytopathologists. It is divided into three sections: the follicular dendritic cell tumor, the interdigitating dendritic reticulum cell tumor and Langerhans cell histiocytosis. Currently used immunohistochemical markers of differentiation are presented and discussed to help in the differential diagnosis of these entities.

24.1 Origin and Classification of Dendritic Cell Tumors

Dendritic cells are major components of essential immune accessory cells necessary for antigen presentation, follicle differentiation and B- and T-cell activation in lymph nodes and other extranodal locations where follicles or germinal centers are found. The dendritic cell compartment is made up of follicular dendritic reticulum cells (FDC), interdigitating dendritic reticulum cells (IDC) and the Langerhans dendritic cell of the skin, epidermis, stomach, esophagus, cervix, vagina and vulva. The first two are generally called nodal reticulum cells because they form a network and supporting structure by which B and T lymphocytes adhere and contact with while they are in the lymph node. FDC generally reside in the follicles of lymph nodes and present antigens to both B and T cells, whereas IDC reside in the interfollicular areas and present antigen to T cells. Known tumors from these cells are therefore named on the basis of finding cells whose immunophenotypic and ultrastructural properties characterize the original cell type (Hollowood et al. 1995; Nayler et al. 1996; Pileri et al. 2002; Kairouz et al. 2007). The Langerhans cells are also T-cell-presenting dendritic cells in epithelial organs, but which can travel through the bloodstream where they are often called "veiled cells", then enter the afferent

L.R. Ylagan (✉)
Washington University Medical School, 660 South Euclid Ave, Box 8118, St. Louis, Missouri 63110, USA
e-mail: lylagan@path.wustl.edu

M.R. Shurin, R.D. Salter (eds.), *Dendritic Cells in Cancer*,
DOI 10.1007/978-0-387-88611-4_24, © Springer Science+Business Media, LLC 2009

lymphatics and reside in the paracortex. Tumors arising from these cells oftentimes have both histiocytic and dendritic properties and have a different immunoprofile than their nodal or germinal center counterparts (Pileri et al. 2002). Hence, tumors arising from these cells in association with epithelia are called Langerhans histiocytosis. They have also been referred to in the literature as histiocytosis X, eosinophilic granuloma or Langerhans cell granulomatosis (WHO).

Follicular and interdigitating dendritic cell tumors commonly arise in lymph nodes. However, organs which may have a robust germinal center formation such as Peyer's patches of the colon, germinal centers in thyroid in cases of chronic lymphocytic or Hashimoto's thyroiditis, tonsils and thymus may give rise to tumors of dendritic cell origin. And in fact, any solid organ by which germinal center formation has been microscopically documented as in cases of chronic pancreatitis, colitis, cirrhosis, gastritis, follicular cervicitis can be a primary foci of extranodal site involvement of a dendritic cell malignancy. Given the variety of location germinal centers may arise, recognition and diagnosis of nascent extranodal dendritic cell tumors then prove difficult.

Over the years since first described by Monda in 1986 (Monda et al. 1986), the basis for their "differing biologic behavior", as is commonly described in the literature, is most likely due to the inherent difficulty in their diagnosis and subcategorization exaggerated by the general under recognition of these entities. Their similar behavior and morphologic appearance to tumors, especially soft tissue sarcomas commonly arising in similar locations, can mimic the more aggressive malignant fibrous histiocytoma or the more indolent inflammatory pseudotumor, hence making this diagnosis and predicting its prognosis even more difficult. It became obvious then that a sub-categorization of dendritic cells was necessary not only for uniformity of nomenclature, but also for predicting biologic behavior in these tumor subtypes.

The International Lymphoma Study Group, in 2002, published a review article based on a panel of 15 antibodies for the subclassification of these tumors of histiocytes and accessory dendritic cells. Based on the morphologic, immunophenotypic and ultrastructural features of these tumors, they were able to classify 93% of the tumors in their study ($n = 63$). Since this is the biggest concentration of tumors of these types, thus far, most of our current diagnosis of these entities is based on this categorization system (WHO).

24.2 Follicular Dendritic Cell Tumor/Sarcoma

The WHO classifies follicular dendritic cell tumor/sarcoma as originating from mesenchymal stem cells. This supposition, however, is currently mainly based on the absence of shared immunophenotypic characteristic of normal monocytes. This designation of sarcoma/tumor is based on the tumor's morphologic feature of spindled cells in whorling, at times interstitial arrangement which on electron microscopy shows long cytoplasmic processes interconnected by desmosomes and oftentimes expression of stromal-derived markers such as vimentin. Special marker of immunoreactivity with CD21 and CD35 with an almost absent staining for other hematolymphoid-derived markers of differentiation characterize this tumor such as in B cells (CD20), T cells (CD3, UCHL1), granulocytes (CD30, CD15) or histiocytes (CD68). They are also not immunoreactive for epithelial markers of differentiation (CK), neural (S-100), smooth muscle (desmin, smooth muscle actin) or vascular markers (CD31, CD34). The tumor is variably positive for EMA (Hollowood et al. 1995) .

Other than primary in the lymph node, extranodal FDC tumors have been described in the mesocolon (Hollowood et al. 1995), pancreas (Hollowood et al. 1995) , tonsil (Nayler et al. 1996), oral cavity (Chan et al. 1994b), spleen (Perez-Ordonez and Rosai 1995) , nasopharynx (Beham-Schmid et al. 1998), thyroid (Galati et al. 1999) and liver (Shek et al. 1996) all with variable degree of aggression. Although previously regarded as a tumor of intermediate behavior, some cases have progressed to liver metastasis even after aggressive radiotherapy (Hollowood et al. 1995; Chan et al. 1997) with abdominal locations carrying a more ominous prognosis (Chan et al. 1994a, 1997). Common sites of metastasis include the liver, peritoneum, pancreas, lymph nodes and lung (Chan et al. 1997). Some cases have been associated with hyaline-vascular Castleman's disease (Chan et al. 1994a, 1997) and a few are found in association with the Epstein–Barr virus (Chan et al. 1997).

24.2.1 Gross Features and Histomorphology

The tumor is generally bulky averaging in size of 6–7 cm, but can reach up to 15 cm in the abdomen (Chan et al. 1997). It is usually a solid slow growing tumor with pushing rather than an infiltrative border. It has very focal areas of necrosis, with mostly fleshy gross consistency. Microscopically, it is described most commonly as a synctitium of spindled cells in whorling, fascicular, diffuse, trabecular, follicular or storiform pattern with very low mitotic index (Chan et al. 1997). There is interspersed within the tumor, mature T lymphocytes which are deemed reactive and non-clonal and some tumors have associated multinucleated giant cells with large pleomorphic nuclei which can resemble a malignant fibrous histiocytoma (MFH). Some cases have distinct perivascular cuffing of lymphocytes (Perez-Ordonez and Rosai 1995, 1998). The spindled cells can have fine granular chromatin without nucleoli, while others have vesicular chromatin with small nucleoli (Fig. 24.1). On electron microscopy, they are described as having very few organelles with long cytoplasmic interdigitating processes bound by desmosomes. This gives the tumor a synctitial look in histology and a solid gross consistency.

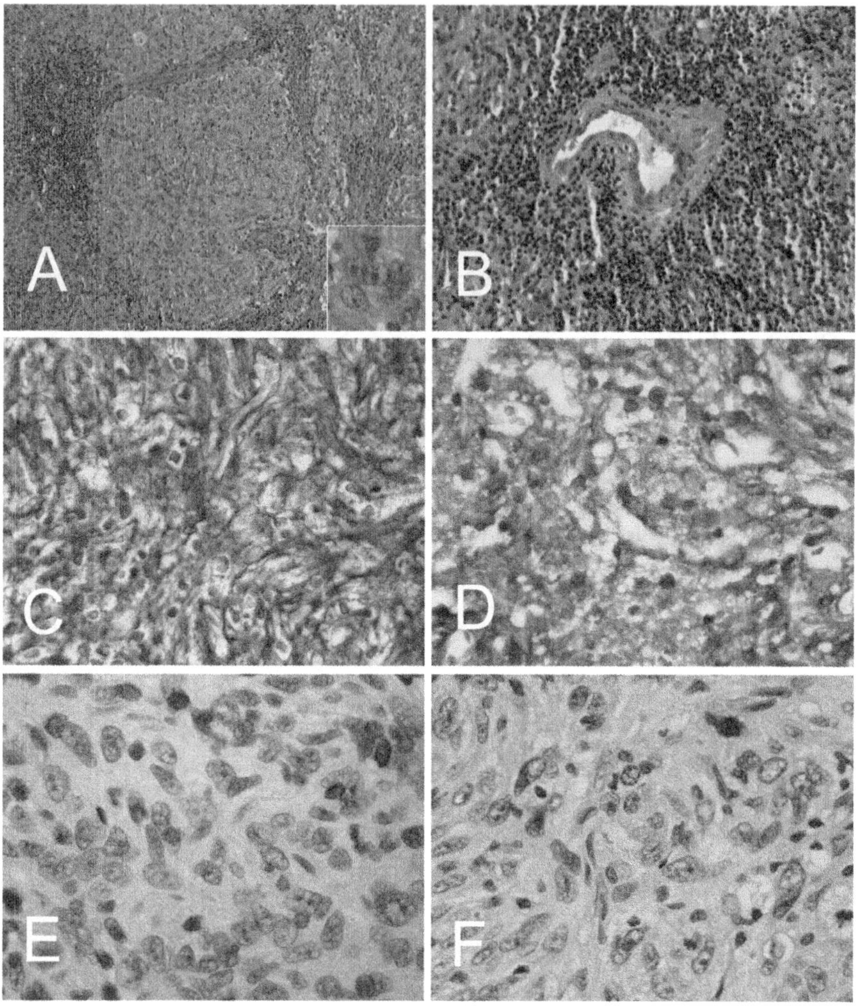

Fig. 24.1 Follicular dendritic cell tumor/sarcoma of the pharynx of a 28-year-old man. (A) Section of tumor in the soft tissue of the pharynx, consisting of tumor cells in storiform configuration resembling a granuloma at low power. It is composed of spindled cells with abundant eosinophilic cytoplasm, vesicular nuclei with prominent nucleoli (inset) (400× original magnification, inset: 1000×). (**B**) Typical perivascular cuffing seen in this tumor (600×). (**C**) CD21 stain is strongly and diffusely immunoreactive as is CD35 stain (**D**). (**E**) CD45 is immunonegative in the tumor cells which highlights the reactive lymphocytes (1000×). (**F**) S-100 stain is negative in this tumor (1000×)

24.2.2 Clinical Behavior and Outcome

Follicular dendritic cell tumors have shown to be more aggressive than was generally thought, at least in the most exhaustive case series of both Chan et al.

($n = 17$) and Perez-Ordonez et al. ($n = 13$) (Perez-Ordonez and Rosai 1995; Chan et al. 1997). Since then a total of the 51 total cases of FDC tumors from 1986 to 2008 have been described in the literature as pure FDC tumors. A meta-analysis by Fonseca et al. showed that these tumors are less aggressive than their interfollicular counterpart or the interdigitating dendritic cell tumor/sarcoma (Fonseca et al. 1998). Although a few patients have experienced a long and indolent course (some going on 9 years), most of these tumors recur within the next 2 years despite a combination of radiotherapy after surgery or aggressive chemotherapy after surgery (Pileri et al. 2002).

Although treatment algorithms are currently not available in this very rare, yet devastating neoplasm, this difference in clinical behavior and outcome makes differentiation between these two tumor cell types the more important. Treatment modalities have included total resections followed by radiation or chemotherapy used for high-grade non-Hodgkin's lymphoma, but recurrences within 2 years after initial diagnosis are usually not survivable by many.

24.3 Interdigitating Dendritic Cell Tumor/Sarcoma

Interdigitating dendritic cell tumors/sarcomas of lymph nodes are primarily nodal in origin and arise from the interdigitating cells of the interfollicular zones or T-cell zones which are the cells responsible for stimulating and presenting antigens to T cells. Unlike follicular dendritic cell tumors/sarcoma, they are thought to be monocyte derived from the bone marrow and may co-express monocyte markers of differentiation.

24.3.1 Gross Features and Histomorphology

Like FDC tumors, IDC tumors can present at extranodal locations as large bulky masses with bulbous pushing borders. A mixture of whorling, storiforming, trabecular and diffuse architecture is also found. In contrast to FDC tumors, however, their long cytoplasmic projections and processes are not attached by desmosomes. Almost all do not express the CD21/CD35 marker, but instead are immunoreactive for monocyte markers CD1a, HLA-DR and are invariably positive for the S-100 protein (Fig. 24.2).

24.3.2 Clinical Behavior and Outcome

IDC tumors are more aggressive in their presentation, and in at least 21 well-documented cases in the literature presented with lymphadenopathy and had earlier time to recurrence than FDC tumors at extranodal sites (Fonseca

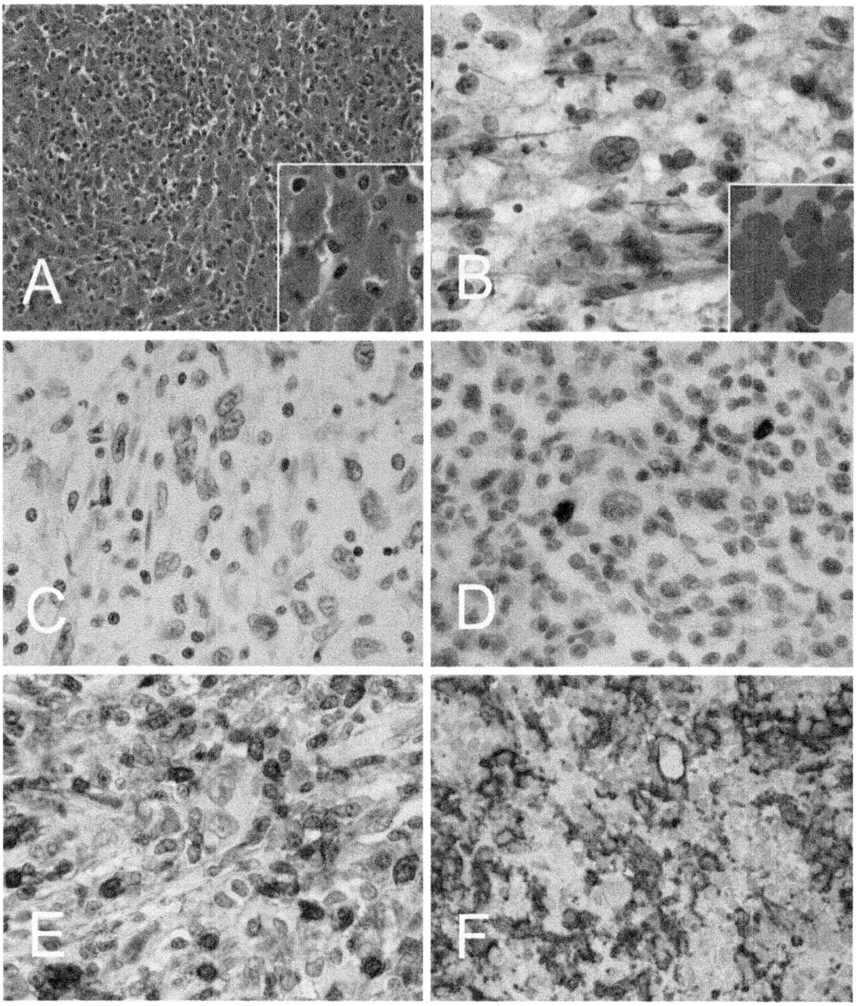

Fig. 24.2 Interdigitating dendritic cell tumor/sarcoma of an intraparotid lymph node of a 66-year-old man. (A) Section of tumor of IDRCT (H&E, 600× original magnification) showing large and pleomorphic high-grade sarcoma cells with bizarre, markedly enlarged nuclei with deep nuclear indentations (inset: 1000×). (**B**) Cytologic features of the same tumor showing enlarged nuclei with convoluted features also seen in Diff Quik stains (inset: 1000×). (**C**) CD21 is immunonegative in the tumor cells as is CD15 (**D**) (1000×). (**E**) CD45 is immunonegative in the tumor cells, but shows reactive T lymphocytes which were immunoreactive for CD45 (1000×). (**F**) CD1a is immunoreactive in the tumor cells (1000×)

et al. 1998; Ylagan et al. 2003). Recurrence within 2 years heralds a more aggressive course without much response to radiation or chemotherapy in most patients.

24.4 Langerhans Cell Histiocytosis/Sarcoma

Langerhans histiocytosis or sarcoma is so named because of its indeterminate cell of origin with both histiocytic and dendritic characteristics. Langerhans cell histiocytosis most commonly occurs in childhood, with a slight male predilection (3:1), more common in whites with an association with acute lymphoblastic leukemia (WHO). Organs of involvement include the flat and long bones: skull, pelvis and femur, but can also involve the skin, liver, lymph nodes and lung (Satter and High 2008). A multifocal unisystem disease is synonymous with Hand–Shuller–Christian disease and a multifocal multisystem disease is synonymous with Letterer–Siwe disease. It has also been associated with adults with a smoking history.

24.4.1 Gross Features and Histomorphology

This tumor usually presents with multifocal involvement of bone and soft tissue. Histologically they are composed of a solid aggregate of histiocytic or foamy cells with the characteristic grooved, indented, folded or lobulated nuclei, with fine granular chromatin and small nucleoli. The most common feature of this neoplasm is its intimate association with eosinophils, so much so that it had been called eosinophilic granuloma at one point. Recognition of these histiocytic cells in close association with eosinophils, lymphocytes, macrophages and giant cells is its identifying feature. They are usually immunoreactive with CD1a and S-100 protein, variably reactive with CD45, CD68 and lysozyme (Fig. 24.3) and on electron microscopy shows the characteristic Birbeck granules (tennis racket-like structures). The cytoplasm has variable numbers of lysosomes and they have no cell junctions.

The summary of immunohistochemical markers used to distinguish these tumors from each other is presented in Table 24.1.

24.4.2 Clinical Behavior and Outcome

Multiple organ involvement heralds a bad prognosis. Patients with unifocal disease progress to multifocality in 10% of cases (Allen and McClain 2007). Some may spontaneously regress, but progression to more than one organ involved heralds a more ominous prognosis. Adult patients who had lung lesions have seen regression of their disease after cessation of smoking.

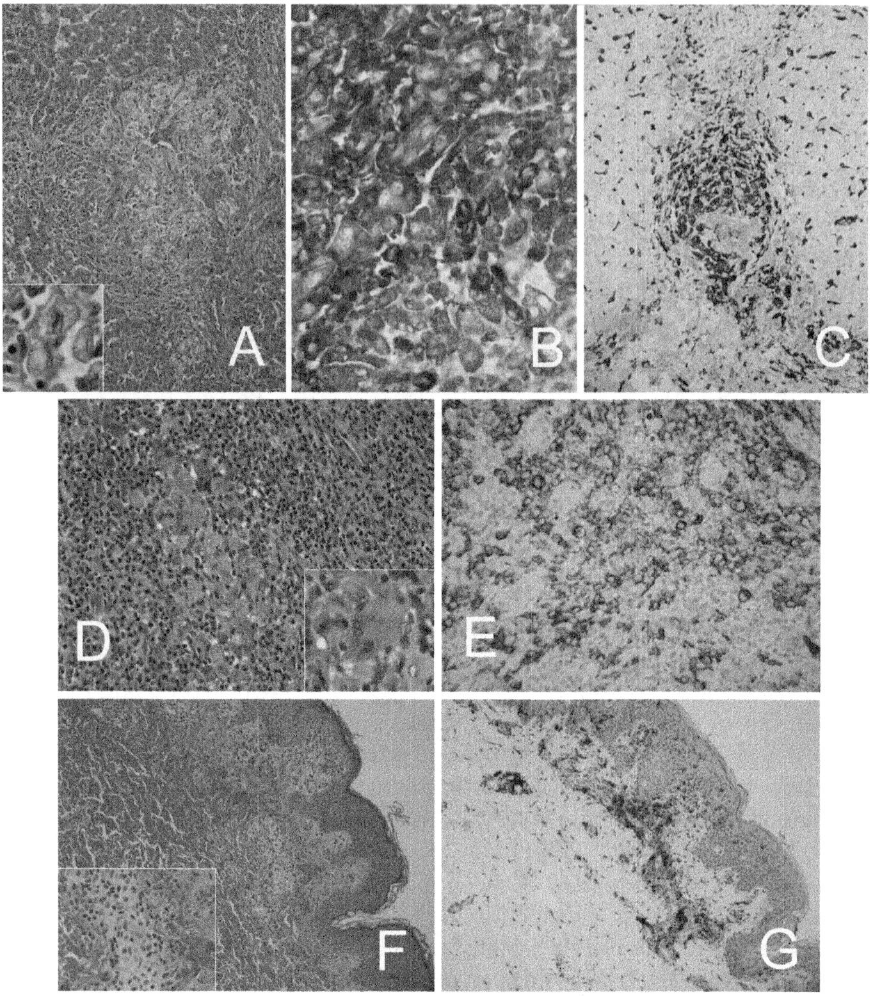

Fig. 24.3 Langerhans cell histiocytosis of liver (A–C), bone (D and E) and skin (F and G). (A) H&E section of explanted liver from a 3-year-old boy with bridging fibrosis and portal tracts that were expanded by an accumulation of foamy histiocytes (400× original magnification, inset: 1000×) which were strongly immunoreactive for CD1a (**B**, 1000×) and CD68 (**C**, 400×). (**D**) Bone marrow curettage specimen from a 6-year-old boy showing abundant accumulation of CD1a-positive cells (**E**, 600×) and eosinophils (**D** inset: 1000×). (**F**) Skin sections from a 4-year-old boy show perivascular nests of foamy cells (400×, inset: 600×). (**G**) These cells were immunoreactive for CD1a (400×) and S-100 (not shown)

Table 24.1 Immunohistochemical markers used in the differential diagnosis in dendritic cell tumors

Markers	FDRCT	IDRCT	LCH	Dilutions	Source of antibodies
CD21	+	–	–	1:40	DAKO*
CD35	+	–	–	1:50	DAKO
CD45	–	–	±	1:400	DAKO
S-100	–	+	+	1:6000	DAKO
CD68	–	–	±	1:2000	DAKO
CD1a	–	+	+	1:40	DAKO
HLA-DR	–	+	–	1:100	DAKO
Lysozyme	–	–	±	1:2000	DAKO
EMA	±	–	–	1:2000	DAKO
CD20	–	–	–	1:4000	DAKO
CD3	–	–	–	1:400	DAKO
CD15	–	–	–	1:50	DAKO
CD30	–	–	–	1:1000	DAKO
CD31	–	–	–	1:40	DAKO
CD34	–	–	–	1:300	DAKO

FDRCT, follicular dendritic reticulum cell tumor; IDRCT, interdigitating dendritic reticulum cell tumor; LCH, Langerhans cell histiocytosis; EMA, epithelial membrane antigen.
* DAKO, Carpinteria, CA, USA.

24.5 Conclusions

Dendritic cell neoplasms are rare tumors that are being recognized with increasing frequency. They were previously classified as lymphomas, sarcomas or histiocytic neoplasms. The World Health Organization classifies dendritic cell neoplasms into several groups, including Langerhans cell histiocytosis, Langerhans cell sarcoma, interdigitating dendritic cell sarcoma/tumor, follicular dendritic cell sarcoma/tumor, dendritic cell sarcoma, not specified otherwise. Pathological, clinical and therapeutic aspects of follicular dendritic, interdigitating dendritic and Langerhans cell tumors require further investigations and analyses. Much still needs to be learned about the most effective adjuvant therapy and the molecular biology of these tumors.

References

Allen, C. E. and McClain, K. L. 2007. Langerhans cell histiocytosis: a review of past, current and future therapies. Drugs Today (Barc) 43:627–643.

Beham-Schmid, C., Beham, A., Jakse, R., Aubock, L. and Hofler, G. 1998. Extranodal follicular dendritic cell tumour of the nasopharynx. Virchows Arch 432:293–298.

Chan, J. K., Fletcher, C. D., Nayler, S. J. and Cooper, K. 1997. Follicular dendritic cell sarcoma. Clinicopathologic analysis of 17 cases suggesting a malignant potential higher than currently recognized. Cancer 79:294–313.

Chan, J. K., Tsang, W. Y. and Ng, C. S. 1994a. Follicular dendritic cell tumor and vascular neoplasm complicating hyaline-vascular Castleman's disease. Am J Surg Pathol 18:517–525.
Chan, J. K., Tsang, W. Y., Ng, C. S., Tang, S. K., Yu, H. C. and Lee, A. W. 1994b. Follicular dendritic cell tumors of the oral cavity. Am J Surg Pathol 18:148–157.
Fonseca, R., Yamakawa, M., Nakamura, S., van Heerde, P., Miettinen, M., Shek, T. W., Myhre Jensen, O., Rousselet, M. C. and Tefferi, A. 1998. Follicular dendritic cell sarcoma and interdigitating reticulum cell sarcoma: a review. Am J Hematol 59:161–167.
Galati, L. T., Barnes, E. L. and Myers, E. N. 1999. Dendritic cell sarcoma of the thyroid. Head Neck 21:273–275.
Hollowood, K., Stamp, G., Zouvani, I. and Fletcher, C. D. 1995. Extranodal follicular dendritic cell sarcoma of the gastrointestinal tract. Morphologic, immunohistochemical and ultrastructural analysis of two cases. Am J Clin Pathol 103:90–97.
James Vardiman (Eds) IARC Press, Lyon 2001:275–301.
Kairouz, S., Hashash, J., Kabbara, W., McHayleh, W. and Tabbara, I. A. 2007. Dendritic cell neoplasms: an overview. Am J Hematol 82:924–928.
Monda, L., Warnke, R. and Rosai, J. 1986. A primary lymph node malignancy with features suggestive of dendritic reticulum cell differentiation. A report of 4 cases. Am J Pathol 122:562–572.
Nayler, S. J., Verhaart, M. J. and Cooper, K. 1996. Follicular dendritic cell tumour of the tonsil. Histopathology 28:89–92.
Perez-Ordonez, B. and Rosai, J. 1995. Follicular dendritic cell tumor: Report of 13 additional cases of a distinctive entity. Annual meeting extracts. Lab Invest. 72:118A.
Perez-Ordonez, B. and Rosai, J. 1998. Follicular dendritic cell tumor: review of the entity. Semin Diagn Pathol 15:144–154.
Pileri, S. A., Grogan, T. M., Harris, N. L., Banks, P., Campo, E., Chan, J. K., Favera, R. D., Delsol, G., De Wolf-Peeters, C., Falini, B., Gascoyne, R. D., Gaulard, P., Gatter, K. C., Isaacson, P. G., Jaffe, E. S., Kluin, P., Knowles, D. M., Mason, D. Y., Mori, S., Muller-Hermelink, H. K., Piris, M. A., Ralfkiaer, E., Stein, H., Su, I. J., Warnke, R. A. and Weiss, L. M. 2002. Tumours of histiocytes and accessory dendritic cells: an immunohistochemical approach to classification from the International Lymphoma Study Group based on 61 cases. Histopathology 41:1–29.
Satter, E. K. and High, W. A. 2008. Langerhans cell histiocytosis: a review of the current recommendations of the Histiocyte Society. Pediatr Dermatol 25:291–295.
Shek, T. W., Ho, F. C., Ng, I. O., Chan, A. C., Ma, L. and Srivastava, G. 1996. Follicular dendritic cell tumor of the liver. Evidence for an Epstein-Barr virus-related clonal proliferation of follicular dendritic cells. Am J Surg Pathol 20:313–324.
WHO. 2001. Tumours of Haematopoietic and Lymphoid Tissues. World Health Organization Classification of Tumours. Elaine Jaffe, Nancy Lee Harris, Herald Stein,
Ylagan, L. R., Bartlett, N. L. and Kraus, M. 2003. Interdigitating dendritic reticulum cell tumor of lymph nodes: case report with differential diagnostic considerations. Diagn Cytopathol 28:278–281.

Chapter 25
Future Directions in Dendritic Cell Research in Cancer

Madhav V. Dhodapkar

One never notices what has been done; one can only see what remains to be done.

Marie Curie (Letter to her brother, 1894)

The most important thing is not to stop questioning.

Albert Einstein (1879–1955)

Abstract Dendritic cells play a critical role in the regulation of both innate and adaptive immunities. Tumors can exploit the biology of these cells to both evade immunity and facilitate their growth. Although much has been learnt about the biology of these cells, much more remains to be discovered. It is likely that several aspects of the host immune response to tumor cells are conditioned by the interactions of dendritic cells and tumor cells. Dendritic cells may also have non-immune effects on the biology and growth of tumor cells. Improved understanding of how tumor cells interact with dendritic cells therefore has implications for both cancer biology and immunity to tumors.

Dendritic cells (DC) were so named for their tree-like projections that allow them to probe and sense the environment and communicate with other cells (Steinman et al. 1975; Steinman and Banchereau 2007). A large body of research over the last two decades has placed these cells as central players in the immune system, capable of interacting with and regulating the function of other immune cells, for both innate and adaptive immunities. Not surprisingly, DC have also been the focus of intensive investigation in cancer, particularly as tools for therapeutic vaccines against cancer (Steinman and Dhodapkar 2001). Attempts to harness the immune-stimulating properties of these cells are also underway in the clinic. Progress (and challenges) in this field has been recently reviewed (Palucka et al. 2007). Here, we will argue that although much has been

M.V. Dhodapkar (✉)
Section of Hematology, Yale University, 333 Cedar Street, Box 208021, New Haven, CT 06510, USA
e-mail: madhav.dhodapkar@yale.edu

M.R. Shurin, R.D. Salter (eds.), *Dendritic Cells in Cancer*,
DOI 10.1007/978-0-387-88611-4_25, © Springer Science+Business Media, LLC 2009

learnt, much more remains to be determined. We will discuss some of the gaps in existing knowledge about the interactions of DC and tumor cells. Improved understanding of the biology of these cells will be needed to effectively harness their properties in the setting of human cancer.

25.1 Location, Location, Location

Tumors in both mice and humans are commonly infiltrated by cells of myeloid origin including DC, macrophages, and myeloid suppressor cells (Talmadge et al. 2007). Several studies have tried to examine the phenotype and activation status of these cells. DC with both myeloid and plasmacytoid phenotype have been identified within tumors. Even within myeloid DC, several subtypes may exist (Banchereau et al. 2000; Ueno et al. 2007). Some of the current confusion relates to the lack of definitive markers for identification of these cells and their subsets within archived human tissues. Location of DC within the tumors may also matter. For example, DC within tumors seem to have an immature phenotype, while cells with more mature phenotype were found near the periphery in human breast cancers (Bell et al. 1999). However, most of these studies only provide static/cross sectional information. It is expected that advances in live imaging will provide opportunities for understanding the dynamics of recruitment, differentiation, mobilization, and fate of tumor-associated DC. Recent studies have begun to elucidate the precursors and differentiation of DC in vivo both under steady state and inflammatory conditions (Jakubzick et al. 2008).

The source of DC may also depend on the tissue. For example, the major precursors to myeloid DC in the spleen in mice seem to be myeloid precursors that are distinct from monocytes. However, monocytes do clearly contribute to the renewal of the DC pool in the steady state in the tissues and under inflammatory conditions (Naik et al. 2006; Shortman and Naik 2007). However, translation of this work to humans remains a challenge. For many of the subsets of splenic DC in mice (such as $CD8\alpha^+$ or $CD8\alpha^-$), the human counterparts remain to be clarified. There may also be important species-specific differences. For example, $Ly6C^{lo}CD115+$ monocytes account for half of murine monocytes, but the corresponding human $CD14^{int}CD16+$ monocytes represent a much smaller subset (Ziegler-Heitbrock 2007). Much remains to be learnt about the biology of myeloid cells within tumors. What are the precursors of DC and their subsets in tumors? How is their recruitment and differentiation regulated? Another important aspect of DC is their migratory properties. For example, alveolar macrophages outnumber DC by over 100-fold within airways. Yet, DC that migrate to lung draining lymphatics outnumber macrophages (Jakubzick et al. 2006).

Due to their migratory properties, DC may also be an important component of the premetastatic niche. In the lymphoid tissues, the specific location of DC (e.g., within the T-cell zone) is of paramount importance. We suspect that the location of DC within tumors is not random, but rather consist of specialized niches or organized structures, wherein their interactions with other cells will be important. For example, resident DC in the lamina propria may play an important role in sampling and responding to gut flora and regulate innate inflammation (Hart et al. 2005; Garrett et al. 2007). Such DC were recently implicated in regulating inflammation in a mouse model of ulcerative colitis, which is known to increase the risk for colon cancer. Similarly, DC in a perivascular location in the bone marrow may be important for survival of long-lived B cells (Sapoznikov et al. 2008). It is notable that DC are often found to be enriched in focal lesions in human B cell tumors such as myeloma (Bahlis et al. 2007). Tumor microenvironment is not uniform but can vary widely with regions of hypoxia, low pH, and altered interstitial pressures due to abnormal vasculature (Grimshaw and Balkwill 2001). All of these aspects may alter the properties of tumor infiltrating DC (and other cells). Improved understanding of the properties of DC in specific tissues may allow novel approaches for regulating inflammation in tissues.

25.2 Function and Plasticity

A critical property of DC is that they are highly specialized in responding to the cues from the microenvironment. The microenvironment within tumors is quite abnormal and includes cytokines, shed molecules, and other tumor derived factors that can provide an immune suppressive signal (Shurin et al. 2006). For example, tumors express cytokines such as IL-6, IL-10, and VEGF that suppress the function of DC in the tumor bed via STAT3 signaling (Kortylewski et al. 2005). Other molecules such as gangliosides shed by tumor cells can also inhibit DC function (Shurin et al. 2001). Other suppressive elements that inhibit DC function include reactive oxygen species, indoleamine dioxygenase (IDO), and arginase-related pathways (Zou 2005). The degree of plasticity in functional properties of tumor-associated DC remains to be clarified. The immune suppressive environment within tumors instructs DC to form suppressive T cells such as FOXP3+ regulatory T cells (Treg) (Banerjee et al. 2006) as well as IL-13 producing T cells (Aspord et al. 2007). In other words, tumors may instruct DC to prevent the activation of antitumor effector T-cell responses. In addition to T cells, DC are also specialized for the activation of innate NK or NKT cells (Munz et al. 2005). An important gap in the present knowledge is the extent to which different subsets of DC interact, both with each other and for activation of T cells in the tumor bed. For example, plasmacytoid DC can produce interferons and membrane bound costimulators that recruit other DC to stimulate immunity to viruses (Yoneyama et al. 2005). DC are not the only antigen-presenting cells in the tumor bed, and the cross talk between DC and other antigen-presenting cells, such as endothelial cells, alternately activated M2 macrophages and B cells in the tumor bed, needs further study.

Most of the current work relates to the ability of DC to activate immune cells, although it is likely that DC also play important roles in the tumor bed that extend beyond the immune system. For example, monocyte-derived human DC were recently shown to be effective for enhancing the clonogenicity of human myeloma (Kukreja et al. 2006). DC and other vascular lymphocytes have also been implicated as potential precursors to new blood vessels or angiogenesis within tumors (Coukos et al. 2005; Noonan et al. 2008). It would be useful to better understand the cellular fates of tumor-associated DC and their contributions to the cellular elements of the microenvironment. We suspect that as the biology of these cells is better understood, their contributions to several aspects of cancer biology will be appreciated and understood at a cellular and molecular level.

25.3 Dendritic Cells as a Focal Point for Immune Regulation

A critical property of DC is their ability to activate both innate and adaptive immune cells (Steinman and Banchereau 2007). The capacity of DC to interact with NK cells and activate them has emerged as an important pathway for the activation of these cells. NK and NKT cells in turn can also deliver an activation stimulus to DC and help link innate and adaptive immunities (Munz et al. 2005). DC have a specialized endocytic system that plays a key role in processing peptides for presentation to CD4+ and CD8+ T cells, self and microbial glycolipids to NKT cells, and intact antigen to B cells. Although the diversity of T-cell response continues to grow (with the most recent addition being inflammatory IL-17 producing T cells), it is likely that DC are critical regulators of the balance of these immune cells (Gutcher and Becher 2007). It is likely that the additional and defined forms of T-cell polarization will be identified and that this will depend on the nature of the signals derived from antigen-presenting cells. An important insight in the recent past has been the recognition of the importance of uptake receptors (Mahnke et al. 2000; Hawiger et al. 2001). Initial studies took uptake of antigens by DC for granted. However, in the past several years, several defined receptors for uptake of antigen have been identified. The molecular specificity of many of these receptors still remains to be fully understood. However, targeting antigens to specific receptors not only yields high efficiency (>100-fold) but also impacts the nature of immune response. For example, the nature of T-cell immunity after FcγR-mediated uptake of immune complexes is determined by the balance of engagement of activating versus inhibitory FcγR (Dhodapkar et al. 2005).

An important property of DC from the perspective of tumor immunity is the property of cross-presentation of exogenous antigens (such as from dying tumor cells) to yield CD8+ killer T cells (Albert et al. 1998; Heath and Carbone

2001). Although DC express several receptors for the recognition of dying cells, the nature of specific receptors involved in the uptake of dying tumor cells in situ remains to be defined. A critical aspect may be the specialized processing of the antigen in the phagosome, limited degradation, and access of antigen to the cytosol in DC (Mellman and Steinman 2001). An emerging concept in this regard relates to immunogenic versus non-immunogenic death of tumors. The distinction between these forms of cell death may relate to the delivery of an activation signal to DC, such as that mediated by exposure of intracellular chaperones (Obeid et al. 2007; Spisek et al. 2007).

The capacity of DC to stimulate immunity is linked to their activation or maturation status. However, DC maturation is a complex process that is still being understood at a molecular level. In particular, how DC integrate different and often opposing signals to stimulate different types of T cells remains to be fully clarified. Future studies will attempt to integrate how signals from diverse subsets of DC are integrated with diverse signals from the same DC (Dhodapkar et al. 2008). Such analyses may also provide novel insights into the need for diverse subsets of DC in the immune response. An important issue is how long-term T-cell memory is generated. Another aspect wherein we suspect DC will be found to play an important role is the regulation of innate immunity and chronic inflammation (Coussens and Werb 2002). There is a growing body of evidence that chronic inflammation may be an early event in carcinogenesis. Understanding the role of DC in this stage of cancer may provide important insights for prevention of cancer.

25.4 Harnessing Dendritic Cells for Immune Therapy of Cancer

The capacity of DC to serve as potent antigen-presenting cells and stimulate tumor-specific immunity has led to several attempts to target these cells for immune therapy or prevention of cancer. Two general approaches are being attempted (Banchereau and Palucka 2005). The approach already being tested in the clinic involves adoptive transfer of DC generated and loaded with antigens ex vivo. Another approach involves targeting DC in situ.

25.4.1 Ex Vivo Dendritic Cell Vaccines

The discovery that precursors from blood monocytes or CD34+ progenitors could be differentiated into DC provided the impetus for testing adoptive transfer of antigen-loaded DC in cancer (Palucka et al. 2007). An advantage of this approach over in vivo targeting is that one can precisely control the delivery of antigen to DC and their activation status. It has been clearly documented that the injection of antigen-pulsed DC in humans can boost immunity to defined antigens and is well tolerated. Although some clinical responses have been observed in these studies, they have been rare, and there is no evidence for improved survival yet. There are several deficiencies of current trials. Most studies have focused on patients with advanced cancer; migration of injected DC to lymph nodes remains suboptimal (de Vries et al. 2005); injected DC often also boost Treg (Banerjee et al. 2006); and true tumor rejection antigens remain unknown (Gilboa 1999). Nonetheless, the approach to optimize adoptive transfer of DC and combine it with other immune modalities remains a promising area of investigation.

25.4.2 *In-Situ Dendritic Cell Targeting*

Another approach is to mobilize DC in situ, either within the tumor or within lymphoid organs. One approach is the injection of tumor cells genetically modified to express GM-CSF, which can recruit patient DC to capture tumor antigens (Soiffer et al. 1998). Another approach, largely attempted in mice, involves delivery of a signal from activated NKT cells to DC (Dhodapkar 2008). Yet another approach is to target antigens to DC via monoclonal antibodies against receptors on DC. One example, now extensively tested in mice, is antibodies against the decalectin uptake receptor, DEC-205 (Mahnke et al. 2005; Charalambous et al. 2006; Bozzacco et al. 2007; Soares et al. 2007). One potential advantage is that large numbers of DC can be targeted in vivo. Key challenges in this area include potential differences in the biology of specific receptors between mice and humans. DC-mediated targeting still remains a field in its infancy and will likely expand as we better understand the biology of these cells and their ability to induce or suppress immunity.

25.5 Summary

A considerable body of evidence over the last decade has demonstrated that DC are not just innocent bystanders, but play an active role in regulating cancer biology and immune resistance against tumors (Steinman and Banchereau 2007). Improved understanding of the roles these cells play and harnessing their immune regulatory properties may have important implications for cancer research, as well as immunology. Efforts to understand the biology of these cells should continue not just in model systems in laboratory animals, but also in defined disease settings in humans.

Acknowledgments MVD is supported in part by funds from NIH, Dana Foundation, and Damon Runyon Cancer Research Fund. I thank Kavita Dhodapkar for critically reading this review.

References

Albert, M. L., Sauter, B. and Bhardwaj, N. 1998. Dendritic cells acquire antigen from apoptotic cells and induce class I-restricted CTLs. Nature 392:86–89.

Aspord, C., Pedroza-Gonzalez, A., Gallegos, M., Tindle, S., Burton, E. C., Su, D., Marches, F., Banchereau, J. and Palucka, A. K. 2007. Breast cancer instructs dendritic cells to prime interleukin 13-secreting CD4+ T cells that facilitate tumor development. J Exp Med 204:1037–1047.

Bahlis, N. J., King, A. M., Kolonias, D., Carlson, L. M., Liu, H. Y., Hussein, M. A., Terebelo, H. R., Byrne, G. E., Jr., Levine, B. L., Boise, L. H. and Lee, K. P. 2007. CD28-mediated regulation of multiple myeloma cell proliferation and survival. Blood 109:5002–5010.

Banchereau, J., Briere, F., Caux, C., Davoust, J., Lebecque, S., Liu, Y. J., Pulendran, B. and Palucka, K. 2000. Immunobiology of dendritic cells. Annu Rev Immunol 18:767–811.

Banchereau, J. and Palucka, A. K. 2005. Dendritic cells as therapeutic vaccines against cancer. Nat Rev Immunol 5:296–306.

Banerjee, D. K., Dhodapkar, M. V., Matayeva, E., Steinman, R. M. and Dhodapkar, K. M. 2006. Expansion of FOXP3high regulatory T cells by human dendritic cells (DCs) in vitro and after injection of cytokine-matured DCs in myeloma patients. Blood 108:2655–2661.

Bell, D., Chomarat, P., Broyles, D., Netto, G., Harb, G. M., Lebecque, S., Valladeau, J., Davoust, J., Palucka, K. A. and Banchereau, J. 1999. In breast carcinoma tissue, immature dendritic cells reside within the tumor, whereas mature dendritic cells are located in peritumoral areas. J Exp Med 190:1417–1426.

Bozzacco, L., Trumpfheller, C., Siegal, F. P., Mehandru, S., Markowitz, M., Carrington, M., Nussenzweig, M. C., Piperno, A. G. and Steinman, R. M. 2007. DEC-205 receptor on dendritic cells mediates presentation of HIV gag protein to CD8+ T cells in a spectrum of human MHC I haplotypes. Proc Natl Acad Sci USA 104:1289–1294.

Charalambous, A., Oks, M., Nchinda, G., Yamazaki, S. and Steinman, R. M. 2006. Dendritic cell targeting of survivin protein in a xenogeneic form elicits strong CD4+ T cell immunity to mouse survivin. J Immunol 177:8410–8421.

Coukos, G., Benencia, F., Buckanovich, R. J. and Conejo-Garcia, J. R. 2005. The role of dendritic cell precursors in tumour vasculogenesis. Br J Cancer 92:1182–1187.

Coussens, L. M. and Werb, Z. 2002. Inflammation and cancer. Nature 420:860–867.

de Vries, I. J., Lesterhuis, W. J., Barentsz, J. O., Verdijk, P., van Krieken, J. H., Boerman, O. C., Oyen, W. J., Bonenkamp, J. J., Boezeman, J. B., Adema, G. J., Bulte, J. W., Scheenen, T. W., Punt, C. J., Heerschap, A. and Figdor, C. G. 2005. Magnetic resonance tracking of dendritic cells in melanoma patients for monitoring of cellular therapy. Nat Biotechnol 23:1407–1413.

Dhodapkar, K. M., Kaufman, J. L., Ehlers, M., Banerjee, D. K., Bonvini, E., Koenig, S., Steinman, R. M., Ravetch, J. V. and Dhodapkar, M. V. 2005. Selective blockade of inhibitory Fcgamma receptor enables human dendritic cell maturation with IL-12p70 production and immunity to antibody-coated tumor cells. Proc Natl Acad Sci USA 102:2910–2915.

Dhodapkar, M. V. 2009. Harnessing human CDld restricted T cells for tumor immunity: progress and challenges. Front Biosci in press.

Dhodapkar, M. V., Dhodapkar, K. M. and Palucka, A. K. 2008. Interactions of tumor cells with dendritic cells: balancing immunity and tolerance. Cell Death Differ 15:39–50.

Garrett, W. S., 1Lord, G. M., Punit, S., Lugo-Villarino, G., Mazmanian, S. K., Ito, S., Glickman, J. N. and Glimcher, L. H. 2007. Communicable ulcerative colitis induced by T-bet deficiency in the innate immune system. Cell 131:33–45.

Gilboa, E. 1999. The makings of a tumor rejection antigen. Immunity 11:263–270.

Grimshaw, M. J. and Balkwill, F. R. 2001. Inhibition of monocyte and macrophage chemotaxis by hypoxia and inflammation–a potential mechanism. Eur J Immunol 31:480–489.

Gutcher, I. and Becher, B. 2007. APC-derived cytokines and T cell polarization in autoimmune inflammation. J Clin Invest 117:1119–1127.

Hart, A. L., Al-Hassi, H. O., Rigby, R. J., Bell, S. J., Emmanuel, A. V., Knight, S. C., Kamm, M. A. and Stagg, A. J. 2005. Characteristics of intestinal dendritic cells in inflammatory bowel diseases. Gastroenterology 129:50–65.

Hawiger, D., Inaba, K., Dorsett, Y., Guo, M., Mahnke, K., Rivera, M., Ravetch, J. V., Steinman, R. M. and Nussenzweig, M. C. 2001. Dendritic cells induce peripheral T cell unresponsiveness under steady state conditions in vivo. J Exp Med 194:769–779.

Heath, W. R. and Carbone, F. R. 2001. Cross-presentation, dendritic cells, tolerance and immunity. Annu Rev Immunol 19:47–64.

Jakubzick, C., Tacke, F., Ginhoux, F., Wagers, A. J., van Rooijen, N., Mack, M., Merad, M. and Randolph, G. J. 2008. Blood monocyte subsets differentially give rise to CD103 + and CD103- pulmonary dendritic cell populations. J Immunol 180:3019–3027.

Jakubzick, C., Tacke, F., Llodra, J., van Rooijen, N. and Randolph, G. J. 2006. Modulation of dendritic cell trafficking to and from the airways. J Immunol 176:3578–3584.

Kortylewski, M., Kujawski, M., Wang, T., Wei, S., Zhang, S., Pilon-Thomas, S., Niu, G., Kay, H., Mule, J., Kerr, W. G., Jove, R., Pardoll, D. and Yu, H. 2005. Inhibiting Stat3 signaling in the hematopoietic system elicits multicomponent antitumor immunity. Nat Med 11:1314–1321.

Kukreja, A., Hutchinson, A., Dhodapkar, K., Mazumder, A., Vesole, D., Angitapalli, R., Jagannath, S. and Dhodapkar, M. V. 2006. Enhancement of clonogenicity of human multiple myeloma by dendritic cells. J Exp Med 203:1859–1865.

Mahnke, K., Guo, M., Lee, S., Sepulveda, H., Swain, S. L., Nussenzweig, M. and Steinman, R. M. 2000. The dendritic cell receptor for endocytosis, DEC-205, can recycle and enhance antigen presentation via major histocompatibility complex class II-positive lysosomal compartments. J Cell Biol 151:673–684.

Mahnke, K., Qian, Y., Fondel, S., Brueck, J., Becker, C. and Enk, A. H. 2005. Targeting of antigens to activated dendritic cells in vivo cures metastatic melanoma in mice. Cancer Res 65:7007–7012.

Mellman, I. and Steinman, R. M. 2001. Dendritic cells: specialized and regulated antigen processing machines. Cell 106:255–258.

Munz, C., Steinman, R. M. and Fujii, S. 2005. Dendritic cell maturation by innate lymphocytes: coordinated stimulation of innate and adaptive immunity. J Exp Med 202:203–207.

Naik, S. H., Metcalf, D., van Nieuwenhuijze, A., Wicks, I., Wu, L., O'Keeffe, M. and Shortman, K. 2006. Intrasplenic steady-state dendritic cell precursors that are distinct from monocytes. Nat Immunol 7:663–671.

Noonan, D. M., De Lerma Barbaro, A., Vannini, N., Mortara, L. and Albini, A. 2008. Inflammation, inflammatory cells and angiogenesis: decisions and indecisions. Cancer Metastasis Rev 27:31–40.

Obeid, M., Tesniere, A., Ghiringhelli, F., Fimia, G. M., Apetoh, L., Perfettini, J. L., Castedo, M., Mignot, G., Panaretakis, T., Casares, N., Metivier, D., Larochette, N., van Endert, P., Ciccosanti, F., Piacentini, M., Zitvogel, L. and Kroemer, G. 2007. Calreticulin exposure dictates the immunogenicity of cancer cell death. Nat Med 13:54–61.

Palucka, A. K., Ueno, H., Fay, J. W. and Banchereau, J. 2007. Taming cancer by inducing immunity via dendritic cells. mmunol Rev 220:129–150.

Sapoznikov, A., Pewzner-Jung, Y., Kalchenko, V., Krauthgamer, R., Shachar, I. and Jung, S. 2008. Perivascular clusters of dendritic cells provide critical survival signals to B cells in bone marrow niches. Nat Immunol 9:388–395.

Shortman, K. and Naik, S. H. 2007. Steady-state and inflammatory dendritic-cell development. Nat Rev Immunol 7:19–30.

Shurin, G. V., Shurin, M. R., Bykovskaia, S., Shogan, J., Lotze, M. T. and Barksdale, E. M., Jr. 2001. Neuroblastoma-derived gangliosides inhibit dendritic cell generation and function. Cancer Res 61:363–369.

Shurin, M. R., Shurin, G. V., Lokshin, A., Yurkovetsky, Z. R., Gutkin, D. W., Chatta, G., Zhong, H., Han, B. and Ferris, R. L. 2006. Intratumoral cytokines/chemokines/growth factors and tumor infiltrating dendritic cells: friends or enemies? Cancer Metastasis Rev 25:333–356.

Soares, H., Waechter, H., Glaichenhaus, N., Mougneau, E., Yagita, H., Mizenina, O., Dudziak, D., Nussenzweig, M. C. and Steinman, R. M. 2007. A subset of dendritic cells induces CD4+ T cells to produce IFN-gamma by an IL-12-independent but CD70-dependent mechanism in vivo. J Exp Med 204:1095–1106.

Soiffer, R., Lynch, T., Mihm, M., Jung, K., Rhuda, C., Schmollinger, J. C., Hodi, F. S., Liebster, L., Lam, P., Mentzer, S., Singer, S., Tanabe, K. K., Cosimi, A. B., Duda, R., Sober, A., Bhan, A., Daley, J., Neuberg, D., Parry, G., Rokovich, J., Richards, L., Drayer, J., Berns, A., Clift, S., Cohen, L. K., Mulligan, R. C. and Dranoff, G. 1998. Vaccination with irradiated autologous melanoma cells engineered to secrete human granulocyte-macrophage colony-stimulating factor generates potent antitumor immunity in patients with metastatic melanoma. Proc Natl Acad Sci USA 95:13141–13146.

Spisek, R., Charalambous, A., Mazumder, A., Vesole, D. H., Jagannath, S. and Dhodapkar, M. V. 2007. Bortezomib enhances dendritic cell (DC)-mediated induction of immunity to human myeloma via exposure of cell surface heat shock protein 90 on dying tumor cells: therapeutic implications. Blood 109:4839–4845.

Steinman, R. M., Adams, J. C. and Cohn, Z. A. 1975. Identification of a novel cell type in peripheral lymphoid organs of mice. IV. Identification and distribution in mouse spleen. J Exp Med 141:804–820.

Steinman, R. M. and Banchereau, J. 2007. Taking dendritic cells into medicine. Nature 449:419–426.

Steinman, R. M. and Dhodapkar, M. 2001. Active immunization against cancer with dendritic cells: the near future. Int J Cancer 94:459–473.

Talmadge, J. E., Donkor, M. and Scholar, E. 2007. Inflammatory cell infiltration of tumors: Jekyll or Hyde. Cancer Metastasis Rev 26:373–400.

Ueno, H., Klechevsky, E., Morita, R., Aspord, C., Cao, T., Matsui, T., Di Pucchio, T., Connolly, J., Fay, J. W., Pascual, V., Palucka, A. K. and Banchereau, J. 2007. Dendritic cell subsets in health and disease. Immunol Rev 219:118–142.

Yoneyama, H., Matsuno, K., Toda, E., Nishiwaki, T., Matsuo, N., Nakano, A., Narumi, S., Lu, B., Gerard, C., Ishikawa, S. and Matsushima, K. 2005. Plasmacytoid DCs help lymph node DCs to induce anti-HSV CTLs. J Exp Med 202:425–435.

Ziegler-Heitbrock, L. 2007. The CD14+ CD16+ blood monocytes: their role in infection and inflammation. J Leukoc Biol 81:584–592.

Zou, W. 2005. Immunosuppressive networks in the tumour environment and their therapeutic relevance. Nat Rev Cancer 5:263–274.

Index

A

Acetylation, of STAT3, 145
α-adrenergic receptors (α–AR), 193
β-adrenergic receptors (β–AR), 193
Aging, 172–176
  allostatic load and, 185
  antigen uptake, 174
  DC maturation/differentiation in old age, 173–174
  DC migration in old age, 174–175
  DC numbers in old age, 172–173
    reduction in LC numbers, 172
    reduction in pDC numbers, 172–173
  T-cell stimulatory properties in old age, 175–176
Allostatic load, 183–185
Alpha-melanocyte-stimulating hormone (α–MSH), 190
Angiogenesis
  DC role in, 247–248
  VEGF role in tumor, 260–261
Angiogenic factors, DC source of, 235–237
Angiogenic properties, of mature DC, 237
ANS response, *see* Autonomic nervous system (ANS) response
Antibody blockade of PD-1 and B7-H1, 358
Antigen-presenting cells (APC), 89, 143, 309
  DC differentiation, 104, 105
Antigens
  loading, 316–318
  selection for therapeutic vaccination, 315–316
  uptake and aging, 174
Antitumor immunity
  induction of, 57
  levels of IL-6 and IL-10, 22
  role of DC in development of, 15–17
  T-cell responses and, 57
  TIDC and, 42, 48
  tumor progression and, 19, 21
APC, *see* Antigen-presenting cells (APC)
Apoptosis, tumor-derived factors and DC, 110–111
Arachidonic acid metabolites, *see* Prostanoids
Autonomic nervous system (ANS) response, 181–183
Avoidance pathway, 19–20

B

B cells
  antibody class switching in, 5
  120G8 antibody and, 82
Biliary carcinoma, TIDC density in, 40
Biological response modifiers (BRM), 337
Blood DC precursors, 1
Breast cancer patients
  intratumoral DC and peritumoral DC in, 64
  T-cell stimulatory activity in, 63
  tumor-induced apoptosis in, 94
Bronchoalveolar lavage fluid, 83

C

Calcitonin gene-related peptide (CGRP), 190–191
Cancer
  DC vaccines and trafficking in, 277–279
  effects senescence of DC in immunotherapy of, 176–177
  ERK and DC dysfunction in, 134
  inflammation and DC function in terminal phase, 166–167
  JAK/STAT3 DC differentiation/ maturation in, 149–150

Cancer (*cont.*)
psychosocial factors and, 182–183
Cancer immunotherapy, *see* Immunotherapy
Cancer patients, circulating DC in, 332
CCL2, role in recruiting monocytes and enriching TAM, 298
CCR7 expression, 34
CD4 expression, 232
CD8 expression, 232
$CD4^+$ T cells
and DC, interaction between, 58
presentation of intracellular antigen to, 3
CD8 + T cells
intracellular antigen to, 3
CEA-specific CTL, 350
Celecoxib, 319
CGRP, *see* Calcitonin gene-related peptide (CGRP)
Chemoattractants, down-regulation of, 19–20
Chemomodulation
cytokine expression in DC, 209–211
effects of, 210
of small Rho GTPase activity in DC, 209
Chemotherapy, regulation of DC function, 206–208
metronomic treatment, 207
MTD for eradication of neoplastic cells, 206
strategies to control toxic effects of, 206–207
tumor microenvironment, 207–208
Cholangiocarcinoma, immunotherapy in, 338
c-jun N-terminal kinases (JNK), 131
Colonic adenocarcinoma
TIDC density in, 39–40, 66
Colony-stimulating factor (CSF), 249
Colorectal carcinoma (CRC)
immunotherapy in, 339–340
CORT, *see* 10^{-6}M corticosterone (CORT)
Corticotrophin-releasing hormone (CRH), 192
COX-1, 162
COX-2, *see* Cyclooxygenase (COX)-2
CPA-7, 151
CpG ODN, *see* Unmethylated cytosine–phosphate–guanine oligodeoxynucleotides (CpG ODN)
CRH, *see* Corticotrophin-releasing hormone (CRH)
Cross-presentation, tumor-derived exosomes and, 122–123
CSF, *see* Colony-stimulating factor (CSF)
CTL, *see* Cytotoxic T lymphocytes (CTL)
CTLA-4, immunoattenuating transmembrane protein, 358
CXC chemokines, role in tumor-induced inflammation, 162
CXCL14, chemokine, expression of, 19–20
Cyclooxygenase (COX)-2, 107, 162
inhibitor, 319
Cyclophosphamide, 318
Cytokine expression, chemomodulation of, 209–211
Cytokine production by DC
factors regulating, 65
IL-10 and IL-12, 66–67
osteopontin, 67
Cytotoxic T lymphocytes (CTL), 5, 310
homing efficiency of, 279

D

Damage-associated molecular pattern molecules (DAMP), 160
DAMP, *see* Damage-associated molecular pattern molecules (DAMP)
DC, *see* Dendritic cells (DC)
DC activation, alteration in, 295–297
DC activity
impact of high-dose *versus* low-dose chemotherapy on, 206–208
modulation by surgery, 201–204
p38 MAPK in, 131–132
DC differentiation
altered, 108–109
cytokines/growth factors affects, 104–105
gangliosides affects, 105–106
IDO role in, 225–226
and JAK/STAT3 in cancer, 149–150
JAK/STAT signaling in, 146–148
in old age, 173–174
p38 MAPK in, 131–132
prostanoids role in, 106–107
tumor antigens and metabolites affects, 108
tumor-derived exosomes and impairment of, 123–125
myeloid cells function, 123–124
TGF-β, 125
DC-directed T-cell fate
factors influencing, 5

DC dysfunction, tumor-induced
  and ERK in cancer, 134
  immune response generated, 107
  JAK/STAT signaling role in, 134–135
  mechanisms, 106
  NF-κB role in, 135–136
  p38 MAPK in, 132–134
    role of TLR, 134
    TCCM function, 132
  tumor-derived factors responsible for, 110
DC function
  aging and alterations in, 172–176
    antigen uptake, 174
    DC maturation/differentiation in old age, 173–174
    DC migration in old age, 174–175
    DC numbers in old age, 172–173
    T-cell stimulatory properties in old age, 175–176
  effects of psychological stress on, 187–190
  IDO role in, 225–226
  and inflammation
    in established tumors, 161–163
    in proto-cancer, 159–161
    during rejection of established tumors, 163–166
    in terminal phase cancer, escaping immune control, 166–167
  JAK/STAT Signaling, 146–148
  mechanisms of stressor-induced modulation of, 190–193
    glucocorticoids, 191–192
    role of sympathetic nervous system, 192–193
  regulation by chemotherapy, 206–208
DC homeostasis
  alterations in, 21
DC maturation
  and JAK/STAT3 in cancer, 149–150
  MAPK-signaling pathways, 131
  NF-κB role in, 135–136
  in old age, 173–174
  p38 MAPK in, 131–132
  process, 232–233
  prostate-specific antigen inhibition and, 108
  tumors interference with, 109
  VEGF inhibition and, 105
DC migration
  metastasis and, 249
  in old age, 174–175
    in vivo lymph node, 177
DC numbers
  alterations in, 295–297
  in old age, 172–173
DC phenotype, alteration, 295–297
DC polarization, tumor microenvironment-associated deviation, 262–263
DC precursors transmigrate, through blood vessel endothelia, 273
DC-SIGN, *see* DC-specific intercellular adhesion molecule 3 (ICAM-3)-grabbing nonintegrin (DC-SIGN)
DC-SIGN expression, *see* DC-specific/intracellular adhesion molecule type 3-grabbing nonintegrin (DC-SIGN) expression
DC-specific intercellular adhesion molecule 3 (ICAM-3)-grabbing nonintegrin (DC-SIGN)
  interaction wit MUC-1, 247
DC-specific/intracellular adhesion molecule type 3-grabbing nonintegrin (DC-SIGN) expression, 175–176
DC-specific transmembrane protein (DC-STAMP), 249
DC-STAMP, *see* DC-specific transmembrane protein (DC-STAMP)
DC subset, 310–312
DC suppression, 319
DC trafficking
  and DC vaccines in cancer, 277–279
  regulation, 271–274
  at tumor site, 274–277
DC vaccination, 318–320
  administrations, 320–321
  evaluating efficacy of, 321–322
DC vaccines
  and DC trafficking in cancer, 277–279
  intratumoral, 279–282
DDC, *see* Dermal DC
Delayed-type hypersensitivity (DTH) response
  psychological stress, 188
Delayed-type hypersensitivity (DTH) skin tests, 322
Dendritic cell neoplasms, 373
Dendritic cells (DC)
  activation of, 4
  alteration in local tumor microenvironment, putative mechanisms of, 45–48
  anatomical location of, 1
  angiogenic properties of mature, 237

Dendritic cells (DC) (*cont.*)
antigen uptake and processing by
CD4+ T cells and MHC, 3
inhibited phagocytosis with VEGF, 59–60
interfere of tumor-derived molecules with, 60
mechanisms of, 4
ovalbumin antigens, 59
tumors suppression, 58
apoptosis in
NO production, 93
peripheral blood DC, 92–93
TGF-β1 role in, 93–94
tumor-derived factors and, 92
tumor-mediated inhibition, 93
in cancer
age-related alterations, 22
in basal cell carcinoma, 15
impaired or re-directed functionality of, 20
LC identification, 13–14
lymph nodes, 17
role in tumor–host interactions, 15–16
SAC and IDC, 14
in cancer patients, 331–332
chemoattractants, down-regulation of, 19–20
chemomodulation of small Rho GTPase activity in, 209
culture of
from CD34+ cells, 332
from CD14+ monocytes, 332, 333
cytokine production by
factors regulating, 65
IL-10 and IL-12, 66–67
osteopontin, 67
definition of, 13
derived from monocyte of cancer patients, 64–65
differentiation
systemic tumor-induced defects of, 62–63
in directing T-Cell responses, 5–6
discovery of
in human skin, 12
immunologic research following, 11
distribution in tumor tissues and regional lymph nodes, 17
electroporation of, 350
endothelial cell properties by, 234–235
entry level deviation by tumors, 258–261
functional characteristics of, 57
functional state of, factors influencing, 21
generated for immunotherapy, quality control required for
maturation, stability, and potency, 335
purity and sterility testing, 334
viability measurement, 334
generation for clinical trials, 333–334
generation of, 312–315
genetic modification of
by viral or non-viral gene transfer vectors, 348–349
from healthy donors and cancer patients, possibility of using, 350–351
in human blood, 58
in immature state, 33
immune regulation by
cross-presentation of exogenous antigens, 378
DC maturation and, 379
NK and NKT cells, 378
for immune therapy of cancer
ex vivo dendritic cell vaccines, 379–380
in-situ dendritic cell targeting, 380
with immunosuppressive properties in tumor microenvironment, 233–234
impact of radiation therapy on, 204–205
interactions with NK cells, 359
JAK/STAT-signaling pathways, 150–151
location within tumors, 376–377
versus macrophages, 249–250
MART-1 peptide in, 205
maturation of, 4–5, 7, 33, 333
expression of accessory molecules, 62
migration of
to lymphoid organs, 60
PgE_2 signaling via EP2 receptor and, 61–62
regulation by cytokines and chemokines, 61
tumor-derived factors affecting, 62
in mouse and humans, 231–232
naturally occurring, 331
NO production in, 93
ovarian cancer-associated immature, 236
pathogenic modifications of, 21
physiologic flow to lymph nodes, 291–193
protumorigenic effects of, 245–249
angiogenesis, 247–248
enhanced tumor growth, 245–246

immunosuppression, 246–247
metastasis, 249
transdifferentiation, 248–249
role in initiating antitumor immune response, 33–34
role in neoplastic process, 35
role in tumor microenvironment, 244–245
source of, 376
source of, for clinical trial, 332
source of angiogenic factors, 235–237
stress impact on, 187
suppressive elements inhibiting, 377
triggering toll-like receptors on, 333
tumor manipulation of biology of, 258
tumor-mediated inhibition of, 18–19
antigen presentation, 63
in breast cancer patients, 94
generation and differentiation of, 91–92
pro-apoptotic factors secretion, 93–94
tumor progression due to abnormalities of, 20
vaccine formulations, 356
Dendritic cell subsets
characterization of, 1
in tumor microenvironment, Tumor microenvironmental DC
Dendritic cell tumors
follicular, *see* Follicular dendritic cell tumor/sarcoma
immunohistochemical markers for differential diagnosis in, 373
interdigitating, *see* Interdigitating dendritic cell tumors/sarcomas
origin and classification of, 365–366
Dendritic cell vaccines, in cancer
apoptotic cholangiocarcinoma cells, 337–338
colorectal carcinoma, 339–340
degree of plasticity in, 377–378
esophageal carcinoma, 336
gastric cancer, 336–337
hepatocellular carcinoma, 338–339
pancreatic cancer, 338
Denileukin difitox, 358
Dermal DC, 1
DEX, *see* Dexamethasone (DEX)
Dexamethasone (DEX), exposure of DC to, 191–192
DTH, *see* Delayed-type hypersensitivity (DTH) skin tests

E

Endometrial carcinoma, TIDC density in, 43
Endothelial cell (EC) properties, by DC, 234–235
ERK, *see* Extracellular signal-regulated kinases (ERK)
Esophageal cancer, role of immunotherapy in, 336
Esophageal SCC, TIDC density in, 39
Established tumors
inflammation and DC functions, 161–163
during rejection, 163–16
tumor-associated inflammation in, 161–163
Exosomes, tumor-derived, 120–122
DC-mediated antigen cross-presentation, 122–123
impairment of DC differentiation, 123–125
myeloid cells function, 123–124
TGF-β, 125
interaction with DC, immune stimulation, 125–126
molecular content of, 121
potential role in stimulating immune response, 121
transport of cytokines/growth factors, 122
Extracellular signal-regulated kinases (ERK), 131
and DC dysfunction in cancer, 134
p38 MAPK role in phosphorylation of, 132
Ex vivo dendritic cell vaccines, 379–380

F

Flow cytometric analysis, of DC in SLN, 299
Flt3 signaling, *see* Fms-related tyrosine kinase 3 ligand (Flt3) signaling
Fms-related tyrosine kinase 3 ligand (Flt3) signaling, 146
Follicular dendritic cell tumor/sarcoma
clinical behavior of, 368–369
clinical features and histomorphology of, 367–368
common sites of, 367
designation of, 366–367
of pharynx, 368
treatment of, 369
WHO classification of, 366
Follicular dendritic reticulum cells (FDC), 365

Forkhead box P3 (FOXP3), 163
Formalin-fixed paraffin-embedded tissue, 35, 37
FOXP3, *see* Forkhead box P3 (FOXP3)

G
Gastric adenocarcinoma, TIDC density in, 39
Gastric cancer, role of immunotherapy in, 336–337
G-CSF, *see* Granulocyte colonystimulating factor (G-CSF)
Genetically modified DC
  antitumor vaccination using, *see* Genetically modified dendritic cell vaccines
  intracellular antigen processing by, 348
Genetically modified dendritic cell vaccines
  antitumor activity of, enhancing
    blockade of tumor-associated immune suppressor factors, 358
    co-transfection of DC, 357
    DC interactions with NK cells, 359
    depletion of T-regulatory cells, 358
    silencing of inhibitory molecules, 359
  in clinical trials, 353–355
    advanced stage small-cell lung cancer, 356
    IL-12, in metastatic gastrointestinal carcinomas, 352, 356
    immune responses and clinical efficacy, correlation between, 357
    metastatic prostate cancer, 352
  pre-clinical models of
    electroporation of DC, 350–351
    immunological tolerance, in HCC patients, 351
    transgenic mice, 350
GILZ, *see* Glucocorticoid-induced leucine zipper (GILZ)
Glucocorticoid-induced leucine zipper (GILZ), 202
Glucocorticoids, effects on DC function, 191–192
GM-CSF, *see* Granulocyte macrophage colony-stimulating factor (GM-CSF)
Graft-vs-host disease (GVHD), 312
Granulocyte colonystimulating factor (G-CSF), 315
Granulocyte macrophage colony-stimulating factor (GM-CSF), 312, 332, 333
  affecting DC differentiation, 104–105
    STAT3 signaling, 146
  immunomodulation of SLN, 299–300
GVHD, *see* Graft-vs-host disease (GVHD)

H
HDAC, *see* Histone deacetylase (HDAC)
Head and neck squamous cell carcinoma (HNSCC), 274
Health, impact of psychosocial factors on, 181–183
Heat shock proteins (HSP), 317
Helicobacter pylori, 159
Hematopoietic stem cells (HSC), 146
Hepatocellular carcinoma (HCC)
  genetically modified DC vaccines role in, 351
  immunotherapy in, 338–339
  TIDC density in, 40
High-mobility group box 1 (HMGB1), 160
Histone deacetylase (HDAC), 145
HLA-DR, 36, 64
HMGB1, *see* High-mobility group box 1 (HMGB1)
HNSCC, *see* Head and neck squamous cell carcinoma (HNSCC)
Homeostasis, 183–185
HPA axis, *see* Hypothalamic–pituitary–adrenal (HPA) axis
HPV infection, 42
HSC, *see* Hematopoietic stem cells (HSC)
HSP, *see* Heat shock proteins (HSP)
Human DC, definition of, 1
Humans, DC in, 231–232
Human telomerase reverse transcriptase (hTERT), 340
Hyperactivation, of STAT3-signaling, 135
Hypothalamic–pituitary–adrenal (HPA) axis, 183, 192

I
IDC, *see* Immature DC (iDC)
IDO, *see* Indoleamine 2,3-dioxygenase (IDO)
IL-4, *see* Interleukin-4 (IL-4)
IL-6
  affecting DC differentiation, 104
    tumor-mediated regulation, 246
  mediated STAT3 activation, 147

IL-10
affecting DC differentiation, 104
role in DC defects in cancer, 246–247
Imiquimod, 164–165
Imiquimod-induced inflammation, 166
Imiquimod-induced NK cell activation, 165–166
Imiquimod-induced TLR activation, 165
Imiquimod-induced tumor rejection, 165
Immature DC (iDC), 258
physiological function of, 58
Immune control, 166–167
Immune reactivity, 294–295
Immune response
effects of stress on T-cell-mediated antigen-specific, 188
exosomes role in stimulating, 121
generated tumor-induced DC dysfunction, 107
Immune-suppressing cytokines, 319
Immune system
stress and, 185–187
stressor impact on, 187
Immunohistochemical markers, 373
Immunohistochemistry, of TIDC, 35
Immunomodulation, reversal of SLN, 299–301
Immunoreceptor tyrosine-based activation motif (ITAM), 249
Immunosuppression
of DC, 246–247
in tumor microenvironment, 233–234
IDO-mediated DC-driven, in tumor environment, 224
induced by surgical stress, 201–204
radiation-induced, 204–205
Immunotherapy, 310–312
DC-based, 315
effect on immune effector cells in SLN, 299
effects senescence of DC in, 176–177
future development of, 340
IDO inhibitors role in DC-targeted, 225
in pancreatic cancer, 338
tumor antigen
in cholangiocarcinoma, 338
in CRC, 339–340
against esophageal cancer, 336
against gastric cancer, 336–337
in HCC, 338–339
Indoleamine 2,3-dioxygenase (IDO), 109
induction in DC, 223–224
inhibitors role in DC-targeted cancer immunotherapy, 225
kynurenine pathway of tryptophan catabolism, 219–221
mediated DC-driven immune suppression in tumor environment, 224
mediated immune suppression, mechanism, 221–223
role in DC function/differentiation, 225–226
Inflammation, tumor-associated, 158–159
established tumors, 161–163
during rejection of, 163–166
in proto-cancer, 159–161
in terminal phase cancer, escaping immune control, 166–167
In-situ dendritic cell targeting, 380
In situ hybridization, 35
Insofar, 185
Interdigitating dendritic cell tumors/sarcomas
clinical behavior and outcome of, 369, 371
histomorphology of, 369
of lymph nodes, 369, 370
Interdigitating dendritic reticulum cells (IDC), 365
Interleukin-4 (IL-4), 312
ITAM, *see* Immunoreceptor tyrosine-based activation motif (ITAM)

J

JAK/STAT-signaling pathways
in cancer, 149–150
in DC differentiation/function, 146–148
in dendritic cells, 150–151
overview of, 144–146
role in tumor-induced DC dysfunction, 134–135
JNK, *see* C-jun N-terminal kinases (JNK)
JSI-124, 150–151

K

Kynurenine pathway, of tryptophan catabolism, 219–221

L

LAMP-hTERT vaccine, 352
Langerhans cell histiocytosis/sarcoma
  clinical behavior and outcome of, 371, 373
  clinical features and histomorphology of, 371
  of liver, bone, and skin, 372
Langerhans cells (LC), 172, 188, 314
  alterations in skin of patients with BCC, 18
  discovery of, 12
  functional studies on, 13–14
  as local cutaneous APC, 15–16
  and monocyte–macrophage system, relationship between, 14
  morphology with dendritic extensions, 2
  S-100 antigen expression, 15
Large amino acid transporter 1 (LAT1), 220, 226
Laryngeal SCC, TIDC in, 38
LAT1, *see* Large amino acid transporter 1 (LAT1)
LC, *see* Langerhans cells (LC)
Lipopolysaccharide (LPS), 173, 193
LM/SNB, *see* Lymphatic mapping and sentinel node biopsy (LM/SNB)
LPS, *see* Lipopolysaccharide (LPS)
Lung carcinomas, DC infiltration in, 64
Lymphatic mapping and sentinel node biopsy (LM/SNB), 295
Lymph nodes
  heterogeneity of immune reactivity in, 294–295
  physiologic flow DC to, 291–293
  two-color flow cytometry, 294–295
Lymphocytopenia, 205

M

MAA, *see* Melanoma-associated antigens (MAA)
Macrophage colony-stimulating factor receptor (M-CSFR), 249
Macrophage inflammatory protein-1α (MIP-1α), 249
Macrophage inflammatory protein (MCP)-1, 208
Macrophages
  versus DC, 249–250
  STAT3-signaling pathways, 148
Macropinocytosis, 3
Malignant effusions, 82
Malignant melanoma, TIDC density in, 43–44
MAPK, *see* Mitogen-activated protein kinases (MAPK)
MAPKK, *see* MAPK kinases (MAPKK)
MAPK kinase kinases (MAPKKK), 130
MAPK kinases (MAPKK), 130
MAPKKK, *see* MAPK kinase kinases (MAPKKK)
MART-1 peptide, 205
Matrix metalloproteinases (MMP)-2, 249
Matrix metalloproteinases (MMP)-9, 249
Mature DC (mDC), 34
Maximum tolerated dose (MTD), in chemotherapy, 206
10^{-6}M corticosterone (CORT), 191, 192
MCP-1, *see* Macrophage inflammatory protein (MCP)-1
M-CSFR, *see* Macrophage colony-stimulating factor receptor (M-CSFR)
Melanoma-associated antigens (MAA), 300
Metastasis, DC migration and, 249
Metastatic prostate cancer, genetically modified DC vaccination in, 352
1-methyl-tryptophan (1-MT), 225
MGC, *see* Multinucleated giant cells (MGC)
Microvesicles, *see* Exosomes
MIIC (MHC class II compartment), HLA-DM and, 3
MIP-1α, *see* Macrophage inflammatory protein-1α (MIP-1α)
Mitogen-activated protein kinases (MAPK)
  DC maturation, 131
  signaling pathways, 130–131
    DC maturation, 131
    diverse cellular functions, 130–131
  type of, 131
MMP-2, *see* Matrix metalloproteinases (MMP)-2
MMP-9, *see* Matrix metalloproteinases (MMP)-9
MO-DC, *see* Monocyte-derived immature DC (MO-DC)
Molecular alteration, in SLN, 297–298
Monocyte-derived immature DC (MO-DC), 173, 174, 175
Mouse DC, 231–232
  definition of, 1
α-MSH, *see* Alpha-melanocyte-stimulating hormone (α–MSH)
1-MT, *see* 1-methyl-tryptophan (1-MT)
MTD, *see* Maximum tolerated dose (MTD)

MUC-1, *see* Mucin 1 (MUC-1) antigen
Mucin 1 (MUC-1) antigen
  interaction with DC-SIGN, 247
Multinucleated giant cells (MGC), 248, 249
Murine B16 melanoma brain tumor model, 350
Murine tumor models, Plasmacytoid DC in cancer, 81–82
Mycobacterium tuberculosis, 173
Myeloid cells, 119
  function of, 123–124
Myeloid-derived suppressor cells (MDSC), 19, 92
Myelopoiesis, 104, 146
Myelosuppression, 201, 206

N
Nasopharyngeal carcinoma, TIDC in, 38
Natural killer (NK) cells, 150
  imiquimod-induced activation of, 165–166
NF-κB, role in DC dysfunction, 135–136
NK cell responses, 359
NK cells, *see* Natural killer (NK) cells
NLR, *see* Nod-like receptors (NLR)
Nod-like receptors (NLR), 160
Non-receptor tyrosine kinases (RTK), 144
Norepinephrine, 192–193

O
Osteopontin, 67
Ovalbumin antigens, 59

P
Paclitaxel, 208
Pancreatic adenocarcinoma, TIDC in, 40
Pancreatic cancer, immunotherapy in, 338
PDC, *see* Plasmacytoid DC (pDC)
Pediatric malignancy, TIDC density in, 44
Peptidoglycan (PGN), 193
P53 gene-modified DC vaccine with chemotherapy, 356
PGN, *see* Peptidoglycan (PGN)
Phagocytosis, 3
  aging and DC, 174
Phosphatase and tensin homolog (PTEN), 174
Physiologic flow, of DC to lymph nodes, 291–293
PIAS, *see* Protein inhibitor of activated STAT (PIAS) family
Plasmacytoid DC (pDC), 2, 7, 172–173
  marker of, 35
  in tumor ascites in patients with ovarian cancer, 19
P38 MAPK
  in DC differentiation/maturation/activity, 131–132
  role in NF-κB–signaling pathways, 135–136
  in tumor-induced DC dysfunction, 132–134
    role of TLR, 134
    TCCM function, 132
Prostanoids, role in DC differentiation, 106–107
Prostate carcinoma, TIDC density in, 41
Protein inhibitor of activated STAT (PIAS) family, 145, 146
Proto-cancer, DC function and inflammation in, 159–161
Protumorigenic effects, of DC, 245–249
  angiogenesis, 247–248
  enhanced tumor growth, 245–246
  immunosuppression, 246–247
  metastasis, 249
  transdifferentiation, 248–249
Psychological stress
  effects on DC function, 187–190
  response to surgery, 202
Psychosocial factors, 181–183
  ANS response, 182
  cancer and, 182–183
PTEN, *see* Phosphatase and tensin homolog (PTEN)
Pulmonary carcinoma, TIDC density and distribution in, 40–41

R
Radiation therapy, impact on DC, 204–205
RANK/RANK-L pathway, 248–249
Reactive oxygen species (ROS), 109
Receptor-mediated endocytosis, of antigens, 3
Renal cell carcinoma, TIDC density in, 42
Rho GTPase activity, chemomodulation in DC, 209
RNAi, *see* RNA interface (RNAi)
RNA interface (RNAi), 319
ROS, *see* Reactive oxygen species (ROS)
RTK, *see* Non-receptor tyrosine kinases (RTK)

S

S-100 antigen, expression of, 15
Semi-mature DC, 6
Senescence, effects of DC in cancer immunotherapy, 176–177
Sentinel lymph node (SLN), 295
  molecular and signaling alterations in, 297–298
  reversal of immunomodulation of, 299–301
Sézary syndrome, 352
SH2 domain, for activation of STAT, 144–145
Signaling alteration, in SLN, 297–298
Signal regulatory protein-α (SIRP–α), 249
Signal transducers and activators of transcription (STAT), 144
  SH2 domain for activation of, 144–145
SIINFEKL peptide, 191
SiRNA knockdown of IDO and SOCS1, 359
SIRP–α, *see* Signal regulatory protein-α (SIRP–α)
SLN, *see* Sentinel lymph node (SLN)
SOCS1
  role in function DC subset, 148
  targeting, 151
SOCS proteins, *see* Suppressor of cytokine signaling (SOCS) proteins
Stability assays, 335
STAT, *see* Signal transducers and activators of transcription (STAT)
STAT proteins
  activation of, 144
  recruitment to receptor, 145
STAT1-signaling pathways, 147
STAT3-signaling pathways
  acetylation, 145
  activity in cancer, 149–150
  in DC differentiation, 146
  for function of macrophages and neutrophils, 148
  hyperactivation, 135
  role in DC development, 147
STAT6-signaling pathwys, 147
Sterility testing, 334
Stress, 183–185
  defined, 183
  and immune system, 185–187
  impact on DC, 187
Stressor
  defined, 183
  impact on immune system, 187
Stressor-induced modulation, of DC function, 190–193
  glucocorticoids, 191–192
  role of sympathetic nervous system, 192–193
Stroma-derived factors, 20
Suppressor of cytokine signaling (SOCS) proteins, 145–146, 147–148, 263
Surgical resection, modulation of DC activity by, 201–204
Survivin, 340

T

TA, *see* Tumor antigens (TA)
TAM, *see* Tumor-associated macrophages (TAM)
TCCM, *see* Tumor culture-conditioning medium (TCCM)
T-cell functional polarization, tumor microenvironment-associated deviation, 262–263
T-cell receptors, 318
T-cell stimulatory properties, in old age, 175–176
TDLN, *see* Tumor-draining lymph nodes (TDLN)
TDO, *see* Tryptophan 2,3-dioxygenase (TDO)
Terminal deoxynucleotidyl transferase biotin-dUTP nick end labeling (TUNEL), 296
TGF-β, *see* Transforming growth factor (TGF)-β
Therapeutic vaccination, antigen selection, 315–316
Therapeutic vaccines, 312–315
Th2 immune responses, 5–6, 65
Thyroid carcinoma, TIDC infiltration in, 39
TLR, *see* Toll-like receptor (TLR)
TLR-3, *see* Toll-like receptor-3 (TLR-3)
TME, *see* Tumor microenvironment (TME)
TNF-α, *see* Tumor necrosis factor (TNF)-α
TNF-receptor, 314
Tolerogenic dendritic cells, generation of, 7
Toll-like receptor (TLR), 134, 160
  aging and expression of, 173–174
  imiquimod-induced activation of, 165
Toll-like receptor-3 (TLR-3), 314
TRAMP-C2, 176
Transdifferentiation, of DC, 248–249
Transforming growth factor (TGF)-β
  as immune suppressor, 297–298
  tumor-derived exosomes and, 125

T-regulatory cells (Treg)
  methods for depletion of, 358
  reduction by DC vaccination in cancer patient, 318
Triggering toll-like receptors on DC, 333–334
Tryptophan catabolism, kynurenine pathway of, 219–221
Tryptophan 2,3-dioxygenase (TDO), 220
Tryptophan starvation, 221–223
Tumor antigen-loading procedures, 336
Tumor antigens (TA), 315
  affecting DC differentiation, 108
  antitumor vaccination using, 347
  in cholangiocarcinoma, 338
  in CRC, 339–340
  in esophageal cancer, 336
  in gastric cancer, 336–337
  in HCC, 338–339
Tumor-associated DC, degree of plasticity in functional properties of, 377
Tumor-associated immune suppressor factors, blockade of, 358
Tumor-associated macrophages (TAM), 149, 163, 259
  CCL2 role in enriching, 298
Tumor culture-conditioning medium (TCCM), 132
Tumor-derived VEGF, 91
Tumor-draining lymph nodes (TDLN)
  antigen-specific immunity in, 258
  blocking of DC maturation/migration into, 261–262
  IDO^+ DC in, 224
  IDO^+pDC in, 222
Tumor growth, enhanced by DC, 245–246
Tumor-infiltrating DC (TIDC)
  in breast carcinoma, 42
  in digestive system malignancy
    colonic adenocarcinoma, 39–40
    gastric adenocarcinoma, 39
    pancreatic adenocarcinoma, 40
  future studies of, 47–48
  in genitourinary malignancy
    prostate carcinoma, 41
    renal cell carcinoma, 42
    transitional cell carcinoma of urinary bladder, 41–42
  in gynecologic malignancy
    endometrial carcinoma, 43
    SCC of uterine cervix, 42–43
  in head and neck malignancy
    laryngeal SCC, 38
    SCC of tongue, 38
    thyroid carcinoma, 39
  identification in cancer using S-100 protein immunostain, 35, 37
  immunohistochemical markers for analysis of, 36
  in malignant melanoma, 43–44
  in pediatric malignancy, 44
  in pulmonary malignancy, 40–41
Tumor manipulation, of DC biology, 258
Tumor metabolites
  affecting DC differentiation, 108
Tumor microenvironmental DC
  $CD4^+$ T cells in, 80
  dysfunction of, 80
  factors affecting distribution of, 78
  factors affecting functions of, 79
  strategies to improve number or function of, 81
  studies of, 82–83
  and tumor-associated DC, 79
Tumor microenvironment (TME)
  blocking maturation/migration of DC, 261–262
  chemotherapy and, 207–208
  compartmentalization of DC, 258–261
  DC with immunosuppressive properties in, 233–234
  deviation of DC/T-cell functional polarization, 262–263
  role of DC in, 244–245
Tumor necrosis factor (TNF)-α, 314
Tumor progression
  DC elimination during, 90–91
  due to abnormalities of DC, 20–21
Tumor site, trafficking of DC at, 274–277
TUNEL, *see* Terminal deoxynucleotidyl transferase biotin-dUTP nick end labeling (TUNEL)
Two-color flow cytometry, lymph nodes, 294–295
Tyrosine phosphatases, 145

U
University of Pittsburgh Cancer Institute, 317
Unmethylated cytosine–phosphate–guanine oligodeoxynucleotides (CpG ODN), 299–300

V

Vascular endothelial growth factor (VEGF)
  affecting DC differentiation
    tumor-derived factor in, 247
  angiogenesis by DC, 235–236
  angiogenic properties of mature DC and, 237
  effect of surgery on, 202–203
  immunosuppressive DC and, 133
  role in tumor angiogenesis alteration of iDC, 260–261
Vasoactive intestinal peptide (VIP), 190
VEGF, *see* Vascular endothelial growth factor (VEGF)
VEGFR-2, *see* VEGF receptor 2 (VEGFR-2)
VEGF receptor 2 (VEGFR-2), 235, 236
VIP, *see* Vasoactive intestinal peptide (VIP)
Vulvar lesions, TIDC density with degree of squamous cell dysplasia in, 43

GPSR Compliance

The European Union's (EU) General Product Safety Regulation (GPSR) is a set of rules that requires consumer products to be safe and our obligations to ensure this.

If you have any concerns about our products, you can contact us on ProductSafety@springernature.com

In case Publisher is established outside the EU, the EU authorized representative is:

Springer Nature Customer Service Center GmbH
Europaplatz 3
69115 Heidelberg, Germany

Batch number: 10755284

Printed by Printforce, the Netherlands